Coffee Science

Coffee Science: Biotechnological Advances, Economics, and Health Benefits highlights the important advances in coffee research and an all-inclusive collection of information on the current status of global coffee production and market, sustainable benefits, novel methods and recent developments in coffee metabolites analysis, advancements in coffee processing technology and improvement of coffee quality by fermentation, solid-liquid extraction methods, and post-harvesting processes to improve the beverage quality and produce coffees with different sensory profiles. The book compiles insights into the biotechnological advances to improve coffee quality. It also describes specialty coffees, which are gaining consumer acceptance and enjoying a good global market.

This book collates work on the influence of various coffee metabolites, such as methylxanthines, phenolic acids, polyphenols, indoleamines and other biogenic amines and coffee diterpenes in human health effects such as cardiovascular diseases, cancer, type 2 diabetes mellitus, Alzheimer's disease, and Parkinson's disease.

This book is a useful resource for scientists, academicians, and professionals all over the world who are engaged in coffee cultivation, research, business, and coffee consumers' health.

Key Features

- Current status on coffee production and the global market
- Novel methods and recent developments in the determination of coffee metabolites
- Advancements in coffee bean processing technology and improvement of coffee quality
- Biotechnological advances to improve coffee quality: the role of molecular markers, tissue culture, transgenic technology, and micro RNAs
- Effects of coffee consumption on human health
- Knowledge contributions from acknowledged experts from across the world.

Coffee Science

Biotechnological Advances, Economics, and Health Benefits

Edited by
Akula Ramakrishna,
Parvatam Giridhar, and
Magdalena Jeszka-Skowron

CRC Press
Taylor & Francis Group
Boca Raton London

CRC Press is an imprint of the
Taylor & Francis Group, an **informa** business

First edition published 2023
by CRC Press
6000 Broken Sound Parkway NW, Suite 300, Boca Raton, FL 33487–2742

and by CRC Press
2 Park Square, Milton Park, Abingdon, Oxon, OX14 4RN

CRC Press is an imprint of Taylor & Francis Group, LLC

Library of Congress Cataloging-in-Publication Data
Names: Ramakrishna, Akula, editor. | Parvatam, Giridhar, editor. | Jeszka-Skowron, Magdalena, editor.
Title: Coffee science : biotechnological advances, economics, and health benefits / Akula Ramakrishna, Giridhar Parvatam, Magdalena Jeszka-Skowron.
Description: First edition | Boca Raton, FL : CRC Press, 2022. | Includes bibliographical references and index.
Identifiers: LCCN 2022002281 (print) | LCCN 2022002282 (ebook) | ISBN 9780367488437 (hardback) | ISBN 9781032303499 (paperback) | ISBN 9781003043133 (ebook)
Subjects: LCSH: Coffee. | Coffee—Breeding. | Coffee—Therapeutic use. | Coffee industry.
Classification: LCC SB269 .C645 2022 (print) | LCC SB269 (ebook) | DDC 633.7/3—dc23/eng/20220218
LC record available at https://lccn.loc.gov/2022002281
LC ebook record available at https://lccn.loc.gov/2022002282

ISBN: 978-0-367-48843-7 (hbk)
ISBN: 978-1-032-30349-9 (pbk)
ISBN: 978-1-003-04313-3 (ebk)

DOI: 10.1201/9781003043133

Contents

Preface

Coffea is an important plantation crop with significant economic, social, and environmental implications. Because there is a surge in coffee research and new trends have developed, the aim of this book is to gather up-to-date information on recent trends in coffee commerce, biotechnology, post-harvest processing, quality improvement, and consequences of coffee consumption on human health. Altogether, the chapters are focused on five different themes. During the preparation of this book, we made sincere efforts to provide quality scientific information on the chosen themes on coffee. The first section begins with a glimpse on the current status of coffee production, the global market, sustainable benefits of the coffee sector, and contemporary topics on coffee research. The second section aptly highlights the latest developments in coffee metabolites analysis. The remaining sections describe advances in post-harvest processing technologies, the intervention of biotechnology for improvement of *Coffea*, and coffee consumption and human health. The chapters, written by the experts in their respective fields from 8 countries, provide vast coverage of areas on the topics, including the latest developments in the field. Each chapter has been presented in a standalone manner for all-inclusive understanding. We strongly believe that the content described here will be very useful to improve the coffee quality and sustainable benefits on the coffee sector. We hope this book will be useful for researchers in academia, professionals in industry, and those all over the world who are involved in coffee science.

As editors, we would like to thank all the authors and coauthors for their timely submissions and cooperation extended during the compilation of the book.

Akula Ramakrishna
Parvatam Giridhar
Magdalena Jeszka-Skowron

Foreword

This book, *Coffee Science: Biotechnological Advances, Economics, and Health Benefits*, edited by Akula Ramakrishna, Parvatam Giridhar, and Magdalena Jeszka-Skowron, provides a valuable window on the current status on global coffee production and market, novel methods and new developments pertaining to coffee metabolites analysis, progressions in coffee processing technology and enhancement of coffee quality by coffee fermentation, the leads of solid-liquid extraction methods, and the implications of post-harvesting processes to enhance beverage quality and produce coffees with different sensory profiles.

The book covers biotechnological advances to improve coffee quality such as molecular markers, tissue culture, transgenic technology, and micro RNA technology and gives an overall glimpse on how best these methods could be used for the improvement of *Coffea*. Coffee consumption has been rising in recent years, and there was significant change in market dynamics owing to the COVID-19 pandemic. This book aptly covers the influence of various coffee metabolites such as methylxanthines, polyphenols, biogenic amines, and coffee diterpenes in human health effects.

I strongly believe the content described here will be very useful for improving coffee quality and sustainable benefits on the coffee sector. I feel honored and privileged to have this opportunity to write a foreword to this book. I am sure this volume will attract scientists, academicians, and professionals from all over the world who are engaged in plant/animal/microbial physiology, biochemistry, biotechnology, life sciences, agriculture science (particularly in coffee science), nutrition, and health. I extend my congratulations to the editors and publisher for this comprehensive, conversant volume of coffee science.

Grażyna Budryn
Faculty of Biotechnology and Food Science
Director of the Institute of Food Technology and Analysis,
Lodz University of Technology, Poland

Acknowledgment

We express our heartfelt gratitude to all the valuable contributors. We understand that the contributors are active scientists and volunteered to share their findings and perspectives with the readers.

We are grateful to Renu Upadhyay, Assistant Commissioning Editor; Jyotsna Jangra, Editorial Assistant; and the team from Taylor & Francis and Balaji K, Project Manager, Apex CoVantage for their help in bringing out the volume in a beautiful and appealing manner.

We are thankful to our families who have encouraged us to take up this task and permitted us to take the time off from the family. We are also thankful to our friends and academic colleagues for their support in the successful accomplishment of the book.

Akula Ramakrishna
Parvatam Giridhar
Magdalena Jeszka-Skowron

Editor biographies

Akula Ramakrishna is currently a Pipeline Enablement Scientist at Bayer Crop Science, Bangalore, India. He holds a master's degree from Sri Krishna Devaraya University, Anantapur, Andhra Pradesh, India. He obtained his PhD in biochemistry from University of Mysore, CSIR-CFTRI, Mysuru. He is a senior research fellow of CSIR, New Delhi. He is involved in various research fields such as plant biotechnology, biochemistry, plant secondary metabolites, food science and technology, vegetable quality analytics, double haploid production in vegetables, and tomato rapid generation advancement. He has published several research papers in leading international journals, as well as reviews and book chapters. He has edited books for CRC Press and two books for Springer. He is also serving as an editorial board member and reviewer for well-known international journals. He has presented over 20 research papers in symposia in India and abroad, as well as a few invited lectures in India. He attended the Fifth International Symposium on Plant Neurobiology held in 2009 in Florence, Italy. He also attended the Technical Community of Monsanto held in 2016 in St. Louis, Missouri. He is a member of the Indian Science Congress Association and Society for Biotechnologists (India). He is a fellow of the Society for Applied Biotechnology, India (2012), a recipient of the Global Vegetable Research Excellence Award (2017), Global Technology Recognition, Rapid Recognition Award, and Special Recognition from the Monsanto Company and Veg R&D champion (2020) from Bayer Crop Science.

Parvatam Giridhar is currently Senior Principal Scientist at CSIR-Central Food Technological Research Institute, Mysuru, Karnataka, India. He holds a master's degree from Kakatiya University, Warangal, Telangana, India. He obtained his PhD from Kakatiya University, Warangal. He was a DAAD Fellow (1995–1996) at University of Bayreuth, Germany, and BOYSCAST Fellow of DST, Government of India (2005–2006) at IRD, Montpellier, France. His research areas of interest are plant tissue culture, secondary metabolites, molecular biology, bioactive metabolites, and food science and technology. Since 2000 he has been engaged with R&D on coffee metabolites, tissue culture, and genomics. He has published several research papers in leading international journals, as well as reviews and book chapters – 190 in all, with 5496 citations, an h-index of 44, and an i10-index of 112. He is an associate editor, editorial board member, and reviewer for well-known international and national journals. He has presented over 50 research papers in conferences in India and abroad. He attended the 21st, 23rd, 24th, and 26th International Conferences on Coffee Science (Association Scientifique International du café [ASIC]) held in 2006 (Montpellier, France), 2010 (Bali, Indonesia), 2012 (San José, Costa Rica) and 2016 (Kunming, China). He also attended scientific conferences in Belgium, Malaysia, and Nepal. He is an elected member of the Plant Tissue Culture Association of India and the National Academy of Sciences, Allahabad, India. He is a fellow of the National Academy of Biological Sciences, India; Association of Food Scientists & Technologists, India; Indian Botanical Society; the Andhra Pradesh Akademi of Sciences, Andhra Pradesh, India; the Academy of Plant Sciences, India; and the Society for Applied Biotechnology, India. He holds nine patents and is a recipient of 12 important awards from various scientific associations and societies of India.

Magdalena Jeszka-Skowron is currently working as an academic teacher and scientist at the Faculty of Chemical Technology, Poznan University of Technology, Poland. She holds a master's degree from Poznan University of Economics at the Faculty of Commodity Sciences, Poland. She obtained her PhD in Food Technology and Nutrition from Poznan University of Life Sciences, Poland. She is involved in various research fields, including food chemistry and science, plant secondary metabolites analysis, food quality, and water and environment sciences.

Her experience in coffee research is connected with analytical methods, especially chromatographic, spectrophotometric, and isotachophoresis analyses, for the determination of bioactive compounds in coffee. She has published 30 research papers in leading international journals, including seven publications with the cooperation of Polish and foreign experts on coffee analysis. She has published a book, a review, and also book chapters. She holds two patents in Food Science. She has presented over 50 research papers in symposia in Poland and abroad, and she was a reviewer over 50 papers. In 2015 she attended the Symposium *Cocotea* – Coffee, Cocoa and Tea in Aveiro, Portugal. In 2015 she received the scholarship Engineer of the Future for the execution of a project titled "Quality of Coffee: Chemical and Toxicological Analysis of Coffee Beans Depending on Their Origin." In 2020 she received a grant

Miniatura from the National Centre of Science, Poland, for the execution of a project titled "The Influence of Fermentation Process of Coffee Beans on Caffeoylquinic Acids and Antioxidant Activity of Coffee Extracts (*Coffea arabica* and *Coffea robusta*)." She is also a lecturer, and one of her lectures for bachelor's degree students is "Instrumental Analysis of Food Products." She teaches laboratory classes in analytical and food chemistry.

Contributors

Dego Yusa Ali
Department of Agricultural Product Processing Technology
Universitas Brawijaya
Malang, Indonesia

Nádia Nara Batista
Department of Biology
Federal University of Lavras
Lavras, Brazil

Monika Beszterda
Department of Food Biochemistry and Analysis
Poznan University of Life Sciences
Poznań, Poland

Ana Paula Pereira Bressani
Food Science Department
Federal University of Lavras
Lavras, Brazil

Maria Emilia Camargo
Universidad de Caxias do Sul
Caxias do Sul, Brasil

David Israel Contreras-Medina
Research and Assistance Center for Technology and Design of the State of Jalisco, Mexico
Guadalajara, Mexico

Philippe Cubry
DIADE
Univ Montpellier, CIRAD, IRD
Montpellier, France

Beata Czarczyńska-Goślińska
Department of Pharmaceutical Technology
Poznan University of Medical Sciences
Poznań, Poland

Alexandre de Kochko
DIADE
Univ Montpellier, CIRAD, IRD
Montpellier, France

Juliana de Paula
Food and Chemistry and Bioactivity Laboratory Coffee
Research Center – NUPECAFE Institute of Nutrition
Federal University of Rio de Janeiro
Rio de Janeiro, Brazil

Disney Ribeiro Dias
Food Science Department
Federal University of Lavras
Lavras, Brazil

Félix O. Díaz-Arango
Caldas University (Universidad de Caldas)
Manizales, Colombia

Elia Socorro Díaz-Nieto
Universidad Autónoma de Querétaro
San Juan del Río, Querétaro, México

Vijayaramu R. Dirisala
Department of Biotechnology
Vignan's Foundation for Science, Technology and Research
Guntur, India

Varun Eranna
SFS Dept
CSIR-CFTRI
Mysore, India

Adriana Farah
Food and Chemistry and Bioactivity Laboratory Coffee
Research Center – NUPECAFE Institute of Nutrition
Federal University of Rio de Janeiro
Rio de Janeiro, Brazil

Adriana S. Franca
Food Science Graduate Program
Universidade Federal de Minas Gerais
Belo Horizonte, Brazil

Robert Frankowski
Institute of Chemistry and Technical Electrochemistry
Poznan University of Technology
Poznań, Poland

Parvatam Giridhar
Plant Cell Biotechnology Department
CSIR-CFTRI
Mysore, India

Anna Maria Jeszka
Institute of International Business and Economics
Poznań University of Economics and Business
Poznań, Poland

Magdalena Jeszka-Skowron
Institute of Chemistry and Technical Electrochemistry
Poznan University of Technology
Poznań, Poland

Mohamad Anuar Kamaruddin
School of Industrial Technology
University Sains Malaysia
Minden, Malaysia

Ngan Giang Khong
National Key Laboratory of Plant Cellular Biotechnology
AGI
Hanoi, Vietnam

Harish Babu Kolla
Department of Biotechnology
Vignan's Foundation for Science, Technology and Research
Guntur, India
and
PCBT Dept.
CSIR-CFTRI
Mysore, India

Hanumohan Reddy Konatham
Department of Biotechnology
Vignan's Foundation for Science, Technology and Research
Guntur, India
and
PCBT Dept.
CSIR-CFTRI
Mysore, India

Ijanu Emmanuel Madu
School of Industrial Technology
University Sains Malaysia
Minden, Malaysia
and
Department of Environmental Science
Federal University Dutse
Dutse, Nigeria

Tunjung Mahatmanto
Department of Agricultural Product Processing Technology
Universitas Brawijaya
Malang, Indonesia

Iychettira Machaiah Mandappa
Coffee Quality Division
Coffee Board
Bengaluru, India

Pierre Marraccini
DIADE
Univ Montpellier, CIRAD, IRD
Montpellier, France
and
National Key Laboratory of Plant Cellular Biotechnology
AGI
Hanoi, Vietnam
and
CIRAD
UMR DIADE
Montpellier, France

Silvia Juliana Martinez
Department of Biology
Federal University of Lavras
Lavras, Brazil

Manoj Kumar Mishra
Plant Tissue Culture and Biotechnology
Plant Biotechnology Centre, CCRI Manasagangothri Campus
Mysore, India

Shibin Mohanan
Department of Botany
Nirmala College
Kerala, India

Soumya Mukherjee
Department of Botany
Jangipur College
University of Kalyani
Kalyani, India

Pushpa Srinivas Murthy
Spices & Flavour Sciences Department
CSIR-CFTRI
Mysore, India

Noorzalila Muhammad Niza
School of Industrial Technology
University Sains Malaysia
Minden, Malaysia

Faris Aiman Norashiddin
School of Industrial Technology
University Sains Malaysia
Minden, Malaysia

Paulus Immanuel Nugroho
PT Santos Jaya Abadi
Kapal Api Global
Jakarta, Indonesia

Leandro S. Oliveira
Food Science Graduate Program
Universidade Federal de Minas Gerais
Belo Horizonte, Brazil

Angela M. Ormaza-Zapata
Faculty of Engineering
Caldas University (Universidad de Caldas)
Manizales, Colombia

Valérie Poncet
DIADE
Univ Montpellier, CIRAD, IRD
Montpellier, France

Yuniar Ponco Prananto
Department of Chemistry
Universitas Brawijaya
Malang, Indonesia

Monika Przeor
Department of Gastronomy Sciences and Functional Foods
Poznan University of Life Sciences
Poznań, Poland

Akula Ramakrishna
Bayer Crop Science
Kallinayakanahalli, India

Niny Z. Rao
Chemistry and Biochemistry Program
Thomas Jefferson University
Philadelphia, PA

Prakash Narayana Reddy
Department of Biotechnology
Vignan's Foundation for Science, Technology and Research
Guntur, India
and
PCBT Dept.
CSIR-CFTRI
Mysore, India

Benjamín A. Rojano
National University of Colombia at Medellín (Universidad Nacional de Colombia sede Medellín)
Medellín, Colombia

Johns Saji
Department of Botany
Nirmala College
Kerala, India

Palakkil Mavilavalappil Sandeep
Coffee Quality Division
Coffee Board
Bengaluru, India

Rosane Freitas Schwan
Department of Biology
Federal University of Lavras
Lavras, Brazil

Abdubaki Mohamed Hussen Shadi
School of Industrial Technology
University Sains Malaysia
Minden, Malaysia

Vaddadi Sridevi
PCBT Dept.
CSIR-CFTRI
Mysore, India

Ewa Stanisz
Institute of Chemistry and Technical Electrochemistry
Poznań University of Technology
Poznań, Poland

Wenny Bekti Sunarharum
Department of Agricultural Product Processing Technology
Universitas Brawijaya
Malang, Indonesia

Sridevi Vaddadi
PCBT Dept.
CSIR-CFTRI
Mysore, India

Anusree Poyyakara Veetil
PCBT Dept.
CSIR-CFTRI
Mysore, India

Tram Vi
DIADE
Univ Montpellier, CIRAD, IRD
Montpellier, France
and
National Key Laboratory of Plant Cellular Biotechnology
AGI
Hanoi, Vietnam

Frank H. Wilkinson
Chemistry and Biochemistry Program
Thomas Jefferson University
Philadelphia, PA

Mohd Suffian Yusoff
School of Civil Engineering
University Sains Malaysia
Pulau Pinang, Malaysia

Brian G. Yust
College of Humanities and Sciences
Thomas Jefferson University
Philadelphia, PA

Agnieszka Zgoła-Grześkowiak
Institute of Chemistry and Technical Electrochemistry
Poznan University of Technology
Poznań, Poland

Section I

Current status of coffee production, global market, contemporary topics in coffee research

1

Current status of coffee production and global marketing: Recent update

Anna Maria Jeszka

CONTENTS

1.1 Characteristics of coffee

Coffee beans are found in the fruits of the coffee tree. They mature for about nine months. Ripe fruits are harvested by hand or by machine. Coffee beans are extracted from the fruits (cherries) in two ways: by soaking in water for a few days or by drying in the sun for several weeks (https://nestle.pl).

Coffee grows in the so-called coffee belt, between the Tropic of Cancer and the Tropic of Capricorn (Figure 1.1). Coffee is a fruit and, like apples, has many varieties. Two varieties of coffee are the best known in the world:

- Arabica, the most popular, accounts for approximately 60% of world production; it has a sweet, delicate taste and is characterized by excellent acidity.
- Robusta contains more caffeine and is more resistant to disease; it has lower acidity, greater bitterness, and a deep body.

Beans purchased from the same producer can be completely different, depending on the rainfall, number of sunny days, and so forth. Coffee is transported from plantations in jute bags to a container ship. Loaded with several tons of coffee beans, the ship sails to one of many European ports. The journey can take up to several months, and the conditions in which the coffee travels can make a difference to its taste. Then the coffee is distributed to major cities on the continent and it goes to roasting plants. This is where the most important process, which gives coffee its flavor and aroma, takes place. The process and its final result are controlled by the roastmaster, who determines the sensory profile and sets the appropriate parameters: time, pressure, and temperature (a difference of even 1 degree Celsius can affect the taste). A series of tests are performed to eliminate errors. Even one bad bean can ruin the entire batch. During roasting, coffee loses about 20% of its weight, so from an initial 10 kg only about 8 kg will remain (Coffee market in Poland 2019).

The coffee industry has been the subject of analysis for a number of years. Consumers are becoming more and more aware and demanding, their tastes are changing, and coffee is drunk both at home and away. Being knowledgeable about coffee, its region of origin, and the roasting and brewing methods has become a symbol of social status. Increasingly, commercial coffee from supermarkets is being replaced with specialty coffee.

1.2 Global coffee production

Coffee is grown in South America, Central America, Asia, Oceania, and Africa. Global coffee production is dominated by several countries. Table 1.1 shows the ten largest coffee producers for the years 2015–2018. It is estimated that the sector employs 70–100 million people, and in 2018 global coffee production was over 10 million tons. Most small and medium-sized coffee producers typically deliver their coffee bags to processing plants, where it is combined with coffee from other producers for bulk processing. These undifferentiated coffees are sold as mass commodities based on New York or London stock prices (Lewin et al., 2004).

Four countries in the world produce nearly 70% of all coffee crops. These include Brazil – the undisputed leader, which in

DOI: 10.1201/9781003043133-2

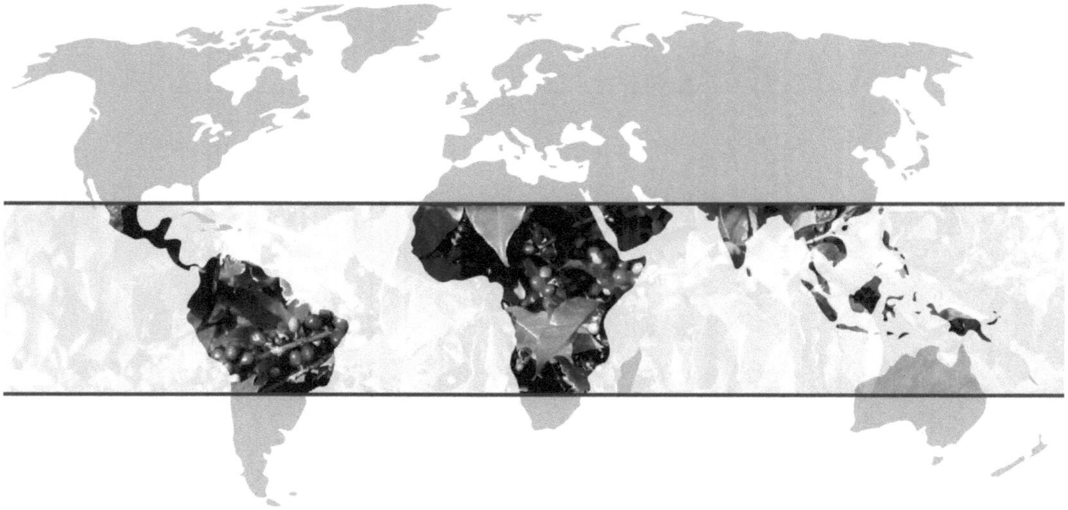

FIGURE 1.1 The coffee belt.

Source: https://www.rainforest-alliance.org/utz/.

TABLE 1.1

Output of the Ten Largest Coffee Producers in the World in 2015–2018, in Thousands of 60-kg Bags

	2015	2016	2017	2018
Brazil	53,850	55,776	55,786	62,944
Vietnam	28,737	25,540	29,688	29,874
Colombia	14,009	14,634	13,824	13,866
Indonesia	12,324	11,369	10,494	9813
Ethiopia	6714	7297	7450	7776
India	5830	6161	5813	5002
Honduras	5786	7457	7560	7328
Uganda	3650	4962	4597	4704
Mexico	2772	3635	4485	4351
Peru	3533	4237	4254	4135

Source: International Coffee Organization.

2018 produced 62,944 bags of coffee – followed by Vietnam with 29,874 bags, Colombia with 13,866, and Indonesia with 9,813.

Figure 1.2 shows the coffee production of ten key exporters in the world in 2015–2018. The leader is Brazil, which in 2018, counting in thousands of 60-kg bags, exceeded 60,000, followed by Vietnam in 2017 and 2018 with production at the level of approx. 30,000. However, in the case of Indonesia, production in the same year fell below 10,000, counting in thousands of 60-kg bags. Uganda, Mexico, and Peru did not exceed 5,000 over the analyzed years. Coffee production in 2016–2018 at a similar level in the case of Honduras and Ethiopia oscillates around 7,000, counting in thousands of 60-kg bags.

Figure 1.3 shows that Brazil, Vietnam, Colombia, and Indonesia are in the lead in terms of the amount of coffee produced. In 2018, Brazil accounted for 37% of the global coffee volume, followed by Vietnam with 18%, Colombia 8%, and Indonesia 6%. The ranking also shows the increasing importance of Ethiopia, which in 2018 had a 5% share of the global coffee market.

1.3 Global coffee consumption

The largest consumer of coffee in the world is the European Union, followed by the United States and Brazil. Among Asian countries, Japan has the highest consumption. Table 1.2 shows the dynamics of coffee consumption in the ten countries with the highest consumption levels in the analyzed years. Europeans are the clear leaders in terms of the amount of coffee drunk.

The International Coffee Organization lists the European Union, the United States, Japan, and Russia among the largest coffee importers. Increasing consumption of coffee is

TABLE 1.2

Global Coffee Consumption in 2017–2020, in Thousands of 60-kg Bags

	2016–2017	2017–2018	2018–2019	2019–2020
European Union	42,567	44,017	46,129	45,667
United States	25,775	26,112	27,759	27,934
Brazil	21,225	21,997	22,250	22,300
Japan	7913	7750	7561	7573
Indonesia	4650	4750	4800	4900
Russia	4638	4324	4506	4627
Canada	3783	3829	3885	3942
Ethiopia	3725	3750	3800	3825
Philippines	3030	3180	3300	3350
Vietnam	2400	2500	2700	2900

Source: International Coffee Organization.

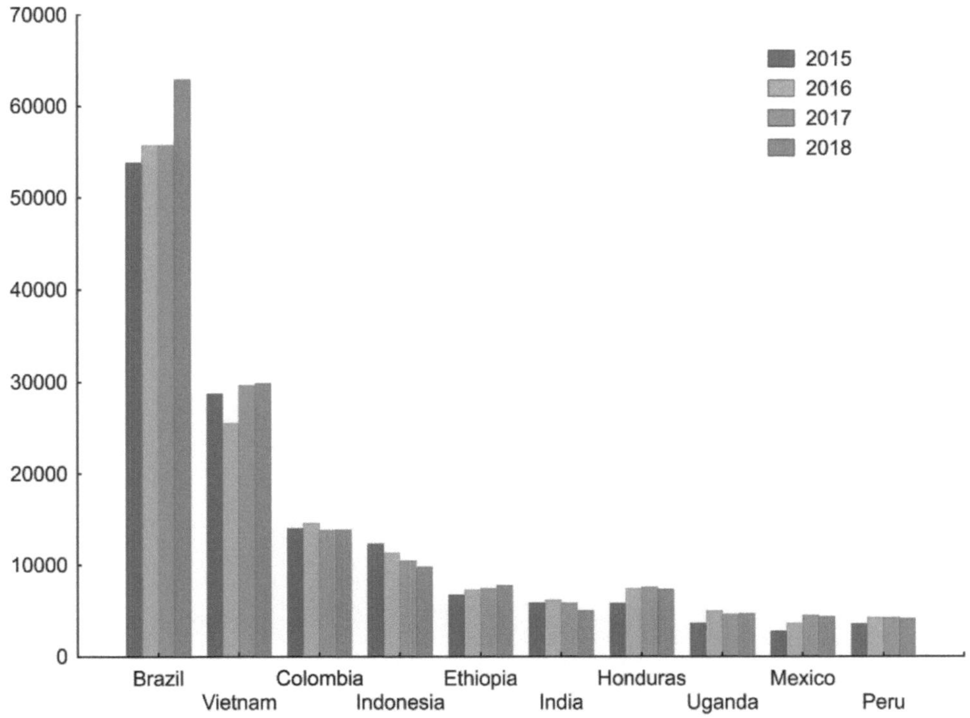

FIGURE 1.2 Global production of the ten largest coffee exporters in thousands of 60-kg bags.

Source: International Coffee Organization.

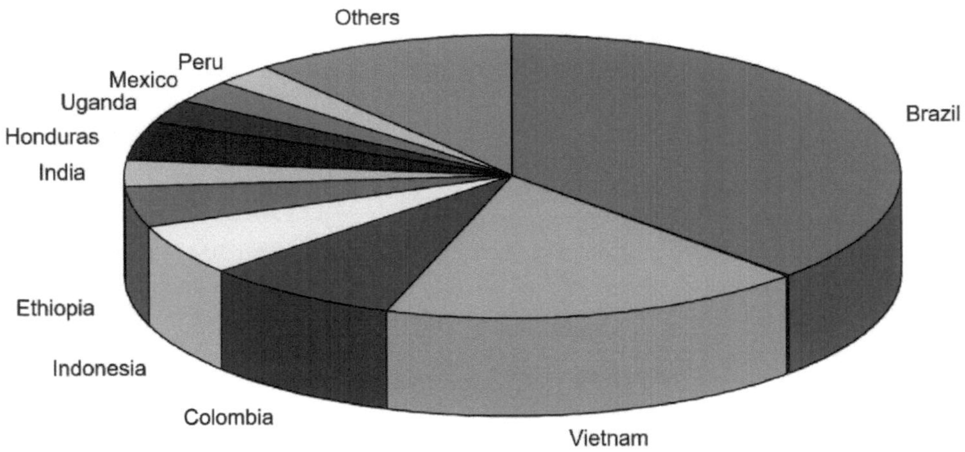

FIGURE 1.3 Global coffee production by country in 2018, in thousands of 60-kg bags (percentage share).

Source: International Coffee Organization.

recorded in Canada and South Korea. Coffee consumption in importing countries in 2019–2020 in thousands of 60-kg bags is shown in Figure 1.4.

The coffee sector is international in nature. As a rule, coffee growers are located in developing countries, whereas the greatest consumption occurs in developed countries. Exceptions include Brazil and Indonesia, which have a significant share in both production and consumption.

1.4 Consumption trends and health benefits of coffee

Coffee is one of the most popular drinks in the world. It has many supporters as well as many opponents. The demand for coffee is growing steadily, but there are some changes in consumption patterns. Depending on the region or country,

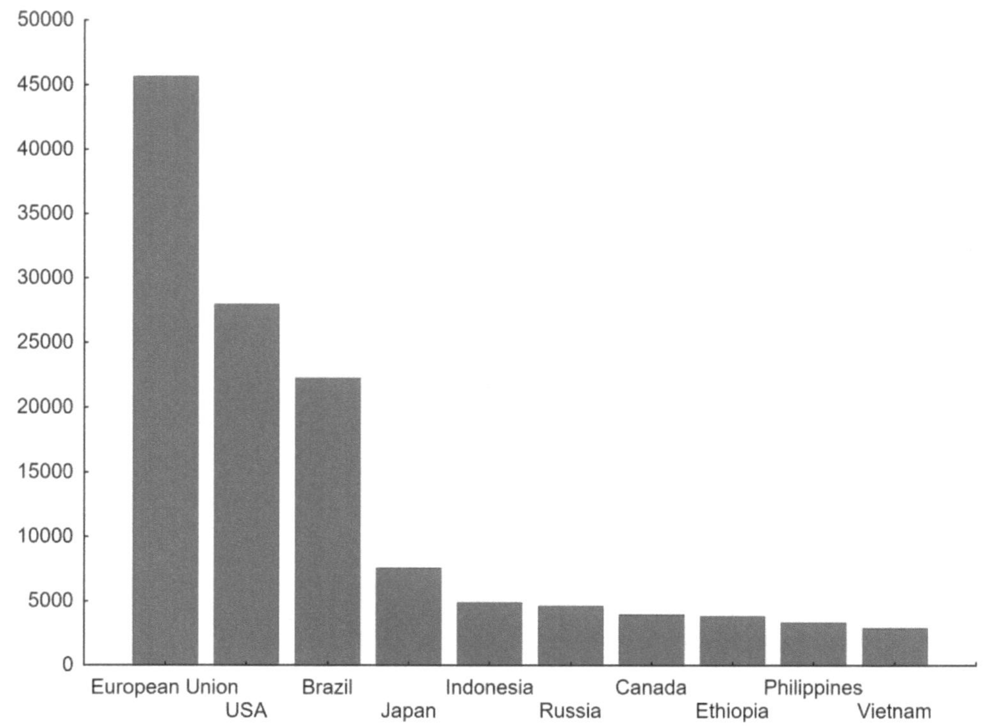

FIGURE 1.4 Coffee consumption in importing countries in 2019–2020 in thousands of 60-kg bags.

Source: International Coffee Organization.

people tend to prefer either coffee or tea, and either instant or ground coffee. In the past, one's status was evidenced by the fact of possessing a coffee machine (which used whole coffee beans, ground coffee, capsules, or pods). Today, such brewing methods such as Chemex, drippers, and AeroPress are back in fashion.

The largest increases in coffee consumption are forecast in China, Russia, and Turkey (Euromonitor 2020), where tea drinking has so far been culturally more important. In mature coffee markets, premiumization of consumption can be observed. This applies to countries where drinking coffee has a long tradition (Japan, the United States, and Western Europe; Euromonitor 2020).

It can be said that the development of the market follows several paths. There are consumers who above all value the ease of preparing a drink; they do not have a problem with having a quick coffee in a paper cup from a vending machine, drinking a canned coffee from the ready-to-drink (RTD) category, or a three-in-one instant coffee. Other consumers want the taste to be reproducible. There was a time when such additives as milk, syrups, or spices were commonly used because the coffee itself was intensely roasted and thus not particularly tasty. The trend of adding milk to coffee led to the emergence of "latte art," which has its own international championships. On the other hand, there are baristas, specialty coffees, and selected beans coming from specific plantations (with a known *terroir*) at particular altitudes that are produced and roasted by craftsmen. Gourmets are able to assess the aroma and, above all, the taste of a freshly prepared coffee brew in terms of a number of criteria, including acidity and body. In the literature, one can find a division into three so-called waves of coffee market development (Samper et al. 2017; Wikierska 2018).

The *first wave* occurred between 1920 and 1960, when coffee as a product became popular. The evolution started with little or no differentiation in terms of the origin of coffee and a market dominated by brands selling in bulk, mainly through chain stores and grocery retail outlets, focusing on standardization and volume. This was the case for most of the twentieth century.

The *second wave* (1960–1990) lasted for the next 30 years or so, and the focus was on differentiation. New players leveraged beverage quality attributes in out-of-home venues such as coffee shop chains. They focused on developing their own intangibles related to new brand experiences and beverage preparation techniques, selling them at higher prices. Second-wave brands have significantly contributed to the expansion of coffee industry value over the past 30 years and sparked interest in a somewhat dormant category.

The *third wave* can be described as "experiential." Currently, the emphasis is on the seller-farmer relationship. The supply chain is shortened. The novelty consists in a departure from coffee shop chains toward those run mainly by independent baristas and coffee shop operators, based on direct trade with coffee producers. This situation creates conditions for product innovation, increasing the role of the quality, value, and origin of coffee beans as product differentiators.

Throughout the development of the coffee market, particularly during the first and second waves, many consumers believed (and many still do) that coffee has a stimulating effect and a positive influence on memory and concentration, but at the same it time raises blood pressure, irritates the stomach, is addictive, and is generally bad for one's health.

In recent decades, many studies and experiments have been conducted to examine the effects of coffee on human health and life. The potential benefits of caffeine consumption are increasingly highlighted (Cornelis 2019). According to recent reports, coffee can prevent many diseases. For example, a meta-analysis showed that drinking two to four cups of coffee a day is likely to be of more benefit than harm to human health, while emphasizing that 300–450 mL is a safe caffeine dose. Studies have shown a positive causal relationships between coffee consumption and a reduced risk of type 2 diabetes, various types of cancer, and metabolic and neurological disorders (Poole et al. 2017). Coffee reduces the risk of diseases that cause brain damage, such as Alzheimer's and Parkinson's, by about 30% (Wierzejska 2017). International reports show that drinking coffee in moderate amounts can reduce the risk of heart disease by about 13%, but this effect does not occur with high coffee consumption (Wu et al. 2009). According to other reports, coffee can improve metabolism and aid weight loss (www.coffeeandhealth.org 2020).

On the other hand, the negative effects of drinking coffee by pregnant women are emphasized. These include low birth weight, premature labor in the first or second trimesters, and even pregnancy loss. Coffee consumption has also been shown to be associated with an increased risk of bone fractures in women (osteoporosis), but not in men (Poole et al. 2017).

As regards instant coffee, attention is paid to the additives and ingredients that are intended to thicken it and improve the taste. For example, nickel may cause allergies, and instant coffee may cause weight gain due to its composition.

1.5 Coffee market segmentation

From the sellers' point of view, coffee can be divided into segments that differ in terms of growth dynamics and development potential in individual countries and regions. The basic segments are the following (*Coffee: The Hidden Crisis Behind the Success Study on Sustainability Within the Coffee Industry* 2019):

1. Whole bean is typical standardized coffee. The increase in consumption is driven by the daily demand for caffeine or the desire to visit coffee shops. This type of coffee accounts for over 90% of coffee available on world markets and it includes the following types:
   - Certified coffees that comply with agronomic, social, and environmental production specifications. These include so-called sustainable certifications, internal labels (such as Nespresso's AAA label and the Starbucks C.A.F.E. Practices label), fair trade certifications, and organic certifications.
   - Origin coffees are those associated with a specific geographic location, which translates into a typical product sensory attributes such as taste and aroma. Classic examples are Colombian, Blue Mountain from Jamaica, and Kopi Luwak from Indonesia.
   - Specialty coffees are coffees of the highest quality produced in limited quantities and traceable to a single origin (*terroir*). They are characterized by a distinctive and superior taste and aroma, which is a result of the specific growing and harvesting conditions, and very few defects.
2. Ground coffee, including capsules and pods which facilitate the preparation of a coffee brew and which are suitable for brewing in coffee machines, and packaged ground coffee.
3. Instant coffee, which still frequently bought because it is easy to brew. There is a risk of instant coffee being replaced with low-cost coffee and energy drinks as well as RTD coffee. Instant coffee is made from a strong infusion of ground coffee, which is sprayed into hot air that turns the droplets into powder or granules. Another method is freeze-drying: the coffee extract is frozen and then dried at a low temperature under vacuum conditions, resulting in coffee in the form of golden crystals, which is more aromatic and of higher quality.
4. RTD coffee, which comprises coffee-based drinks. This type of coffee is most popular in Japan, but it is also gaining ground in European markets.

In the coffee market, a slowdown in sales volume growth is observed, but at the same time there is a sharp increase in sales value (Euromonitor 2020). Consumers may be drinking less coffee, but they spend more money on it. This trend is called premiumization.

Starbucks coffee shops make an increasing contribution to the growth in sales value. The partnership between Nestle and Starbucks (since 2018) is a very clear move (at least in part) to increase their share in the US retail coffee market, where Starbucks accounts for over 13% of retail coffee spending. In markets such as Indonesia and China, where overall coffee consumption is low, out-of-home channels could well remain the primary way of consuming coffee. It is quite possible that coffee shops will remain the primary places for coffee consumption in key growth markets such as China (Euromonitor 2020).

Specialty coffee is a premium coffee in higher-income markets. It is coffee from small producers for people who expect new and diverse flavors and origins. These are Italian-style espresso-based drinks but also coffee based on innovations, such as cold brewing. It is an opportunity for less mechanized production markets (e.g., Nicaragua).

The global coffee capsule market is entering a new era. The convenience of brewing one cup from a Nespresso, Keurig, or other capsule system has driven unprecedented value increases in wealthier markets, even as overall coffee retail sales have slowed or even reversed in some markets.

Capsule systems have become particularly popular in the Netherlands, where Philips and Douwe Egberts/Sara Lee probably started the capsule boom in the early 2000s. In the

Netherlands private label capsules now account for more than a quarter of the category, the highest proportion in the world (Euromonitor 2020).

Pods are also becoming more and more popular, mainly in Germany and Scandinavian countries, where in this case special attention is paid to ecology and reducing waste (*Coffee in Western Europe* 2020).

1.6 Coffee supply chain and value creation

The coffee supply chain begins with small farmers and growers who grow coffee trees and harvest the fruit (cherries). The cherries are then processed using the dry or the wet method to extract the coffee beans, which are green at this stage. Next, the beans are purchased by local agents and exporters. In the supply chain, coffee beans sometimes change hands several times.

The coffee supply chain typically comprises three general stages:

Stage 1: Agricultural steps – growing and harvesting coffee
Stage 2: Processing and preparation of green coffee beans, and production steps including roasting and grinding
Stage 3: Retailing, consumption, and disposal.

In fact, when it comes to "ordinary" coffee, which goes to world markets, and specialty coffee, there are two different types of supply chains. The first supply chain consists of conventional links and intermediaries. The coffee beans are transported to foreign markets and go to roasting facilities. The coffee supply chain is dominated by big players such as large consortia that roast and sell the coffee. Appropriately roasted beans are packaged, and are dispatched in retail or bulk packaging to consumers: individual customers, the hotel industry, institutions, restaurants, bars, and coffee shops, including coffee shop chains. The second supply chain is shortened to a minimum: the coffee beans go directly from the grower to a roasting plant, bypassing "world trade" channels on the stock exchanges in New York and London. In this type of supply chain, the relationship between the producer and the barista is important, as it is the latter who personally makes decisions regarding the choice of raw beans, their roasting, and brewing the beverage to satisfy customers.

A difference in practice between the largest roasters and smaller roasters is that the former generally rely upon traders to generate their supply of coffee rather than dealing directly with producers or producer groups, whereas the smaller ones, particularly specialty roasters, tend to secure their supplies directly from producers and their organizations (Byrnes et al. 2016).

Publications on the subject usually focus on the first type of supply chain because most of the coffee on the world's markets follows this type of chain. It is often emphasized that coffee producers (i.e., coffee growers) obtain rather low margins in relation to other participants of the supply chain. Particular criticism is directed at big companies, roasters that transform green coffee beans into roasted whole beans or instant coffee. This process, according to some, causes coffee prices to rise disproportionately. An answer to the question of why roasting plants have the greatest share in profits is provided by Cole and William (2019, chap. 32). According to these authors, this has to do with the capital committed to coffee roasting, as well as considerable expertise in terms of combining different varieties and batches of coffee so that the taste and aroma will be consistent and appreciated by customers.

It is worth looking at price fluctuations and the proportion of the price that coffee growers receive. This difficult issue has been raised by Fairtrade activists:

> While coffee is clearly profitable for food companies, it's a very different story for coffee farmers. The share of the retail value of coffee retained by the producer has fallen over the decades – in the 1970s, producers retained an average of 20 per cent of the retail price of coffee sold in a shop. When oversupply caused prices to crash to historic lows during the coffee crisis of 1994–2004, research found coffee growers received just 1–3 per cent of the price of a cup of coffee sold in a café in Europe or North America and 2–6 per cent of the value of coffee sold in a supermarket. Farmers are also the worst affected by the notorious volatility of world coffee prices. In recent years the price of Arabica coffee has swung from a 30-year low of 45 cents a pound in 2001 to a 34-year high of almost 309 cents in 2011. Between May 2011 and December 2013, prices fell by 65% as a result of the Euro crisis and oversupply of coffee. And after drought in Brazil saw prices rise briefly in 2014, prices fell back to a range of 115 to 145 cents a pound in 2015 and 2016.

Price instability is a very big problem for coffee producers. Climate change is another major issue, especially for mechanized coffee farms in Brazil and Vietnam.

The cost of roasted coffee includes the cost of green coffee beans, the cost of roasting, and the cost of shrinkage due to roasting (estimated at 10%–20%). Roasting costs are divided into two main categories: direct labor and overhead. The overhead costs include (1) superintendence, (2) power, (3) lighting and heating, (4) fuel and supplies, (5) repairs and maintenance of machinery and equipment, (6) depreciation of machinery and equipment, (7) premises (without depreciation), (8) depreciation of premises, (9) insurance, (10) rent, and (11) sundry roasting and processing costs.

In short, a share of income along the coffee supply chain is received by each of the following categories (Byrnes et al. 2016):

- Growers that pick and process cherries receive a "farm gate price".
- Processed cherries are traded at a "factory gate price."
- The green beans that are available to export by intermediaries are sold at free on board (FOB) prices.

- The beans shipped to consuming countries are exchanged for cost, insurance, and freight (CIF) prices.
- The beans are sold at wholesale prices.
- Roasters sell the processed beans at factory gate prices.
- Distributors sell the coffee to the end consumers at retail prices.

The value chain differs from the supply chain in that the value chain requires allocating value to each stage of the supply chain. Value chain analysis estimates the distribution of total income or surplus along the coffee chain. Surplus is the net retained profit at each stage of the coffee chain. It is possible to estimate the retained values as a percentage of the retail price. Retained value describes the income that remains within each producing and consuming country. In consuming countries, it is measured by subtracting the cost of imported green coffee from the retail price, which is equivalent to the total value added in a given consuming country. In producing countries, the total value added is estimated by subtracting the producer's price from export unit value (Byrnes et al. 2016).

The total income along the entire coffee chain comes from the money spent by consumers on purchasing coffee products. A study by Professor John M. Talbot divides the total income per pound of coffee into four parts: (1) the total income of coffee growers; (2) the incomes of processors, traders, exporters, and the state in producing countries; (3) the incomes of coffee manufacturers, wholesalers, retailers, and the state in consuming countries; and (4) a residual category including shipping costs, the profits of shippers and financiers, and weight losses that occur during the shipping and roasting of the coffee. The share of margin along the coffee chain is allocated in the following order (William et al. 2019):

- Producers, 20%
- Local intermediaries, 11%
- International traders, 8%
- Exporters, 6%
- Insurance and freight fees, 2%
- Roasters, 30%
- Retailers, 15%
- Taxes, 6%.

The largest players trading in green coffee are the so-called brokers, the largest trading consortia on the green coffee (raw material) market. The top four of these companies control more than half of the global coffee trade (Goldstein 2018):

1. Neumann Kaffee Gruppe (NKG), based in Hamburg, controls about 10% of the total coffee demand, with a vast international structure with about 50 subsidiaries in 27 countries. NKG is vertically integrated, including three large estates totaling about 7500 ha. The group owns production facilities, including 27 dry mills and seven wet mills. NKG runs a vertically integrated supply chain, from farming and farm management; export; grading; import and trade of industrial Y-grade volumes, specialty coffee, and instant/soluble coffee; services of semi-finished products, logistics, and storage; and risk management and finance. NKG is also involved in specialty coffee trading for the US market under its company InterAmerican Coffee, which is Fairtrade and Rainforest Alliance certified and operates from its offices in Houston, San Diego, and Providence. In early 2018, Neumann acquired a majority stake in Atlas Coffee Importers, a specialty coffee dealer based in Seattle, shifting its focus to maintaining relationships with quality roasters as opposed to focusing on the "commodity coffee" volume.
2. Ecom Agroindustrial, a Switzerland-based coffee trading company with annual sales of nearly $20 billion, is currently one of the world's largest coffee traders. It began trading coffee in 1935 with its first coffee office in São Paulo, Brazil. Ramon Esteve, the sixth generation of the family, continues to be involved in the company's management.
3. Olam, which began operating in 1989 and has achieved a strong position in certain commodity markets, including edible nuts, palm oil, and spices. Olam's global coffee growth strategy includes setting up new plantations in Laos, Zambia, Tanzania, and Brazil. Following market trends, Olam has also increased its presence in the specialty coffee market as Olam Specialty Coffee in the Americas, and Schluter for the UK/EU market, which it acquired in 2016 for $7.5 million and which will represent Olam's specialty coffee arm in Europe.
4. Volcafe Ltd./ED&F Man: Volcafe operates as the coffee trading arm of ED&F Man, one of the world's largest commodity trading firms. Volcafe has been trading coffee since 1851 and prides itself on being capable of handling both small, single-origin premium coffees and mass-market coffee blends. The company is vertically integrated and sources from over 15 different countries through its coffee trading hubs.

   Other top traders include:
5. Louis Dreyfus Company (LDC; also Louis Dreyfus or just Dreyfus) operates from its coffee trading centers in Geneva (Switzerland), Santos (Brazil), Wilton (United States), Nairobi (Kenya), and Singapore. LDC is one of the largest coffee traders in the world and operates actively in the coffee business via strong origin sourcing from locations such as Dalat and Lamdong (Vietnam), Villanueva (Honduras), and Uganda.
6. Sucafina is a leading coffee trader founded in 1977. Its main offices are located in Geneva. Sucafina progressively developed toward more sophisticated commodity trading, hence the name SU (sugar), CA (coffee), and FINA (financials). Sucafina specialties include 32CUP, a specialty green coffee distributor and Beyers Koffie, a private-label roaster, both based in Belgium.

7. Mercon is a green coffee trading specialist that has been in the coffee trading business for over 50 years. Mercon's headquarters are located in the Netherlands, but the company also runs operations in nine other countries: Brazil, Guatemala, Honduras, Netherlands, Nicaragua, Panama, Spain (Barcelona), the United States, and Vietnam (Ho Chi Minh City). Its main business lies in import/export, but it also controls some plantations in Vietnam. Additionally, Mercon runs a high-quality specialty coffee division called Mercon Premium Estates, which sources single-origin coffees from such countries as Nicaragua, Vietnam, Guatemala and Honduras.
8. ADM Investors Services (Archer Daniels Midland): ADM's presence in the coffee market is limited to being a market maker/broker in options and futures for the most important coffee traders in the world.

1.6.1 Roasters and distributors

It is also worth looking at other intermediaries in the coffee supply chain, those who receive the largest value from the coffee value chain in retail markets. For a large industrial coffee buyer, the main challenge is to ensure a regular supply of coffee beans. The role of coffee processing companies in the supply chain is roasting and packaging coffee and distributing it to the retail market. The following is a list of the largest companies who buy most of the raw coffee beans in the world, process it, and distribute it on the market under their own brands (Goldstein 2018):

1. Nestle SA (Switzerland): Nestle is one of the world's leading coffee companies with about +20% of the total retail market. It controls some of the world's best-known coffee brands such as the instant coffee Nescafe and the premium coffee capsule system Nespresso. In 2017, Nestle bought Blue Bottle Coffee, a US specialty coffee brand, for $425 million. In 2018, Nestle agreed to pay Starbucks $7.15 billion in exchange for gaining rights to exclusively retail coffee and tea under the Starbucks brand.
2. Kraft Heinz (United States): Kraft Heinz was born as a result of the merger between Kraft Foods and Heinz. It has annual sales in the region of $26 billion, and its shares trade on the NASDAQ as KHC. One of the largest roasters and distributors in the world, Kraft Heinz owns brands such as Maxwell House, Gevalia, General Foods International, Baker's, Cafe Collection, Yuban, Sanka, and Tassimo.

   Mondelez International, an American corporation, is one of the world's strongest coffee distributors and ranks either first or second in 20 different markets. After a fusion with Heinz in 2012, Mondelez, headquartered in Deerfield, Illinois, took over the snack and food brands of the former Kraft Foods brands umbrella. In 2015, the company divested most of the coffee business to focus on its core snack businesses. The newly created joint venture is called Jacobs Douwe Egberts. The company controls some of the world's most popular coffee brands, including Jacobs, Kenco, Carte Noire, Tassimo, Saimaza, Kaffee Hag, Jaques Vabre, Gevalia, Splendid, Maxwell House, and Grand Mere.
3. Jacobs Douwe Egberts (Netherlands): In 2015, Mondelez and D.E. Master Blenders announced the creation of a joint venture that fully combined their coffee businesses. The new company is called Jacobs Douwe Egberts, based in the Netherlands, and has over €5 billion in revenue from coffee-related sales. The reasoning behind this consolidation was to unite forces against Nestle.
4. Keurig Dr. Pepper (United States): In July 2018, Keurig Green Mountain announced a merger with Dr. Pepper Snapple Group. The new company is called Keurig Dr. Pepper and aims at becoming a leading player in the hot and cold beverages segment. With a combined annual revenue in excess of $11 billion and with about 25,000 employees, Keurig Dr. Pepper holds a portfolio of 125 brands including roasted coffee brands such as Green Mountain Coffee Roasters, The Original Coffee Shop, Van Houtte, and Barista Bros.
5. Tchibo (Germany): Founded in 1949 in Hamburg, Tchibo is a popular German coffee retailer with over 700 direct shops in Germany and over 300 in the rest of the world.
6. Lavazza (Italy): In 1895, Luigi Lavazza founded the eponymous iconic Italian coffee maker located in the northern Italian city of Torino. Almost 125 years later, the company – now run by third- and fourth-generation family members – achieved a yearly revenue of almost $1.5 billion. In October 2018, Lavazza bought Mars's coffee business, including its Flavia and Klix vending systems.
7. Strauss Coffee (Israel): Strauss Coffee, a top-ten coffee company in the world, is growing fast in emerging markets with its 16 coffee brands and about 6,000 employees. Traded on the Tel Aviv Stock Exchange, Strauss Coffee controls brands such as BeanZ, Elite Coffee, MK Cafe, Ambassador, Doncafe, Amigo, Chernaya Karta, Fort, and Chorna Karta. Strauss Coffee runs its purchasing and green coffee trading hub from Zug, Switzerland.
8. Starbucks (United States): Starbucks is one of the world's most famous coffee firms. As one of the truly international coffee franchises, the company focuses on darker roasts of Arabica-only beans. In 2018, it was announced that Starbucks and Nestle reached an agreement regarding the use of the Starbucks brand in the sales of coffee and tea. Nestle will from now onwards be able to distribute roasted beans under Starbucks branding.

1.7 A sustainable coffee supply chain

The global coffee sector experiences strong fluctuations in coffee prices, causing farmers to abandon growing the plant when the price does not cover production costs. In addition, there are rising production costs, the need to ensure food security, and the poverty of the people who work on coffee plantations. The changing climate is also a challenge.

For the entire sector to develop, it is essential to improve the situation at the initial stage of the supply chain (i.e., the working conditions of the farmers responsible for the production of coffee beans). For several decades, a number of international organizations have been trying to help those farmers and their families. The most important thing is that the people who work hard on plantations receive support in the cultivation of "black gold" as well as remuneration that is fair in relation to the value created for other participants in the supply chain.

Fair trade organizations collaborate mainly with small producers' cooperatives. Their basic guiding principles include the following:

- Fair wages for workers;
- Safe and healthy working conditions;
- Ensuring the equality of women and strengthening their economic position;
- No discrimination or child and slave labor;
- Respecting workers' freedom of association and right of collective bargaining;
- Investing the Fairtrade social premium in development projects for local communities, such as building and running health centers, schools, drinking water supplies, childcare programs, and training;
- Using environmentally friendly cultivation and production methods that preserve the quality of land, water, and air and protecting the health of agricultural workers and, consequently, consumers. (Fairtrade Fundation, 2020)

The majority of fair trade producers also adhere to organic production standards, which is confirmed by appropriate certificates.

Fair trade importers ought to ensure the following:

- Fair local prices for products and fair employee wages;
- Purchasing directly from organizations of small producers in southern countries, bypassing intermediaries;
- Long-term contracts and advance payments for producer groups, which create the basis for sustainable livelihoods and economic security and help them develop their business activities;
- Transferring the Fairtrade premium to producer groups for development projects in their local communities;
- Technical assistance in the form of training, access to market information, and so forth.

Many roasters that retail their products under their own brands make sure that the coffee comes from sustainable sources. For example, the following information can be found on the website of Sagafredo, which uses Organic, Fairtrade, UTZ, and Rainforest Alliance certification.

The Organic certificate confirms that the raw materials come from supply chains in which the principles of organic farming have been implemented. For companies operating in Europe, the rules formalizing the characteristics of organic farming are set out in the legislation of the European Community, while in the United States such rules are developed by the National Organic Program (NOP).

Fairtrade certification guarantees that products bearing the Fairtrade mark have been produced respecting the rights of workers in Asia, Africa, and Latin America.

The UTZ standard is based on such principles as transparency of business practices, traceability of raw materials and compliance with the conventions of the International Labor Organization.

The Rainforest Alliance standard is based on the following principles: protection of local flora and fauna, protection of ecosystems, protection of soil, protection of water resources, fair treatment of workers and good working conditions, good community relations, and integrated waste and pesticide management, monitoring, and disposal.

The Rainforest Alliance, founded in 1987, focuses on the protection of the environment and the preservation of rainforests in the countries where coffee is grown. The organization also monitors the working conditions of coffee plantation workers. Products from plantations that meet the requirements of safety, social justice and environmental protection receive the Rainforest Alliance Certified seal. This certification means that the cultivation methods used on the coffee plantation comply with the standards of economic, social, and environmental sustainability and are controlled by the Rainforest Alliance and its partner organizations in the Sustainable Agriculture Network (SAN). Farmers and their families experience the positive effects on their farms through fair wages, improved working conditions, and larger crops and better quality beans.

In plantations where coffee is grown sustainably, no pesticides or chemical fertilizers are used, the soil is fertile and gives higher yields, and both the soil and climate are protected.

Certified coffee cultivation means that the coffee plantations use sustainable growing methods, which is monitored in accordance with certification standards. The criteria that farmers must comply with include environmental protection, social justice, and economic issues. To be certified, plantations must meet all the core criteria and at least 80% of the total number of 99 criteria.

Fairtrade coffee producers introduce solutions that increase the quality and efficiency of crops. In return, they can count on a fair income that guarantees covering the cost of sustainable coffee production. Small farmers receive at least the Fairtrade minimum price for their products as well as the so-called Fairtrade premium. For the last 25 years, the Fairtrade price has been higher than the price on world markets, allowing farmers not only to cover the cost of sustainable coffee cultivation but also to live in dignity and support their families.

Organic coffee beans usually are typically retailed as specialty coffee. They are of single origin and high quality. Currently, large international coffee consortia are also interested in the margins that can be achieved in this segment; consequently, organic coffee can more and more often be found in retail stores and supermarket chains.

Most organic food stores offer various types of Fairtrade coffee, often with organic certification; Aldi, ground and whole-bean coffee under Markus and Moreno private labels;

Carrefour, among others, coffee under the Carrefour BIO own brand; Kaufland, among others, coffee under the K-Bio own brand; Lidl, coffee under the Bellarom's own brand; and Tchibo, two types of Fairtrade Barista whole-bean and ground coffee.

The UTZ organization was established in the 1990s, and the UTZ standard operates through two sets of guidelines:

1. The Code of Conduct (for producers/farmers)
2. The Chain of Custody (for the remaining supply chain participants).

The UTZ Code of Conduct for producers includes training for farmers with regard to, among other things, improving soil quality, crop diversification, and alternative methods of crop protection and increasing yields. The purpose of these activities is to ensure better crops and increase the farmers' income. In order to obtain UTZ certification, producers must comply with good agricultural practices, ensure appropriate working conditions, and minimize any negative effects of agricultural production on the natural environment.

The UTZ Certified mark means that the producer complies with the standards specified in the Code of Conduct, has been audited by a designated certification body, and is regularly inspected by UTZ program inspectors and auditors. During an audit, documentation relating to planting and the use of fertilizers and pesticides is checked. This includes such information as harvest dates, information about the land plot, and yields from specific areas. Inspectors also check documentation related to the employment of workers, including contracts, work records, and payrolls. In this system, each quantity of produce sold is registered in the UTZ electronic product identification system (the Good Inside Portal).

Plantation audits also include the monitoring of working conditions, facilities, and activities in terms of good agricultural practices. Anonymous interviews are conducted with employees with regard to such issues as gender or ethnic discrimination and worker abuse. Additionally, the program sets out specific requirements with regard to child protection, mainly through banning child labor. The UTZ system aims to ensure that all children of farmers and farms employees can grow up in a healthy and safe environment and that they receive educational opportunities.

The UTZ certificate for raw material processors confirms that the product produced by them is physically and/or administratively related to the raw material from a UTZ-certified farmer.

The audit covers all stages of production, from the purchase of raw materials to the delivery of the finished product to the customer:

> In addition to the qualitative identification of the UTZ certified raw material, the quantity of the final product in relation to the quantity of the raw material purchased is also checked. The standards define detailed requirements concerning the management of the raw material, the documentation as well as the labelling and electronic registration processes.

In order to obtain the UTZ certificate, a processor must purchase a specific amount of UTZ-certified raw material.

The fair trade movement is well established in the coffee industry. The most famous certificates are Rainforest Alliance, UTZ, and Fairtrade, but there also some internal certificates issued by individual companies such as Starbucks. Consumer awareness regarding the ethical aspects of manufacturing products and the origin of raw materials plays an increasingly important role. This is significant for processing companies, which seek to increase their competitiveness by having their products certified.

1.8 Conclusions

Coffee is an important commodity in international trade. The manner of coffee consumption is changing: even though the same varieties of coffee continue to be traded around the world, consumers have now started to pay attention to the coffee's origin, the region where the coffee is grown, the harvesting and processing methods, and the quality selection. The method of roasting, the degree of grinding, and the type of brewing are also becoming increasingly important to consumers. From a global point of view, coffee production and consumption continue to grow, although the latter seems to be close to its upper limit. The dynamics of growth varies depending on the country and segment: coffee drunk in coffee shops, specialty and origin coffees, coffee in capsules and pods, and instant coffee.

In the global supply chain of standardized coffee used for the coffee blends commonly available on the retail market, the lowest profits are invariably achieved by farmers and the largest ones by commercial coffee roasters, places where the beans are processed and packaged to later be dispatched to retail chains, coffee shops, or the hospitality industry under various brands. While specialty coffee has always had its faithful supporters, now also global players, including department stores and roasters, are becoming more and more interested in this segment. They are concerned about the sustainability of the supply chain and cooperate with regard to sustainable cultivation and fair trade. The sales of certified sustainable coffee have an increasing share in global coffee sales.

REFERENCES

Byrnes, W., Perez N.C., & Khodakarami N. 2016 (on-line).

Coffee in Western Europe. 2020. Market Research Report.

Coffee: The Hidden Crisis Behind the Success Study on Sustainability Within the Coffee Industry. 2019. Max Havelaar France, The French Fair Trade Platform, the "Rethinking Value Chains" Collective, BASIC.

Cole, R.T., & Byrnes, W. 2019. Publisher: LexisNexis.

Cornelis, M. *Nutrients*, 2019;*11*(2):416.

Fairtrade Foundation. fairtrade.org.uk (accessed 26 January 2020).

Global coffee: understanding threats to growth in 2020 and beyond. Euromonitor. 2020.

Goldstein, D. *Who Moves the Coffee Markets? Meet the World's Largest Green Coffee Traders*. 2018. commoditytrading.guru

International Coffee Organization (ICO) www.ico.org (accessed 30 January 2020).
Lewin, B., Giovannucci D., & Varangis P. *SSRN Electronic Journal,* 2004.
nestle.com (accessed 26 January 2020).
Poole, R., Kennedy O. J., Roderick P., Fallowfield J. A., Hayes P. C., & Parkes J. *BMJ*, 2017;j5024.
Rynek kawy w Polsce – szybko rośnie segment kawy specialty (Coffee market in Poland – specialty coffee segment is growing rapidly). www.wiadomoscihandlowe.pl/artykuly/rynek-kawy-w-polsce-szybko-rosnie-segment-kawy-spe,58361 [in Polish] (accessed 26 January 2020)
Samper, L.F., Giovannucci D., & Vieira L. M. WIPO. 2017.
UTZ – UTZ Certified. utz.org (accessed 27 January 2020).
Wierzejska, R. *Arch Med Sci.* 2017;*3*:507–514.
Wikierska, A. Trzecia fala kawy (Third wave of coffee). *Kaworama.* www.kaworama.pl/ciekawostki/trzecia-fala-kawy/ 2018. [in Polish] (accessed 26 January 2020)
Wu, J., Ho S.C., Zhou C., Ling W., Chen W., Wang C., & Chen Y. *Int. J. Cardiol.* 2009;*137*(3):216–225.

2

Scientific and technological knowledge: A review of sustainable benefits on the coffee sector

David Israel Contreras-Medina, Maria Emilia Camargo, and Elia Socorro Díaz-Nieto

CONTENTS

2.1 Science and technology's role in the world

Science and technology are concepts that have been recognized as important drivers for social and economic progress (United Nations Educational, Scientific and Cultural Organization, 2012; National Research Council, 2006) and as vital elements for providing solutions about how to eradicate poverty (United Nations Educational, Scientific and Cultural Organization, 2017), and, in the present day, they are being considered in the transition toward a sustainable world (United Nations, 2013; Clark and Dickson, 2003).

Science, defined as the careful study of the structure and behavior of the physical world, has been combined with many kinds of tools for practical purposes based on a set of principles, laws, and methods for building new knowledge (Cambridge University Press, 2020a; Andersen and Hepburn, 2015). This conjunction was significantly important since the hunting age, where in taking advantage of the physical world our ancestors built sharp stones and bows and arrows to obtain food; in the agricultural age, they created modest technologies to work the land; during the Industrial Revolution, the combination of technologies to substitute for manual jobs increased the volume of production; and into the digital age, where technologies such as sensors, robots, and virtual systems are being used to enhance management, streamline the industrial process, and increase sustainability (Sachon, 2017).

Despite these advancements, in today's world the use and benefits of science and technology have manifested heterogeneously. In developed countries, the activities related to scientific and technological knowledge support industries for social, economic, and environmental sustainable development; whereas in developing countries, the productive sectors, mainly agricultural, present important gaps between science and technology activities, manifesting in social, economic, and environmental lags and resulting in unsustainable development in these regions (United Nations, 2018a, 2018b).

Agriculture is the backbone of economies in nations of Asia, Latin American, the Caribbean, and Africa, with around 25 million people and their families dedicated to farming activities (International Coffee Organization, 2019; International Monetary Fund, 2020; Food and Agriculture Organization, 2013). This includes the production and consumption of corn, rice, and coffee (among others; Food and Agriculture Organization, 2017) in both developing and developed countries.

In comparison with the most alternative economic activities, coffee growing and trading contributes in a positive way to social and environmental issues (International Coffee Organization, 2020), from its contribution to carbon sequestration in the environment (see Birkenberg and Birner, 2018) to preventing and fighting cancer (McNutt and He, 2019), or as potential feedstock for biodiesel production using coffee waste (Kamil et al., 2020).

To underscore the fact that the science and technology has greatly contributed to the development of human beings, the social, economic, and environmental balance for sustainable practices is promoted by almost all countries in the world. In this sense, and in order to explore the social, economic, and environmental benefits created by scientific and technological knowledge for agriculture, it is imperative to explore the helpful or positive effects (Cambridge University Press, 2020b) in the context of sustainable coffee activity throughout the world.

2.2 Sustainable development evolution

Sustainable development is defined as "the development that meets the needs of the present without compromising the ability of the future generations to meet their own needs" (Brundtland, 1987). One of the first antecedents came in 1864, in studies related to the relationship of human activities with

DOI: 10.1201/9781003043133-3

the environment, which already focused on deforestation without reforestation, the excessive use of natural resources without compensation, and the evident bad relationship between man and the environment (Lowenthal, 2015). This manifestation recognized the need to promote the care and conservation of natural spaces, the requirement to find a balance in the use and exploitation of natural resources, and the demand to harmonize society with the natural environment (Goldstein et al., 2013).

Given the warning about the damage to the environment and the subsequent risk to economic development, the Stockholm Summit was held in 1972, at which several countries began to show concern for the environment (Flavin and French, 2002). However, it was not until 1983, with the creation of the Brundtland Commission, that the concepts of economy, environment, and society began to interrelate (Alfaro Barbosa et al., 2014). From this, it was recognized that it is not possible to conceive of economic growth without a sustainable environment, and that economic and social activities are basic principles for sustainable development (Larrouyet, 2015).

The interrelationship between social, economic, and environmental features began to become evident through national, regional, and international action plans at the 1992 Rio de Janeiro Earth Summit (da Fonseca et al., 2012). Then, in the year 2015, these plans were revisited and now are the pillars of the 2030 Agenda for Sustainable Development and of the 17 Goals for people and for the planet (United Nations, 2020).

In the context of sustainability, the social dimension implies a system of collective organization that alleviates poverty, the economic dimension involves a system of production for satisfying consumption levels without compromising future needs, and the environmental dimension is related to the conservation of ecosystems, biodiversity, and climate change (United Nations, 2015; Littig and Griessler, 2005; Lobo et al., 2015). The integration and balance of these social, economic, and environmental dimensions is the commitment of almost all the economies in the world to achieving sustainable development (United Nations, 2015).

In achieving social, economic, and environmental balance and facing the challenges of increasing population, it is necessary to accelerate the transition to sustainable agriculture. This action will provide economic and social opportunities and protect the primary ecosystem on which everyone in the world depends (Food and Agriculture Organization, 2014).

2.3 Scientific and technological knowledge for sustainable coffee production

The benefits of science and technology have been probed in the past, but today, with the introduction of sustainability, the worldwide challenges multiply. This review describes and links 52 scientific studies and technological actions examining the benefits (helpful good effects, or routes intended to help) of the social, economic, and environmental dimensions in the worldwide coffee context, in the period following the establishment of the 2030 Sustainable Development Goals.

In this sense, technological actions for strengthening coffee industry have called on different routes intended to help. For example, in Africa (specifically in Ethiopia, which accounts for 32% of coffee exports), market volatility, evident in the price fluctuations arising from national economic policy and international markets for coffee products, makes the sector vulnerable. Consequently, platforms have been developed in which buyers and sellers come together to trade and guarantee quality deliveries and payments, and the Ethiopian Intellectual Property Office (EIPO) mechanism was founded to help obtain a higher percentage of the benefits from the country's coffee growers (Mohan, 2016; Gashaw and Kibret, 2018). In Kenya, where six million people make their living on coffee, challenges such as managing knowledge and the lack of stable production capacity for coffee growers and cooperatives affects its vigor to develop competitive advantages and access to global viable markets. In this sense, actions related with soft technologies (ST; Jin, 2002) such as certification standards, production capacity, and delivery have been developed for increasing the competitive advantage of the sector (Chiuri and Taku, 2015). In Rwanda, which has low accessibility to markets, actions as agricultural agreements by contract (ST) are the key advantages for obtaining market access, price stability, and technical assistance, which ensure increased productivity through intensity (Garcia, 2018). In Latin America and the Caribbean, the challenges for increasing the yield and the quality of coffee crops from Mexico are focused on supervising coffee growth variables through precision agriculture tools such as greenhouses and equipment for transmission devices that allow the monitoring of coffee crops and accessing data via Web tools (Contreras-Medina et al., 2019). Panama monitors the races of coffee rust through a research platform, and Peru uses solar dryers in their coffee-growing activities (Inter-American Institute for Cooperation on Agriculture, 2018). In Asia, specifically in Vietnam, the introduction of an automatic harvester has helped achieve productivity growth (Texeira et al., 2019). The intercropping model (ST) for income diversification for reducing coffee price volatility is also promoted in Vietnam (Food and Agriculture Organization, 2019) (see TABLE 2.1).

In terms of scientific studies, the benefits, effects, or routes are traced in different topics. In the environmental dimension, the lines explored by scientific studies include using coffee waste as a potential feedstock, showing savings with performance comparable to regular diesel and lowered pollutant emissions. Other study revealed the influence of climate change on coffee production, with a yield increase in some Brazilian states reaching 2439.64 kg ha^{-1} (during El Niño conditions) and a decrease to 325.96 kg ha^{-1} (during La Niña conditions).

Another study was related to the production of biodiesel from recycled coffee waste, recommending the inclusion of pelleted biomass for viability. Other research explored the potential of spent coffee grounds as a commercial feedstock, revealing 40 novel components and alternative products that could be produced based on coffee. Research has also explored the environmental impacts of biodiesel production from spent coffee grounds, concluding that the implementation of a plant will be feasible and profitable when production is on the order of 10,000 tons per yield. An additional study described programs related to certification and sustainability criteria.

Another study identified water conservation, the use of retention barriers, and water harvesting as the most important

TABLE 2.1

Benefits, Effects, or Routes Related with Technological Actions

Technological actions	Benefits, effects, or routes intended to help	Region
Platform	Articulation of sellers with buyers	Ethiopia
Certification standards	Increasing competitive advantage	Kenya
Contract agreement	Improving market access	Rwanda
Precision and Web tools	Monitoring coffee crops	Mexico
Solar dryers	Monitoring coffee activities	Panama and Peru
Automatic harvester	Increasing production	Vietnam
Intercropping model	Reducing price volatility	Vietnam

Sources: Mohan et al. (2016); Gashaw and Kibret (2018); Chiuri and Taku (2015); Garcia Thomas (2020); Contreras-Medina et al. (2019); IICA (2018); Texeira et al. (2019); Food and Agriculture Organization (2019).

factors in terms of the adoption of sustainable practices. A review studied options for productively utilizing spent coffee grounds, resulting in a set of possibilities as oil, biodiesel, for sugar recovery, as bioethanol and solid biofuel, for phenolic compounds, as antioxidants, composting, fertilizer, for preventing and fight cancer cells, as material in the construction industry, as plastic, as bricks and as an absorptive filter. Another study analyzed the kinetics of polyphenolic compounds, concluding that solvent, Soxhlet, and ultrasonic-assisted extraction have great potential for extracting polyphenolic compounds due to its simplicity, high reproducibility, high efficiency, and low energy requirements.

An investigation related to the interaction of pest biology and climate/CO_2, recommending the integration of GIS technology, social networking, and visual assessments into traditional pest management practices. The study reviewed the adoption of in-house certification by coffee farms, concluding that farmers with in-house certification are more diligent about their environmental conduct than those without the certification. The research was concerned with the impact of a shade coffee certification program of forest degradation, finding that the certification program is effective in alleviating forest degradation; additionally, the price per kilogram increased by 15%–20%.

Another study explored the potential of coffee husks as feedstock to wood, revealing that coffee husks could contribute 24 GWh of energy, decreasing deforestation and environmental degradation. Another group analyzed the environmental and economic effects of organic certification, concluding that it has significant effects on the adoption of farm practices that are associated with positive environmental outcomes, decreasing input costs and yields, but it does not have a significant effect on labor cost or income (see TABLE 2.2).

Regarding the economic dimension, one study compared environmental conditions to capital for resilient strategies implementation, revealing that financial capital, technical assistance and transport are important for crop diversification; financial and literacy knowledge capital are needed for rainwater collection system; literacy knowledge is vital for pest management; financial capital and knowledge are needed for the management of seed banks and nurseries; and financial and human knowledge are necessary for solar dryers.

Another study analyzed the Arabica coffee–macadamia intercropping system, which showed an increase of almost 150% over monocropped coffee systems. New research relates to the valorization of spent coffee grounds, exposing high solid yields through torrefaction experimental practices. Another study was concerned with technical efficiency and the sources of inefficiency in small-scale coffee farming, concluding that the chemical fertilizer and labor are the most important factors in coffee production. Other research covered certification as a governance mechanism on coffee sector, resulting in the need to improve efficiency, productivity, subsidies, and technical assistance to yield economic benefits.

In this sense, the study about collective actions to improve and revitalize the coffee sector detected that the Fairtrade certification, regular collaboration, an enhanced governance system and the actions related with transport, are the drivers for group success. Another investigation exposed the determinants of agricultural price distortions, establishing that inflation, public debt, mobile phone usage, and institutional quality are the key factors that affect price distortions, but registering that mobile phone usage can reduce it 0.22%. Another study evaluated the sustainability of coffee farms, determining that inputs and the improvement of technical efficiency could be the drivers to improve the level of sustainability.

In the same aspect, another research explored the cost of sustainable coffee production considering web-based application, resulting in a app to input key variables related to the production cost. Another study assessed the impact of the cooperative–small farmer structure on the specialty coffee sector, concluding that cooperatives have a positive impact at a better price (20% higher) than those in just a semi-processed chain, although the profits are not shared with farmers. Another study is related to the comparison of agro-ecosystems, recommending the integration of the banana intercropping system to obtain additional benefits of fruits, firewood, and timber provided by bananas and shade trees.

Another research was concerned with the identification of inputs that influence smallholders' ability to produce high-quality coffee and thereby to participate in a cooperative, concluding that the economic benefit on participants is in forms of enhanced access to credit and other financial services from the cooperative. Another investigation was related to harvesting coffee mechanization, resulting in three lines of the coffee grower's ability to use mechanization: by own machines, by outsourcing contracts, and by the use of both. Another research was related to the impact of women association on

TABLE 2.2
Environmental Benefits, Effects, or Routes Related with Scientific Studies

Scientific studies	Benefits, effects, or support routes	Author(s)
Coffee waste for biodiesel feedstock	Feedstock saving, comparable performance to regular diesel, lowered pollutant emissions	Kamil et al. (2020)
The influence of El Niño and La Niña on coffee production	Yield increase to 2439.64 kg ha^{-1} (with El Niño conditions) and 325.96 kg ha^{-1} (with La Niña) (limited to some Brazilian states)	Silva et al. (2020)
The production of biodiesel from recycled coffee waste	Biomass pellets including spent coffee grounds that make biodiesel production viable	Kamil et al. (2019a)
The potential of spent coffee grounds as a commercial feedstock	Forty novel components and alternative products that could be produced from coffee	Massaya et al. (2019)
Environmental impacts of biodiesel production from spent coffee grounds and the implementation in a plant	Spent coffee grounds for biodiesel production will be feasible and profitable when the plant's production capacity reaches 10,000 tons per yield	Kamil et al. (2019b)
Certification program review	Description of programs related with certification and sustainability criteria	Verburg et al. (2019)
The adoption of sustainable practices on smallholder producers	Water conservation, the use of retention barriers for preventing soil erosion, and water harvesting are practices that will become increasingly important in the region	Bro et al. (2019)
Options for productively utilizing spent coffee grounds	Oil, biodiesel, sugar recovery, bioethanol and solid biofuel, phenolic compounds, antioxidants, composting, fertilizer, to prevent and fight cancer cells, create material in the construction industry, plastic, bricks, and absorbent filters	McNutt and He (2019)
Analyzing the kinetics of polyphenolics compounds	Solvent, Soxhlet, and ultrasonic-assisted extraction have great potential for the extraction of polyphenolic compounds	Caballero-Galván et al. (2018)
Interaction of pest biology and climate/CO_2	Integration of GIS technology, social networking, and visual assessments into traditional pest management practices	Ziska et al. (2018)
Adoption of in-house certification by coffee farms	Farmers with in-house certifications are more diligent about their environmental conduct	Giuliani et al. (2017)
Impact of a shade coffee certification program of forest degradation	Certification program is effective in alleviating forest degradation, and the price per kilogram is increased by 15%–20%	Takahashi and Todo (2017)
The potential of coffee husks as feedstock to wood	If utilized in a viable manner, coffee husks could contribute 24 GWh of energy, decreasing deforestation and environmental degradation	Miito and Banadda (2017)
Environmental and economic effects of organic certification	Certification has significant effects on the adoption of farm practices	Ibanez and Blackman (2016)

Sources: Kamil et al. (2019a, 2019b, 2020); Silva et al. (2020); Massaya et al. (2019); Verburg et al. (2019); Bro et al. (2019); McNutt and He (2019); Caballero-Galván et al. (2018); Ziska et al. (2018); Giuliani et al. (2017); Takahashi and Todo (2017); Miito and Banadda (2017); Ibanez and Blackman (2016).

coffee production, resulting in a perception that women feel they don't see immediate benefits with Fairtrade organic production.

Another study explored Fairtrade and organic certification, revealing that Fairtrade-certified farmers had a 29% higher yield than non-certified, and 31% higher than the organically certified farmers; however, the authors registered that coffee certification has no significant effects on the total income of certified farmers. Another research was related with the comparison between socioeconomic and ecological conditions and the history of policies that directly and indirectly influence social, economic and environmental sustainability of coffee production in two countries, concluding that the interventions must focus on strengthening of the social and economic development via supporting the cooperatives in Nicaragua, while interventions in India must focus on environmental goals and cooperatives support.

Another study explored the impact of climate change on productivity and quality, revealing the adaptation process of coffee production is according rain distribution. Other research was related with the effects of torrefaction on benzene, toluene, ethylbenzene, and xylene yields and production cost, concluding that under certain conditions, torrefaction is recommended to reduce production cost, transport cost, and oxygen content. One more investigation exposed the main pathways in the adoption of sustainability standards, determining that the structure of coffee institutions, distribution of land, and certifications are the main pathways for voluntary sustainability adoption (see TABLE 2.3).

Relating to the social dimension, one study explored the determinants for paying for coffee produced by a cooperative, indicating that the coffee consumer will pay 51% more for such coffee. Another study exposed the truth behind smallholder participation on sustainable practices, revealing that sustainable farming is more profitable and cost-effective than conventional coffee production. Other research was about the adaptation drivers in a climate change context, concluding to the household head's awareness of climate variability, social networks, and involvement in policy formulation as the major drivers.

Another study related to the determinants of smallholder sustainability coffee production, concluding that the participation of younger farmers, the diversification of incomes, and the payment for coffee in the moment are the key for sustainable coffee production. Other research explored the challenges

TABLE 2.3

Economic Benefits, Effects, or Routes Related with Scientific Studies

Scientific studies	Benefits, effects, or support routes	Author(s)
Comparison of environmental conditions and capital necessary for resilient strategies implementation	Crop diversification financial capital, technical assistance, and transport are important; for rainwater collection system financial and literacy knowledge capital are needed; for pest management literacy knowledge is vital; financial capital and knowledge are vital for management of seed banks and nurseries on a large scale, and for solar dryers financial and human knowledge are necessary	Shapiro-Garza et al. (2020)
Arabica coffee–macadamia as intercropping system	Economic benefit of intercropping was almost 150% higher than achieved with monocropped coffee	Perdoná and Soratto (2020)
Valorization of spent coffee grounds	High solid yields through torrefaction experimental practices	Vakalis et al. (2019)
Estimating technical efficiency and sources of inefficiency in small-scale coffee farming	Chemical fertilizer and labor as the most important factors in the production of coffee	Ngango and Kim (2019)
Certification as a governance mechanism on coffee sector	Improving efficiency, productivity, subsidies, and technical assistance to obtain economic benefit	Hajjar et al. (2019)
Collective actions to improve and revitalize the coffee sector	The Fairtrade certification, regular collaboration with partners and enhanced governance system are the actions related with transport as drivers for success	Sengere et al. (2019)
Determinants of agricultural price distortions (focus on mobile phone usage)	Inflation, public debt, mobile phone usage, and institutional quality are the key factors that affect price distortions, registering that the mobile phone usage can reduce it 0.22%	Nsabimana and Amuakwa-Mensah (2018)
Sustainability level of coffee farms	Inputs and the improvement of technical efficiency could be the drivers for improving sustainability	Gaitán-Cremaschi et al. (2018)
Web-based application to explore sustainable coffee production	Web-based application to input key variables of the production cost	Leshed et al. (2018)
Impact of an cooperative–small farmer structure in the specialty coffee sector	Positive impact of cooperatives	Vicol et al. (2018)
Comparison of agro-ecosystems	Integration of banana intercropping system to obtain additional benefits of fruits, firewood, and timber provided by bananas and shade trees	Rahn et al. (2018)
The identification of inputs that influence smallholders' ability to produce high-quality coffee and thereby to participate in a cooperative	Economic benefit on participants is to access to credit and other financial services from the cooperative	Hernández-Aguilera et al. (2018)
Transaction of harvesting mechanization between coffee growers and specialized service providers	Three lines of coffee grower's ability to use the harvesting mechanization: by own machines, by outsourcing contracts, and by the use of both	Magalhaes de Oliveira and Zylbersztajn (2018)
The impact of women association on coffee production	Women feel they don't see immediate benefits with Fairtrade organic production	Lyon et al. (2017)
Fairtrade and organic certification	Fairtrade-certified farmers had 29% higher yield than non-certified, and 31% higher than organically certified farmers	Jena et al. (2017)
A comparison, in two countries, between socioeconomic and ecological conditions and the history of policies that directly and indirectly influence social, economic, and environmental sustainability of coffee production	In Nicaragua, interventions must focus on strengthening of the social and economic development via support the cooperatives, whereas in India they must focus on environmental goals and support of cooperatives	Mithöfer et al. (2017)
The potential impact of climate change on productivity and quality	The adaptation of coffee production areas is according to rain distribution	Chemura et al. (2015)
Effects of torrefaction on benzene, toluene, ethylbenzene, and xylene yields and production cost	Torrefaction is recommended to reduce production cost, transport cost, and oxygen content	Chai et al. (2016)
Pathways in the adoption of sustainability	The structure of coffee institutions, distribution of land and certifications are the main pathway for sustainability adoption	Grabs et al. (2016)

Sources: Shapiro-Garza et al. (2020); Perdoná and Soratto (2020); Vakalis et al. (2019); Ngango and Kim (2019); Hajjar et al. (2019); Sengere et al. (2019); Nsabimana and Amuakwa-Mensah (2018); Gaitán-Cremaschi et al. (2018); Leshed et al. (2018); Vicol et al. (2018); Rahn et al. (2018); Hernández-Aguilera et al. (2018); Magalhaes de Oliveira and Zylbersztajn (2018); Lyon et al. (2017); Jena et al. (2017); Mithöfer et al. (2017); Chemura et al. (2015); Chai et al. (2016); Grabs et al. (2016).

faced during certification process of a cooperative, evidencing that the conjunction of PAS 2060 certification and Fairtrade for reduction of transaction cost and energy expenses are key drivers to solve the challenges. One more investigation exposed the role of cooperatives in food security, finding that they don't have a significant impact on food expenditure and income obtained by their members.

Another research explored the impact of Rainforest Alliance (RA) certification on economic benefits, concluding that economic benefits from RA certification only come from a price effect and not from a yield effect. Another study exposed the socioeconomic benefits under coffee agroforestry, resolving that the diversification of agroforestry implies an increasing income. Another investigation explored paying for shade-grown coffee, concluding with the purchasing of coffee 10% over the regular price.

Additional studies were developed for evaluating the impact of coffee cooperatives, concluding that the members have important economic benefits to their participants and to entire community. Another study was about the role of Fairtrade certification on the capacity-building and incomes of coffee farm, resolving that the Fairtrade certification has a positive impact on small farmers' income, expecting to improve the livelihood and the integration in new export markets. One more study was related with the transactions of coffee farmers, revealing that they prefer the economic rather than social motives in their business dealings (see TABLE 2.4).

2.4 Conclusion

The scientific and technological knowledge in the context of coffee sector has been in evolution. From the use of contract agreements, the consideration of standards certification, or the creation of platforms, to the studies related with the reutilization of coffee waste, crop diversification and the determinants for paying a specialty coffee, the benefits, effects or routes intended to help, are present in the context of the coffee sector.

The scientific studies and technological actions reflect the benefits, effects, or routes intended to help, in the lines of coffee waste for biodiesel production, take advantage of El Niño and La Niña conditions for yield increase, the consideration of certifications for obtaining a better price, the adoption of the intercropping system and collective strategies for achieving economic benefit, the introduction of technology for reduce the price distortions, the determinants for paying a "specialty coffee," however, to the authors' knowledge, just Coopedota cooperative balancing social, economic and environmental elements by including Fairtrade for reduction of transaction cost and energy expenses, better incomes for reducing poverty, based on PAS 2060 standard certification. This review is in line with the UN recognition that "the ending of poverty must go hand-in-hand with strategies that build economic growth and address a range of social needs, while tackling climate change and environmental protection" (UN, 2020).

TABLE 2.4

Social Benefits, Effects, or Routes Related with Scientific Studies

Scientific studies	Benefits, effects, or support routes	Author(s)
Determinants for paying a "specialty coffee" producing by cooperative coffee members	The coffee consumer pays 51% more for a coffee produced by a cooperative	Ufer et al. (2019)
The truth behind an increasing number of smallholder farmers participating in sustainable coffee farming in Vietnam	Sustainable farming is actually more profitable and cost-effective than conventional farming	Hung Anh et al. (2019)
Adaptation drivers in a climate change context	Household head's awareness of climate variability, social networks, and involvement in policy formulation as the major drivers	Mulinde et al. (2019)
Determinants of smallholder sustainability coffee production	The participation of younger farmers, the diversification of incomes, and the payment for coffee in the moment are the keys for sustainable coffee production	Wairegi et al. (2018)
The challenges faced during the certification process of a cooperative	The conjunction of PAS 2060 certification and Fairtrade for reduction of transaction cost and energy expenses	Birkenberg and Birner (2018)
The role of cooperatives in food security	Coffee cooperatives were found to have no significant impact on food expenditure and income obtained by their members	Shumeta and D'Haese (2018)
Impact of Rainforest Alliance (RA) certification on economic benefits	Economic benefits from RA certification only come from a price effect and not from a yield effect	Mitiku et al. (2018)
The exploration of socioeconomic benefits under coffee agroforestry	The diversification of agroforestry implies an increasing income	Yasin et al. (2018)
To pay for shade-grown coffee	The purchasing of coffee 10% over regular price	Klimas and Webb (2018)
To evaluate the impact of cooperatives	Economic benefits for members and for the community	Mojo et al. (2017)
The role of Fairtrade certification on the capacity-building and incomes of coffee farms	Fairtrade certification has a positive impact on small farmers' incomes	Jena and Grote (2017)
Personalized transactions of coffee farmers	Coffee farmers prefer the economic rather than social motives	Gelaw et al. (2016)

Sources: Ufer et al. (2019); Hung Anh et al. (2019); Mulinde et al. (2019); Wairegi et al. (2018); Birkenberg and Birner (2018); Shumeta and D'Haese (2018); Mitiku et al. (2018); Yasin et al. (2018); Klimas and Webb (2018); Mojo et al. (2017); Jena and Grote (2017); Gelaw et al. (2016)

In this sense, future scientific studies and technological actions could be in the line of balancing social, economic, and environmental elements, considering economic growth, reducing social needs, based on environmental protection. Greater collective effort is needed for achieving sustainable agriculture!

2.5 Acknowledgment

The authors appreciate the support of the National Council for Science and Technology from Mexico (Consejo Nacional de Ciencia y Tecnología [CONACYT]), specially to the program "Catedras CONACYT"; to Research and Assistance Center for Technology and Design of State of Jalisco A.C. (Centro de Investigación y Asistencia en Tecnología y Diseño del Estado de Jalisco A.C. CIATEJ), to Universidad de Caxias do Sul (University of Caxias do Sul), and to Universidad Autonoma de Querétaro (Autonomous University of Queretaro), for their support in the development of this review.

REFERENCES

Alfaro Barbosa Juan Manuel, Limón Rodríguez Benjamín, Martínez Turanzas Gustavo Ángel. *Ambiente y sustentabilidad por una educación ambiental.* Grupo Editorial Patria. 2014.

Andersen Hanne, Hepburn Brian. *Scientific Method.* Stanford Encyclopedia of Philosophy Archive. 2015.

Birkenberg Athena, Birner Regina. *The world's first carbon neutral coffee: Lessons on certification and innovation from a pioneer case in Costa Rica.* Journal of Cleaner Production. 2018; 189: 485–501.

Bro Aniseh, Clay Daniel, Ortega David, Lopez Maria. *Determinants of adoption of sustainable production practices among smallholder coffee producers in Nicaragua.* Environment, Development and Sustainability. 2019; 21: 895–915.

Brundtland GH. Report of the World Commission on environment and development "our common future". 1987; 1–300.

Caballero-Galván Ashley Sthefania, Restrepo-Serna Daissy Lorena, Ortiz-Sánchez Mariana, *Cardona-Alzate Carlos Ariel. Analysis of extraction kinetics of bioactive compounds from spent coffee grounds (Coffea arábica).* Waste and Biomass Valorization. 2018; 9: 2381–2389.

Cambridge University Press. *Science.* Cambridge Dictionary. 2020a.

Cambridge University Press. *Benefit.* Cambridge Dictionary. 2020b.

Chai Li, Saffron Christopher M, Yang Yi, Zhang Zhongyu, Munro Robert W, Kriegel Robert M. *Integration of decentralized torrefaction with centralized catalytic pyrolysis to produce green aromatics from coffee grounds.* Bioresource technology. 2016; 201: 287–292.

Chemura Abel, Kutywayo Dumisani, Chidoko Pardon, Mahoya Caleb. *Bioclimatic modelling of current and projected climatic suitability of coffee (Coffea arabica) production in Zimbabwe.* Regional environmental change. 2015; 16: 473–485.

Chiuri, Taku. Kenya Scholars & Studies Association Annual Conference. 2015.

Clark William C, Dickson Nancy M. Sustainability science focuses on the dynamic interactions between nature and society. Proceedings of the National Academy of Sciences. 2003; 100: 8059-8061.

Clark WC, Dickson NM. Proc. Nation. Aca. Sci. 2003; 100: 8059–8061.

Contreras-Medina David Israel, Sánchez Osorio Ever, Olvera Vargas Luis Alberto, Romero Romero Yair. *Technology roadmapping architecture based on knowledge management: case study for improved indigenous coffee production from Guerrero, Mexico.* Journal of Sensors. 2019; 2019: 1–17.

da Fonseca Gustavo AB. *De Rio a Rio: 20 años promoviendo la economía verde.* Washington, DC: Global Environment Facility. 2012.

Flavin, French. *La Situación del Mundo 2002: Informe Anual Del Worldwatch Institute Sobre Progreso Hacia una Sociedad Sostenible* Icaria Editorial. 2002.

Food and Agriculture Organization. *Statistical Yearbook of the Food and Agriculture of United Nations.* FAO. 2013.

Food and Agriculture Organization. *Building a common vision for sustainable food and agriculture.* FAO. 2014.

Food and Agriculture Organization. *El Estado Mundial de la Agricultura y la Alimentación.* FAO 2017.

Food and Agriculture Organization. *Greening Coffee Production in Vietnam.* FAO. 2019.

Gaitán-Cremaschi Daniel, Van Evert Frits K, Jansen Don M, Meuwissen Miranda P, Oude Lansink Alfons G. *Assessing the sustainability performance of coffee farms in Vietnam: a social profit inefficiency approach.* Sustainability. 2018; 10: 2–23.

García Thomas. Países productores. Coffee geek. 2018.

Gashaw Bizualem Assefa, Kibret Saron Mebratu. *The role of Ethiopian commodity exchange (ECX) in crop value chain development in Ethiopia.* Int J Bus Econ Res. 2018; 7: 183–190.

Gelaw Fekadu Speelman, Stijin Van Huylenbroeck Guido. *Farmers' marketing preferences in local coffee markets: Evidence from a choice experiment in Ethiopia.* Food policy. 2016; 61: 92–102.

Giuliani Elisa, Ciravegna Luciano, Vezulli Andrea, Kilian Bernard. *Decoupling standards from practice: the impact of in-house certifications on coffee farms' environmental and social conduct.* World Development. 2017; 96: 294–314.

Goldstein Natalie. *Conservation Heroes.* New York: John Muir-Infobase Learning. 2013.

Grabs Janina, Kilian Bernanrd, Hernández Daniel Calderon, Dietz Thomas. *Understanding coffee certification dynamics: A spatial analysis of voluntary sustainability standard proliferation.* International Food and Agribusiness Management Review. 2016; 19: 31–56.

Hajjar R, Newton P, Adshead D, Bogaerts M, Maguire-Rajpaul VA, Pinto LFG, McDermott CL, Milder JC, Wollenberg E, Agrawal A. *Scaling up sustainability in commodity agriculture: Transferability of governance mechanisms across the coffee and cattle sectors in Brazil.* Journal of Cleaner Production. 2019; 206: 124–132.

Hernández-Aguilera J Nicolas, Gómez Miguel I, Rodewald, Aamanda D, Rueda Ximena, Anunu Colleen, Bennet Ruth, van Es Harold M. *Quality as a driver of sustainable agricultural value chains*: The case of the relationship coffee model. Business Strategy and the Environment. 2018; 27: 179–198.

Hung Anh Nguyen, Bokelmann Wolfgang, Do Thi Nga, Van Minh Nguyen. *Toward sustainability or efficiency: the case of smallholder coffee farmers in Vietnam*. Economies, 2019;7:1–25.

Ibanez Marcela, Blackman Allen. *Is eco-certification a win–win for developing country agriculture? Organic coffee certification in Colombia*. World Development. 2016; 82: 14–27.

Inter-American Institute for Cooperation on Agriculture, IICA. *Annual report 2017*. IICA. 2018.

International Coffee Organization. *Coffee Development Report 2019* Overview. ICO. 2019.

International Coffee Organization, ICO. *Developing a sustainable coffee economy*. 2020.

International Monetary Fund. *Debt Relief Under the Heavily Indebted Poor Countries (HIPC) Initiative*. IMF. 2020.

Jena P, Grote U. *Fairtrade certification and livelihood impacts on small-scale coffee producers in a tribal community of India*. Applied Economic Perspectives and Policy. 2017; 39: 87–110.

Jena PR, Stellmacher T, Grote U. *Can coffee certification schemes increase incomes of smallholder farmers? Evidence from Jinotega, Nicaragua*. Environment, Development and Sustainability. 2017; 19: 45–66.

Jin Z. *Soft technology: the essential of innovation*. Futures Research Quarterly. 2002; 18: 1–24.

Kamil M, Ramadan KM, Olabi AG, Shanableh A, Ghenai C, Al Naqbi AK, Awad OI, Ma X. *Comprehensive evaluation of the life cycle of liquid and solid fuels derived from recycled coffee waste*. Resources, Conservation and Recycling. 2019a; 150: 104446.

Kamil M, Ramadan KM, Awad OI, Ibrahim TK, Inayat A, Ma X. *Environmental impacts of biodiesel production from waste spent coffee grounds and its implementation in a compression ignition engine*. Science of the Total Environment. 2019b; 675: 13–30.

Kamil M, Ramadan KM, Olabi, Ghani AG, Al-Ali EI, Ma X, Awad OI. *Economic, technical, and environmental viability of biodiesel blends derived from coffee waste*. Renewable Energy. 2020; 147: 1880–1894.

Klimas CA, Webb E. *Comparing stated and realized preferences for shade-grown vs. conventionally grown coffee*. International Journal of Consumer Studies. 2018; 42: 76–92.

Larrouyet MC. *Desarrollo sustentable: origen, evolución y su implementación para el cuidado del planeta*. Repositorio Institucional Digital de Acceso Abierto de la Universidad Nacional de Quilmes RIDAA, 2015.

Leshed Gilly, Rosca Masha, Huang Michael, Mansbach Liza, Zhu Yicheng, Hernández-Aguilera Juan Nicolás. *CalcuCafé: Designing for Collaboration Among Coffee Farmers to Calculate Costs of Production*. Proceedings of the ACM on Human-Computer Interaction. 2018; 2: 1–26.

Lyon S, Mutersbaugh T, Worthen H. The triple burden: *The impact of time poverty on women's participation in coffee producer organizational governance in Mexico*. Agriculture and Human Values. 2017; 34: 317–331.

Littig B, Griessler E. Int. J Sustain. Develop. 2005; 8: 65–79.

Lobo Maria-Jesús, Pietriga Emmanuel, Appert Caroline. *An evaluation of interactive map comparison techniques*. Proceedings of the 33rd annual ACM conference on human factors in computing systems. 2015; 3573–3582.

Lowenthal D. *Marsh's Man and Nature at 150*. The George Wright Forum. 2015; 32: 227–237.

Magalhaes de Oliveira G, Zylbersztaju D. *Make or buy: the case of harvesting mechanization in coffee crop in Brazil*. International Food and Agribusiness Management Review. 2018; 21: 895–914.

Massaya J, Pereira AP, Mills-Lamptey B, Benjamin J, Chuck CJ. *Conceptualization of a spent coffee grounds biorefinery: a review of existing valorisation approaches*. Food and Bioproducts Processing. 2019; 118: 149–166.

McNutt J. He QS. *Spent coffee grounds: A review on current utilization*. Journal of industrial and engineering chemistry. 2019; 71: 78–88.

Miito GJ, Banadda N. A *short review on the potential of coffee husk gasification for sustainable energy in Uganda*. F1000Research. 2017; 6: 1–8.

Mithöfer D, Méndez VE, Bose A, Vaast P. *Harnessing local strength for sustainable coffee value chains in India and Nicaragua: Reevaluating certification to global sustainability standards*. International Journal of Biodiversity Science, Ecosystem Services & Management. 2017; 13: 471–496.

Mitiku F, Nyssen J, Maertens M. *Certification of Semi-Forest Coffee as a Land-sharing Strategy in Ethiopia*. Ecological Economics. 2018; 145: 194–204.

Mohan S, Gemech F, Reeves A, Struthers J. *The welfare effects of coffee price volatility for Ethiopian coffee producers*. Qualitative Research in Financial Markets. 2016; 8: 288–304.

Mojo D, Fischer C, Degefa T. *The determinants and economic impacts of membership in coffee farmer cooperatives: recent evidence from rural Ethiopia*. Journal of Rural Studies. 2017; 50: 84–94.

Mulinde C, Majaliwa JGM, Twinomuhangi R, Mfitumukiza D, Komutunga E, Ampaire E, Assimwe J, Asten PV, Jassogne L. *Perceived climate risks and adaptation drivers in diverse coffee landscapes of Uganda*. NJAS-Wageningen Journal of Life Sciences. 2019; 88: 31–44.

National Research Council. *The fundamental role of science and technology in international development: an imperative for the US Agency for international development*. National Academies Press. 2006.

Ngango J, Kim SG. *Assessment of Technical Efficiency and Its Potential Determinants among Small-Scale Coffee Farmers in Rwanda*. Agriculture. 2019; 9: 1–12.

Nsabimana A, Amuakwa-Mensah F. *Does mobile phone technology reduce agricultural price distortions? Evidence from cocoa and coffee industries*. Agricultural and Food Economics. 2018; 6: 6–20.

Perdoná MJ, Soratto RP. *Arabica coffee–macadamia intercropping: Yield and profitability with mechanized coffee harvesting*. Agronomy Journal. 2020; 112: 429–440.

Rahn E, Liebig T, Ghazoul J, van Asten P, Läderach P, Vaast P, Sarmiento A, Garcia c, Jassogne L. *Opportunities for sustainable intensification of coffee agro-ecosystems along an altitudinal gradient on Mt. Elgon, Uganda*. Agriculture, Ecosystems & Environment. 2018: 263: 31–40.

Sachon M. *Cuando las personas y maquinas trabajan juntos. Los cinco puntales de la cadena de valor en la industria 4.0.* IESE Insight. 2017; 2: 15–22.

Sengere RW, Curry GN, Koczberski G. *Forging alliances: Coffee grower and chain leader partnerships to improve productivity and coffee quality in Papua New Guinea.* Asia Pacific Viewpoint. 2019; 60: 220–235.

Shapiro-Garza E, King D, Rivera-Aguirre A, Wang S, Finley-Lezcano J. *A participatory framework for feasibility assessments of climate change resilience strategies for smallholders: lessons from coffee cooperatives in Latin America.* International Journal of Agricultural Sustainability. 2020; 18: 21–34.

Shumeta Z, D'Haese M. *Do Coffee farmers benefit in food security from participating in coffee cooperatives? Evidence from Southwest Ethiopia Coffee Cooperatives.* Food and nutrition bulletin. 2018; 39: 266–280.

Silva KA, de Souza, Rolim G, Valeriano TTB. *Influence of El Niño and La Niña on coffee yield in the main coffee-producing regions of Brazil.* Theoretical and Applied Climatology. 2020; 139: 1019–1029.

Takahashi R, Todo Y. *Coffee certification and forest quality: evidence from a wild coffee forest in Ethiopia.* World Development. 2017; 92: 158–166.

Texeira M, Nguyen P, Symmes JC. *With technology and machinery, Brazil and Vietnam tighten their grip on coffee trade.* The Japan times. 2019.

Ufer D, Lin W, Ortega DL. *Personality traits and preferences for specialty coffee: Results from a coffee shop field experiment.* Food Research International. 2019; 125: 108504.

United Nations Educational, Scientific and Cultural Organization, UNESCO. Consultations of the Director-General with Member States. 2012.

United Nations Educational, Scientific and Cultural Organization, UNESCO. Science, Technology and Innovation Policy. Science Policy. 2017.

United Nations, UN. *Science, technology and innovation, and the potential of culture, for promoting sustainable development and achieving the Millennium Development Goals.* Report of the Secretary-General. 2013.

United Nations, UN. *Transformar nuestro mundo: la agenda 2030 para el Desarrollo Sostenible.* 2015.

United Nations, UN. Closing the technology gap in least developed countries. The Magazine of the United Nations. 2018a.

United Nations, UN. *Least developed nations lagging behind on attaining sustainable development, speakers say, as second committee debates countries in special situations.* Meetings Coverage & Press Releases. 2018b.

United Nations, UN. *17 goals to transform our world. Losing the technology gap in least developed countries.* 2020.

Vakalis S, Moustakas K, Benedetti V, Cordioli E, Patuzzi F, Loizidou M, Baratieri M. *The "COFFEE BIN" concept: centralized collection and torrefaction of spent coffee grounds.* Environmental Science and Pollution Research. 2019; 26: 35473–35481.

Verburg R, Rahn E, Verweij P, van Kuijik M, Ghazoul J. *An innovation perspective to climate change adaptation in coffee systems.* Environmental Science & Policy. 2019; 97: 16–24.

Vicol M, Neilson J, Hartatri DFS, Cooper P. *Upgrading for whom? Relationship coffee, value chain interventions and rural development in Indonesia.* World Development. 2018; 110: 26–37.

Wairegi L, Bennett M, Nziguheba G, Mawanda A, de los Rios C, Ampaire E, Jassogne L, Pali P, Mukasa D, van Asten PJ. *Sustainably improving Kenya's coffee production needs more participation of younger farmers with diversified income.* Journal of Rural Studies. 2018; 63: 190–199.

Yasin H, Kebebew Z, Hundera K. *Woody Species Diversity, Regeneration and Socioeconomic Benefits Under Natural Forest and Adjacent Coffee Agroforests at Belete Forest, Southwest Ethiopia.* Ekológia (Bratislava). 2018; 37: 380–391.

Ziska L, Bradley BA, Wallace RD, Bargeron CT, LaForest JH, Choudhury RA, Garret K, Vega FE. *Climate change, carbon dioxide, and pest biology, managing the future: coffee as a case study.* Agronomy. 2018; 8: 1–21.

3

Contemporary topics in coffee research: A glance

Parvatam Giridhar, Akula Ramakrishna, and Magdalena Jeszka-Skowron

CONTENTS

3.1 Introduction

Coffee is a high-profile, commercially important agricultural commodity and has a special place in the lives of many people in the world. It has an especially pivotal role for those in coffee-producing countries. According to the contemporary reports, the compound annual growth rate (CAGR) for coffee is 4.22% for 2020–2025, with Europe the largest marketplace for consumers and South America the fastest-growing market (Roughan, 2020). Like several other alternative commercial commodities, the COVID-19 pandemic had a noticeable impact on coffee in the year 2020, as global production was down 2.2% compared to the previous year, and a 5% drop was recorded for Arabica coffee (ICO September 2020 Report). Conjointly, a vital impact was detected on the coffee market because of COVID-19 on coffee exporting countries, and because of this, coffee prices are volatile due to shock in demand and supply.

3.2 Various issues pertaining to sustainability in coffee production

3.2.1 Environmental issues and policies

It is sensible to understand that as per the current trend, mindful coffee consumers are on the upsurge, however, coffee yields were down and pests were up, mainly attributed to the impact of climate variation (Pham et al., 2019). This has a major influence on the coffee industry (Tavares et al., 2018) and also environmental impacts throughout the supply chain. Under the current context of the global situation pertaining to environmental safety concerns, there is a great attention in both industry and academics on sustainability of coffee production, organic shade-grown coffee, bird-friendly coffee, and the coffee trade (Samper and Quiñones-Ruiz, 2017). There is much apprehension and concern about the cost and quality of coffee from an industry perspective. Although various governmental and non-governmental organizations are associated with a defined coffee policy in this aspect, a lot of work remains to be done to minimize the disparity pertaining to coffee issues and social aspects (www.coffeeresearch.org/politics/overview.htm, accessed 1 April 2021). In a recent report, Bianco (2020) highlighted various challenges and opportunities pertaining to climate change adaptation, coffee, and corporate social responsibility. Also it is the clarion call for more collaboration and research around the adaptation issue for the coffee industry.

3.2.2 Socioeconomic aspects

Efforts were made to bring good economic prospects to farmers through advancement or refining opportunities, by providing a sustainability platform for exotic, time-tested single-origin coffees and by that facilitate offering geographical indications (GIs; Samper and Quiñones-Ruiz. 2017). In

DOI: 10.1201/9781003043133-4

coffee commerce, price volatility is a big concern. Most of the time, green coffee fetches relatively small and unpredictable prices, which is one of the foremost challenge growers face. Some environmental, social, and economical encounters related to coffee growers and coffee industry were elaborated by Samper and Quiñones-Ruiz (2017). Few case studies are reported on sustainable coffee management; one highlighted various challenges and problems faced by the Coorg coffee industries in India (Kavya Dechamma and Mishra, 2018) and Vietnam (Nguyen, and Sarker, 2018).

3.3 Current trends in coffee marketing and growing trends in specialty coffee

Significant knowledge is offered on varied ongoing efforts to guarantee economic viability and sustainability of coffee growers, issues pertain to coffee marketing, inclinations in the single-serve, ready-to-drink, and specialty coffee markets such as signature blends, distinct origin, solitary farm, or estate. Such information helps us to comprehend the trends, prospects, and risks associated with coffee market (www.cbi.eu/market-information/coffee/trends, accessed 1 April 2021).

3.3.1 Caffeine-fortified energy drinks

Energy drinks were introduced in the 1980s and have become so popular among the public and athletes. Caffeine-fortified energy drinks are now a multibillion-dollar global industry (Petit et al., 2012; Alia et al., 2019). However, significant reports express apprehension over the safety and efficacy of these drinks and their related adverse effects on the health of consumers (Gutiérrez-Hellín and Varillas-Delgado, 2021; Manchester et al., 2017; Higgins et al., 2018). This area needs to be explored further so that science-reinforced and proof-based interpretations can be documented to offer recommendations, regulatory actions, and so forth (De Sanctis et al. 2017). It is quite fascinating to grasp the sway of new trends in consumption on the coffee market. Particularly the relation among main megatrends, consumer drifts and product developments on consumers' behavior is deliberated vitally because of its anticipated profound influence on the coffee market. In particular, natural and sustainable practices geared toward health and nutrition add a lot to consumer behavior (Maciejewski and Mokrysz, 2019).

3.3.2 Ready-to-drink coffee

Although many consumers still purchase mainstream coffee blends, nowadays ready-to-drink coffee is among the quickest growing market segments in Europe, whereas in the United States there is decent demand for single-cup brewing equipment for domestic use. According to a recent report by Technavio Research (www.businesswire.com/news/home/20200904005195/en/Research-Report-Specialty-Coffee-Shops-Market-2020–2024-Increasing-Consumption-of-Coffee-to-Boost-the-Market-Growth-Technavio, accessed 1 April 2021), the specialty coffee shop market is poised to grow by $80.78 billion during 2020–2024, moving ahead at a CAGR of nearly 13% during the projection period. The increasing consumption of coffee will offer enormous growth prospects.

Those who have observed the topical trends in types of coffee-based drinks have definitely witnessed coffee's transformation from a beverage to an experience. Coffee is a non-alcoholic beverage, but now it is no more just a "hot" beverage. Hitherto, we have only been familiar with green coffee beans being roasted and then ground to make coffee, but the latest trend is Nitro coffee, which is a cold brew infused with nitrogen and then released through a pressurized valve. Another latest preference is "coffee-leaf tea."

3.3.3 Specialty coffees

"Specialty coffee refers to the single-origin green coffee beans of the highest quality, presenting certified labels such as organic coffee, fair trade, and sustainable production." High-quality green coffee beans have little or no physical defects and show unique attributes in coffee cupping upon roasting, scoring the highest points (De Luca et al., 2016; Tolessa et al., 2016; Alcantara et al., 2021).

Many varieties in the elite class of premium single-origin coffees possess overall great flavor: Jamaican Blue Mountain, Hawaiian Kona, Indonesian Toraja, and Tanzanian Peaberry, as well as Ethiopian coffees such as Yirgacheffe, Sidamo, Harar/Mocca, and Nekemte.

Another unique coffee is Gesha, a coffee native to Ethiopia, which now is planted in Panama and other countries. It is one of the most coveted coffee varieties and also among the most expensive (www.cbi.eu/market-information/coffee/trends).

Premium and exotic coffee involve unique methods of processing. One well-known type of coffee is Kopi Luwak, determined the most expensive coffee in the world (Marcone and Alrifai, 2019). In a natural bioprocess, coffee cherries pass through the gastrointestinal tract of mammals, birds, and bats. Ivory coffee is usually procured from Arabica coffees obtained from the feces of the Asian elephant (*Elephas maximus*). These digested coffees place premium and exotic coffees on the raising market (Raveendran and Murthy, 2021).

The new trends of the best-quality coffees for consumer needs indicate development of craft coffees, ready-to-consume hot and cold brewed coffees. These trends, as well as coffee with new flavors and ingredients, help improve health conditions (Maciejewski and Mokrysz, 2019).

3.4 Influence of coffee on consumer health

Though coffee has been cultivated for more than 350 years, its consumption and its consequences on health did not receive attention until the 1970s, and in the following decade the beneficial and harmful effects coffee drinking were deliberated (Troyer, 1984). A good number of claims and comprehensive reports on coffee and the influence of caffeine on consumer health are well documented (Cornelis, 2019). Various insights on medical and psychological aspects of coffee drinking further increased our knowledge in view of the recent findings on the benefits of coffee on health. Coffee may offer some protection against Parkinson's disease, type 2 diabetes, liver disease and cancer, heart attack, and stroke (Kolahdouzan et al., 2017; Hong et al., 2020). Most of these beneficial claims are based on investigations on experimental animals and a few in human volunteers.

3.5 Biotechnology progressions in coffee

Biotechnological advancement of coffee is another important area, wherein researchers are making sustainable efforts to improve the qualitative and quantitative traits of coffee (Sridevi and Giridhar, 2008; Avinash Kumar et al., 2016). A few to mention are breeding, marker based selection, germplasm improvement through traditional methods, improved genetic transformation methods (Mishra and Slater, 2012; Avinash Kumar et al., 2018), abiotic stress factors influence on caffeine (Avinash Kumar et al., 2016) and WRKY transcription factors (Giridhar et al., 2012), metabolic engineering and molecular biology interference such as antisense technology, RNAi methods to regulate caffeine biosynthesis (Shibin et al., 2014; Fernandes-Brum et al., 2017), and transcriptional factors involved in caffeine metabolism (Giridhar et al., 2012) and TFs identification in pW1a promoter (Petitot et al., 2013). The current focus is on employing genome editing technologies for the improvement of coffee (Breitler et al., 2018).

3.6 Exploring potential of coffee waste

Generating wealth from coffee waste in both the pre- and post-harvesting stages is a fascinating area of investigation. Various studies emphasized the possibility of producing biogas (Akinbomi et al., 2014; Corro et al., 2014; Ulsido and Li, 2016; Ulsido et al., 2016), alcohol (Gouvea et al., 2009), and the reutilization of cherry husks as biosorbents for cyanide (Gebresemati et al., 2017), for the removal of heavy metals from aqueous solutions (Oliveira et al., 2007), for the removal of dyes from aqueous solutions (Ahalya et al. 2014), and as a biosorbent for de-fluorination of water (Getachew et al., 2015). Such applications are not limited to these; they also find application as fuel briquettes or pellets (Cubero-Abarca et al., 2014) and in the production of pectinases and cellulases (Boccas et al., 1994). Recent studies on spent coffee grounds triggered lot of curiosity among researchers and also entrepreneurs and have opened new avenues. Being considered as an underutilized source of biomass, spent coffee grounds are likely a rich spring of industrially important reducing sugars and extractable bioactives (Scully et al. 2016). Highly worthy engineering studies on process optimization and selection of appropriate enzymes for processing lignocellulose biomass are reported (Scully et al. 2016). This area needs to be explored further to gain greater insight on developing the process chain for the production of bioactive polyphenols, caffeine, extractable lipids, and mainly fermentable sugars.

3.7 Quality of coffee

The quality of coffee is the most important factor for consumers, distributors, and producers. For example, the strict legal requirements for coffee safety, which are connected with traceability and hygiene, can be found in the European Union. Contaminations of coffee beans, which include pesticides (maximum residue level for each pesticide), mycotoxins, mold, and *Salmonella* bacteria, are constantly controlled.

Traditional sensory analyses of coffee beans and brews are the most commonly used, but nowadays there are also many novel instrumental methods to evaluate coffee's quality attributes (Jeszka-Skowron et al., 2015). Marker compounds such as kahweol and 16-O-methylcafestol (16-OMC) dominate in Robusta samples (Martellossi et al. 2005; Barbin et al., 2014; Górnaś et al., 2014; Bertone et al., 2016; Gunning et al., 2018; DIN 10779:2011–03). PCR testing of the extracted DNA for amplification and quantitation of Arabica and Robusta species can be used to check the adulterants (Combes et al., 2018). To identify and authenticate specialty coffees such as Kopi Luwak, the discriminant markers or specific markers of adulterants have been found using a metabolomics approach (Jumhawan et al., 2013, 2015; Burns and Walker, 2020; Wang et al., 2020).

3.8 Conclusion

During the last few decades, noteworthy progress has been recorded on various aspects of coffee, from cultivation, development of high-quality varieties, and post-harvest processing of coffee beans. With the advent of the latest technologies, whether it is biotechnology, instrumentation, nanotechnology-based methods, or even artificial intelligence–assisted techniques, bright avenues are still open. In the light of the previous discussion, there are abundant opportunities for researchers to pursue investigations on various aspects from pre-harvest, post-harvest processing, marketing, and trade, as well as on coffee health.

3.9 Acknowledgments

The authors are grateful to CSIR-CFTRI, Mysore, for the opportunity to carry out research and development on coffee for more than a decade. That precise experience helped in preparation of this manuscript. This work was funded by the National Science Centre – NCN (Poland) under the program MINIATURA 4, grant number: 2020/04/X/NZ9/00736.

REFERENCES

Ahalya N, Chandraprapha MN, Kanamadi RD. Adsorption of fast Green on to coffee husk. *J Chem Eng Res*. 2014; 2: 201–207.

Akinbomi J, Brandberg T, Sanni SA, Taherzadeh MJ. Development and dissemination strategies for accelerating biogas production in Nigeria. *BioResources*. 2014; 9: 5707–5737.

Alcantara GMRN, Dresch D, Melchert WR. Use of non-volatile compounds for the classification of specialty and traditional Brazilian coffees using principal component analysis. *Food Chem.* 2021; 360: 130088. doi:10.1016/j.foodchem.2021.130088.

Alia AO, Petrunich-Rutherford ML. Anxiety-like behavior and whole-body cortisol responses to components of energy drinks in zebrafish (*Danio rerio*). https://pubmed.ncbi.nlm.nih.gov/31497403/ *Peer J.* 2019 August 20; 7: e7546. doi:10.7717/peerj.7546. eCollection 2019.

Avinash K, Simmi PS, Giridhar P. Cell wall remodelling involving galactomannan de-branching influence *Agrobacterium tumefaciens*-mediated transformation of *Coffea canephora* somatic embryos. *Plant Cell Tiss and Org Cult.* 2018; 134: 369–382.

Avinash K, Simmi PS, Nandini PS, Giridhar P. Developing sustainable disease resistance in coffee: Breeding vs. transgenic approaches. In: *Plant Pathogen Resistance Biotechnology*, First Edition. Edited by David B. Collinge. © 2016 John Wiley & Sons, Inc. Published 2016 by John Wiley & Sons, Inc. pp. 217–243.

Barbin DF, de Souza Madureia Felicio AL, Sun DW, Nixdorf SL, Hirooka EY. Application of infrared spectral techniques on quality and compositional attributes of coffee: An overview. *Food Res Int.* 2014; 61: 23–32.

Bertone E, Venturello A, Giraudo A, Pellegrino G, Geobaldo T. Simultaneous determination by NIR spectroscopy of the roasting degree and Arabica/Robusta ratio in roasted and ground coffee. *Food Control.* 2016; 59: 683–689.

Bianco GB. Climate change adaptation, coffee, and corporate social responsibility: Challenges and opportunities. *Int J Corporate Soc Responsibility.* 2020; 5: 3. doi:10.1186/s40991-020-00048-0

Boccas F, Roussos S, Gutierrez M, Serrano L, Viniegra GG. Production of pectinase from coffee pulp in solid-state fermentation system – selection of wild fungal isolate of high potency by a simple 3-step screening technique. *J Food Sci and Tech.* 1994; 31: 22–26.

Breitler JC, Dechamp E, Campa C *et al.* CRISPR/Cas9-mediated efficient targeted mutagenesis has the potential to accelerate the domestication of *Coffea canephora*. *Plant Cell Tiss Organ Cult.* 2018; 134: 383–394. doi:10.1007/s11240-018-1429-2

Burns TB, Walker MJ. Critical review of analytical and bioanalytical verification of the authenticity of coffee. *J AOAC Int.* 2020; 103: 283–294.

Combes MC, Joët T, Lashermes P. Development of a rapid and efficient DNA-based method to detect and quantify adulterations in coffee (Arabica versus Robusta). *Food Control.* 2018; 88: 198–206.

Cornelis MC. The impact of caffeine and coffee on human health. *Nutrients.* 2019; 11: 416. doi:10.3390/nu11020416.

Corro G, Pal U, Cebada S. Enhanced biogas production from coffee pulp through deligninocellulosic photocatalytic pretreatment. *Energy Sci & Engineering.* 2014; 2: 177–187.

Cubero-abarca R, Moya R. Use of coffee (*Coffea arabica*) pulp for the production of briquettes and pellets for heat generation. *Ciência e Agrotecnologia.* 2014; 38: 461–470.

De Luca S, De Fillipis M, Bucci R, Magrì AD, Magrì AL, Marini F. Characterization of the effects of different roasting conditions on coffee samples of different geographical origins by HPLC-DAD, NIR and chemometrics. *Microchem J.* 2016; 12: 348–361.

De Sanctis V, Soliman N, Soliman AT, Elsedfy H, Di Maio S, El Kholy M, Fiscina B. Caffeinated energy drink consumption among adolescents and potential health consequences associated with their use: A significant public health hazard. *Acta Biomed.* 2017; 88(2): 222–231.

DIN 10779:2011-03: Analysis of coffee and coffee products – Determination of 16-O-methyl cafestol content of roasted coffee – HPLC-method, German Institute for Standardization, Berlin, Germany.

Fernandes-Brum CN, Rezende PM, Cherubino Ribeiro TH, Ricon de Oliveira R, Cunha de Sousa Cardoso T, Rodrigues do Amaral L, de Souza Gomes M, Chalfun-Junior A. A genome-wide analysis of the RNA-guided silencing pathway in coffee reveals insights into its regulatory mechanisms. *PLOS One.* 2017. doi:10.1371/journal.pone.0176333

Gebresemati M, Gabbiye N, Sahu O. Sorption of cyanide from aqueous medium by coffee husk: Response surface methodology. *J Applied Res Tech.* 2017; 15: 27–35.

Getachew T, Hussen A, Rao VM. Defluoridation of water by activated carbon prepared from banana (*Musa paradisiaca*) peel and coffee (*Coffea arabica*) husk. *Int J Environ Sci Tech.* 2015; 12: 1857–1866.

Giridhar P, Avinash K, Simmi PS, Ravishankar GA. Differential expression of WRKY transcriptional factors in endosperm tissues during stress and ontogeny of fruits of *Coffea canephora* with respect to caffeine biosynthesis. *Proceedings of 24th International Conference on Coffee Science*, November 11–16, 2012, San Jose, Costa Rica, pp. 522–526.

Górnaś P, Siger A, Pugajeva I, Czubinski J, Waśkiewicz A, Polewski K. New insights regarding tocopherols in Arabica and Robusta species coffee beans: RP-UPLC-ESI/MS^n and NP-HPLC/FLD study. *J Food Comp Anal.* 2014; 36: 117–123.

Gouvea, Torres C, Franca AS, Oliveira LS, Oliveira ES. Feasibility of ethanol production from coffee husks. *In Biotech Letters.* 2009; 31: 1315–1319.

Gunning Y, Defernez M, Watson AD, Beadman N, Colquhoun IJ, Le Gall G, Philo M, Garwood H, Williamson D, Davis AP, Kemsley EK. 16-O-methylcafestol is present in ground roast Arabica coffees: Implications for authenticity testing. *Food Chem.* 2018; 248: 52–60.

Gutiérrez-Hellín J, Varillas-Delgado D. Energy drinks and sports performance, cardiovascular risk, and genetic associations; Future prospects. *Nutrients.* 2021; 24; 13(3): 715. doi:10.3390/nu13030715.

Higgins JP, Babu K, Deuster PA, Shearer J. Energy drinks: A contemporary issues paper. *Curr Sports Med Rep.* 2018; 17(2): 65–72.

Hong CT, Chan L, Chyi-Huey Bai C-H. The effect of caffeine on the risk and progress on of Parkinson's disease: A meta-analysis. *Nutrients.* 2020; 12(6): 1860.

Jeszka-Skowron M, Zgoła-Grześkowiak A, Grześkowiak T. Analytical methods applied for the characterization and the determination of bioactive compounds in coffee. *Eur Food Res Technol.* 2015; 240: 19–31.

Jumhawan U, Putri SP, Yusianto, Bamba T, Fukusaki E. Application of gas chromatography/flame ionization detector-based metabolite fingerprinting for authentication of Asian palm civet coffee (Kopi Luwak). *J Biosci Bioeng.* 2015; 120: 555–561.

Jumhawan U, Putri SP, Yusianto, Marwani E, Bamba T, Fukusaki E. Selection of discriminant markers for authentication of

Asian palm civet coffee (Kopi Luwak): A metabolomics approach. *J Agric Food Chem*. 2013; 61: 7994–8001.

Kavya Dechamma KM, Mishra KD. Challenges and problems faced by coffee industry – A case study of coorg coffee industries v/s other mechanised industry attracting youth. *International J of Mech Engin and Technol*. 2018; 9: 452–459.

Kolahdouzan M, Hamadeh MJ. The neuroprotective effects of caffeine in neurodegenerative diseases. *CNS Neurosci. Ther*. 2017; 23: 272–290. doi:10.1111/cns.12684.

Maciejewski G, Mokrysz S. New trends in consumption on the coffee market. *Zeszyty Naukowe Polityki Europejskie, Finanse i Marketing* 2019; 22: 132–144.

Manchester J, Eshel I, Marion DW. The benefits and risks of energy drinks in young adults and military service members. *Mil Med*. 2017; 182(7): e1726–e1733. doi:10.7205

Marcone MF, Alrifai O. Origins and compositional analysis of novel foods: Kopi Luwak coffee and bird's nest soup. *Comprehensive Biotechnology (Third Edition)*, 2019; 4: 739–759.

Martellossi C, Taylor EJ, Lee D, Graziosi G, Donni P. DNA extraction and analysis from processed coffee beans. *J Agric Food Chem*. 2005; 53: 8432–8438.

Mishra MK, Slater A. Recent advances in the genetic transformation of coffee. *Biotechnol. Res Int*. 2012. doi:10.1155/2012/580857

Nguyen GNT, Sarker T. Sustainable coffee supply chain management: a case study in Buon Me Thuot City, Daklak, Vietnam. *Int J Corporate Soc Responsibility*. 2018; 3: 1 doi:10.1186/s40991-017-0024-x

Oliveira WE, Franca AS, Oliveira LS, Rocha SD. Untreated coffee husks as biosorbents for the removal of heavy metals from aqueous solutions. *J Hazardous Materials*. 2007; 152: 1073–1081.

Petit A, Levy F, Lejoyeux M, Reynaud M, Karila L. Energy drinks: An unknown risk. *Rev Prat*. 2012; 62(5): 673–678. PMID: 22730801 Review. French?

Petitot AS, Barsalobres-Cavallari C, Ramiro D, Albuquerque Freire E, Etienne H, Fernandez D. Promoter analysis of the WRKY transcription factors CaWRKY1a and CaWRKY1b homoeologous genes in coffee (*Coffea arabica*). *Plant Cell Rep*. 2013; 32(8): 1263–1276. doi:10.1007/s00299-013-1440-3.

Pham Y, Reardon-Smith K, Mushtaq S. *et al*. The impact of climate change and variability on coffee production: A systematic review. *Climatic Change*. 2019; 156: 609–630. doi:10.1007/s10584-019-02538-y

Raveendran A, Murthy PS. New trends in specialty coffees – "the digested coffees". *Crit Rev Food Sci Nutr*. 2021; 28: 1–7.

Roughan G. The key coffee industry trends for 2021 & beyond. Posted October 29, 2020. https://www.unleashedsoftware.com/blog/the-key-coffee-industry-trends-for-2021-beyond. Accessed on April 1, 2021.

Samper LF, Quiñones-Ruiz XF. Towards a balanced sustainability vision for the coffee industry. *Resources*. 2017; 6: 17. doi:10.3390/resources6020017

Scully DS, Jaiswal AK, Abu-Ghannam N. An investigation into spent coffee waste as a renewable source of bioactive compounds and industrially important sugars. *Bioengineering (Basel, Switzerland)*. 2016; 3(4): 33.

Shibin M, Satyanarayana KV, Sridevi V, Kalpashree G, Giridhar P, Chandrashekar A, Ravishankar GA. Evaluating the effect and effectiveness of different constructs with a conserved sequence for silencing of *Coffea canephora* N-methyltransferases. *J. Plant Biochem. Biotechnol*. 2014; 23: 399–409.

Sridevi V, Giridhar P. Recent trends in Coffee biotechnology towards quality improvement – a review. *Indian J Bot Res*. 2008; 4: 5–12.

Tavares PD, Giarolla A, Chou SC, Silva AJD, Lyra AD. Climate change impact on the potential yield of Arabica coffee in Southeast Brazil. *Reg Environ Chang*. 2018; 18: 873–883. doi:10.1007/s10113-017-1236-z

Tolessa K, Rademaker M, De Baets B, Boeckx P. Prediction of specialty coffee cup quality based on near infrared spectra of green coffee beans. *Talanta*. 2016; 150: 367–374.

Troyer RJ. Coffee drinking an emerging social problem? *Social Problems*. 1984; 31: 403–416.

Ulsido MD, Li M. Effect of organic matter from coffee pulp compost on yield response of chickpeas (*Cicer arietinum* L.) in Ethiopia. *Eng Rural Dev*. 2016; 1339–1347.

Ulsido MD, Zeleke G, Li M. Biogas potential assessment from a coffee husk: An option for solid waste management in Gidabo watershed of Ethiopia. *Eng Rural Dev*. 2016; 16: 1348–1354.

Wang X, Lim L-T, Fu Y. Review of analytical methods to detect adulteration in coffee. *J AOAC Int*. 2020; 103(2): 295–305.

Section II

Coffee metabolites analysis

4

The bioactive compounds and substances with toxic potential in coffee products: The properties and the methods of analysis

Magdalena Jeszka-Skowron and Beata Czarczyńska-Goślińska

CONTENTS

4.1 Introduction

Coffee is one of the most popular drinks of Europeans and Americans. This beverage is drunk in homes and restaurants for its flavor and stimulative properties. The first coffee house was opened in Mecca in the fifteenth century, and it can be assumed that the consumption of coffee is still growing. Coffee is a complex composition of more than 1000 compounds, caffeine, chlorogenic acids, and other bioactive compounds and elements. After roasting, and also with the degree of roasting, the composition of coffee beans changes. Proteins, polysaccharides, chlorogenic acids, and trigonelline content decreases, and other substances such as volatile compounds are formed and nicotinic acid levels increase (Farah et al. 2006a; Jeszka-Skowron et al. 2020). Apart from bioactive compounds increased and formed after roasting, compounds with toxic potential, such as acrylamide, can also be formed. In green coffee beans (as well as coffee brews) and supplements, which are more and more popular due to their anti-obesity potential, the mycotoxins can be determined. There is still growing evidence that drinking 2–4 cups of coffee per day is associated with better health because coffee possesses the potential of contributing to the reduction in stroke cases (21% lower; Shao et al. 2021) and to a lower risk of mortality (Abe et al. 2019). Therefore, the constant quality control of commercial coffee beans, brews, coffee leaves, and their infusions should be monitored. In this chapter, the bioactive compounds and potentially toxic compounds found in coffee and coffee products, which are rarely analyzed, are described.

4.2 Main compounds in coffee beans, instant coffees, and coffee leaves

The coffee (*Coffea* of the family Rubiaceae) is an evergreen shrub or a small tree that grows natively in the mountains of southwest Ethiopia (Ukers 1922; see Figure 4.1). More than 100 *Coffea* species grow in tropical and subtropical areas in the equatorial regions. *Coffea arabica* L., *C. canephora* L. (commonly known as *C. robusta*), and *C. liberica* represent the most significant coffees in the world, but the most commercialized are *C. arabica* L. and *C. robusta* (Toci et al. 2016). After the roasting process, and also after the steaming process of coffee beans, physical changes and chemical

DOI: 10.1201/9781003043133-6

reactions are observed, such as the Maillard and Strecker reactions and production of the brown pigments from monosaccharides, and some compounds are degraded (Clarke 2003). The differences in quantity of compounds and elements in Arabica and Robusta coffees and green and roasted beans are shown in Table 4.1.

The phenolic compound profiles in coffee are affected by coffee variety and cultivar, climatic conditions (altitude, soil, fertilization, sun exposure, temperature, rainfall), harvesting, and post-harvest conditions (fermentation, washing, drying, and storage; Alonso-Salces et al. 2009; Jeszka-Skowron et al. 2015, 2016). The bioactive compounds in coffee brews show positive health effects on the human body, specifically in terms of diabetes, cancer, and cardiovascular, gastrointestinal, liver, and neurodegenerative disorders (Cano-Marquina et al. 2013; Gokcen and Sanlier 2017). Specially prepared coffees such as Kopi Luwak and coffee leaf infusions are more and more popular not only in their countries of origin but also in Europe.

Instant coffee (soluble) is a product of coffee bean brew after drying. Spray-drying and freeze-drying processes are

TABLE 4.1

Main Compounds, Bioactive Compounds, and Elements in Green and Roasted *C. arabica* and *C. robusta* Beans

Compounds/elements (%)	Arabica		Robusta	
	Green	Roasted	Green	Roasted
Water	8.5–12.0	1.5–5.0	8.5–12.0	1.5–5.0
Proteins/peptides	10.0–11.0	7.5–10.0	11.0–15.0	7.5–10.0
Carbohydrates/fiber (total)	45.1–58.1	36.3–42.5	54.1–64.4	42.3–43.9
Polysaccharides	34–44	31–33	47–55	37
Lipids (total)	14.0	17.9	7.0	11.0
Diterpene esters	0.5–1.2	0.9	0.2–0.8	0.2
Elements (total, approx. 30)	3.0–4.2	4.5	4.4–4.5	4.7
Potassium	1–2	1–2	1–2	1–2
Magnesium	0.001–0.003	0.001–0.003	0.0025–0.006	0.0025–0.006
Melanoidins	–	23–25	–	23–25
Caffeine	0.9–1.3	0.9–1.3	1.5–2.5	1.4–2.5
Trigonelline	0.6–2.0	0.2–1.2	0.6–0.9	0.3–0.7
Nicotinic acid (mg 100 g^{-1})	<LOD–0.481	0.16–30.4	<LOD–0.672	0.014–25.0
Nicotinamide (mg 100 g^{-1})	<LOD–0.088	<LOD–0.202	<LOD–0.167	<LOD–0.190
Chlorogenic acids	7.6–8.6	0.4–2.9	5.2–7.1	0.4–5.9
Chlorogenic acid lactones	–	0.070–0.110	–	0.170
Acrylamide (μg 100 g^{-1})	–	16.2–37.4	–	18.8–70.8

Notes: Content varies according to cultivar, agricultural practices, climate, soil composition, and methods of analysis. LOD = limit of detection.

Sources: Trugo and Macrae (1984); Clarke (2003); Farah et al. (2006a, 2006b); Farah and Donangelo (2006); Fischer et al. (2001); Jeszka-Skowron et al. (2020); Jeon et al. (2019); Perrone et al. (2008b); Mojska and Gielecińska (2013); Bagdonaite et al. 2008).

FIGURE 4.1 Coffee fruit and leaves.

Source: Photo by Piotr Werner.

commonly used in the coffee industry. The finished product should contain a similar quantity of compounds to roasted coffee brew, mainly from Robusta coffee beans (Table 4.2). Therefore, a poorer sensory quality of instant coffee has been observed. Also, the addition of freeze-dried green coffee extracts to freeze-dried soluble coffee has resulted in an increase of the level of chlorogenic acid in the new, combined product (Corso et al. 2016).

In addition, there are various methods to obtain better quality coffee, such as fermentation of coffee beans. Kopi Luwak coffee is one of the most expensive coffees in the world (ICO 2021; Smith 2014). It is a unique coffee due to its specific treatment: the internal fermentation inside the gut of the Asian palm civet (*Paradoxurus hermaphroditus*), a small mammal native to southern and northern Asia, especially Java, Sumatra, and Sulawesi. The animal eats the best-flavored berries/cherries and excretes them (Jumhawan et al. 2016). After the civet's feces are collected, the beans are extracted, cleaned, wet-fermented, and sun-dried (Figure 4.2). The main compounds and other bioactive substances and elements of Arabica and Robusta Kopi Luwak coffees are shown in Table 4.2.

TABLE 4.2

Main Compounds, Bioactive Compounds, and Elements in Instant Coffees and Kopi Luwak Arabica and Robusta Seeds

Compounds/elements (%)	Instant coffee	Kopi Luwak Arabica seeds		Kopi Luwak Robusta green seeds
		green	roasted	
Water	2.7–3.5 (max. 5%)	12.0	5.0	11.7
Protein	12.2	12.5–13.9	12.7–14.3	14.5
Polysaccharides	50–75			58.4
Lipids	0.2–1.6	12.0–12.7	14.7–15.1	12.0
Elements (total): Potassium Magnesium	8.8–10 3.5 0.3			3.2[1] 1.82 0.002
Caffeine	3.1	1.1–1.2	1.0–1.2	
Nicotinic acid (niacin) (mg 100 g^{-1})	3.6–28			
Tryptophan	0.03			
Chlorogenic acids	0.87–2.69	3.6–4.8	0.6–1.2	
Isoflavones	0.0009			
Acrylamide (mg 100 g^{-1})	0.096–5.46	–		–

Notes: [1] Elements described as ash. Blanks indicate values not determined yet.

Sources: Smith (1985); USDA (2010); Muzaifa et al. (2020); Jeszka-Skowron et al. (2020); Perrone et al. (2008a); Jeon et al. (2019); Surma et al. (2017); Mojska and Gielecińska (2013).

FIGURE 4.2 Kopi Luwak coffee after cleaning.

Source: Photo by Piotr Werner.

TABLE 4.3
Chemical Composition of Coffee Leaves and Coffee Leaves Prepared as Infusions

Compound/element (%)	Arabica coffee leaves		Robusta coffee leaves		Processed coffee leaves (infusions)					
					White tea		Green tea		Black tea	
	Young leaves	Mature leaves	Young leaves	Mature leaves	Young leaves	Mature leaves	Young leaves	Mature leaves	Young leaves	Mature leaves
Moisture		5.5–7.8								
Protein		14.4–19.0								
Carbohydrates		51.0–64								
Fiber		17.1–20								
Fat		4.5–12.5								
Elements (total as ash):		8.8–12.4								
Ca		0.124–0.175								
Fe		0.0028–0.093								
Mg		0.0095–0.080								
Caffeine	1.49	0.60	1.00–1.32	0.28–0.75	0.77	0.58	0.71–0.89	0.54–0.69	0.88	0.52
Trigonelline	1.17	0.30	0.48–0.79	0.42–0.60	0.66	0.51	0.52–0.67	0.43–0.54	0.68	0.481
Theobromine	0.081	0.005	0.017–0.41	0.0045–0.081						
Nicotinic acid	0.005[1]									
Chlorogenic acids[2]	7.35	2.21	3.57–4.69	1.82–2.38	0.26	0.89	0.84–2.41	1.08–1.48	0.56	0.11
Rutin					0.25	0.20	0.16–0.33	0.17–0.21	0.13	0.04
Mangiferin	1.47	0.22	nd	nd	0.23	0.28	0.27–0.79	0.27–0.40	0.26	0.037
Isomangiferin	0.108	0.011			0.04	0.04	0.05–0.14	0.04–0.06	0.045	0.005

Notes: [1] Not reported as young or mature leaves; [2] chlorogenic acids as a sum of 3-, 4- and 5-caffeoylquinic acids; 3,5-diCQA = 3,5-dicaffeoylquinic acid; 3,4-diCQA = 3,4-dicaffeoylquinic acid; 4,5-diCQA = 4,5-dicaffeoylquinic acid; nd = not detected (below limit of detection [LOD]).

Sources: Campa et al. (2012); Chen et al. (2018, 2019); Monteiro et al. (2020); Silva et al. (2020); Woldesenebet (2005).

The coffee leaves are also the rich source of many bioactive compounds, not only caffeine, trigonelline, and chlorogenic acids but also mangiferin, isomangiferin, nicotinic acid, carotenoids, and others (Table 4.3). The dried coffee leaves are consumed as a substitute for tea in Indonesia and Ethiopia, as well as in Jamaica, India, Java, Sumatra, and South Sudan (Campa et al. 2012; Chen 2019). Drinking such tea could influence the positive health processes in the human body: anti-inflammatory, antioxidant, antihypertensive, antibacterial, and antifungal properties (Campa et al. 2012; Campa et al. 2017; Jyotshna et al. 2016). The level of ent-kaurane diterpenoid derivatives should be monitored due to their toxic effect on organisms. In the latest report from the European Food Safety Authority (EFSA; European Union), no safety concern was raised to place the infusion from coffee leaves (Arabica and/or Robusta) on the European market (EFSA 2020).

4.3 Bioactive compounds

4.3.1 Phytoestrogens from phenolic group of compounds

Phenolic compounds are plant metabolites that are related to the defense responses in the plant. They also possess other functions concerning attractive substances to accelerate pollination, coloring for camouflage and defense against herbivores, and antibacterial and antifungal activities (Acamovic and Brooker 2005; Edreva et al. 2008). These compounds can be determined in all parts of plants (i.e., fruit, nuts, seeds, leaves, bark, and roots; Naczk and Shahidi, 2004, 2006; Wanasundara et al. 1996). In the group of phenolic compounds, there are few classes of compounds, usually as follows: flavonoids, tannins, phenolic acids (hydroxybenzoic and hydroxycinnamic acids), lignans, stilbenes, and coumarins (Jeszka-Skowron et al. 2021). The coffee beans (almost 90% of all phenolic compounds) and leaves contain mainly chlorogenic acids. However, compounds such as flavonoids (epigallocatechin gallate and rutin) can also be determined (Chen 2019; Jeszka-Skowron et al. 2021; Król et al. 2020). Flavonoids can be classed as flavones, flavanones, isoflavones, flavonols, flavanonols, flavan-3-ols, anthocyanidins, and anthocyanins (acyl glycosides and glycosides of anthocyanidins; Jeszka-Skowron et al. 2021). Isoflavones and lignans were also found in coffee samples (Figure 4.3; Alves et al. 2010a; Angeloni et al. 2018, 2020). These compounds belong to the phytoestrogens group and are used for prevention of hypercholesterolemia, carcinogenesis, and osteoporosis, and they can relieve menopausal symptoms (Murkies et al. 1998).

The estrogen-like structure of isoflavones (3-phenylchromen-4-one core structure) is differed by the substituents such as methoxy, hydroxy, and glucosides. Some compounds such as genistein (4′,5,7-trihydroxyisoflavone), daidzein (4′,7-dihydroxyisoflavone), and formononetin (7-hydroxy-4′-

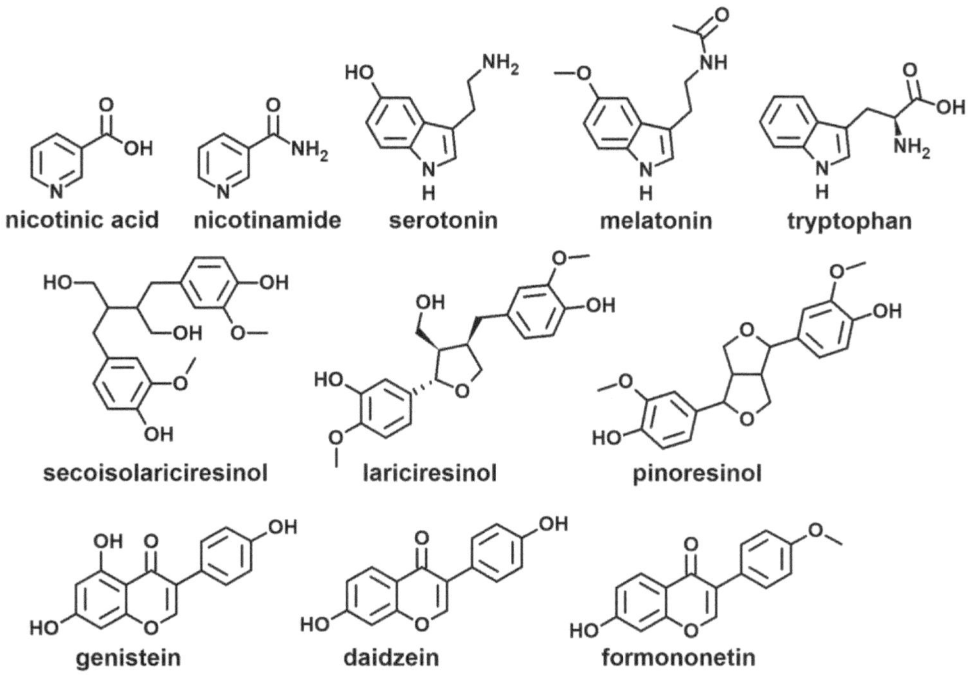

FIGURE 4.3 Chemical structures of nicotinic acid, nicotinamide, main phytoestrogens, serotonin, melatonin, and tryptophan in coffee.

methoxyisoflavone) are the main compounds determined in the coffees. Instant coffee possesses about 9 µg g^{-1} of total isoflavones (Kuhnle et al. 2008). These compounds reveal weak estrogenic and antioxidant activity (in vitro and in vivo; Ruiz-Larrea et al. 1997; Mu et al. 2009; Ungar et al. 2003). Also, isoflavones showed the preventive activities in hormone-related cancers such as prostate, breast, bowel, cardiovascular diseases, osteoporosis, and menopause symptoms (Cotterchio et al. 2006; Kits 1987; Kumar et al. 2002; Rice-Evans 2004). It was found that coffee has estrogenic properties due to isoflavone content (Kitts 1987; Lucero et al. 2001).

The functioning of lignan compounds in plants is probably due to their role in defense against various pathogens and pests, and also in plant growth. Enterodiol and enterolactone, the metabolites of lignans, are formed by gastrointestinal microbiota, called mammalian lignans or enterolignans, found in high concentration in plasma and urine samples, and show many biological activities (antioxidant, estrogenic, and antiestrogenic) with effects on the hormone metabolism (Adlercreutz and Mazur 1997; Landete 2012).

4.3.2 Analysis of phytoestrogens

Lately, the determination of the phytoestrogens in coffee samples has been performed with the use of liquid chromatography (LC) coupled with tandem mass spectrometry (MS; Angeloni et al. 2018, 2020). Other methods can be used, such as high-performance liquid chromatography (HPLC; Alves et al. 2010b) or gas chromatography–mass spectrometry (GC-MS; Mazur et al. 1996, 1998; Thompson et al. 2006). Also, an automated solid phase extraction (SPE) and LC-MS/MS were applied (Kuhnle et al. 2007). Isoflavones in plants occur mainly as glycosides or glycoside esters (Grynkiewicz et al. 2005). Therefore, the detection of the aglycones (free aglycones of isoflavones) and their quantification need to be after acid hydrolysis (Müllner and Sontag 1999) or enzymatic procedures (Oomah 2002). Mazur et al. (1998) found that lignin secoisolariciresinol (SECO) in commercially available coffees was on average 561 ± 103 µg 100 g^{-1}, and matairesinol (MAT) was below the limit of detection (LOD) using the GC-MS method. The determination of phytoestrogens was performed by Thompson et al. (2006), who analyzed the commercial coffee samples for isoflavones (genistein, daidzein, glycitein, and formononetin), lignans (SECO, MAT, pinoresinol, lariciresinol [LARI]), and coumestan (coumestrol) with the use of time-consuming SPE preparation steps on samples and then analyzed them on GC-MS apparatus.

Recently, Angeloni et al. (2018) have used three different extraction methods ("dilute and shoot", acidic hydrolysis, and enzymatic digestion) for the determination of three lignans (SECO, MAT, and LARI) using LC-MS/MS in espresso coffee from capsules from five regions: Brazil, Colombia, El Salvador, Ethiopia, and India. They noted that both SECO (27.9–52.0 µg L^{-1}) and LARI (5.3 to 27.8 µg L^{-1}) were found in every coffee sample in contrary to matairesinol, which was not possible to detect in coffee samples (below LOD). Enzymatic hydrolysis was developed and used to avoid long clean-up and preparation steps, and separation time was 4 min. In 2020, Angeloni et al. used the aforementioned method to determine lignans (SECO, MAT, and LARI) and isoflavones (genistin, daidzin, daidzein, genistein, biochanin A, and formononetin) in green coffee samples in one analysis after optimization of seven extraction methods. The best performing process was a double extraction composed of base hydrolysis in methanol and enzymatic digestion with clara-diastase (a mixture of enzymes containing α-amylase, cellulase, invertase, peptidase, phosphatase, and sulfatase); the method showed recoveries at 74%–94% levels and the highest total concentration of all the compounds studied. Twenty-five different green coffee samples possess total lignans in the following ranges: 286.5–8131.8 μg kg^{-1} and isoflavones: 3.4–300.0 μg kg^{-1}. MAT was also below the LOD in all samples. Coffee samples from Ethiopia were a good source of lignans.

The differences between coffee species, roast degree, and brewing procedure in isoflavone content were also described (Alves et al. 2010a). They reported that the total isoflavone amount was sixfold higher in Robusta coffees than in Arabica coffees, mainly due to formononetin level. After roasting, the content of isoflavones decreased, whereas their extractability increased (especially for formononetin). The total content of isoflavones in espresso coffee (30 mL) varied from 40 μg (100% Arabica) to 285 μg (100% Robusta). In espresso coffee (30 mL) prepared from commercial blends, average amounts of genistein, daidzein, and formononetin were 6, 17, and 78 μg, respectively. The differences between brewing methods revealed that total content of isoflavones ranged from espresso (average 170 μg in 30 mL) and press-pot (average 130 μg in 60 mL) to mocha coffees (average 360 μg in 60 mL), and amounts were similar to those of filtered coffee (180 μg in 120 mL).

4.3.3 Nicotinic acid and nicotinamide

The main alkaloids determined in coffee are methylxanthines, and caffeine is usually the dominating compound from this group. The next important alkaloid found in coffee is trigonelline (N-methyl nicotinic acid). The highest level of this compound can be found in green coffee beans and young coffee leaves (Tables 4.1 and 4.3). After the roasting of coffee beans, apart from volatile compounds, nicotinic acid (NA), N-methylpyridinium (NMP), and nicotinamide (NAM) are also formed (Trugo and Macrae 1989). Nicotinic acid and nicotinamide are usually called niacin (vitamin B_3), a semi-essential vitamin that is formed from tryptophan in the human body, with approximately 60 mg of tryptophan being equivalent to 1 mg nicotinic acid (Figure 4.3). Safe therapy in humans is less than 3 g per day (Knip et al. 2000). In contrast, Zhou et al. showed that long-term nicotinamide overload could cause insulin resistance and oxidative stress, and can even lead to diabetes (Zhou et al. 2009; Zhou et al. 2010). The EFSA has recommended the determination of urinary metabolites: N^1-methylnicotinamide (NMNAM) and N^1-methyl-2-pyridone-5-carboxamide (2-Py) to measure an individual's niacin intake (EFSA 2014). In a human intervention study, urinary levels of NA were established to increase 2 h after consuming 350 mL coffee (Lang et al. 2010). Also, an intervention study of the excretion of major niacin metabolites in the urine of volunteers after ingestion of 500 mL "regular coffee" containing 34.8 μmol L^{-1} nicotinic acid and 0.58 μmol L^{-1} nicotinamide were determined (Kremer et al. 2018). Other metabolites (NMNAM, 2-Py, and nicotinuric acid) were also found. The results revealed that coffee could be a source of niacin in the diet.

Depending on the time of the roasting process, the content of nicotinic acid ranges from about 10 mg 100 g^{-1} in weak, light-roasted commercial coffees (American coffee) to 40 mg in heavy, dark-roasted coffees (i.e., Italian coffee; Adrian and Frangne 1991). Also, the different extraction techniques, decoction, infusion, and pressure methods used for coffee brewing influence nicotinic acid content (Table 4.4).

4.3.4 Analysis of nicotinic acid and nicotinamide

Recently, the bitter compounds in coffee brews have been determined using multivariate analysis of the data obtained from the analytical instrument and electronic taste sensor experiments (Fujimoto et al. 2020). At first, the investigation of bitter compounds (from 30 compounds) in coffee brews using a taste sensor and measurements from LC-MS/MS, liquid chromatography–photodiode array (LC-PDA), and liquid chromatography–ultraviolet detector (LC-UV) were performed (after fractionation of coffee brews by liquid-liquid extraction). A taste sensor analyzed the bitterness response value of each fraction. Candidate bitter compounds in the coffee brews were identified regarding their variable importance in projection and by a coefficient of projection to latent structure regression (PLS-R) analysis. "PLS-R analysis proposed that nicotinic acid, L-lactic acid, and nicotinamide contributed to the bitterness of the coffee brews." Nicotinic acid, L-lactic acid, and nicotinamide added to the coffee brews increased the bitterness response value of brews compared to those brews without them (Fujimoto et al. 2020).

In 1959, niacin was microbiologically determined with the use of *Lactobacillus arabinosus* as the test organism (Bressani and Navarrete 1959). Nowadays, HPLC and LC-MS are the dominant techniques in determining nicotinic acid and nicotinamide in coffee samples. HPLC is usually used to determine nicotinic acid and other compounds, mainly trigonelline and others (Caprioli et al. 2014; Casal et al. 2000; Liu et al. 2012). The analysis of nicotinamide in coffees is very rare. In the last decades, LC-DAD-MS^n is more common, with additional advantages such as compound identification and quantification on lower levels (lower LODs), therefore nicotinamide can be determined (Jeszka-Skowron et al. 2020; Rodrigues and Bragagnolo 2013). The isotopically labeled standard d_4-nicotinic acid was spiked to every coffee sample, and fast 4-min LC-ESI-MS/MS analysis was conducted (Buckel et al. 2019). Nicotinic acid and d_4-NA were detected in multiple reaction monitoring modes; MS transitions were m/z 124→80 as a qualifier and m/z 124→78 as quantifier and the isotopically labeled analogue m/z 128→84, respectively. The quantitative determination of trigonelline, nicotinic acid, and nicotinamide during 10 min was used for the analysis in complex matrix samples (roasted coffee and human plasma) with the LC-MS/MS-SIDA (SIDA – stable isotope dilution analysis; Lang et al. 2008). In addition to the chromatographic methods, a chromatographic low-pressure flow injection system comprising a short length monolithic column (coated with the cationic surfactant didecyldimethylammonium bromide [DDAB]) using amperometric detection was developed by Santos and Rangel (2015). This approach involves simple configuration, low cost, fast and easy assembling, and only brewing and filtration of the sample without any additional sample clean-up procedures.

Perrone et al. (2008a) found that the amount of nicotinic acid in green coffee beans was below the LOD, in comparison

TABLE 4.4

Nicotinic Acid in Coffee Brews Depending on Degree of Roasting and Preparation Method

Coffee preparation	Roasting process	Preparation temperature/ pressure and time		Nicotinic acid ($\mu g\ 100\ mL^{-1}$)	Reference
Standard "drip" coffee maker (filter paper)[1]	3 – intensity[2]	nr		1031–1142	Buckel et al. 2019
Automatic coffee maker[1]	3 – intensity[2]	nr		998–1082	Buckel et al. 2019
Cold brew[1]	3 – intensity[2]	4 °C	24 h	991–1034	Buckel et al. 2019
Pod machine – espresso beans[1]	5 – intensity[2]	nr		1512–1544	Buckel et al. 2019
Traditional filter – classic beans[1]	3 – intensity[2]	40°C	nr	401–539	Buckel et al. 2019
Traditional filter – classic beans[1]	3 – intensity[2]	60°C	nr	734–804	Buckel et al. 2019
Traditional filter – classic beans[1]	3 – intensity[2]	80°C	nr	812–883	Buckel et al. 2019
Traditional filter – classic beans[1]	3 – intensity[2]	90°C	nr	890–924	Buckel et al. 2019
Traditional filter – classic beans[1]	3 – intensity[2]	100°C	nr	896–925	Buckel et al. 2019
Coffee maker – green beans[1]	No roasting process	Hot water	nr	50–300	Minamisawa et al. 2004
Coffee maker – roasted beans[1]	Light roasted beans	Hot water	nr	300–1950	Minamisawa et al. 2004
Coffee maker – roasted beans[1]	Medium roasted beans	Hot water	nr	1100–2700	Minamisawa et al. 2004
Coffee maker – roasted beans[1]	"City" roasted beans[3]	Hot water	nr	650–2400	Minamisawa et al. 2004
Coffee maker – roasted beans[1]	"Full city" roasted beans[3]	Hot water	nr	3050–4450	Minamisawa et al. 2004
Coffee maker – roasted beans[1]	French roasted beans[3]	Hot water	nr	2800–3850	Minamisawa et al. 2004
Capsule coffee maker (65 capsules from 5 different brands in Italy)	nr	60.0–78.5°C/15–20 bars	nr	1000–4000	Angelino et al. 2018
Standard filter basket – ground Arabica[4]	Medium dark (182°C and 13 min)	93 ± 2°C/9 bars	nr	10,700–19,100	Khamitova et al. 2020
Standard filter basket – ground Robusta[4]	Medium dark (182°C and 15 min)	93 ± 2°C/9 bars	nr	7100–14,100	Khamitova et al. 2020
Lower size filter basket – ground Arabica[5]	Medium dark (182°C and 13 min)	93 ± 2°C/9 bars	nr	8400–16,700	Khamitova et al. 2020
Lower size filter basket – ground Robusta[5]	Medium dark (182°C and 15 min)	93 ± 2°C/9 bars	nr	6400–29,600	Khamitova et al. 2020

Notes: [1] 100% Arabica beans; [2] taste intensity (3–5) – degree of roasting from medium to heavy roasted beans; [3] "city" type of roasting is the standard roast (New York style), "full city" is deeper roast than city, and French is the deepest grade, roasted at 230°C; [4] filter basket – standard, less than 300 μm-sized filter basket for Italian espresso coffee; [5] filter basket – 180 μm-sized filter basket; nr – not reported.

to the regular and decaffeinated commercial ground roasted coffees, which contained 9.6–30.4 mg 100 g^{-1} of coffee dry weight, and instant coffees, which contained 34.2–36.3 mg 100 g^{-1} of coffee dry weight. On the other hand, Jeszka-Skowron et al. (2020) revealed that Arabica and Robusta green coffee brews contained nicotinic acid from 3.20 to 4.81 μg g^{-1} and 4.01 to 6.72 g μg^{-1}, respectively, depending on the coffee origin, and nicotinamide from 0.31 to 0.88 μg g^{-1} and from 0.30 to 1.66 μg g^{-1}, respectively. After the light roasting process, the level of NA changed in Arabica coffees from 470% to 2305% and Robusta coffees from 326% to 1765%; and the level of NAM changed in Arabica coffees from 150% to 407% and in Robusta coffees from 114% to 322%.

Rodrigues and Bragagnolo (2013) noted that the level of this compound in regular and decaffeinated commercial ground roasted coffees was higher and 48.7–133.8 mg 100 g^{-1} of coffee dry weight, and instant/soluble coffees with also decaffeinated samples were 19.7–43.8 mg 100 g^{-1} of dry coffee weight. These results were different because of coffee origin, roasting time, preparation procedure, and the method of analysis. The comparison of the nicotinic acid level in μg in 100 mL of differently prepared coffees is presented in Table 4.4. "The nicotinic acid is almost completely extracted in the preparation of the beverage for drinking, the amount in a cup of white coffee (200 mL, half 10% coffee decoction and half milk) between 1 and 2 mg" (Hughes and Smith 1946). The nicotinic acid value in instant coffees was found at 0.01% on average (Liu et al. 2012; Perrone et al. 2008a). The trigonelline contents, nicotinic acid, and nicotinamide in coffee brews can depend on coffee species, variety, geographical origin, and roasting conditions (Casal et al. 2000; Jeszka-Skowron et al. 2020; Ky et al. 2001).

In roasted Arabica coffees, the nicotinic acid level varied from 19 to 29 mg 100 g^{-1} (Kitzberger et al. 2014). The method used for nicotinic acid determination was described by Alves et al. (2006). The extraction was prepared with the use of water: acetonitrile (95:5 v/v). After filtration, dilution, and further filtration through a nylon membrane (0.45 mm), a reverse-phase Spherisorb ODS1 column (250 mm × 4.6 mm i.d., 5 μm) was applied. Compounds were eluted in a gradient of acetic acid 5% (v/v) (A) and acetonitrile (B) as follows: 0–5 min, 5% of B; 5–10 min, 5%–13% of B (linear); and 13–35 min, 13% of B. A flow rate of 0.7 mL min^{-1} and an oven temperature of 25°C were used. Nicotinic acid and trigonelline were detected at 264 nm, caffeine was detected at 272 nm, 5-CQA was detected at 325 nm, and identification was based on retention time, co-elution, and UV spectra. In addition, the fast LC-MS method (6 min of analysis) for simultaneous determination of caffeine, trigonelline, nicotinic acid, and sucrose was applied (Perrone et al. 2008a). The HPLC-VWD (variable wavelength detector) method showed linearity ($R^2 > 0.9985$) and recoveries (71%–92%) in 25 min of analysis. LODs and LOQs (limit of quantification) were in the range of 0.03–0.06 and 0.1–0.2 mg L^{-1}, respectively (Caprioli et al. 2014).

The latest research showed the salting-out assisted liquid-liquid extraction (SALLE) method to investigate how effectively caffeine, trigonelline, and nicotinic acid are extracted from green and roasted coffee samples (Chirfa et al. 2020). To extract and/or preconcentrate analytes by SALLE, 0.5 g of ground raw or roasted coffee sample was taken into a beaker and then dissolved in 25 mL distilled water and boiled in a water bath at 80–90°C. After cooling, the filtration process was performed. Subsequently, 1 mL of the filtrate and 9 mL deionized water containing 5 mmol L^{-1} tetrabutylammonium hydrogensulfate was taken into a Falcon tube. After gentle shaking of the mixture, 2 mL acetonitrile and 5 g $(NH_4)_2SO_4$ were added. The mixture was again shaken until the salt was completely dissolved. To further enhance homogenization, it was mixed by the vortex shaking, then centrifuged to obtain phase separation. Then 1 mL of the supernatant was taken with micropipette into a 25 mL beaker and dried in an oven at 85–90°C. The residue was reconstituted with 2 mL of deionized water, then filtered and injected into an HPLC instrument. The smaller volumes of less toxic organic solvents were used.

4.3.5 Biogenic amines

Coffee contains not only phenolic compounds, alkaloids, and vitamins but also other bioactive compounds such as the amines serotonin and melatonin as well as the amino acid – tryptophan (Figure 4.3). The first two substances belong to the important group of compounds called biogenic amines (BAs), which are formed as a result of decarboxylation of amino acids when roasting the coffee beans. They can also appear when microbial decarboxylases are formed during fermentative processes, suggesting inappropriate storage or low-quality defective fermented seeds and transamination of aldehydes and ketones (Dias et al. 2012). Biogenic amines can be divided into polyamines (putrescine, spermine, and spermidine), catecholamines (tyramine), imidazoles (4-methylimidazole [4-MeI]), and indoleamines (serotonin and melatonin; Baranowska and Płonka 2015). Polyamines are involved in plants' growth and development, coffee tissues, and fruit (Sridevi et al. 2009). All BAs are involved in cellular processes in the human organism, generated in the brain, and act as neurotransmitters. Serotonin (5-hydroxytryptamine) controls blood pressure, smooth muscle stimulation, and sensory threshold increase, and inhibits gastric secretion (Halász and Baráth 2002). Other amines, including putrescine, spermidine, and cadaverine, can act as free radical scavengers, while tyramine has a remarkable antioxidant activity that depends on amino and hydroxy groups. The primary function of BAs is their impact on flavor and taste of food (Halász et al. 1994; Papageorgiou et al. 2018).

It is essential to monitor the amount of these compounds in food, coffee, and coffee bean supplements because serotonin, histamine, 2-phenylethylamine, or tyramine could be toxic in high concentrations. For tyramine, no observed adverse effect level (NOAEL) limits between 200 mg per oral consumption and 600 mg per person per meal is reported for persons without monoamine oxidase inhibitor (MAOI) drug treatment in the EU (EFSA 2011). There are many physiological effects of high doses of biogenic amines, including human allergic reactions and hypotension (histamine, putrescine, cadaverine) or hypertension (tyramine), nausea, headache, rash, dizziness, cardiac palpitation, emesis, intracerebral hemorrhage, and anaphylactic shock syndrome (Özdestan 2014). Biogenic amines can also be precursors of nitrosamines, which can cause carcinogenic and mutagenic changes in the organism.

4-Methylimidazole (4-MeI) is a neurotoxic agent that may induce symptoms in the central nervous system, such as seizures, tremors, and lack of motor coordination (Chan et al. 2008; Sivertsen et al. 2009). The International Agency of Research on Cancer reported that 4-MeI is a potentially carcinogenic compound to humans (group 2B; IARC 2011). The Office of Environmental Health Hazard Assessment within the California Environmental Protection Agency recommends that the no significant risk level (NSRL) of 4-MeI should be less than 29 μg per day (OEHHA 2012).

4.3.6 Analysis of biogenic amines

The coffee leaves contain the polyamines – ornithine, putrescine, spermine, threonine, triethanolamine, and tyramine (Martins et al. 2014). Some other BAs can be found in coffee beans and brews. The differences between green, American roasted (300°C for 6 min), and French roasted coffee (300°C for 12 min) were presented with the use of HPLC equipment (Cirilo et al. 2003). Serotonin was the dominating biogenic amine in green coffee beans (1.13 ± 0.15 mg 100 g^{-1}) followed by putrescine (1.03 ± 0.09 mg 100 g^{-1}), spermidine (0.60 ± 0.07 mg 100 g^{-1}) and spermine (0.44 ± 0.08 mg 100 g^{-1}). After roasting, the levels of putrescine and spermine decreased (below LOD). There was also a reduction of spermidine and serotonin. There are differences between French and American roasted coffees: French roasted coffees possess higher amounts of serotonin and spermidine, and also, agmatine is formed.

On the other hand, putrescine is the dominant compound in green coffee beans, followed by spermine and spermidine (Casal et al. 2002, 2004). Cadaverine, histamine, serotonin, and tyramine are also found. The level of serotonin in the roasted coffee brew is from <LOD to 1.4 mg in 100 mL and melatonin content 0.006–0.008 mg in 100 mL in blends of Arabica and Robusta. In the brewed Turkish prepared coffees from the Turkish market, the mean value of serotonin was 6.864 mg L^{-1}, tyramine 4.532 mg L^{-1}, cadaverine 4.387 mg L^{-1}, tryptamine 3.656 mg L^{-1}, and 2-phenylethylamine 0.8904 mg L^{-1}. Spermine, spermidine, and histamine were not detected (below LOD; Özdestan 2014). The level of total BAs in brewed coffee from the Italian market (100% Arabica and blends of Robusta and Arabica) was from 3.04 mg L^{-1} to 9.88 mg L^{-1} and usually, all eight BAs were

determined (2-phenylethylamine, putrescine, cadaverine, histamine, tyramine, spermidine, serotonin, and spermine); instant coffees usually possessed a lower level of BAs.

> Beverages prepared by espresso, capsule, and pod machines had the lowest BAs contents, as a result of the thermal and physical stress imposed on ground coffee by these methods, while mocha contained the highest BAs amounts owing to lower pressure and longer brewing time (Restuccia et al. 2015).

The putrescine and spermine level was higher in wet-processed coffee beans than in entire cherry dried beans (dry processed; Casal et al. 2004). On the other hand, the overall level of amines was detected in a lower amount in wet-processed beans. "This process was found to have a more uniform drying period and avoided the formation of amines like histamine, tyramine, and cadaverine" (Dias et al. 2012).

Apart from serotonin, the serotonin 5-*O*-β-glucoside and tryptophan-N^1-glucoside were determined in the green coffee beans (Servillo et al. 2016). Both substances were identified by their product ion spectra and by comparison with synthesized authentic standards. Serotonin 5-*O*-β-glucoside was found in all varieties of Robusta (3.1 ± 0.4 mg kg^{-1}) and Arabica (5.4 ± 0.8 mg kg^{-1}) coffee species. On the other hand, tryptophan-N^1-glucoside was at high levels in beans of Robusta (45.4 ± 5.2 mg kg^{-1}) and at significantly lower levels in those of Arabica species (0.8 ± 0.2 mg kg^{-1}) ($p < 0.0001$). In the medium roasted coffee brew, 1.4–2.5 mg tryptophan in 50 mL was determined; the Arabica coffees possess higher amounts of it (Martins et al. 2010). Melatonin amounts in green coffees are usually at very low levels of 39.0 ± 6.50 pg g^{-1} (Kocadağlı et al. 2014).

The determination of amines is usually very laborious due to the derivatization steps and need of preconcentration (Papageorgiou et al. 2018). The BAs are extracted by trichloroacetic acid (TCA). After an ion-pair clean-up procedure with *bis*-2-ethylhexylphosphate, BAs are converted to the corresponding dansyl derivatives for HPLC-DAD analysis at 254 nm (Casal et al. 2002, 2004). In addition, 4-methylimidazole determination in an instant and other substitutes of coffees was developed (Cunha et al. 2016). In the method, based on Casal et al. (2002) procedure, methanol was used as extrahent, then chloroform. The mixture was mixed and centrifuged. The aliquot of the aqueous phase was derivatized with a mixture of acetonitrile, isobutanol, pyridine, and two portions of *bis*-2-ethylhexylphosphate. Then aqueous sodium bicarbonate solution with isooctane was added; after mixing, the top layer was transferred to an autosampler vial, and 1 μL was injected into the GC-MS system. Mass spectrometric parameters were the following: electron impact ionization with 70 eV energy; ion source temperature, 230°C; MS quadrupole temperature, 150°C; and solvent delay, 2 min. The MS system was set in the selective ion monitoring (SIM) mode, and 4-MeI was quantified based on peak area using one target and three qualifier ions. The 4-MeI with the double peak at RT 8.66 min and RT 8.84 min, characteristic qualifier ions were m/z 82, 109, 81, and quantifier ion m/z 182. A single peak 157 was obtained at RT 8.20 min for the internal standard, with characteristic qualifier ions m/z 95, 123, 81, and quantifier ion m/z 196.

The latest study showed the simultaneous determination of free amino acids and monoamines, diamines, and polyamines in green coffee beans supplements and infusions with the use of ultra high-performance liquid chromatography (UHPLC) with UV and fluorescence detectors (Macheiner et al. 2021). Separation of the target analytes was achieved using an ethylene bridged hybrid (BEH)–based particle diol-bonded surface C18 column (50 mm × 2.1 mm i.d., 1.7 μm). Gradient elution was conducted at 35°C, using 0.1 mol L^{-1} sodium acetate in water, adjusted to a pH of 4.80 with acetic acid (solvent A) and acetonitrile (solvent B). This method was based on the procedure obtained by Fiechter and Mayer (2014) for the determination of amino acids. Due to the amounts of spermidine and histamine, the level of biogenic amines in supplements should be monitored.

4.4 Toxic compounds in coffee

4.4.1 Acrylamide

Although acrylamide (AA) is mainly associated with a contaminant appearing in potato, cereal, and bakery products as a result of their heat treatment (Krishnakumar and Visvanathan 2014; Tareke et al. 2002; Svensson et al. 2003; Surdyk et al. 2004; Borda et al. 2011; Riboldi et al. 2014), coffee has become one of its main dietary sources due to the roasting process (Zamani et al. 2017; Figure 4.4). It has been reported to contribute substantially to the total dietary intake of acrylamide both in the United States and in the Nordic countries of Sweden, Norway, and Denmark. Roasted and ground coffee is usually consumed in the form of a brew after the addition of hot water and subsequent filtration. AA is formed in food during heat processing at above 120°C (Claus et al. 2008; Krishnakumar and Visvanathan 2014). During the roasting process, coffee beans are subject to high temperatures in the range of 220–250°C. Many thermally driven reactions may occur under such conditions, and the significant intermediaries of AA such as acrolein, acrylic acid, and 3-aminopropionamide (3-APA) occur. However, the basic reaction of AA formation is always considered the Maillard reaction, induced by the condensation of free amino acid (asparagine) and reducing sugars (glucose and fructose) or alternatively reactive carbonyls (including α-dicarbonyls, n-aldehydes, and 2-oxo acids; Yaylayan et al. 2003; Zyzak et al. 2003; Stadler et al. 2002; Friedman 2003). Therefore, the levels of carbohydrates and asparagine in coffee seem to be important in terms of AA's potential formation. The studies on correlations between their concentrations and amounts of AA provided after roasting show only a weak positive correlation of amounts of asparagine and AA and no correlation with the amounts of reducing sugars (Lantz et al. 2006; Guenther et al. 2007).

Intake of acrylamide can lead to neurotoxicity, carcinogenicity, reproductive toxicity, and immunotoxicity. Studies on animals revealed three neurological effects: ataxia, hind-limb foot splay, and skeletal muscle weakness. Human exposure to AA in the workplace environment resulted in impaired nerve function or disrupted nervous system. However, a document issued by FAO/WHO (2002) indicated no risk of neuropathy

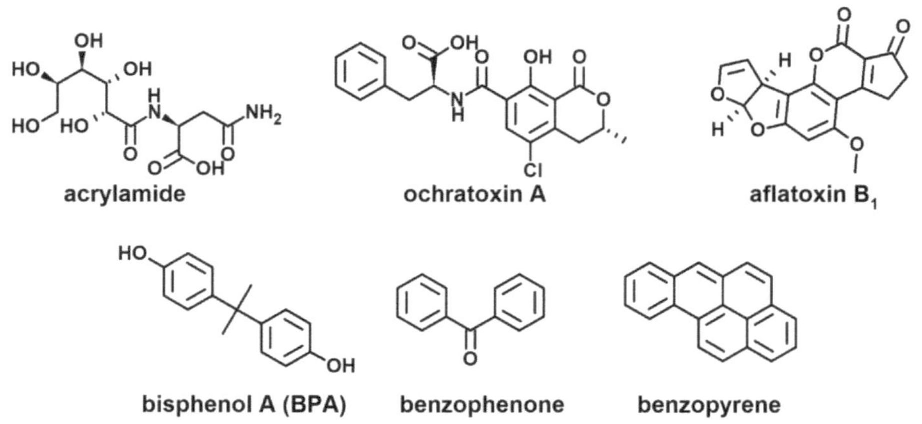

FIGURE 4.4 The chemical structures of acrylamide, mycotoxins, and estrogenic chemicals in coffee.

or fertility in normal conditions caused by intake of AA from food due to the consumption of much lower doses than these affecting human health. AA has a similar structure to carcinogen agents, and it is considered a probable human carcinogen. A working group of 23 scientists from ten countries teamed with the IARC to evaluate the potential carcinogenicity of drinking coffee. The research was based on more than 1000 observational and experimental studies to analyze long-term studies on rodents and compared both tumor-promoting and cancer-preventing activity. They concluded that overall coffee drinking was considered unclassifiable concerning its carcinogenicity to humans. As far as animals were concerned, the scientists found the gained evidence insufficient for the recognition of coffee as being a carcinogen (Loomis et al. 2016).

The genotoxicity of AA and its metabolite glycidamide resulted from the ability to cause DNA damage. Moreover, AA turned out to affect both the female and male reproductive systems of rodents. It reduced progesterone and testosterone levels in the serum of animals (Ali et al. 1983; Wei et al. 2014; Zamani et al. 2017; Nehlig and Cunha 2020). The assumed threat of AA to human health depends strongly on many various factors, such as the ways of exposure and distribution in the human organism, its elimination, and the action affected by the components of a diet (Shimamura et al. 2017).

4.4.2 Analysis of acrylamide

The methods developed so far to determine AA in coffee and coffee products are based mainly in LC-MS/MS with variations in the clean-up conditions employed. Different analytical methods have been described since the first reports of the occurrence of AA in food in 2002. Liquid chromatography–mass spectrometry (LC-MS) with the use of an isotope-labeled internal standard and GC-MS seemed to be one of the most popular methods (Gielecińska and Mojska 2013; Soares et al. 2007).

The key issue in the LC-MS methods is the extraction and clean-up steps, including proper selection of solid-phase extraction (SPE) cartridges to remove interfering compounds from the rich matrix. The significant number of LC-MS methods are based on a triple quadrupole instrument (MS/MS) and multiple reaction monitoring for selectivity enhancement. For ion production, most often electrospray ionization (ESI) or, less frequently, atmospheric pressure chemical ionization (APCI; Aguas et al. 2006). In MS/MS detection, the use of multiple reaction monitoring significantly enhances the selectivity of the method (Lineback et al. 2012).

Mojska and Gielecińska (2013) developed and validated the LC-MS/MS method employing the SPE procedure to investigate the influence of the roasting process of coffee on acrylamide levels. They tested 28 ground roasted coffee samples, 11 instant coffees, and three coffee substitutes (grain coffees). There were 14 brands of coffee samples from eight manufacturers. Among the roasted coffee samples were six Arabica species, 10 Robusta species, and 12 mixtures of both. Nine of 11 instant coffee samples were also mixtures of Arabica and Robusta. Coffee substituents were from roasted cereals. Coffee samples were mixed with water and hexane and centrifuged. After removal of the organic layer, the aqueous phase was incubated in an ultrasonic bath and cooled to room temperature. The Carrez I solution was prepared by dissolving potassium hexacyanoferrate in water, whereas Carrez II was a zinc sulfate water solution. Both these solutions were added to the sample, which was purified on a carbon SPE column previously conditioned with methanol and water. The methanol eluate was evaporated to dryness and reconstituted in a water and methanol mixture. Such samples were filtered and analyzed with LC-MS/MS. Isotope labeled 2,3,3-d_3 acrylamide was used as an internal standard. The authors noted that the roasting process of coffee impacts the levels of acrylamide in natural coffee. The highest AA concentration was determined for coffee substitute (818 μg kg^{-1}), then for instant coffee (358 μg kg^{-1}) and roasted coffee (179 μg kg^{-1}). No significant differences between Arabica and Robusta coffee samples were revealed.

Time-consuming GC-MS methods are less frequently applied than LC-MS ones. Due to the low volatility and polarity of AA, derivatization by bromination is required in most GC-MS methods. Later, the excess bromine needs neutralization, and the apolar derivative is extracted by liquid-liquid extraction into an organic solvent (Lineback et al. 2012).

Soares et al. (2007) developed a GC-MS method with two main purification steps to determine AA in coffee and coffee products. They dissolved the coffee sample in water and shook it with absolute ethanol. After the addition of internal standards and shaking, the sample was cooled and centrifuged. The solution was transferred to a centrifugal flask and acidified. Carrez I and Carrez II solutions were added, and after the centrifugation, the solution was filtrated and its volume was reduced in a rotary evaporator and placed into a vial. The column for extraction was layered with C18 sorbent covered with a polyethylene frit. After column conditioning with methanol and then with water, the sample solution was applied to the column. In the procedure for instant coffee, a previous SPE cleaning with a C18 column was conducted. Due to the need for derivatization, calcinated KBr was added to the eluate, then HBr for acidification, and finally saturated bromine solution. The addition of sodium thiosulfate decomposed the excess bromine. The extract was saturated with NaCl, and the derivatized analyte was isolated by liquid-liquid extraction into an organic solvent, ethyl acetate/n-hexane 4:1 (v/v). The volume was reduced under nitrogen, and anhydrous Na_2SO_4 was added. After centrifugation of the sample solution, the upper layer was subject to evaporation to the 0.5 mL volume for injection. The solutions were injected into a gas chromatograph with MS detection. AA was investigated in 26 coffee sample (roasted beans, instant coffees, and coffee blends). The concentrations of AA were in the range of 11.4–36.2 $\mu g\ L^{-1}$ in ground roasted coffee, 47.4–95.2 $\mu g\ L^{-1}$ for instant coffee, and the highest 200.8–229.4 $\mu g\ L^{-1}$ for coffee blends with cereals, which indicated a significant increase of AA resulting from the presence of roasted cereals (Soares et al. 2007).

4.4.3 Mycotoxins

The presence of mycotoxins in coffee is of great concern, as coffee is one of the world's most popular beverages nowadays. Mycotoxins are toxic secondary metabolites produced by certain filamentous fungi infecting crops in the field or in processing, storage, transportation, and finally in the processed foods. They do not affect the normal growth and reproduction of a fungus, but they induce biochemical and pathological changes in many living organisms (Pleadin et al. 2019). The term *mycotoxin* originates from the Greek *mycos* (fungus) and the Latin *toxicum* (poison; Paterson et al. 2001; Vieira and Casal 2015; Richard 2007). A few classes of mycotoxins can be defined, from which are produced aflatoxins and sterigmatocystin (mainly by *Aspergillus* spp.) and ochratoxins (mainly by *Penicillium* spp.); these mycotoxins have been found in coffee. The mold growth production of mycotoxins depends on many factors, including region of origin, different climatic, crop processing, and storage conditions (Romani et al. 2000). So far there are three identified naturally occurring ochratoxins (A, B, and C). Ochratoxin A (OTA) is the most prevalent and the most deleterious toxin of this group (Vieira and Casal 2015; Figure 4.4). OTA, the metabolite of *Aspergillus ochraceus*, has proven toxicity to all animals species studied, doing harm to kidneys and acting as an immunosuppressant, a liver toxin, a carcinogen, and a teratogen (Paterson et al. 2001; Bennett and Klich 2003). In all nonruminant mammals poisoned with OTA, tubular degeneration, interstitial fibrosis, and decreased renal function were observed (Vieira and Casal 2015). This substance has been found in many products such as barley, oats, rye, and wheat. Moreover, it may be present in wines made from grapes contaminated with *A. carbonarius* (Bennett and Klich 2003). OTA is supposed to inhibit the enzyme involved in the synthesis of the phenylalanine tRNA complex. Additionally, it inhibits mitochondrial ATP production and induces lipid peroxidation. Its carcinogenicity in humans is associated with the induction of oxidative stress or oxidative DNA lesions coupled with direct DNA adducts. There are observations of DNA damage and ROS production in various cell types, including human kidney (HK-2), proximal tubular cells and colonic, adenocarcinoma cell line (CaCo-2), as well as peripheral blood mononuclear cells and hepatoma cell lines (Leitao 2019; Arbillaga et al. 2007; Guerra et al. 2005; Schaaf et al. 2002; Cui et al. 2013). OTA has been found in animal tissues and even in human milk. In Scandinavian countries, it is associated with an endemic Danish disease, porcine nephropathy. OTA is also linked with dangerous poultry disease. There are some hypotheses that ochratoxins, besides other agents, are involved in human Balkan nephropathy, occurring endemically in populations of some regions of Romania, Bulgaria, and the former Yugoslavia. The disease can also lead to urinary tract tumors. Although its etiology is unknown, it is often named an ochratoxicosis. According to some speculations, ochratoxin might be a risk factor for testicular cancer. Human exposure to ochratoxin seems broad, as it has been detected in blood and serum during studies in Canada, Sweden, Germany, and the former Yugoslavia. Also, both ochratoxin and aflatoxin were detected in the urine of children from Sierra Leone (Bennett and Klich 2003). The presence of OTA in coffee poses a health threat and still raises controversy. Although the UE Commission Regulation No. 105/2010 imposed legal limits for amounts of OTA in roasted coffee and instant coffee only, some countries use own statutory legal limits for the amount of OTA in green coffee (European Union 2010). Italy permits 8 $\mu g\ kg^{-1}$ of OTA in green coffee, while the limit in Greece is 20 $\mu g\ kg^{-1}$ (Vieira and Casal 2015).

Among mycotoxins, there is a group called aflatoxins which are produced by many species of *Aspergillus*, especially during improper storage conditions of coffee beans. Aflatoxins are stable and resistant to quite high temperatures with little degradation. These compounds are mutagenic, carcinogenic, teratogenic, immunotoxic, nephrotoxic (B_1 and G_1), and hepatotoxic. The most potent carcinogenic aflatoxin is aflatoxin B_1 (Figure 4.4), which is produced by *A. flavus* and *A. parasiticus*. Aflatoxins G_1, G_2, and B_2, are usually less toxic and do not occur in the absence of B_1 (Vieira and Casal 2015; Bennett and Klich 2003; Al-Abdalall et al. 2014).

4.4.4 Analysis of mycotoxins

Mycotoxins are mainly determined by the methods allowing analysis of a single compound (mainly OTA) or a group of these contaminants. The length and quality of the analytical procedure depend on sample size, analyte isolation, and purification. Therefore, efficient clean-up and preconcentration steps based on immunoaffinity columns (IAC) and SPE should be elaborated. According to the present trend in mycotoxin analysis, IACs are becoming more frequently applied as a clean-up and enrichment technique for such a complex matrix as coffee (Monaci and Palmisano 2004). They feature high recovery and can be used for purification of samples contaminated with various mycotoxins. After preconditioning, the determined compounds bind selectively to the antibodies on the column. While rinsing, most of the possible interfering substances are removed due to no interactions of matrix components. The analytes can be eluted by antibody denaturation (Rahmani et al. 2009).

Contamination of green coffee beans with OTA is quite frequent all over the world, as confirmed by Romani et al. (2000). They tested 162 samples of green coffee, including 84 from Africa, 60 from Central and South America, and 18 from Asia. Each coffee sample was first ground and subjected to an isolation procedure with chloroform and phosphoric acid. After liquid-liquid partition into bicarbonate solution, SPE was conducted. OTA was eluted with ethyl acetate–methanol–acetic acid. An HPLC system with fluorescence detection (at an excitation wavelength of 333 nm and an emission wavelength of 460 nm) was used for analysis. It turned out that 106 out of 162 coffee samples were contaminated with OTA (at levels ranging from <LOD to 48 μg kg^{-1}). The highest concentrations of OTA and the highest frequency of this mycotoxin were found in African samples. In further studies, Romani et al. (2003) decided to check the influence of the roasting process on the OTA concentration in coffee beans. For this purpose, they prepared four various coffee samples and submitted them to roasting with an air temperature of 450°C and different roasting times to achieve a light, medium, or dark degree of roasting. Both the green and roasted coffee samples were used for the experiment. Ground coffee samples were mixed with methanol and 3% aqueous sodium hydrogen carbonate. Then they were filtered and diluted with phosphate-buffered saline (PBS) and finally applied to the IAC. After washing the column with water, OTA was eluted with methanol. The eluate was evaporated to and redissolved in the mixture of acetonitrile–sodium acetate/acetic acid (19:1 v/v). The quantitative analysis was performed with an HPLC system equipped with a fluorescence detector. The comparison of the OTA content in coffee samples before and after roasting confirmed the assumption that its level was reduced during the roasting process. Its efficiency depended on the severity of the thermal process. Moreover, the most significant OTA decrease was observed in the highest contaminated samples after each roasting level. For one sample, a decreased OTA concentration was observed only after roasting to a dark level. The conclusion could be drawn that OTA reduction resulting from roasting is related to its thermal destruction, and the process efficiency is influenced by the roasting conditions and initial OTA concentration. After subjecting samples to roasting parameters typical of an Italian espresso coffee brew, reductions of >90% of the OTA levels were found, both for samples of high and low mycotoxin contamination.

Vaclavik et al. (2013) developed and evaluated an analytical method for screening of 34 mycotoxin substances in 50 commercial dietary products containing green coffee beans. The targeted compounds included ochratoxin A and B; aflatoxin B_1, B_2, G_1, G_2; trichothecenes A and B (deoxynivalenol, deoxynivalenol-3-glucoside, nivalenol, zearalenone, HT-2 toxin, T-2 toxin, neosolaniol, 3-acetyldeoxynivalenol, and diacetoxyscirpenol), alternariol; alternariol–methyl ether; tentoxin; fumonisin B_1, B_2, B_3; enniantin A, A_1, B, B_1; beauvericin; and citrinin, cyclopiazonic acid, mycophenolic acid, penicillic acid, penitrem A, roquefortin C, gliotoxin, and sterigmatocystin. Analytes were isolated from the matrices using a QuEChERS-like procedure (quick, easy, cheap, effective, rugged, and safe). The authors used standard additions to overcome problems resulting from matrix effects. The contents of different dosage forms were homogenized and shaken with water containing 2% formic acid. After 20 min of incubation, acetonitrile was added and shaken. In the next stage of the QuEChERS protocol, anhydrous $MgSO_4$ and NaCl were added to the samples, shaken, and centrifuged. The acetonitrile layer was filtered and diluted 1:1 with deionized water. The samples were analyzed with a UHPLC system coupled with a hybrid triple quadrupole/linear ion trap mass spectrometer equipped with a TurboIon electrospray (ESI) ion source. Additional accurate mass measurements and high-resolution (HR) MS/MS spectra with the use of a benchtop quadrupole-orbitrap tandem mass spectrometer were additionally used for confirmation of identities of reported compounds. Studies revealed the occurrence of 4 out of 34 mycotoxins. Ochratoxin A, ochratoxin B, fumonisin B_1, and mycophenolic acid were determined in concentration ranges of <1.0 (LOD)–136.9 μg kg^{-1}, <1.0 (LOD)–20.2 μg kg^{-1}, <50.0 (LOD)–415.0 μg kg^{-1}, and <5.0 (LOD)–395.0 μg kg^{-1}, respectively, and the highest incidence rate for OTA (36%). Ochratoxin B co-occurred only with OTA and was present in 16 of 18 positive products. Fumonisin B_1 was detected in 10% of products. What was interesting is that in the two most contaminated products, all detected mycotoxins were found at similar concentrations (ochratoxin A, 136.9 and 128.0 μg kg^{-1}; ochratoxin B, 20.2 and 17.0 μg kg^{-1}; fumonisin B_1, 387.0 and 415.0 μg kg^{-1}; mycophenolic acid 227.3 and 217.0 μg kg^{-1}), which could indicate their origin from the same raw material (Vaclavik et al. 2013).

Reichert et al. (2018) carried out studies on the development and validation of an analytical method for the simultaneous extraction of a vast array of pesticides and mycotoxins from raw coffee beans followed by LC-ESI-MS/MS. The samples of ground raw coffee were slurried with water and spiked with two separate standard solutions. Both standard solutions, one with 131 pesticides and a second one with 35 mycotoxins, were divided into three groups of different concentrations. For the isolation procedure, QuEChERS technique was optimized. Different approaches were investigated – acidification of acetonitrile with acetic acid or formic acid, extraction with or without buffer, and with or without clean-up of the extracts before LC-ESI-MS/MS analysis. Before the clean-up stage, seven dispersive SPE sorbents and their mixtures were tested, and the best results were obtained for C18, Graphitized Carbon Black, Chlorofiltr, and the mixture of Oasis HLB and Chlorofiltr. The method selected for validation was based on employing acetonitrile containing 1% acetic acid as an extraction solvent. Additionally, buffering, and performing any clean-up steps were skipped as such procedure was cheaper, easier, faster, and allowed for obtaining comparable results to that with the clean-up step. The twofold dilution of extracts with methanol was carried out before LC-ESI-MS/MS analysis to improve the chromatographic performance and decrease matrix effects (Reichert et al. 2018).

Jeszka-Skowron et al. (2017) determined the mycotoxin levels in green coffee of different origin: *C. arabica* beans from Brazil, Rwanda, China, Laos, and Peru, and *C. robusta* beans from Vietnam. They applied an ultra-performance liquid chromatography (UPLC) system coupled to a tandem quadrupole (TQ) mass detector to detect ochratoxin A, aflatoxins (B_1, B_2, G_1, and G_2), and also ergosterol. This steroid substance is a marker indicating the fungal presence, and it was found in every sample tested. Aflatoxin B_1 appeared to be the dominant mycotoxin in coffee bean samples, with the highest level at 17.45 ng g^{-1} in Laotian green beans of Robusta coffee. Ochratoxin A was determined in four samples at levels ranging from 1.27 to 4.34 ng g^{-1}. Robusta beans from Vietnam and Uganda and Arabica beans from Laos and China contained the highest ochratoxin A, but the concentrations were consistent with the legal limits. All determined mycotoxins were found in Chinese Arabica beans. Fungi potentially able to produce this toxin, such as *A. oryzae, A. foetidus, A. tamarii, Alternaria* sp., and *P. citrinum* were isolated. An important conclusion was drawn that ergosterol cannot be taken into consideration for predicting contamination of green coffee with mycotoxins. Ground coffee samples were immersed overnight in the mixture of acetonitrile and water, then filtrated, diluted with water and filtered through a paper filter. The IAC was used to wash the filtrate put on its top. The washing solution contained NaCl, $NaHCO_3$, Tween-20, and water. For elution of OTA, methanol was applied. Before extraction of aflatoxins, ground coffee samples were immersed overnight in the mixture of methanol and water. Then they were filtrated through a paper filter and diluted with phosphate buffer saline. After application of the filtrate on top of IAC, the column was washed with distilled water. For the elution of aflatoxins from the column, methanol was used. Finally, the samples were evaporated and reconstituted with methanol for the analysis of OTA and aflatoxins (Jeszka-Skowron et al. 2017).

Napolitano et al. (2007) determined OTA in green and roasted coffee and corresponding by-products from seven different geographic regions using HPLC-FLD. They prepared a mixture of ground coffee samples and acetonitrile. Such suspension was centrifuged, and the supernatant was combined with PBS solution. For the sample, clean-up IAC was used. The diluted supernatant solution was applied for washing the IAC, and for its cleaning PBS solution and distilled water were passed to elute OTA with methanol. After resuspending the residue in HPLC mobile phase and its filtering, the HPLC analysis with fluorometric detection was performed. The excitation wavelength was at 333 nm, and the emission wavelength was at 460 nm. In the case of samples containing more than 100 ng kg^{-1} of OTA, the HPLC analysis coupled with tandem MS and equipped with a turbo ion spray source was performed in multiple reaction monitoring (MRM) to confirm the structures. The highest level of OTA was determined for Costa Rica green coffee (13 µg kg^{-1}), while the lowest contamination was found for the Ethiopian coffee sample (3.8 µg kg^{-1}). For all coffee brands, OTA concentration was the highest for green coffee samples, lower for roasted coffee, and the lowest for coffee beverages. Therefore it was noted that both roasted and ground coffee and soluble coffee can be considered rather minor sources of OTA due to reduction of its level during coffee processing (Napolitano et al. 2007). Bessaire et al. (2019) developed and validated the method for the simultaneous analysis of 31 mycotoxins in green coffee. They investigated the presence of a vast range of mycotoxins in 71 green coffee beans belonging to Arabica or Robusta species collected from nine countries. For sample preparation, ground coffee was dissolved in water and homogenized with acetonitrile containing 1% acetic acid. Then QuEChERS salt mixture (consisting of $MgSO_4$ and NaCl) was added, shaken, and centrifuged. The obtained supernatant was split into two parts. Each portion was cleaned up by different protocols. In one protocol, the supernatant was agitated with n-hexane to de-fat the sample and centrifuged. After removing the hexane phase, the lower acetonitrile phase was left for evaporation. In the latter protocol, the portion of supernatant was passed through an SPE cartridge suitable for mycotoxins, and the eluate was left for evaporation. The dried samples in both protocols were reconstituted in methanol. After the addition of water, mixing, and sonicating, the extracts were finally centrifuged to enable precipitation of possible impurities. The extracts were analyzed separately with the LC system coupled with MS detection with a turbo ionization source. The matrix-matched calibration curve approach was applied for measurements of concentrations. Low limits of quantification (0.25–150 µg kg^{-1}) were reached. Most often present and in the highest concentrations were ochratoxin A, fumonisin B_2, sterigmatocystin, beauvericin, and enniatin A with maximum levels at 12.2, 76.3, 19.8, 4.4, and 1.7 µg kg^{-1}, respectively. Aflatoxin B_1 was rarely found, with the maximum level at 1.2 µg kg^{-1}. Sample contamination with multiple mycotoxins was seldom and usually concerned two various substances at most. Comparing the contamination of coffee species, mycotoxins occurred more frequently in Robusta than in Arabica. However, no health threat caused by contaminated coffee consumption due to low levels of mycotoxins and their rare occurrence should be expected (Bessaire et al. 2019).

Twarużek et al. (2020) investigated the levels of mold contamination and mycotoxin in commercial green coffee products and dietary supplements with green coffee extract. They tested 34 samples from green coffee products, including 16 samples of raw beans, 15 samples of ground coffee, and three samples of instant coffee, as well as 19 samples in the form of capsules, two tablets, and one sachet. The spread-plate method was applied to count the total mold. Sample preparation consisted of sample extraction with acetonitrile: water and addition of PBS solution. After applying solvent to the column, a mixture of methanol: CH_3COOH and then purified water were passed through the column. The eluate was subjected to HPLC coupled with a fluorescence detector. They identified molds in 94% of green coffee beans, 100% of the ground and instant coffee samples, and 55% of dietary supplements. Citrinin was found in none of the samples. OTA was detected in 40% of ground coffee samples and in 67% of instant coffee samples, and only in one sample of raw green coffee beans. This contaminant was also found in 58% of the supplement capsules, excluding tablets and sachets. According to the outcomes of the study, the OTA contamination of raw material (e.g., green coffee beans) was lower than that of the processed products. This could suggest the occurrence of contamination during the production process. The mean concentration of OTA in green coffee beans (0.4 ng g^{-1}) was significantly lower than in processed counterparts coffee (3.28 ng g^{-1} for ground coffee samples and 4.09 ng g^{-1} for instant coffee samples; Twarużek et al. 2020).

4.4.5 Estrogenic chemicals in coffee

Food contact materials (FCMs) are one of the sources of human exposure to endocrine disrupting chemical (EDC) substances resulting from their leaching into food, including coffee (Ćwiek-Ludwicka and Ludwicki 2014; Muncke 2011). According to the US Environmental Protection Agency, EDC is "an exogenous agent that interferes with synthesis, secretion, transport, metabolism, binding action, or elimination of natural blood-borne hormones that are present in the body and are responsible for homeostasis, reproduction, and developmental process." There is increasing public concern on the possible health threat posed by the omnipresent EDCs (Diamanti-Kandarakis et al. 2009; Encarnação et al. 2019). These xenobiotics mimic or antagonize natural hormones by interacting with various receptors, for example to estrogen receptors (ERs), androgen receptors (ARs), progesterone receptors, thyroid receptors (TRs), and retinoid receptors. EDCs can influence fetal development, or their action can be manifested in adulthood or in senility (Shanle and Xu 2011; Diamanti-Kandarakis et al. 2009). Interestingly, chronic exposure to low amounts of combinations of these chemicals more often causes human disorders. They can also have an impact on nonnuclear steroid hormone and non-steroid hormone receptors and orphan receptors and affect hormonal and homeostatic systems in various ways. There are many reports indicating their harmful effects on female reproductive development and function, including polycystic ovarian syndrome (PCOS), premature ovarian failure, decreased ovarian reserve, aneuploidy, granulosa steroidogenesis, reproductive tract anomalies, uterine leiomyomas (fibroids), and endometriosis. EDCs are also associated with susceptibility to breast cancer during vulnerable periods of life. These xenobiotics are also suspected of being able to alter male reproductivity, but this issue still requires far more research. EDC exposure may bring about prostate carcinogenesis, disruption of reproductive neuroendocrine and thyroid function, obesity, and metabolic disorders (Diamanti-Kandarakis et al. 2009; Vandenberg et al. 2007).

Xenobiotics migrating from packaging into food are usually components of packaging. Both these substances and their side or degradation products may leach into the food or beverage contents.

Bisphenol A (BPA) is a popular industrial chemical widely used in the production of synthetic polymers such as epoxy resins. It is incorporated into the inner lining of metal food cans, beverages, and bottles for drinking water. The leaching of BPA depends on various factors such as packaging damage, storage conditions, and high temperatures (Almeida et al. 2018). BPA plays a role in the pathogenesis of various endocrine dysfunctions and is considered a hazardous chemical, especially due to the continuous daily exposure and its bioaccumulation (Konieczna et al. 2015). It causes an increase of oxidative mediators and the reduction of antioxidant enzymes, which affects the function of mitochondria, cell signaling pathways, and leads to apoptosis (Meli et al. 2020). Contact with high levels of BPA can be correlated with metabolic disorders, cardiovascular diseases, polycystic ovarian disease, or low sperm count (Fenichel et al. 2013). There are also limited data on the association between susceptibility to BPA and incidence of breast cancer, prostate cancer, and meningioma. Its prenatal and fetal exposure in rodents may alter brain development and result in behavior disorders, and impact functions of the liver, gut, kidneys, adipose tissue, endocrine pancreas, mammary glands, and reproductive tract (EFSA 2015; Meli et al. 2020; Fenichel et al. 2013).

Benzophenone (BP), a widely used chemical in FCMs, acts as a photoinitiator in printing inks for paper or plastic packaging. It activates the radical polymerization of inks, speeds the drying process, and improves the curing efficiency. Due to its small dimensions, the benzophenone molecule can easily migrate from the printed part into the contents (Muncke 2011; Sanches-Silva et al. 2008; Anderson and Castle 2003). According to animal studies, benzophenone, as EDC-exerted carcinogenic activity, increasing incidences and/or severities of non-neoplastic lesions in the kidney and liver of rats and in the liver, kidney, nose, and spleen of mice (Anonym 2006).

Polycyclic aromatic hydrocarbons (PAHs) are substances qualified as endocrine disruptors that can also occur in food due to food processing, similarly to acrylamide, and are found in coffee. PAHs are a group of organic compounds formed as a result of various process techniques involving incomplete combustion or high-pressure processes, such as coffee roasting (ATSDR 2009), roasting conditions and parameters, and equipment used to affect the level of PAH formation. The highest contamination level was determined in dark coffee blends that had been subjected to higher degree of roasting and for a longer time (Singh et al. 2016). Considering the health risks PAHs pose, the most significant is their possible carcinogenicity, mainly for lung, skin, and bladder (ATSDR 2009). According to the number of aromatic rings in the molecule, two groups of PAHs, light and heavy (more toxic), can be differentiated (Ciecierska et al. 2019).

4.4.6 Analysis of estrogenic chemicals

Sakaki et al. (2020) performed studies to assess the concentrations of estrogenic chemicals in capsules and in French press coffee and to check the degree of exposure to these compounds from the typical consumption. They tested 22 brands of coffee in capsules and 15 brands of French coffee, all of which were brewed. For sample preparation, they conducted liquid-liquid extraction. The samples were spiked with 4-hydroxy-biphenyl-d_9 surrogate and with the nine targeted ECs including caprolactam, BPA, BPS (bisphenol S), BPF (bisphenol F), benzophenone, 4-nonylphenol (4-NP), dimethyl terephthalate (DMTP), dibutyl phthalate (DBP), and di(2-ethylhexyl) phthalate (DEHP). After the addition of ethyl acetate, each sample was centrifuged and the top organic layer removed. The remaining emulsion was subjected to evaporation for 60 min. Finally, methanol and the internal standard 1-naphthol-d_3 were added. For the quantitative analysis of EDs, ultra-performance liquid chromatography with tandem mass spectrometry (UPLC-MS/MS) was applied. In capsule coffee, five out of nine ECs were detected. BP turned out to be the prevailing contaminant of the capsule coffee that was found in six samples. In the French press coffee, BPA and BPF were identified. BP, 4-NP, and DBP were not found in any French press coffee, and there were no DMTP, BPS, and caprolactam in either type of coffee. Although the coffee was contaminated with ECs, both in capsules and in French press brands, the levels of contaminants were considered relatively low regarding human safety (Sakaki et al. 2020).

Kang and Kondo (2002) tested the leaching of BPA from the internal coating of cans into two types of coffee – one from a decaffeinated instant coffee and the latter one from non-decaffeinated instant coffee. They dissolved coffee in water at 90°C. After 20 min, they cooled it to room temperature and transferred the sample into a can. After heating at 121°C for 30 min, the sample was left at room temperature for 3 h. For the extraction, they applied methanol and water to the cartridge, then sample and finally water for washing the cartridge. From the dried cartridge, they eluted BPA with acetonitrile-water solution. The samples were analyzed with the HPLC system, and fluorescence detection and the leaching of BPA into the can content was confirmed. The mean value of BPA obtained for the sample of decaffeinated instant coffee was 66.2 ng mL^{-1} and for the sample of non-decaffeinated instant coffee 84 ng mL^{-1}.

Ciecierska et al. (2019) investigated the contamination of green coffee beans (Arabica and Robusta) with 19 different PAHs before and after the mild roasting process, while a temperature of 125–135°C for 25–26 min was applied. Ground samples of coffee beans with anhydrous sodium sulfate were placed in an Erlenmeyer flask with hexane/acetone solution mixture into an ultrasonic bath for 30 min. The solution was filtered and subjected to evaporation in a rotary vacuum evaporator. After dissolving the remaining solution in 0.5 mL of cyclohexane, the clean-up procedure on the silica gel column was carried out, and elution with cyclohexane was performed. The concentrated eluate was dissolved in 5 mL of cyclohexane/ethyl acetate and filtered. After the final purification procedure employing size exclusion chromatography (SEC) on a silica gel column, the extract was dissolved in acetonitrile after solvent evaporation. Samples were analyzed with HPLC-FLD/DAD. For verification of the results, GC-MS was performed. The contamination of roasted coffee samples (4.29–16.17 μg kg^{-1} of PAHs) turned out to be lower than that of raw coffee beans (8.66–76.63 μg kg^{-1} of PAHs). It was concluded that mild roasting parameters contributed to the reduction of the PAH contamination level, which strongly depended on the roasting conditions.

4.5 Conclusions

Coffee consumption in moderation may be beneficial for human health due to antioxidant properties, positive effects on cardiovascular system, improvement of cognitive performance, regulation of blood sugar, or even cancer prevention. However, its overconsumption can lead to irritability, digestion problems, sleep disorders, high blood pressure, or dehydration. The source of origin, the proper packaging, storage of coffee beans, and their processing conditions are of great importance for the coffee quality and its possible contamination. Some toxic substances may appear in coffee products due to fungi infection of coffee beans before and after harvest, their roasting parameters, and also migration of chemical packaging components to the packed contents. The level of both bioactive substances and possible contaminants of coffee should be controlled to meet the requirements for pure and safe coffee products. HPLC is the most commonly applied analytical method for determination of these compounds. LC coupled with MS/MS detection with variations in the clean-up conditions is widely recommended in case of very low amounts of analytes.

4.6 Acknowledgments

This work was also supported by the Poznan University of Medical Sciences statutory funding 502-01-03314429-03439. This research was funded by the National Science Centre – NCN (Poland), under the program MINIATURA 4, grant number: 2020/04/X/NZ9/00736.

REFERENCES

Abe S. K., Saito E., Sawada N., Tsugane S., Ito, H., Lin Y., Tamakoshi A., Sado J., Kitamura Y., Sugawara Y., Tsuji I., Nagata C., Sadakane A., Shimazu T., Mizoue T., Matsuo K., Naito M., Tanaka K., Inoue M., & Research Group for the Development and Evaluation of Cancer Prevention Strategies in Japan. *Prev Med.* 2019; 123: 270–277.

Acamovic T. and Brooker J.D. *Proc Nutr Soc.* 2005; 64: 403–412.

Adlercreutz H. and Mazur W. *Ann Med.* 1997; 29: 95–120.

Adrian J. and Frangne R. *Adv Exp Med Biol.* 1991; 289: 49–59.

Agency for Toxic Substances and Disease Registry (ATSDR) 2009. https://www.atsdr.cdc.gov/csem/pah/docs/pah.pdf (accessed January 21, 2021)

Aguas P.C., Fitzhenry M.J., Giannikopoulos G., Varelis P. *Anal Bioanal Chem.* 2006; 385(8): 1526–1531.

Al-Abdalall A.H.A., Al-Talib E.A. *Can J Appl Sci.* 2014; 8(2): 2825–2836.

Ali S.F., Hong J.S., Wilson W.E., Uphouse L.L., Bondy S.C. *Arch Toxicol.* 1983; 52: 35–43.

Almeida S., Raposo A., Almeida-Gonzalez M., and Carrascosa C. *Compr Rev Food Sci Food Saf.* 2018; 17(6): 1503–1517.

Alonso-Salces R. M., Serra F., Reniero F., Heberger K. *J Agric Food Chem.* 2009; 57: 4224–4235.

Alves S. T., Dias R. C. E., Benassi M. T., Scholz M. B.S. *Quim Nova.* 2006; 29: 1164–1168.

Alves R.C., Almeida I.M.C., Casal S., Oliveira M.B.P.P. *J Agric Food Chem.* 2010a; 58: 3002–3007.

Alves R.C., Almeida I.M.C., Casal S., Oliveira M.B.P.P. 2010b. *Food Chem.* 2010b; 122: 914–919.

Anderson W.A. and Castle L. *Food Addit Contam.* 2003; 20(6): 607–618.

Angeloni S., Navarini L., Sagratini G., Torregiani E., Vittori S., Caprioli G. *Int J Food Sci Nutr.* 2020; 71: 193–200.
Anonym. *Natl Toxicol Program Tech Rep Ser.* 2006; 533: 1–264.
Arbillaga L., Azquet A., Ezpeleta O., Lopez de Cerain A. *Mutagenesis* 2007; 22: 35–42.
Bagdonaite K., Derler K., Murkovic M. *J Agric Food Chem.* 2008; 56: 6081–6086.
Baranowska I. and Płonka J. *J. Food Anal Methods.* 2015; 8: 963–972.
Bennett J.W. and Klich M. *Clin Microbiol Rev.* 2003; 16(3): 497–516.
Bessaire T., Perrin I., Tarres A., Bebius A., Reding F., Theurillat V. *Food Control.* 2019; 96:59–67.
Borda D. and Alexe P. *Rom J Food Sci.* 2011; 1(1): 3–15.
Bressani R. and Navarrete D.A. *J Food Sci.* 1959; 24: 344–351.
Buckel L., Kremer J.I., Stegmüller S., Richling E. *Proc.* 2019; 11:13.
Campa C., Mondolot L., Rakotondravao A., Bidel L.P., Gargadennec A., Couturon E., La Fisca P., Rakotomalala J.J., Jay-Allemand C., Davis A.P. *Ann Bot.* 2012; 110: 595–613.
Campa C. and Petitvallet A. 2017; *Achieving sustainable cultivation of coffee.* Cambridge: Burleigh Dodds Science Publishing Limited, pp. 237–258.
Cano-Marquina A., Tarín J.J. Cano A. *Maturitas* 2013; 75: 7–21.
Caprioli G., Cortese M., Maggi F., Minnetti C., Odello L., Sagratini G., Vittori S. *Int J Food Sci Nutr* 2014; 65: 465–469.
Casal S., Oliveira M.B., Ferreira M.A. *Food Chem.* 2000; 68: 481–485.
Casal S., Oliveira M.B.P.P., Ferreira M.A. *J Liq Chromatogr Relat Technol.* 2002; 25: 2535–2549.
Casal S., Mendes E., Rui Alves M., Alves R.C., Beatriz, M., Oliveira, P.P., Ferreira, M.A. *J Agric Food Chem.* 2004; 52: 6188–6192.
Chan P.C., Hills G.D., Kissling G.E., Nyska A. *Arch Toxicol.* 2008; 82: 45–53.
Chen X.M. *Crit Rev Food Sci Nutr* 2019; 59: 1008–1025.
Chen X.M., Ma Z., Kitts D.D. *Food Chem.* 2018; 249: 143–153.
Chen X., Mu K., Kitts D.D. *Food Chem.* 2019; 271: 248–258.
Chirfa G., Merdassa Y., Gure A. *J Anal Test.* 2020; 4: 298–306.
Ciecierska M., Derewiaka D., Kowalska J., Majewska E., Drużyńska B., Wołosiak R. *J Food Sci Technol.* 2019; 56(2): 737–745.
Cirilo M. P. G., Coelho A. F. S., Araújo C. M., Gonçalves F. R. B., Nogueira F. D., Glória M. B. A. *Food Chem.* 2003; 82: 397–402.
Clarke R.J. 2003; *Encyclopedia of Food Science and Nutrition*, 2nd edition, Oxford: Academic Press, Vol. 3.
Claus A., Carle R., Schieber A. *J Cereal Sci.* 2008; 47(2): 118–133.
Commission Regulation (EU) No 105/2010 of 5 February 2010 amending Regulation (EC) https://eur-lex.europa.eu/LexUriServ/LexUriServ.do?uri=OJ:L:2010:035:0007:0008:EN:PDF
(accessed January 21, 2021)
Corso M. P., Vignoli J. A., Benassi M. *J Food Sci Technol.* 2016; 53: 1380–1388.
Cotterchio M., Boucher B. A., Manno M., Gallinger S., Okey A., Harper P *J Nutr.* 2006; 136: 3046–3053.
Cui J., Liu, J., Wu, S., Wang Y., Shen H., Xing L., Wang J., Yan X., Zhang X. *Arch Toxicol.* 2013; 87: 1829–1840.
Cunha S. C., Senra L., Cruz R., Casal S., Fernandes J. O. *Food Control* 2016; 63: 15–20.
Ćwiek-Ludwicka K. and Ludwicki J.K. *Rocz Panstw Zakl Hig.* 2014; 65(3): 169–177.
Diamanti-Kandarakis E., Bourguignon J.P., Giudice L.C, Hauser R., Prins G.S., Soto A.M., Zoeller R.T., Gore A.C. *Endocr Rev.* 2009; 30(4): 293–342.
Dias E., Pereira R.G.F.A., Borém F.M., Mendes E., De Lima R.R., Fernandes J.O., Casal S. *J Agric Food Chem.* 2012; 60: 4120–4125.
Edreva A., Velikova V., Tsonev T., Dagnon S., Gürel A.L., Aktas L. *Gen Appl Plant Physiol.* 2008; 34: 67–78.
EFSA. *J EFSA.* 2011; 9: 2393.
EFSA. *J EFSA.* 2014; 12: 3759.
EFSA. *J EFSA.* 2015; 13: 3978.
EFSA. 2020; EFSA Supporting Publication: EN-1783.
Encarnação T., Pais A.A., Campos M.G., Burrows H.D. *Sci Prog.* 2019; 102(1): 3–42.
FAO/WHO. 2002. https://apps.who.int/iris/handle/10665/67372 (accessed January 21, 2021).
Farah A. and Donangelo C.M. *Braz J Plant Physiol.* 2006; 18: 23–36.
Farah A., de Paulis T, Moreira D.P., Trugo L.C., Martin P.R. *J Agric Food Chem.* 2006a; 54: 374–381.
Farah A., Monteiro M.C., Calado V., Franca A.S, Trugo L.C. et al. *Food Chem.* 2006b; 98: 373–380.
Fenichel P., Chevalier N., Brucker-Davis F. *Ann. Endocrinol.* (Paris) 2013; 74(3): 211–220.
Fiechter G. and Mayer H.K. *Curr Anal Chem.* 2014; 10(3): 366–374.
Fischer M., Reimann S., Trovato V., Redgwell R. J. *Carbohydr Res.* 2001; 330: 93–101.
Friedman M. *J Agric Food Chem.* 2003; 51(16): 4504–4526.
Fujimoto H., Narita Y., Iwai K., Hanzawa T., Kobayashi T., Kakiuchi M., Ariki S., Wu X., Miyake K., Tahara Y., Ikezaki H., Fukunaga T., Toko K. *Food Chem.* 2020; 342: 128228.
Gielecińska I. and Mojska H. *Rocz Panstw Zakl Hig.* 2013; 64: 85–90.
Gokcen B.B. and Sanlier N. *Crit Rev Food Sci Nutr* 2017; 30: 1–13.
Grynkiewicz G., Ksycinska H., RamzanJ., Zagrodzka J. *Acta Chromatogr.* 2005; 15: 31–65.
Guenther H., Anklam E., Wenzl T., Stadler R.D H. *Food Addit Contam.* 2007; 24 Suppl 1: 60–70.
Guerra M.C., Galvano F., Bonsi L., Speroni E., Costa S., Renzulli C., Cervellati R. *Br J Nutr.* 2005; 94: 211–220.
Halász A., Baráth Á., Simon-Sarkadi L., Holzapfel W. *Trends Food Sci Technol.* 1994; 5: 42–49.
Halász A. and Baráth A. *Food Science and technology COST 917 Biogenically active amines in food,* pp. 131–141, EC Publication, Luxembourg, 2002.
Hughes E.B. and Smith R.F. *J. Chem. Technol. Biotechnol.* 1946; 65: 284–286.
IARC 2011. http://monographs.iarc.fr/ENG/Monographs/vol101/mono101-015.pdf (accessed February 02, 2021).
ICO. 2021. International Coffee Organization. www.ico.org (accessed January 01, 2021).
Jeon J.S., Kim H.T., Jeong I.H., Hong S.R, Oh M.S., Yoon M.H., Shim J.H., Jeong I.H., Abd El-Aty A.M. *J Adv Res.* 2019; 17: 85–94.

Jeszka-Skowron M., Zgoła-Grześkowiak A., Grześkowiak T. *Eur Food Res Technol.* 2015; 240: 19–31.
Jeszka-Skowron M., Stanisz E., De Peña M.P. *LWT – Food Sci Technol.* 2016; 73: 243–250.
Jeszka-Skowron M., Zgoła-Grześkowiak A., Waśkiewicz A., Stępień Ł., Stanisz E. *J Sci Food Agric.* 2017; 97: 4022–4028.
Jeszka-Skowron M., Frankowski R., Zgoła-Grześkowiak A. *LWT – Food Sci Technol.* 2020; 125: 109344.
Jeszka-Skowron M., Zgoła-Grześkowiak A., Grześkowiak T. 2021; *Analytical methods in the determination of bioactive compounds and elements in food,* pp. 31–82, Springer Nature Switzerland AG.
Jumhawan U., Putri S.P., Yusianto, Bamba T., Fukusaki E. *Biosci Bioeng.* 2016; 122: 79–84.
Jyotshna, Khare P., Shanker K. *Biofactors.* 2016; 42: 504–514.
Kang J.H. and Kondo F. *Food Addit Contam.* 2002; 19(9): 886–890.
Khamitova G., Angeloni S., Fioretti L., Ricciutelli M., Sagratini G., Torregiani E., Vittori S., Caprioli G. *Food Res Int* 2020; 133: 109220.
Kitts D.D. *J Toxicol Environ Health Sci.* 1987; 20: 37–49.
Kitzberger C.S.G., dos Santos Scholz M. B., de Toledo Benassi M. *Food Res Int* 2014; 61: 61–66.
Knip M., Douek I., Moore W., Gillmor H. A., McLean A. E. M., Bingley P. J., Gale E. *Diabetologia* 2000; 43: 1337–1345.
Kocadağlı T., Yılmaz C., Gökmen V. *Food Chem.* 2014; 153: 151–156.
Konieczna A., Rutkowska A., Rachoń D. *Rocz Panstw Zakl Hig.* 2015; 66(1): 5–11.
Kremer J. I., Gömpel K., Bakuradze T., Eisenbrand G., Richling E. *Nutr Food Res* 2018; 62: e1700735.
Krishnakumar T. and Visvanathan R. *Int. J. Food Process. Technol.* 2014; 5: 344.
Król K., Gantner M., Tatarak A., Hallmann E. *Eur Food Res Technol.* 2020; 246: 33–39.
Kuhnle G.G., Dell'Aquila C., Low Y.-L., Kussmaul M., Bingham S.A. *Anal Chem.* 2007; 79: 9234–9239.
Kuhnle G.G.C., Dell'Aquila C., Aspinall S.M., Runswick S.A., Mulligan A.A., Bingham S.A. *J Agric Food Chem.* 2008; 56: 7311–7315.
Kumar N. B., Cantor A., Allen K., Riccardi D., Cox C. E. *Cancer* 2002; 94: 1166–1174.
Ky C.-L., Louarn J., Dussert S., Guyot B., Hamon S., Noirot M. *Food Chem.* 2001; 75: 223–230.
Landete J.M. *Food Res Int* 2012; 46: 410–424.
Lang R., Yagar E. F., Eggers R., Hofmann T. *J Agric Food Chem.* 2008; 56: 11114–11121.
Lang R., Wahl A., Skurk T., Yagar E. F., Schmiech L., Eggers R., Hauner H., Hofmann T. *Anal Chem.* 2010; 82: 1486–1497.
Lantz I., Ternite´R., Wilkens J., Hoenicke K., GuentherH., van der Stegen G. *Mol Nutr Food Res.* 2006; 50(11): 1039–1046.
Leitao A.L. *Beverages.* 2019; 5: 36.
Lineback D.R., Coughlin J.R., Stadler R.H. *Annu Rev Food Sci Technol.* 2012; 3: 15–35.
Liu H., Shao J., Li Q., Li Y., Yan H.M., He L. *J AOAC Int.* 2012; 95: 1138–1141.
Loomis D., Guyton K.Z., Grosse Y. *Lancet Oncol.* 2016; 17(7): 877–878.
Lucero J., Harlow B. L., Barbieri R. L., Sluss P., Cramer D. W. *Fertil Steril.* 2001; 76: 723–729.
Macheiner L., Schmidt A., Mayer H.K. *Eur Food Res Technol.* 2021; 247: 85–99.
Martins A.C.C.L. and Gloria M.B.A. *Food Chem.* 2010; 118: 529–533.
Martins S. C., Araujo W. L., Tohge T., Fernie A. R. and F. M. DaMatta. *PLoS One.* 2014; 9: e94862.
Mazur W., Fotsis T., Wahala K., Ojala S., Salakka A., Adlercreutz H. *Anal Biochem.* 1996; 233: 169–180.
Mazur W., Wähälä K., Rasku S., Salakka A., Hase T., Adlercreutz H. *Br J Nutr.* 1998; 79: 37–45.
Meli R., Monnolo A., Annunziata C., Pirozzi C., Ferrante M.C. *Antioxidants* 2020; *(Basel)* 9(5): 405.
Minamisawa M., Yoshida S., Takai N. *Anal Sci.* 2004; 20: 325–328.
Mojska H. and Gielecińska I. *Rocz Panstw Zakl Hig.* 2013; 64(3): 173–181.
Monaci L. and Palmisano F. *Anal Bioanal Chem.* 2004; 378(1): 96–103.
Monteiro A., Colomban S., Azinheira H., Guerra-Guimarães L., Do Céu Silva M., Navarini L., Resmini M. *Antioxidants.* 2020; 9: E6.
Mu H., Bai Y.-H., Wang S.-T., Zhu Z.-M., Zhang Y.-W. *Phytomedicine.* 2009; 16: 314–319.
Muncke J. *J Steroid Biochem Mol Biol.* 2011; 127(1–2): 118–127.
Murkies A.L., Wilcox G., Davis S.R. *J Clin Endocrinol Metab.* 1998; 83: 297–303.
Muzaifa M., Hasni D., Febriani, Patria A., Abubakar A. *IOP Conf. Ser. Earth Environ. Sci.* 2020; 425: 012001.
Müllner C. and Sontag G. *Fresenius J. Anal. Chem.* 1999; 364: 261–265.
Naczk M. and Shahidi F. *J. Chromatogr A.* 2004; 1054: 95–111.
Naczk M. and Shahidi F. *J Pharm Biomed Anal.* 2006; 41: 1523–1542.
Napolitano A., Fogliano V., Tafuri A., Ritieni A. *J Agric Food Chem.* 2007; 55: 10499–10504.
Nehlig A. and Cunha R.A. *Nutrients.* 2020; 12: 3141.
OEHHA 2012. http://oehha.ca.gov/prop65/law/adopt020812.html (accessed February 02, 2021).
Oomah B.D. 2002; *Methods of analysis for functional foods and nutraceuticals*, pp. 1–64, Boca Raton, FL: Taylor & Francis Group LLC.
Özdestan Ö. *Food Res Int.* 2014; 61: 167–175.
Papageorgiou M., Lambropoulou D., Morrison C., Kłodzińska E., Namieśnik J., Płotka-Wasylka J. *TrAC Trends Anal Chem.* 2018; 98: 128–142.
Paterson R.R.M., Baker P.S., van der Stegen G.H.D. 2001; *Coffee Futures. A source book of some critical issues confronting the coffee industry.* pp. 16–25. The Commodities Press.
Perrone D., Donangelo C.M., Farah A. *Food Chem.* 2008a; 110: 1030–1035.
Perrone D., Farah A., Donangelo C.M., de Paulis T., Martin P.R. *Food Chem.* 2008b; 106: 859–867
Pleadin J., Frece J., Markov K. *Adv Food Nutr Res.* 2019; 89: 297–345.
Rahmani A., Jinap S, Soleimany F. *Compr Rev Food Sci Food Saf.* 2009; 8(3): 202–251.
Reichert B., de Kok A., Pizzutti I.R., Scholten J., Cardoso C.D., Spanjer M. *Anal Chim Acta.* 2018; 1004: 40–50.
Restuccia D., Spizzirri U.G., Parisi O.I., Cirillo G., Picci N. *Food Chem.* 2015; 175: 143–150.

Riboldi B.P., Vinhas A.M., Moreira J.D. *Food Chem.* 2014; 157: 310–322.
Rice-Evans C. *Free Radic Biol Med.* 2004; 36: 827–828.
Richard J.L. *Int. J. Food Microbiol.* 2007; 119(1–2): 3–10.
Rodrigues N.P. and Bragagnolo N. *J Food Compos Anal.* 2013; 32: 105–115.
Romani S., Sacchetti G., Chaves Lopez C., Pinnavaia G.G., Dalla Rosa M. *J Agric Food Chem.* 2000; 48(8): 3616–3619.
Romani S., Pinnavia G.G., Dalla Rosa M. *J Agric Food Chem.* 2003; 51(17): 5168–5171.
Ruiz-Larrea M.B., Mohan A.R., Paganga G., Miller N.J., Bolwell G.P., Rice-Evans C.A. *Free Radic Res.* 1997; 26: 63–70.
Sakaki J.R., Melough M.M., Provatas A.A., Perkins Ch., Chun O.K. *Toxicol Rep.* 2020; 7: 1020–1024.
Sanches-Silva A., Pastorelli S., Cruz J.M., Simoneau C., Paseiro-Losada P. *J Food Sci.* 2008; 73(2): C92-C99.
Santos J.R. and Rangel A.O. *Food Chem.* 2015; 187: 152–158.
Schaaf G.J., Nijmeijer S.M., Maas R.F., Roestenberg P., de Groene E.M., Fink-Gremmels J. *Biochim Biophys Acta.* 2002; 1588: 149–158.
Servillo L., Giovane A., Casale R., Cautela D., D'Onofrio N., Balestrieri M.L., Castaldo D. *LWT-Food Sci Technol.* 2016; 73: 117–122.
Shanle E.K. and Xu W. *Chem Res Toxicol.* 2011; 24: 6–19.
Shao C., Tang H., Wang X., He J. *J Stroke Cerebrovasc Dis.* 2021; 30: 105452.
Shimamura Y., Lio M., Urahira T., Masuda S. *J Sci Food Agric.* 2017; 97: 2419–2425.
Silva F.L.F., Nascimento G.O., Lopes G.S., Matos W.O., Cunha R.L., Malta M.R., Trevisan M.T.S. *Food Res Int.* 2020; 134: 109221.
Singh L., Varshney JG, Agarwal T. *Food Chem.* 2016; 199: 768–781.
Sivertsen T., Nygaard A.K, Mathisen G., Fonnum F. *Toxicol Mech Methods.* 2009; 19: 214–218.
Smith A.W. 1985; *Coffee.* pp. 31–33. London: Elsevier Applied Science.
Smith I.B. 2014; pp. 1–116. Ltd, Middlesex, United Kingdom: IBS Publishing.
Soares C., Cunha S., Fernandes J. *Food Addit Contam.* 2007; 23(12): 1276–1282.
Sridevi V., Giridhar P., Ravishankar G.A. *Acta Physiol. Plant.* 2009; 31: 757–764.
Stadler R.H., Blank I., Varga N., Robert F., Hau J., Guy P.A., Robert M.-C., Riediker S. *Nature.* 2002; 419(6906): 449–450.
Surdyk N., Rosén J, Andersson R, Aman P. *J Agric Food Chem.* 2004; 52(7): 2047–2051.
Surma M., Sadowska-Rociek A., Cieślik E., Sznajder-Katarzyńska K. *Microchem J.* 2017; 131: 98–102.
Svensson K., Abramsson L., Becker W., Glynn A., Hellenäs K. E., Lind Y., Rosén J. *Food Chem Toxicol.* 2003; 41(11): 1581–1586.
Tareke E., Rydberg P., Karlsson P., Eriksson S., Törnqvist M. *J Agric Food Chem.* 2002; 50(17): 4998–5006.
Thompson L.U., Boucher B.A., Liu Z., Cotterchio M., Kreiger N. *Nutr Cancer.* 2006; 54: 184–201.
Toci A.T., Farah A., Pezza H.R., Pezza L. *Crit Rev Anal Chem.* 2016; 46: 83–92.
Trugo L.C. and Macrae R. *Food Chem.* 1984; 15: 219–227.
Trugo L.C. and Macrae R. *Arch Latinoam Nutr.* 1989; 39: 96–107.
Twarużek M., Kosicki R., Kwiatkowska-Giżyńska J., Grajewski J., Ałtyn I. *Toxicon.* 2020; 188: 172–177.
Ukers W. 1922; pp. 5–9. *The Tea and Coffee Trade Journal.* New York.
Ungar Y., Osundahunsi O.F., Shimoni E. *J Agric Food Chem.* 2003; 51: 4394–4399.
USDA. 2010. National Nutrient Database for Standard Reference. NDB No: 14218.
Vaclavik L., Vaclavikova M., Begley T.H., Krynitsky A.J., Rader J.I. *J Agric Food Chem.* 2013; 61: 4822 – 4830.
Vandenberg L.N., Hauser R., Marcus M., Olea N., Welshons W. *Reprod Toxicol.* 2007; 24: 139–177.
Vieira T.S. and Casal S. 2015; *Coffee in Health and Disease Prevention* pp. 225–233. Academic Press.
Wanasundara U.N., Amarowicz R., Shahidi F. *Food Res Int.* 1996; 28: 525–530.
Wei Q., Li J., Li X., Zhang L., Shi F. *Reprod Toxicol.* 2014; 46: 121–128.
Woldesenebet A. Msc thesis, Center for Food Science and Nutrition, 2005; Addis Abeba University.
Yaylayan V.A., Wnorowski A., Perez Locas C. *J Agric Food Chem.* 2003; 51(6): 1753–1757.
Zamani E., Shokrzadeh M, Fallah M, Shaki F. *Pharm Biomed Res.* 2017; 3(1): 1–7.
Zhou S.S., Li D., Sun W.P., Guo M., Lun Y.Z., Zhou Y.M., Xiao F.C., Jing L.X., Sun S.X., Zhang L.B., Luo N.., Bian F.N., Zou W., Dong L.B., Zhao Z.G., Li S.F., Gong X.J., Yu Z.G., Sun C.B., Zheng C.L., Jiang D.J., Li Z.N. *World J Gastroenterol.* 2009; 15: 5674–5684.
Zhou S.S., Li, D., Zhou, Y. M., Sun, W. P., Liu, X. X., Lun, Y.Z. *Acta Physiol Sinica.* 2010; 62: 86–92.
Zyzak D.V., Sanders R.A., Stojanovic M., Tallmadge D.H., Eberhart B.L., Ewald D.K., Gruber D.C., Morsch T.R., Strothers M.A., Rizzi G.P. Villagran M.D. *J Agric Food Chem.* 2003; 51(16): 4782–4787.

5

Elemental analysis of coffee products

Ewa Stanisz

CONTENTS

5.1 Introduction

The chemical composition of coffee beans is influenced, among others, by species type, geographical origin, the method of cultivation (e.g. organic or conventional fertilized soils) or processing after harvesting the beans (during industrial production and beverage preparation; Cruz et al. 2015; Petracco 2005; Pohl et al. 2013). The two most important species of coffee in economic terms are Arabica coffee (*Coffea arabica*) and Robusta coffee (*C. canephora*). Two other species that are grown on a much smaller scale are Liberica coffee (*C. liberica*) and Excelsa coffee (*C. dewevrei*). When considering coffee analysis, in most cases Arabica and Robusta are studied, as they are most often used for preparation of coffee brews (infusions), and the demand for the others is not commercially significant (Vázquez and Vélez 2015; Cruz et al. 2015).

Arabica is generally considered to be the nobler variety due to the range of aromas produced by roasting the beans and the appreciated sour taste. Robusta contains more caffeine and has a stronger and more bitter taste. From the agricultural point of view, Robusta coffee trees are more resistant and less demanding than Arabica trees in various aspects, including the ideal growing climate and the use of pesticides. All these factors make Arabica beans more expensive compared to other coffee species (Clarke and Vitzthum 2001).

Coffea sp. plants are grown on soils with an acid pH (4.2–5.1) in more than 70 tropical and subtropical countries (Vázquez and Vélez 2015; Jarošová et al. 2014; Clarke and Vitzthum 2001). The reports of the International Coffee Organization (ICO) – the organization dealing with the trade in this product, inter alia – show that global coffee production in 2019 was calculated at 167.18 million bags (with a bag estimated as 60 kg). Arabica output decreased by 8.6% to 94.29 million bags, while Robusta output rose by 1.9% to 72.88 million bags. The decrease in the production is attributed primarily to the reduction in Brazil, as it was an off-year for its Arabica production (ICO 2019). However, Brazil remains the largest producer of coffee beans. Alone, this country produced 58 million bags (in 2019). One of the reasons for dominance of Brazil in the world is that its climate and temperature are ideal for growing this plant. Coffee export is also an important part of the Vietnamese economy, with a share value of 30.75 million bags. Furthermore, coffees from Colombia, Indonesia, and Ethiopia are very well known to amateurs of this drink. The list of world coffee producers also includes countries such as India, Honduras, Mexico, Guatemala, and Uganda.

With such a large group of coffee producers, the use of geographical indications allows them to obtain market recognition and often a premium price. Thus, the development of new and increasingly sophisticated procedures for determining the geographical origin of coffee products is highly desirable for consumers, agricultural farmers, retailers, and international authorities (Luykx and van Ruth 2008).

The ability to determine the geographic origin of coffee using analytical approaches is based on the composition of plants grown in a particular geographic area, which tends to reflect the elements present in the soil (Schmidt et al. 2009; Tagliaferro et al. 2007; dos Santos et al. 2010). Thus the elemental composition of

DOI: 10.1201/9781003043133-7

coffee may provide a unique marker enabling characterization of coffee according to country of producer (Carter et al. 2015; Bitter et al. 2020; Santato et al. 2012; Valentin and Watling 2013). Coffee assimilates numerous elements from the soil where it is grown in similar ratios as is found in the soil. Therefore, the elemental analysis allows to establish the so-called fingerprint, which is the arrangement of the concentrations of elements characteristic for a given origin of beans (Rodrigues et al. 2009, 2011a). The problem becomes more serious when the considered types of coffee are more expensive or more desirable (e.g., because of the origin or method of processing). Furthermore, attempts may be made to place on the market mixtures containing cheaper beans or falsify data on product labels.

For years, coffee samples have been subjected to sensory tests. Therefore, qualified testers may have recognized and assigned samples (e.g., in terms of their origin and market value). Unfortunately, even experienced experts are not able to recognize a large number of coffee origins, including subtle differences between them (Krivan et al. 1993). Thus attempts for coffee characterization have been made and various ingredients are used for this purpose; for example, alkaloids, chlorogenic acids, soluble solids in green beans (Bicho et al. 2011), volatile components (Cheong et al. 2013), and elements (Carter et al. 2015; Bitter et al. 2020; Santato et al. 2012; Valentin and Watling 2013).

5.2 Elemental composition of coffee products

Nowadays analytical procedures have determined about 60 different elements in green and roasted coffee beans, as well as in coffee infusions, instant coffee, and spent coffee grounds. These elements may be divided into three main groups: macroelements (Ca, K, Mg, Na, S, P); microelements (Cl, Co, Cr, Cu, Fe, Mn, Mo, Ni, Se, Sr, and Zn); and trace elements (Al, As, B, Ba, Br, Cd, Hg, Pb, and Sn). The group of toxic elements (e.g. Al, As, Cd, Cr, Ni, Pb, Hg), which have a negative impact on human health, is also very often determined in coffee products (Pohl et al. 2013; Jarošová et al. 2014).

In general, green coffee has a total element content of approximately 5% m/m (Şemen et al. 2017). This value depends mainly on the growing origin of the coffee, which is closely related to the soil type, coffee variety, field practices, climate, and processing (Debastiani et al. 2019; Pohl et al. 2013). Additionally, soil and water pollution and the use of some fertilizers and/or pesticides increase the total content of metals in soil, and thus in crops (dos Santos et al. 2009; Şemen et al. 2017). Inorganic contamination may also appear during the production and storage processes of coffee (dos Santos and de Oliveira 2001; Pohl et al., 2013).

Research on the elemental composition of coffee is conducted to establish the safety and quality of consuming infusions in the context of the presence of toxic metals (dos Santos and de Oliveira 2001; dos Santos et al. 2010; Árvay et al. 2019) and applied plant additives, such as chicory and barley (Oliveira et al. 2012). The concentration of the elements is also important when considering the daily requirement for a given nutrient with moderate coffee consumption (Pohl et al. 2018; Suseela et al. 2001; Şemen et al. 2017; Oliveira et al. 2012) as well as the element profile related to the geographical origin of beans (Bitter et al. 2020; Oliveira et al. 2015; Krvan et al. 1993; Valentin and Watling 2013). Because of the widespread consumption of this product, the chemical composition and hence quality of coffee has to be known and kept under strict control (Pohl et al. 2013).

5.2.1 Beans

As previously mentioned, the chemical composition of coffee beans is mainly related to origin of the crops, coffee species, the soil conditions, and method of cultivation (dos Santos et al. 2009). The coffee bean processing procedures should also be taken into account (Árvay et al. 2019; Van Cuong et al. 2014).

Elemental analyses have been widely applied to coffee beans to determine their geographic origin (Krivan et al. 1993; Anderson and Smith 2002; Liu et al. 2014; Worku et al. 2019; Habte et al. 2016; Mohammed et al. 2019; Bitter et al. 2020; Rodrigues et al. 2011a). According to some authors, Mn may be good discriminator between the coffee production regions (e.g., Colombia, Costa Rica, Cuba, El Salvador, Mexico, Nicaragua, Panama, Papua New Guinea; Krivan et al. 1993). Other authors differentiated between origins with a particular contribution of Cu, Na, Mn, and Fe (Anderson and Smith 2002; Cruz et al. 2015). The feasibility of using isotopic compositions of coffee beans to identify their geographic origins was presented by Liu et al. (2014) and Rodrigues et al. (2009, 2011a). Rodrigues et al. (2009) proposed multivariate analysis of isotopes ($\delta^{13}C$, $\delta^{15}N$, $\delta^{18}O$) and the percentage composition of carbon and nitrogen in coffee beans from 20 different geographic origins. For discriminating between coffees according to their geographical origin at the level of individual countries, the use of the $^{87}Sr/^{86}Sr$ isotope ratio coupled with $\delta^{18}O$ was also presented (Rodrigues et al. 2011a). Liu et al. (2014) classified the coffee beans based on concentrations of Rb, Sr, and Ba. However, according to the authors, the isotope ratios of B and Sr turned out to be excellent indicators of geographic origin.

In addition to the determination of geographic origin, the correct definition of the coffee species and its authenticity is another topic in coffee research. Arabica coffees are generally considered to be of better quality and thus of higher market prices (Santato et al. 2012; Cruz et al. 2015). Early studies reported that Arabica usually reveals higher amounts of Mn (2.5–6.0 mg 100 g^{-1}) when compared with Robusta (1.0–3.3 mg 100 g^{-1}; Clarke 1985). Later, the content of P, Mn, and Cu was proposed as Arabica and Robusta roasted coffee discriminators. Additionally, these descriptors enabled the resolution of Arabica/Robusta mixtures with a prediction error of about 7% (Martín et al. 1999).

The results of research on the influence of roasting degree on concentration of elements in green and roasted coffee showed that the studied concentrations ranged from 1447.97 to 1342.10 mg 100 g^{-1} for K; from 768.22 to 1259.44 μg g^{-1} for Ca; from 10.35 to 13.15 μg g^{-1} for Mn; and from 17.38 to 20.97 μg g^{-1} for Cu in green and roasted coffee beans, respectively (Van Cuong et al. 2014). All determined elements were the lowest value in green coffee and then increased with increasing roasting level and reached the highest value in roasting temperature of 250°C. Additionally, Mg concentration ranged from 682.70 to 3647.73 μg g^{-1}; Fe, 37.20 to 53.44 μg g^{-1}; Zn, 5.97–6.89 μg g^{-1}; and Pb, 2.18–15.04 μg 100 g^{-1}. For green coffee and roasted coffee beans, K, Mg, and Ca were the elements with the highest concentration, while Zn and Pb were found at lower concentrations. The roasting process did not eliminate any of the elements determined. Except for Fe, the other remaining elements had the lowest values in green coffee (Van Cuong et al. 2014). Another study shows that roasting degrees may affect the concentration of elements in resulting infusions (Árvay et al. 2019).

The mineral composition of coffee beans is also connected with environmental and agronomical conditions (Cruz et al. 2015). Soil characteristics are naturally diversified, varying from volcanic soils with high K contents (e.g., Indonesia and Central America) to those rich in basalt and granite with low K values (e.g., Brazil, West Africa, and India; Cruz et al. 2015; Anzueto et al. 2005). The soil elements effect the mineral composition of coffee beans, providing the basis for the development of markers for geographical discrimination, as detailed earlier. Long-term agricultural use of inorganic fertilizers increases the total content of heavy metals in cultivated soils, and thus in coffee plants and beans (dos Santos et al. 2009; Anzueto et al. 2005). A study presented by dos Santos et al. (2009, 2010) suggested that Cd, Cu, Zn, and other toxic elements contained in fertilizers used in the traditional and technological coffee farms, as well as in organic coffee farms, may cause an increase in the content of these metals in coffee plants.

However, regardless of the origin of the product the and coffee species, K is the element with the highest value, followed by Ca, P, Mg, S, and Na (Pohl et al. 2013; dos Santos and de Oliveira 2001; Suseela et al. 2001; Vega-Carrillo et al. 2002; Tagliaferro et al. 2006; Zaidi et al. 2006; Grembecka et al. 2007; Ashu and Chandravanshi 2011). Among minor elements, Fe, Cu, Mn, Zn, and Co are usually reported by the authors (Ashu and Chandravanshi 2011; Stelmach et al. 2015; Debastiani et al. 2014; Santato et al. 2012; Rodrigues et al. 2011b; Şemen et al. 2017; Filho et al. 2007; Martín et al. 1998; Krivan et al. 1993). The presence of major elements in coffee products (K, Ca, P, Mg, S, and Na) in different coffee products is essential because these elements may provide a certain nutritional value, especially considering the world popularity of coffee (Grembecka et al. 2007; Ashu and Chandravanshi 2011; Oliveira et al. 2012). The product can also be regarded as a source of some minor elements (Cr, Cu, Fe, Mn, and Zn) that have a significant impact on human health (dos Santos and de Oliveira 2001; Grembecka et al. 2007; Ashu and Chandravanshi 2011; Oliveira et al. 2012). However, the contribution of coffee elements in dietary reference intakes (DRIs) is not significant and can be estimated at several percent (Onianwa et al. 1999; Suseela et al. 2001; Zaidi et al. 2006; Oliveira et al. 2012). Importantly, some trace elements present in coffee beans, and thus in the infusions, can negatively affect human health (e.g., Al, Cd, Ni, Sb, Sn, Pb; Pohl et al. 2013). Therefore, concentrations of these elements in the beans and final product as well as in the production process should be strictly controlled using analytical methods. Such approaches may also allow the estimation of the pollution levels introduced at the production stages and prove the quality of the product in accordance with food legislation regulations (dos Santos and de Oliveira 2001; Suseela et al. 2001; Anderson and Smith 2002; De Nadai Fernandes et al. 2002; Vega-Carrillo et al. 2002; Ashu and Chandravanshi 2011).

The concentrations of selected elements in green coffee samples (based on literature data) are presented in Table 5.1.

TABLE 5.1

Concentrations of Selected Elements in Green Coffee Samples

Element	Concentrations ($\mu g\ g^{-1}$)	References
Al	1.16–45.28	Santato et al. 2012; Rodrigues et al. 2011b; Şemen et al. 2017
As	0.01–0.02	Santato et al. 2012; Şemen et al. 2017
B	5.38–14.62	Santato et al. 2012; Rodrigues et al. 2011b; Şemen et al. 2017
Ba	0.29–10.2	Santato et al. 2012; Rodrigues et al. 2011b; Şemen et al. 2017; Martín et al. 1998; Krivan et al. 1993
Be	0.001	Santato et al. 2012
Ca	401–8366	Stelmach et al. 2015; Debastiani et al. 2014; Santato et al. 2012; Şemen et al. 2017; Filho et al. 2007; Martín et al. 1998; Krivan et al. 1993
Cd	0.001–0.026	Santato et al. 2012; Şemen et al. 2017
Co	0.02–0.33	Santato et al. 2012; Rodrigues et al. 2011b; Krivan et al. 1993; Şemen et al. 2017
Cr	0.02–10.32	Santato et al. 2012; Rodrigues et al. 2011b; Krivan et al. 1993
Cu	6.95–76.9	Stelmach et al. 2015; Debastiani et al. 2014; Santato et al. 2012; Rodrigues et al. 2011b; Şemen et al. 2017; Filho et al. 2007; Martín et al. 1998; Krivan et al. 1993
Hg	0.12	Şemen et al. 2017
K	5505–21,570	Debastiani et al. 2014; Santato et al. 2012; Martínet al. 1998; Krivan et al. 1993; Şemen et al. 2017
Li	0.001–0.042	Santato et al. 2012
Mg	648–2476	Stelmach et al. 2015; Debastiani et al. 2014; Santato et al. 2012; Rodrigues et al. 2011b; Şemen et al. 2017; Filho et al. 2007; Martín et al. 1998; Krivan et al. 1993
Mn	3.69–154	Stelmach et al. 2015; Debastiani et al. 2014; Santato et al. 2012; Rodrigues et al. 2011b; Şemen et al. 2017; Filho et al. 2007; Martín et al. 1998; Krivan et al. 1993
Mo	0.07–0.44	Santato et al. 2012; Rodrigues et al. 2011b; Şemen et al. 2017
Ni	0.03–3.97	Santato et al. 2012; Rodrigues et al. 2011b; Şemen et al. 2017
P	182–2200	Debastiani et al. 2014; Santato et al. 2012; Martín et al. 1998; Şemen et al. 2017
Pb	0.003–0.30	Santato et al. 2012; Rodrigues et al. 2011b; Şemen et al. 2017
Sb	0.8×10^{-4} to 0.004	Santato et al. 2012; Şemen et al. 2017
Se	0.003–0.15	Santato et al. 2012; Şemen et al. 2017
Sn	0.054–13.24	Santato et al. 2012; Şemen et al. 2017
Sr	1.12–71	Debastiani et al. 2014; Santato et al. 2012; Şemen et al. 2017; Martín et al. 1998; Krivan et al. 1993
U	0.001–0.002	Santato et al. 2012; Şemen et al. 2017
Zn	3.62–61.3	Debastiani et al. 2014; Santato et al. 2012; Rodrigues et al. 2011b; Şemen et al. 2017; Filho et al. 2007; Martín et al. 1998; Krivan et al. 1993

5.2.2 Infusions

During the brewing process, only a fraction of the total elements present in the beans goes into the brew. For this reason, apart from the concentration of an element in the infusion, the degree of extraction from the beans is also often determined as a percentage (Pohl et al. 2013, 2016, 2018; Szymczycha-Madeja et al. 2015, 2016). Additionally, not all elements present the same degree of extraction (Jaganyi and Madlala 2000). High extractabilities can be observed for K, Mg, Ca, Na, and Mn, while only up to 12% Cu is extracted or a maximum of 25% for Fe (Cruz et al. 2015; Pohl et al. 2014b; Ashu and Chandravanshi 2011; Grembecka et al. 2007). In the case of instant coffee, it is usually assumed that the total content of an element in the infusion is equal to the content in the solid product (Clarke and Vitzthum 2001)

The degree of minerals extraction depends on the nature of the element, its speciation forms, and the degree of bonding to the matrix. The amount of minerals in infusion is related also to the roasting degree, grinding grade, coffee/water ratio, and pressure and temperature of brew preparation (Illy 2005; Petracco 2005; Cruz et al. 2015). Water, being the important ingredient of coffee brews, also plays a role in the quality of coffee and in the total mineral content of the product (Cruz et al. 2015). Thus its pH, hardness, and chemical contamination should be controlled in the brewing process (Illy, 2005). Mean leachabilities (%) of selected elements into instant coffee and ground coffee brews are presented in Figure 5.1.

For most popular espresso coffee brews, the mineral content, in general, varies between 7000 and 7200 mg L^{-1}, with K amounts in the range of 2900–3200 mg L^{-1}. These values are higher than those reported for other coffee beverages prepared by different methods due to the concentrated character of espresso (Petracco 2005; Cruz et al. 2015). Selected mineral contents in commercial espresso coffees are presented in Figure 5.2.

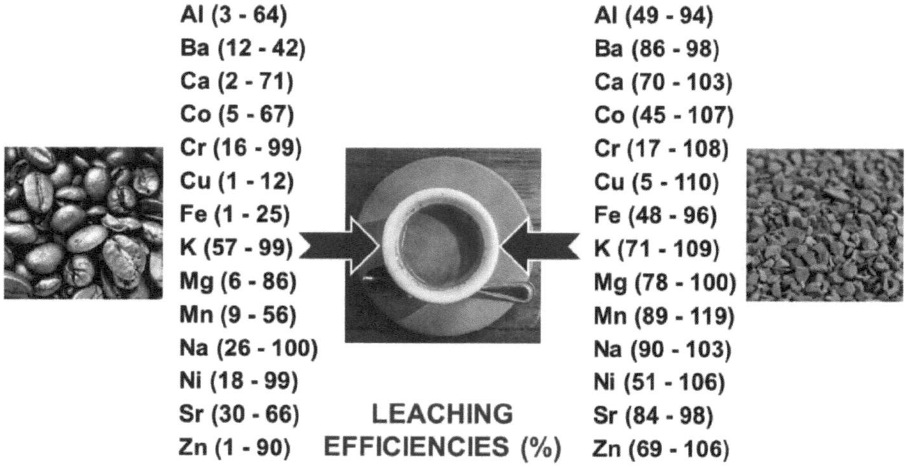

FIGURE 5.1 Mean leachabilities (%) of selected elements into ground coffee (*left*) and instant coffee (*right*) brews.

Sources: Szymczycha-Madeja et al. (2016); Pohl et al. (2014b); Welna et al. (2013); Stelmach et al. (2013); Ashu and Chandravanshi (2011); Fraňková et al. (2009); Grembecka et al. (2007).

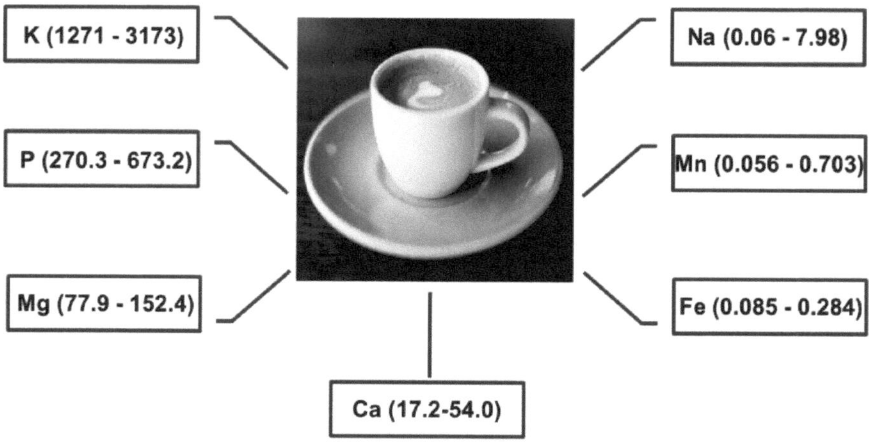

FIGURE 5.2 Selected mineral contents (mg 100 g^{-1}) in commercial espresso coffees (bracketed values are the minimum and maximum presented for coffees from different continents).

Sources: Oliveira et al. (2015); Illy (2005); Petracco (2005).

Similarly to coffee beans, infusions of ground and instant coffee contain the most K (491–681 µg mL^{-1}). In general, the average content of macro- and microelements is slightly higher in instant coffees compared to infusions of ground coffees, which is related to the technological processes of producing coffee in the instant form and the appropriate additives (Pohl et al. 2013, 2016, 2018; Szymczycha-Madeja et al. 2015, 2016). For example, the Ca content is 20.9 µg mL^{-1} in ground coffee, and in instant coffee 63.3 µg mL^{-1}; Mn content is 294 × 10^3 µg mL^{-1} in ground coffee, and in instant coffee 565 × 10^3 µg mL^{-1}. According to the literature data, the Fe content in ground coffee infusions is relatively low, at 87.1 × 10^3 µg mL^{-1}, while in soluble coffee it can exceed 573 × 10^3 µg mL^{-1}. However, the concentration of Cu does not exceed 8.74 × 10^3 µg mL^{-1} in infusions of ground coffee and 5.05 × 10^3 µg mL^{-1} in infusions of instant coffee. Zn is present in ground coffee infusions at the level of 33.3 × 10^3 µg mL^{-1} and it ranges from 42.5 × 10^3 µg mL^{-1} to 107 × 10^3 µg mL^{-1} in instant coffee (see Tables 5.2 and 5.3).

5.2.3 Instant coffee

Instant coffee is the dried soluble product obtained from roasted coffee. It is provided to consumers in powder or granule form for immediate makeup in hot water. The name by which this product is generally known is "instant coffee," although the synonym "soluble coffee" is also used (Clarke 2003).

The instant coffee market has increased over recent years, mainly due to the low price of the product and the ease of preparation. Mixtures of instant coffee with soluble cereal extracts are also popular due to the possible reduction of caffeine ingestion and economic reasons. In the soluble coffee industry, Robusta beans are usually preferred, which is related to their higher total soluble solids value and lower market price (Cruz et al. 2015; Clarke 2003; Zaidi et al. 2005; Vega-Carrillo et al. 2002; Oliveira et al. 2012)

In the production of soluble coffee brews, most minerals are extracted into the concentrated drink. Therefore, the elements content in the final product depends mostly on the green coffee initial composition and the extraction efficiency. Additionally, some minerals form less or more stable complexes with the matrix, and some are naturally more water soluble than others, which can affect the production process and preparation steps (Cruz et al. 2015; Stelmach et al. 2016; Szymczycha-Madeja et al. 2015, 2016). Water characteristics are presumed to influence the instant coffee mineral content, especially in the case of Ca and Na amounts (Cruz et al. 2015). Concentrations (µg mL^{-1}) of selected elements in brews of instant coffee are presented in Table 5.3.

5.2.4 Spent coffee grounds

The coffee technological production process includes several steps, where huge amounts of residues are generated. In countries that are major coffee producers, coffee husks and pulp are obtained after dry or wet hulling of the coffee cherries (Cruz et al. 2015, 2012). The milling of dried beans and the roasting of green coffee beans are also the source of biowaste, which contributes to environmental pollution (Campos-Vega et al.

TABLE 5.2

Concentrations (µg mL^{-1}) of Selected Elements in Brews of Ground Coffee

Element	Concentration (µg mL^{-1})	
	Range	Mean[1] (CV, %)
Al × 10^3	6.90–28.7	13.5 (55.9)
Ba × 10^3	16.2–62.7	30.3 (53.1)
Ca	14.6–32.4	20.9 (25.7)
Co × 10^3	0.9[2]–6.70	3.85 (35.8)
Cr × 10^3	0.4[2]–2.76	1.48 (34.8)
Cu × 10^3	3.52–14.2	8.74 (45.1)
Fe × 10^3	74.4–129	87.1 (20.8)
K	463–518	491 (3.6)
Mg	34.2–44.0	40.2 (9.1)
Mn × 10^3	181–433	294 (38.2)
Na	225–1510	694 (59.8)
Ni × 10^3	2.0[2]–34.2	15.2 (79.3)
Sr × 10^3	57.3–129	82.2 (32.5)
Zn × 10^3	21.2–52.2	33.3 (31.3)

Notes: [1] Mean values with coefficients of variance (CVs) in brackets; [2] detection limit (DL).

Source: Data from Szymczycha-Madeja et al. (2016).

TABLE 5.3

Concentrations (µg mL^{-1}) of Selected Elements in Brews of Instant Coffee

Element	Concentration (µg mL^{-1})	
	Range	Mean[1] (CV, %)
Al × 10^3	21.3–208	65.2 (88.5)
Ba × 10^3	33.0–213	91.5 (65.2)
Ca	50.3–80.0	63.3 (13.5)
Co × 10^3	0.9[2]–23.7	11.1 (73.3)
Cr × 10^3	1.78–3.42	2.70 (19.4)
Cu × 10^3	0.895–13.6	5.05 (74.5)
Fe × 10^3	268–1333	573 (58.1)
K	605–750	681 (7.8)
Mg	60.0–83.9	71.6 (9.8)
Mn × 10^3	294–950	565 (43.4)
Na	0.501–66.6	28.8 (72.3)
Ni × 10^3	2.0[2]–44.4	30.8 (92.7)
Sr × 10^3	118–328	181 (39.0)
Zn × 10^3	42.5–107	67.5 (35.3)

Notes: [1] Mean values with coefficients of variance (CVs) in brackets; [2] detection limit (DL).

Source: Data from Szymczycha-Madeja et al. (2016).

2015). Spent coffee grounds (SCGs) are generated by the soluble coffee industry and after beverage preparation at homes or coffee shops (Cruz et al. 2012). SCGs contain polysaccharides, oligosaccharides, lipids, aliphatic acids, amino acids, proteins, alkaloids, phenolics, lignins, and some minerals. Therefore, many studies have focused on investigating the composition of SCGs and their secondary applications (Cruz et al. 2012; Mussatto et al. 2011; Kovalcik et al. 2018; Campos-Vega et al., 2015).

In general, industrial coffee residues contain small amounts of soluble compounds, with a mineral content in the range of 0.3 to 1 g 100 g^{-1}, on a dry basis, with K as its major constituent (approx. 350 mg 100 g^{-1}), followed by P, Mg, Ca, Al, Fe, Mn, Cu, and Zn (Mussatto et al. 2011). The composition of the spent coffee grounds depends on the extraction efficiency during the brewing procedure, but it indicates a relatively high mineral content (Campos-Vega et al. 2015). According to Cruz et al. (2012) for espresso spent coffee grounds, selected mineral content (K, Mg, P, Ca, Na, Fe, Mn, and Cu) varied from 2.5 to 882.4 mg 100 g^{-1} for Cu and K, respectively. The other determined elements were characterized by the following concentrations (in mg 100 g^{-1}): Mg, 220.1 ± 134.1; P, 153.4 ± 50.3; Ca, 34.9 ± 12.2; Na, 20.1 ± 15.0; Fe, 4.6 ± 2.1; and Mn 2.7 ± 1.0 (Cruz et al. 2012).

5.3 Methods of coffee elemental analysis

Nowadays, elemental food composition data are important to both consumers and health professionals, and recent food labeling legislation has additionally highlighted this requirement (Andrade Korn et al. 2008; Trugo 2003). Therefore, there is increasing demand for analytical methodologies enabling industry and nutrition science to determine the quality of coffee, its origin, and safety in terms of the elements content (Andrade Korn et al. 2008). A number of experimental works have been focused on elemental composition of different coffee samples and/or coffee brews (Stelmach et al. 2015; Szymczycha-Madeja et al. 2016; Alves da Silva et al. 2017; Schmidt et al. 2009; Pigozzi et al. 2018). It is understandable that great attention is given to hazardous or toxic elements (Alves da Silva et al. 2017; Ashu and Chandravanshi 2011), because high concentrations of several heavy metals have been found in some coffee samples (Grembecka et al. 2007; Santos and de Oliveira 2001; Alves da Silva et al. 2017; Ashu and Chandravanshi 2011).

5.3.1 Preparation of coffee samples

Sample preparation is still one of the most important steps of the analytical procedure. If efficient sample preparation/digestion is conducted, homogeneity of the sample solution is ensured, the negative impact of the matrix components is limited, and thus reliable analytical results are provided. On the other hand, it is usually a time-consuming step that may introduce negative changes in the composition of the sample (e.g., loss of analytes or contamination). Therefore, the development of new methods for determination of elements in coffee must go through the introduction of reliable methodologies for the preparation of samples (Ferreira et al. 2010; Trindade et al. 2020).

The preparation of coffee samples before elemental analysis may consist of several steps. They depend on whether the analysis involves green or roasted beans, commercial instant or ground coffee, its infusions, and other by-products obtained during the production.

Before preparation, pre-treatment stages for green coffee beans include washing with tap water followed by detergent, diluted acid solution (e.g., HCl), and finally deionized water for removing soil dust (dos Santos et al. 2009; Pohl et al. 2013). Next, they can be dried in laboratory ovens at temperatures of 60–103°C and then pulverized in cryogenic mills or ground and powdered using rotor-type mills or grinders (Krivan et al. 1993; Filho et al. 2007; Oleszczuk et al. 2007; Bertrand et al. 2008). Commercially available roasted beans are usually manually or mechanically grinded using mortars or different kind of mills (ball, knife, and cryogenic) and sieved to select required particle size fractions. Commercial instant coffees (Szymczycha-Madeja et al. 2016; dos Santos and de Oliveira 2001; Suseela et al. 2001; Vega-Carrillo et al. 2002; Ribeiro et al. 2003; Fernandes et al. 2005; Zaidi et al. 2006; Grembecka et al. 2007; Oliveira et al. 2012) or ground roasted coffee samples are usually subjected to further stages of the elemental analysis without any special pre-treatment before the preparation step (Fraňková et al. 2009; Pohl et al. 2018; Szymczycha-Madeja et al. 2016; Stelmach et al. 2016).

5.3.1.1 Preparation of solid coffee products

The determination of trace elements in coffee, using atomic and mass spectrometry techniques, requires extensive sample preparation consisting of digestion, dissolution, or/and extraction stage prior to instrumental analysis (Andrade Korn et al. 2008). The methodology mostly aims at reducing matrix effects originating from organic compounds and releasing elements in their simplest form (Filho et al. 2007; Castro et al. 2009; Pohl et al. 2018). In general, two main types of digestion procedures have been used for this purpose: dry ashing in the atmosphere of air (Suseela et al., 2001; Grembecka et al., 2007; Bertrand et al. 2008) and wet oxidative digestion (Pigozzi et al. 2018; Oleszczuk et al. 2007; Rodrigues et al. 2011a; Martín et al. 1999; dos Santos et al. 2010).

Dry ashing is one of the traditional techniques used in routine food analysis (Hoenig 2003; Akinyele and Shokunbi 2015). Usually, the procedure is performed by calcination at atmospheric pressure in programmable muffle furnaces, but the microwave furnaces especially adapted to achieve elevated temperatures are also available (Hoenig 2003, 2005). In the usual high-temperature ashing, fresh or dried coffee samples are weighed into suitable ashing vessels (quartz, porcelain, platinum) and placed in the furnace. The temperature is progressively elevated, following a convenient heating program to attain maximum constant temperature, and then maintained for several hours. For coffee sample preparation, the commonly utilized temperatures are in the range of 350–550°C (Suseela et al. 2001; Bertrand et al. 2008; Grembecka et al. 2007; Onianwa et al. 1999). The resulting inorganic residue (ash) is then dissolved in small portions of acid solutions, such as HNO_3 and $HClO_4$ (Suseela et al. 2001), concentrated HCl (Grembecka et al. 2007; Bertrand et al. 2008), or HNO_3 (Onianwa et al. 1999). Thus, the coffee sample is diluted to a controlled volume and analyzed. Additionally, the preconcentration of trace elements in the final solution is achieved, which is useful when low concentrations of elements are determined. The ash is also completely free of organic matter, which is required for some detection techniques (Hoenig 2003, 2005; Pohl et al. 2013; Akinyele and Shokunbi 2015). However, running the process in an open system may lead to loss of some trace analytes or contamination of the sample during preparation.

Nowadays, wet oxidative digestion techniques are used more frequently than dry ashing to prepare coffee samples prior to elemental analysis. They involve the chemical degradation of sample matrix in solution, usually with a combination of acids to increase solubility (Twyman 2005). If the procedure takes place in an open system (Magalhães et al. 1999; Martín et al. 1999; Anderson and Smith 2002; Ribeiro et al. 2003; Filho et al. 2007), unfavorable changes in the sample composition may occur due to volatilization of analytes, adsorption onto the vessel walls, and contamination of the sample with substances in the air. In general, the use of open vessels allows the preparation of higher sample masses (e.g., 1–2 g), but the efficiency of decomposition is limited to the boiling temperature of solvents (Ferreira et al. 2010). In case of hot plates and heating blocks applied for open-vessel systems, samples are heated at 80–120°C for 4–12 h (Valentin and Watling 2013; Ribeiro et al. 2003) or at much higher temperatures (approx. 180°C) for 1–4 h (Anderson and Smith 2002; Filho et al. 2007). In some procedures reported in the literature, Kjeldahl apparatus (Gure et al. 2018; Ashu and Chandravanshi 2011) and alkaline solubilization of the samples in tetramethylammonium hydroxide (TMAH; Ribeiro et al. 2003) were used. For conventional open systems heating on heating blocks, water baths or laboratory ovens are used. The process of increasing the temperature can also be assisted by the influence of microwave radiation, which significantly shortens the time of conversion.

The use of closed systems, especially microwave pressurized vessels (Fernandes et al. 2005; Filho et al. 2007; Oleszczuk et al. 2007; Santos et al. 2008; Castro et al. 2009; Fraňková et al. 2009; dos Santos et al. 2010), enables isolating the digestion reaction from the surroundings (Andrade Korn et al. 2008) and therefore to reduce both contamination and sample loss (Twyman 2005). In general, pressurized (approx. 40 bars) closed vessels provide fast and complete decomposition of coffee samples under controlled temperatures with relatively low volumes of reagents. Additionally, microwave heating has several advantages over conventional heating, as the energy is generated in the digestion mixture and not transferred by conduction (Flores 2014). Among the key benefits of microwave-assisted digestion are short digestion time and a significant reduction in the consumption of aggressive reagents to obtain high efficiency of digestion (Andrade Korn et al. 2008; Flores 2014). However, using high-pressure systems the amount of sample digested is lower in comparison to open systems (Ferreira et al. 2010).

It should be emphasized that the type of acid used in the preparation procedure may have a significant impact on the detection system (Andrade Korn et al. 2008). It is well known that nitric acid is the most desirable reagent in all techniques of atomic spectrometry. Sometimes (e.g., in ICP-OES), the occasional signal suppression is observed in its presence. However, no serious analytical problems are encountered with concentrations up to 10% in most atomic spectrometric techniques. Hydrogen peroxide, which is added in many coffee mineralization procedures, is also rarely responsible for analytical problems (Arruda 2007) The presence of hydrochloric acid is not a problem in the analysis utilizing ICP-OES. Nevertheless, in electrothermal AAS determinations, this reagent is forbidden due to the possible formation of volatile and difficult to dissociate chlorides which may cause significant interferences (Welz and Sperling 1999). The use of sulfuric acid is usually avoided in the spectroscopic method despite its effectiveness in mineralization of coffee matrix. This is mainly due to its high viscosity, thus complications in the sample introduction by nebulization may occur (FAAS, ICP-OES, ICP-MS; Andrade Korn et al. 2008).

In the course of wet decomposition of coffee samples, the most commonly used are mixtures of concentrated HNO_3 and 30% m/m H_2O_2 solutions at different ratios (2:1, 3:1, 7:1, 10:1, or 12:1; dos Santos et al. 2010; Castro et al. 2009; Fernandes et al. 2005; Filho et al. 2007; Oleszczuk et al. 2007; Santos et al. 2008; Ribeiro et al. 2003; dos Santos and de Oliveira 2001). Additionally, mixtures of concentrated HNO_3 and HF solutions (Krivan et al. 1993), HNO_3 and H_2SO_4 solutions (Martín et al. 1999), and HNO_3 and $HClO_4$ solutions (dos Santos et al. 2009; Ashu and Chandravanshi 2011) were used for the decomposition of coffee.

Wet oxidative digestion and dry ashing of coffee samples may be relatively time-consuming and often suffer from systematic errors arising from incomplete mineralization, contamination, and losses of volatile analytes (Anthemidis and Pliatsika 2005). Thus, direct solid sample introduction (DSSI) via a graphite furnace tube (Oleszczuk et al. 2007) or slurry sampling (SS; Magalhães et al. 1999; Anthemidis and Pliatsika 2005; Trindade et al. 2020) may overcome the aforementioned problems by combining matrix destruction and recording the analytical signal in a one procedure. In general, slurry sampling is considered to have certain advantages over direct solid sampling, because it is possible to change the slurry concentration by simple dilution. Therefore, combination of the advantages of solid and liquid sampling is achieved. Additionally, aqueous standards may be used for calibration. However, the stabilization of the slurry, its homogeneity, particle size, and sedimentation also have to be considered (Andrade Korn et al. 2008).

For direct solid introduction, small portions of coffee samples (approx. 0.03–0.5 mg) are weighted onto graphite platforms and introduced into graphite tubes for a thermal proceeding (Oleszczuk et al. 2007). Slurries of coffee samples may be prepared by adding a diluted HNO_3 solution to the solid sample (30–50 mg, <105 μm) and heating to approximately 90°C to increase extraction efficiency (Magalhães et al. 1999). Then a diluted solution of Triton X-100 is added, and ultrasonication is conducted, resulting in complete homogenization of solid sample particles (Magalhães et al. 1999; Trinidade et al. 2020). The online formation of sample slurries also has been reported (Anthemidis and Pliatsika 2005).

In the course of neutron activation analysis (NAA), coffee beans are washed to remove any surface contamination with distilled and deionized water, then dried with hot air (Zaidi et al. 2006). After washing, the samples are first freeze-dried (Zaidi et al. 2006) or oven-dried (Tagliaferro et al. 2006), ground to fine powder, and placed in special polyethylene or quartz vials for analysis (Tagliaferro et al. 2006; Krivan et al. 1993; Zaidi et al. 2006). It should be highlighted that NAA requires extra care due to the radioactivity induced on the samples in the preparation step.

Selected examples of sample preparation techniques prior to elemental analysis of coffee products are presented in Table 5.4.

TABLE 5.4

Selected Procedures of Coffee Samples Preparation prior to Their Elemental Analysis (Determined Elements and the Detection Techniques Are Also Presented)

Analytes	Sample preparation[1]	Detection technique	References
Cd, Co, Cr, Cu, Fe, Ni, Pb, Zn	Dry ashing: GC/550°C; ash + HNO_3 (1 mol L^{-1})	FAAS	Onianwa et al. 1999
Ca, Cd, Cr, Cu, Fe, K, Mg, Mn, Ni, Pb, Sr, Zn	Dry ashing: 2.0 g SC/350°C; ash + 2 mL HNO_3 + 1 mL $HClO_4$ (microwave digestion)	FAAS	Suseela et al. 2001
Ca, Cd, Co, Cr, Cu, Fe, K, Mg, Mn, Na, Ni, P, Pb, Zn	Dry ashing: 6 g SC + 150 mL hot water; filtration; evaporation to dryness/540°C; ash + 1.5 mL 36.5% HCl + drop 65% HNO_3 5 g GC/540°C; ash + 1.5 mL 36.5% HCl + drop 65% HNO_3	FAAS	Grembecka et al. 2007
B, Ca, Cu, Fe, K, Mg, Zn	Dry ashing: 0.5 g GC/500°C/2 h; ash + 2 mL 6 M HCl/drying + 2 mL 0.5 M HCl; residue/500°C/2 h; ash + HF +2 mL 0.5 M HCl	ICP-OES	Bertrand et al. 2008
Cd, Pb, Cu, Cr, Ni, Zn	Infusions GC: 12 g + 100 mL water (95–100°C); then 25 mL of infusion concentrated in the oven 60°C to the final volume 2.5 mL Wet digestion: infusions GC and GC (solid) +18 mL (mixture HNO_3: $HClO_4$, 3:1)/heated 190°C	FAAS	Pigozzi et al. 2018
Ca, Cd, Cr, Co, Cu, Fe, K, Mg, Mn, Ni, Pb, Zn	Wet digestion, Kjeldahl apparatus: 0.5 g GC + mixture of 5 mL HNO_3 (70%) and 1.5 mL $HClO_4$ (70%)/270°C/4 h/hot plate	FAAS	Gure et al. 2018
Ca, Co, Cu, Fe, K, Mn, Mg, Na, Zn	Wet digestion, Kjeldahl digestion block: 0.5 g GC + 5 mL HNO_3 70% + 1.5 mL $HClO_4$ 70%/4 h/350°C Infusions: 25 mL after evaporation + 4 mL HNO_3 + 1 mL $HClO_4$/4 h/300°C	FAAS	Ashu and Chandravanshi 2011
Ba, Ca, Cu, Fe, K, Mg, Mn, Na, P, Sr, Zn	Wet digestion, hot plate: 1.0 g GC + H_2SO_4 + HNO_3 (conc. 1:4)	ICP-OES	Martín et al. 1999
Ca, Cu, Fe, K, Mg, Mn, Na, P, Se, Sn, Zn	Alkaline solubilization (heating block): 0.35 g SC + 0.5 mL TMAH 25% (w/v)/10 min/80°C	ICP-OES	Ribeiro et al. 2003
Ag, Al, As, Ba, Be, Bi, Ca, Cd, Ce, Co, Cr, Cs, Cu, Dy, Er, Eu, Fe, Ga, Gd, Ge, Hf, Ho, K, La, Li, Lu, Mg, Mn, Mo, Na, Ni, Nb, Nd, P, Pb, Pd, Pr, Rb, S, Sb, Sc, Se, Sm, Sn, Sr, Ta, Te, Ti, Tb, Th, Tl, Tm, U, V, W, Y, Yb, Zn, Zr	Wet digestion, water bath: 1 g GC + 5 mL HNO_3/4 h/room temperature; then 90°C/12 h; then + 2 mL H_2O_2; evaporated + 20% HNO_3 (up to 25 mL)	ICP-OES ICP-MS	Valentin and Watling 2013
Co, Cu, Mn	Direct solid sampling: 0.03–0.5 mg GC; weighed directly onto the SS platforms and introduced into the graphite furnace Microwave-assisted wet digestion: 0.5 g GC + 7 mL HNO_3 65% + 1 mL H_2O_2 30%; first pre-digestion 1 h; then digestion: total time 25 min. (four-step temperature program), max. 1000 W/180°C.	ETAAS ICP-OES	Oleszczuk et al. 2007

Analytes	Sample preparation[1]	Detection technique	References
Al	Slurry sampling: 30–50 mg GC (<105 μm) + 20 mL 0.2% (v/v) HNO_3; heating (hot plate) 80–90°C/15 min; then + 20 mL 10% (v/v) Triton X-100; sonication/10 min	ETAAS	Magalhães et al. 1999
Al, Ca, Co, Cr, Cu, Fe, Mg, Mn, Ni, Zn	Slurry sampling, online formation: 0.01 g GC (<70 μm) + 0.5% (v/v) Triton X-100 in 1% (v/v) HNO_3 Wet digestion, conventional heating: 0.3 g GC + 3 mL HNO_3 65% (m/m); pressurized bomb (120–130°C/2 h)	ICP-OES/FAAS/GFAAS	Anthemidis and Pliatsika 2005
Ca, Fe, Zn	Slurry sampling: 0.1 g (<100 μm) + 40 mL Triton X-100 5% (v/v) + 5 mL (mixture 1:1; 1.0 mol L^{-1} HCl /1.0 mol L^{-1} HNO_3); sonication/10 min	FAAS	Trindade et al. 2020
Al	Microwave-assisted wet digestion: 0.3 g GC + 10 mL HNO_3 70%; total time 40 min (two-step temperature program); max 200°C, 195 psi, 800 W	ICP-OES	Fraňková et al. 2009
Al, Ca, Cu, Fe, K, Mg, Mn, Na, Pb, S, Se, Si, Sn, Sr, Zn	Microwave-assisted wet digestion: 0.25 g GC + 3 mL HNO_3 concentration + 1 mL H_2O_2 concentration; total time: 19 min (five-step temperature program); max 700 W. Infusions: 1.0 g GC + 10 mL boiling water/15 min/centrifugation/acidification with HNO_3 to 0.014 mol L^{-1}	ICP-OES	Fernandes et al. (2005)
Ca, Cd, Cr, Cu, Fe, K, Mg, Mn, Na, Ni, Pb, Zn	Microwave-assisted wet digestion: 0.1 g GC + 2 mL HNO_3 65% (v/v) + 1 mL H_2O_2 30% (m/m) + 6 mL of ultrapure water; total time 45 min (two-step temperature program)	FAAS/ICP-OES	dos Santos et al. 2010
Ba, Cu, P	Microwave-assisted wet digestion: 0.25 g GC + 7 mL 3.5 mol L^{-1} HNO_3 + 1 mL H_2O_2 30% (w/w); total time: 33 min (four-step procedure, 750 W, 35 bar) Wet digestion, conventional heating (muffle furnace): 0.2 g + 2 mL HNO_3 3.5 mol L^{-1} + 1 mL H_2O_2 30% (w/w); pressurized bomb (12 h/120°C)	ICP-OES	Castro et al. 2009
Ca, Cu, Fe, Mg, Mn, Zn	Wet digestion, conventional heating (block): 0.4 g GC + 10 mg V_2O_5[2] + 3 mL HNO_3 65% (v/v) + 2 mL H_2O_2 30% (m/m); 1 h/180°C Microwave-assisted wet digestion: 0.4 g GC + 3 mL HNO_3 65% (v/v) + 2 mL H_2O_2 30% (m/m); total time 18 min (four-step temperature program); max 1000 W.	FAAS	Filho et al. 2007
Ca, Cd, Cr, Cu, Fe, Mg, Mn, Ni, Pb, Zn	Microwave-assisted wet digestion: 0.2 g GC + 3 mL of HNO_3 (67%) + 1 mL H_2O_2 (30%); total time 20 min (four-step temperature program); max 600 W.	FAAS/ GFAAS/ ICP-MS	Jarošová et al. 2014
Al, Ba, Ca, Cd Co, Cr, Fe, K, Mg, Mn, Na, Ni, Pb, Sr	Infusions: 6.0 g GC + 200 mL water (100°C)/10 min/filtration Infusions: 6.0 g SC + 250 mL water (90–95°C) dissolving and mixing/centrifugation 10.0 g of infusion (GC or SC)/acidification with HNO_3 to 0.25 mol L^{-1}	FAAS /ICP-OES	Szymczycha-Madeja et al. 2016
Ca, Fe, Mg, Mn	Infusions: 2.5 g SC + 100 mL water (95°C)/centrifugation/without any pre-treatment step.	FAAS	Stelmach et al. 2016
Al, As, B, Ba, Be, Bi, Ca, Cd, Co, Cr, Cu, Hg, K, Li, Mg, Mn, Mo, Ni, P, Pb, Se, Sr, Sn, Sb, Th, U, Zn	Microwave-assisted wet digestion: 0.5 g GC + 7 mL HNO_3 65% (v/v) + 1 mL H_2O_2 30% (v/v); total time 28 min (four-step temperature program); max 1600 W, 500 psi, 180°C. Infusions GC: acidification with HNO_3 2%	ICP-MS	Şemen et al. 2017
Al, Ba, Ca, Cu, Fe, Mg, Mn, Ni, Sr, Zn	Infusions GC or SC: acidification with HNO_3 to 1.3 mol L^{-1}	ICP-OES	Pohl et al. 2018

(*Continued*)

TABLE 5.4 (Continued)

Analytes	Sample preparation[1]	Detection technique	References
Ba, Br, Ca, Co, Cr, Cs, Fe K, La, Mn, Na, Rb, Sc, Zn	Direct analysis for NNA: 70–200 mg GC was sealed in polyethylene capsules or quartz tubes before irradiations Wet digestion, conventional heating for AAS: 200 mg GC + 3 mL HNO_3 65% (m/v) + 0.1 mL HF 40% (m/v) pressurized vessel/160°C/4 h	NAA/FAAS/ GFAAS	Krivan et al. 1993
Ba, Br, Ce, Cl, Co, Cr, Cs, Eu, Fe, Hg, In, K, Mn, Na, Rb, Sb, Sc, Se, Yb, Zn	Direct analysis: Coffee beans were first freeze-dried, then ground, sealed in polyethylene or quartz capsules for irradiation SC was sealed in polyethylene or quartz capsules for irradiations	NAA	Zaidi et al. 2006
Br, Ca, Co, Cs, Fe, K, Na, Rb, Sc, Zn	Direct analysis: Coffee beans were dried (60°C), then ground; 400 mg was placed in polyethylene vials for irradiations	NAA	Tagliaferro et al. 2006
Fe, La, Sc, Sm, Th	Infusions GC: 15 g + 400 mL water (espresso brewing technique 94°C, 1.5 MPa); then frozen at −10°C, freeze-dried for 72 h, and ground; 500 mg was placed in special polyethylene capsules for irradiations	NAA	Tagliaferro et al. 2007

Notes: [1] Technique used, sample mass, reagents, digestion time (if given by the authors); [2] used as catalyst; TMAH = tetramethylammonium hydroxide; GC = ground coffee; SC = soluble coffee; FAAS = flame atomic absorption spectrometry; ICP-OES = inductively coupled plasma optical emission spectrometry; ICP-MS = inductively coupled plasma mass spectrometry; ETAAS = electrothermal atomic absorption spectrometry; GFAAS = graphite furnace atomic absorption spectrometry; NAA = neutron activation analysis.

5.3.1.2 Preparation of coffee infusions before elemental analysis

To prepare a coffee infusion (brew) for analytical purposes, most commonly hot (approx. 100°C) de-ionized water is added to a coffee portion (ground roasted or instant). After several minutes, the resulting brew is filtered or centrifuged and may be further analyzed (Pohl et al. 2018; Szymczycha-Madeja et al. 2016; Tagliaferro et al 2007).

The style of brewing may affect the composition of coffee product, therefore the preparation technique of the drink must be considered. Boiling (e.g., Turkish coffee), steeping (e.g., instant or coffee bags), dripping, and pressure methods (e.g., mocha pot or espresso) may be a source of metal contamination of the drink (e.g. Al or Ni; Fraňkova et al. 2009; Windisch et al. 2020; Magalhães et al 1999; Berg et al. 2000).

Coffee infusions without special preparation stage are analyzed quite often. Then only dilution (Fraňkova et al. 2009; Oliveira et al. 2012) or acidification for stabilization of the infusion are performed. Ground coffee infusions may be acidified with HCl at 1% (Oliveira et al. 2015) or with HNO_3 at 0.25 mol L^{-1} (Szymczycha-Madeja et al. 2016), 1.3 mol L^{-1} (Pohl et al. 2014b; Stelmach et al. 2013) or 0.014 mol L^{-1} (Fernandes et al. 2005). Instant coffee after dissolution in water may be also acidified with HCl at 1% (Oliveira et al. 2012) or with HNO_3 at 1.0 mol L^{-1} (Pohl et al., 2014a), 0.25 mol L^{-1} (Szymczycha-Madeja et al., 2016), or without it (Stelmach et al. 2016).

To obtain information about the total concentrations of elements released or dissolved in infusions, sometimes it is necessary to apply the methods discussed earlier for solid coffee samples (Section 5.3.1.1). Infusions of ground coffee may be decomposed in mixtures of concentrated HNO_3 and 30% H_2O_2 (Welna et al. 2013) or concentrated HNO_3 and $HClO_4$ (Ashu and Chandravanshi, 2011). Some authors propose evaporation of infusion to dryness followed by dry ashing and dissolution of ash in solutions of HNO_3 or HCl (Onianwa et al. 1999; Grembecka et al. 2007). Infusions of instant coffee may be wet digested in mixtures of concentrated HNO_3 and 30% H_2O_2 (dos Santos and de Oliveira 2001; Welna et al. 2013) or dry ashed (Grembecka et al. 2007). For NNA analysis, portions of coffee infusion are frozen, freeze-dried and ground. The resulting powder is placed in special polyethylene or quartz vials for analysis (Tagliaferro et al. 2007).

Selected examples of sample preparation techniques prior to elemental analysis of coffee infusions are presented in Table 5.4.

5.3.2 Detection techniques used in the elemental analysis of coffee

Several spectroscopic techniques have been described in the literature for the determination of elements in coffee products. Flame atomic absorption spectrometry (FAAS; Trindade et al. 2020; Ashu and Chandravanshi, 2011; Grembecka et al. 2007; Pigozzi et al. 2018; Stelmach et al. 2016; Szymczycha-Madeja et al. 2016); electrothermal (or graphite furnace) atomic absorption spectrometry (ET(GF) AAS; Magalhães et al. 1999; Oleszczuk et al. 2007; Anthemidis and Pliatsika 2005; Jeszka-Skowron et al. 2016); inductively coupled plasma optical emission spectrometry (ICP-OES; Szymczycha-Madeja et al. 2016; Bernard et al. 2008; Martín et al. 1999; dos Santos et al. 2010; Valentin and Watling 2013); inductively coupled plasma mass spectrometry (ICP-MS; Habte et al. 2016; Şemen et al. 2017; Valentin and Watling 2013; Jarošová et al. 2014); neutron activation analysis (NAA; Tagliaferro et al. 2006, 2007; Krivan et al. 1993; Zaidi et al. 2006) have been the most widely applied techniques for this purpose. Selected examples of detection techniques used in the elemental analysis of coffee products are presented in Table 5.4.

Atomic absorption spectrometry is an analytical technique based on the absorption of ultraviolet-visible radiation by free atoms in the gaseous state. In general, liquid samples (coffee after digestion) or solid samples (in the case of solid sampling) are atomized through rapid heat application in flame or graphite furnace. The analyte concentration is then quantified from the amount of light absorption (Yeung et al. 2017). It is a relatively simple approach and has been the most widely used form of atomic spectroscopy in coffee analysis for many years. FAAS was presented for determination of different major (e.g., Ca, K, Mg and Na), minor (e.g. Fe, Mn and Zn), and trace (e.g. Cd, Co, Cr, Cu, Ni and Pb) elements in coffee. Unfortunately, FAAS is considered not sensitive enough to quantify some important trace elements (nutrients but also toxic) in coffee products (Grembecka et al. 2007; Pohl et al. 2013). Nerveless, electrothermal AAS allows for lower quantification limits and was used generally to determine individual elements (e.g., Al, Ni as contaminants in coffee; Windisch et al. 2020; Magalhães et al. 1999; Fraňkova et al. 2009; Berg et al. 2000), which is related to the time-consuming determinations and their single-element character. Similar to FAAS, samples of coffee have to be digested before analysis. However, ETAAS enables, apart from neutron activation analysis (NAA), direct analysis of solid coffee samples with minimal sample preparation by direct solid (Oleszczuk et al. 2007) or slurry sampling (Magalhães et al. 1999; Anthemidis and Pliatsika 2005; Trindade et al. 2020), which is an advantage compared to ICP-OES or ICP-MS.

ICP-OES is very often applied in the elemental analysis of coffee samples. In the past, the name ICP-AES was commonly used for this technique; currently, the abbreviation ICP-OES is preferred. The principle of the technique is based on the fact that excited electrons emit energy at a given wavelength as they return to ground state (Yeung et al. 2017; Boumans 1987). The basic feature of this process is that each element emits light at wavelengths appropriate to its chemical nature. The intensity of light emitted at the selected wavelength is proportional to the concentration of this element in the analyzed sample. The technique is mainly utilized for liquid samples, therefore, a flow injection (FI) system for an online sample slurry formation of solid samples without a decomposition step also has been proposed to improve and shorten the coffee analysis (Anthemidis and Pliatsika 2005).

In comparison to AAS, ICP-OES is an emission spectrometric technique, thus enabling faster multi-element determinations, lower detection limits, and higher sensitivities. It operates in a wider linear dynamic ranges and allows of quantification of elements, such as P and S (Anderson and Smith 2002; Fernandes et al. 2005). Using this technique, macronutrients (e.g., Ca, K, Mg, Na, S, and P), micronutrients (e.g. Co, Cr, Cu, Fe, Mn, Mo, Ni, Se, and Zn), and nonessential (e.g., Ba, Sr, Si and V) and toxic elements (e.g., Al, Sn, Cd, and Pb) were determined in coffee products.

Because most elements exist predominantly as singly charged ions in the ICP, it is effectively used as an ionization source for mass spectrometry (Olesik 1991; Smith and Thakur 2017). Inductively coupled plasma mass spectrometry allows extremely low detection limits at the single part per trillion (ppt) levels, enhanced multielement capabilities, and the ability to quantify individual isotopes of the elements (Yeung et al. 2017). In this technique, solid coffee samples are digested and aspirated into the ICP torch in the same manner as in ICP-OES. However, instead of having an optical system and a device for separating and detecting radiation of specific wavelengths, the ICP-MS uses a mass spectrometer to separate and detect ions of the elements directly based on their unique mass-to-charge (m/z) ratios (Yeung et al. 2017; Smith and Thakur 2017).

The technique is particularly suitable for the classification of coffee, because trace element concentrations and isotope ratios of heavy elements are among the most useful parameters determining the nature and quality of coffee (Aceto 2016; Georgiou and Danezis 2015). ICP-MS was applied by Rodrigues et al. (2011a) to Sr isotope fingerprinting of green coffee beans. It has also been shown that isotopic analysis of C, N, and O is an effective method in verifying the geographical origin for specialty coffees on a local geographical scale in Brazil (Barbosa et al. 2014; Worku et al. 2019).

Many works on the subject can be found in the literature concerning multi-element profiling (Gonzálvez and de la Guardia 2013), isotopic analysis (Rodrigues et al. 2011a, 2013; Zhao et al. 2014), and both elemental and isotopic analysis (Drivelos and Georgiou 2012; Georgiou and Danezis 2015).

Instrumental neutron activation analysis (INNA), sometimes referred to as simply NAA, has also been utilized for the elemental analysis of coffee (Krivan et al. 1993; De Nadai Fernandes et al. 2002; Vega-Carrillo et al. 2002; Tagliaferro et al. 2006, 2007; Zaidi et al. 2006). The technique is based on conversion of stable atomic nuclei into radioactive nuclei by irradiation with neutrons and subsequent detection of the radiation emitted by the radioactive nuclei and its identification. Samples are usually encapsulated in polyethylene or some other suitable packaging and irradiated. Coffee may be analyzed without pre-treatment, dissolution, or decomposition, eliminating measurement errors due to incomplete digestion or loss of volatile elements (Hamidatou et al. 2013; Cunningham and Stroube 1987). The analysis is non-destructive, therefore samples can be re-analyzed by the NAA or other reference techniques if needed (Tagliaferro et al. 2006, 2007; Krivan et al. 1993; Zaidi et al. 2006).

5.4 Conclusions

Coffee is one of the most popular drinks in the world with great commercial and social importance. Research in the field of coffee analysis, conducted by food industry, governmental agencies, or universities, requires determination of coffee ingredients and impurities as well as their characteristics. Coffee consumption is constantly increasing due to its physiological effects, taste, and aroma. Therefore, consumer demands as well as international regulations and coffee industry requirements constantly pose new challenges to chemists dealing with the analysis of this product. The results of their work are essential for monitoring coffee composition and to ensure the quality and safety of the coffee supply.

The nature of the coffee samples and the specific reason for the analysis strongly influence the choice of analytical methods. The success of the entire analytical procedure depends on the proper preparation of the coffee sample, reliably conducted analysis, and appropriate interpretation of the data. Because of the numerous challenges that arise during the analysis of coffee, it seems necessary to improve existing methods and develop new methods of analyzing this food product.

In routine elemental analysis of coffee, new sample preparation techniques will be developed to reduce this step as much as possible. In this field, laser ablation, used for direct elements determination in solid samples with minimal sample pre-treatment, seems to be a promising approach. The advantages of this technique are direct characterization of solids, no chemical procedures for dissolution, and thus reduction of contamination and sample loss. Additionally, ICP-MS appears to be the most promising detection technique in coffee analysis. Thus its combination with laser ablation may create the possibility of quick and multi-element analysis of coffee products.

5.5 Acknowledgments

This work was supported by Polish Ministry of Science and Higher Education.

REFERENCES

Aceto, M. 2016. 137–164. Publisher: Amsterdam: Elsevier.

Akinyele, I. O. and Shokunbi, O. S. *Food Chem.* 2015;173:682–684.

Alves da Silva, S., Queiroz Mendes, F., Rodrigues Reis, M., Passos, F. R., de Carvalho, A. M. X, Rodrigues de Oliveira Rocha, K. and Garcia Pinto, F. *Afr. J. Agric. Res.* 2017;12: 221–228.

Anderson, K. A. and Smith, B. W. J. *Agric. Food Chem.* 2002; 50: 2068–2075.

Andrade Korn, M. das G., da Boa Morte, E. S., Batista dos Santos, D. C. M., Teixeira Castro, J., Tiago Pereira Barbosa, J., Paixão Teixeira, A., Pires Fernandes, A., Welz, B., Piler Carvalho dos Santos, W., Batista Guimarães Nunes dos Santos, E. and Korn, M. *Appl. Spectrosc. Rev.* 2008;43:67–92.

Anthemidis, A. N. and Pliatsika, V. G. J. *Anal. At. Spectrom.* 2005;20:1280–1286.

Anzueto, F., Baumann, T. W., Graziosi, G., Piccin, C. R., Söndahl M. R. and van der Vossen, H. A. M. 2005. 21–86. Publisher: Amsterdam: Elsevier Academic Press.

Arruda, M. A. Z. 2007. Publisher: New York: Nova Science Publishers.

Árvay, J., Šnirc, M., Hauptvogl, M. Bilčíková, J., Bobková, A., Demková, L., Hudáček, M., Hrstková, M., Lošák, T., Král, M., Kováčik, A. and Štefániková, J. *Biol. Trace Elem. Res.* 2019;190:226–233.

Ashu, R. and Chandravanshi, B. S. *Bull. Chem. Soc. Ethiop.* 2011;25:11–24.

Barbosa, R. M., Batista, B. L., Varrique, R. M., Coelho, V. A., Campiglia, A. D. and Barbosa Jr., F. *Food Res. Int.* 2014;61:246–251.

Berg, T., Petersen, A., Pedersen, G. A., Petersen, J., and Madsen, C. *Food Addit. Contam.* 2000;17:189–196.

Bertrand, B., Villarreal, D., Laffargue, A., Posada, H., Lashermes, P., and Dussert, S. J. *Agric. Food Chem.* 2008;56: 2273–2280.

Bicho, N. C., Leitão, A. E., Ramalho, J. C. and Lidon, F. C. *Eur. Food Res. Technol.* 2011;233:303–311.

Bitter, N. Q., Fernandez, D. P., Driscoll, A. W., Howa, J. D. and Ehleringer, J. R. *Food Chem.* 2020;320:126602.

Boumans, R. W. J. M. 1987. Publisher: Weinheim: John Wiley & Sons.

Campos-Vega, R., Loarca-Piña, G., Vergara-Castañeda, H. A. and Oomah, B. D. *Trends Food Sci. Tech.* 2015;45:24–36.

Carter, J. F., Yates, H. S. A. and Tinggi, U. J. *Agric. Food Chem.* 2015;63:5771–5779.

Castro, J. T., Santos, E. C., Santos, W. P. C., Costa, L. M., Korn, M., Nóbrega, J.A. and Korn, M. G. K. *Talanta.* 2009;78:1378–1382.

Cheong, M. W., Tong, K. H., Ong, J. J. M., Liu, S. Q., Curran, P. and Yu, B. *Food Res. Int.* 2013;51:388–396.

Clarke, R. J. 1985, pp. 42–82. Publisher: Dordrecht, The Netherlands: Springer.

Clarke, R. J. 2003. 1493–1498. Publisher: Amsterdam: Elsevier Academic Press.

Clarke, R. J. and Vitzthum, O. G. 2001. Publisher: Blackwell Science Ltd.

Cruz, R., Cardoso, M. M., Fernandes, L. Oliveira, M., Mendes, E., Baptista, P., Morais, S. and Casal, S. *Agric. Food Chem.* 2012;60:7777–7784.

Cruz, R., Morais, S. and Casal, S. 2015. 549–558. Publisher: Amsterdam: Elsevier Academic Press.

Cunningham, W. C. and Stroube, W. B. *Sci. Total Environ.* 1987;63:29–43.

Debastiani, R., dos Santos, C. E. I., Yoneama, M. L., Amaral, L. and Dias, J. F. *Nucl. Instrum. Methods Phys. Res. B.* 2014;318:202–206.

Debastiani, R., dos Santos, C. E. I., Ramos, M. M., Sobrosa Souza, V., Amaral, L., Yoneama, M.L. and Ferraz Dias, J. *Food Res. Inter.* 2019;119:297–304.

De Nadai Fernandes, E. A., Tagliaferro, F. S., Azevedo-Filho, A. and Bode, P. *Acred. Quall Assur.* 2002;7:378–387.

dos Santos, E. J. and de Oliveira, D. J. *Food Compos. Anal.* 2001;14:523–531.

dos Santos, J. S., Santos, M. L. P., Conti, M. M., dos Santos, S. N. and de Oliveira, E. *Food Chem.* 2009;115:1405–1410.

dos Santos, J. S., Santos, M. L. P. and Conti, M. M. *J. Braz. Chem. Soc.* 2010;21:1468–1476.

Drivelos, S. A. and Georgiou, C. A. *Trends Anal. Chem.* 2012;40:38–51.

Fernandes, A. P., Santos, M. C., Lemos, S. G., Ferreira, M. M. C., Nogueira, A. R. A. and Nóbrega, J. A. *Spectrochim. Acta Part B At. Spectrosc.* 2005;60:717–724.

Ferreira, S. L. C., Miró, M., da Silva, E. G. P., Matos, G., Reis, P., Brandao, G., Santos, W., Duarte, A., Vale, M. and Araujo, R. *Appl. Spectrosc. Rev.* 2010;45:44–62.

Filho, V. R. M., Polito, W. L. and Neto, J. A. G. *J. Braz. Chem. Soc.* 2007;18:47–53.

Flores, É. M. M. 2014. Publisher: Amsterdam: Elsevier.

Fraňková, A., Drábek, O., Havlík, J., Száková, J. and Vaněk, A. *J. Inorg. Biochem.* 2009;103:1480–1485.

Georgiou, C. A. and Danezis, G. P. 2015. 132–224. Publisher: Amsterdam: Elsevier.

González, A. and de la Guardia, M. 2013. 51–76. Publisher: Amsterdam: Elsevier.
Grembecka, M., Malinowska, E. and Szefer, P. *Sci. Total Environ.* 2007;383:59–69.
Gure, A, Chandravanshi, B. S. and Godeto, T. W. *Bull. Chem. Soc. Ethiop.* 2018;32:27–38.
Habte, G., Hwang, I. M., Kim, J. S. Hong, J. H., Hong, Y. S., Choi, J. Y., Nho, E. Y., Jamila, N., Khan, N. and Kim, K. S. *Food Chem.* 2016;212:512–520.
Hamidatou, L., Slamene, H., Akhal, T., and Zourane, B. 2013. Publisher: IntechOpen.
Hoenig, M. 2003. 235–255. Publisher: Amsterdam: Elsevier.
Hoenig, M. 2005. 131–136. Publisher: Amsterdam: Elsevier.
ICO. 2019. *International Coffee Organization.* www.ico.org (accessed November 13, 2020).
Illy, A. 2005. 1–20. Publisher: Amsterdam: Elsevier Academic Press.
Jaganyi, D. and Madlala, S. P. J. *Sci. Food Agric.* 2000;80:85–90.
Jarošová, M., Milde, D., and Kuba, M. *Czech J. Food Sci.* 2014;32:354–359.
Jeszka-Skowron, M., Stanisz, E., and De Peña, M. P. *LWT.* 2016;73:243–250.
Kovalcik, A., Obruca, S., and Marova, I. *Food Bioprod. Process.* 2018;110:104–119.
Krivan, V., Barth, P. and Morales, A. F. *Mikrochim. Acta.* 1993;110:217–236.
Liu, H., You, C., Chen, C., Liu, Y. and Chung, M. *Food Chem.* 2014;142:439–445.
Luykx, D. M. A. M. and van Ruth, S. M. *Food Chem.* 2008;107:897–911.
Magalhães, C. E. C., Lima, E. C., Krug, F. J. and Arruda, M. A. Z. *Microchim. Acta.* 1999;132:95–100.
Martín, M. J., Pablos, F., and González, A. G. *Anal. Chim. Acta.* 1998;358:177–183.
Martín, M. J., Pablos, F. and González, A. G. *Food Chem.* 1999;66:365–370.
Mohammed, F., Guillaume, D., Dowman, S. and Abdulwali, N. *Microchem. J.* 2019;145:173–179.
Mussatto, S. I., Carneiro, L. M., Silva, J. P. A., Roberto, I. C., Teixeira, J. A. *Carbohydr. Polym.* 2011;83(2):368–374.
Olesik, J. W. *Anal. Chem.* 1991;63:12A–21A.
Oleszczuk, N., Castro, J. T., da Silva, M. M., Korn, M. G. A., Welz, B. and Vale, M. G. R. *Talanta.* 2007;73:862–869.
Oliveira, M., Casal, S., Morais, S., Alves, C., Dias, F., Ramos, S., Mendes, E., Delerue-Matos, C. and Oliveira, M. B. P. P. *Food Chem.* 2012;130:702–709.
Oliveira, M., Ramos, S., Delerue-Matos, C. and Morais, S. *Food Chem.* 2015;177:330–338.
Onianwa, P. C., Adetola, I. G., Iwegbue, C. M. A., Ojo, M. F. and Tella, O. O. *Food Chem.* 1999;66:275–279.
Petracco, M. 2005. 290–315. Publisher: Amsterdam: Elsevier Academic Press.
Pigozzi, M. T., Passos, F. R., Mendes, F. Q. O. and Horta, G. H. *J. Food Qual.* 2018;2018:1–7.
Pohl, P., Stelmach, E., Welna, M. and Szymczycha-Madeja, A. *Food Anal. Methods.* 2013;6:598–613.
Pohl, P., Stelmach, E. and Szymczycha-Madeja, A. *Food Chem.* 2014a;163:31–36.
Pohl, P., Stelmach, E. and Szymczycha-Madeja, A. *Food Anal. Methods.* 2014b;7:676–682.
Pohl, P., Szymczycha-Madeja, A., Stelmach, E. and Welna, M. *Talanta.* 2016;160:686–693.
Pohl, P., Szymczycha-Madeja, A. and Welna, M. *Food Chem.* 2018;263:171–179.
Ribeiro, A. S., Moretto, A. L., Arruda, M. A. Z. and Cadore, S. *Microchim. Acta.* 2003;141:149–155.
Rodrigues, C., Maia, R., Miranda, M., Ribeirinho, M., Nogueira, J. and Máguas, C. J. *Food Compos. Anal.* 2009;22:463–471.
Rodrigues, C., Máguas, C. and Prohaska, T. *Eur. Food Res. Technol.* 2011a;232:361–373.
Rodrigues, C., Brunner, M., Steiman, S., Bowen, G. J., Nogueira, J. M., Gautz, L., Prohaska, T. and Máguas, C. J. *Agric. Food Chem.* 2011b;59:10239–10246.
Rodrigues, C., Maia, R., Lauteri, M., Brugnoli, E. and Máguas, C. 2013. 77–99. Publisher: Amsterdam: Elsevier.
Santato, A., Bertoldi, D., Perini, M., Camin, F. and Larcher, R. J. *Mass Spectrom.* 2012;47:1132–1140.
Santos, W. P. C., Hatje, V., Lima, L. N., Trignano, S. V., Barros, F., Castro, J. T. and Korn, M. G. A. *Microchem. J.* 2008;89:123–130.
Schmidt, C. A. P., Miglioranza, E., Nagashima, G. and Grecco, F. *Cienc. Rural.* 2009;39:1591–1594.
Şemen, S., Mercan, S., Yayla, M. and Açıkkol, M. *Food Chem.* 2017;215:92–100.
Smith, J. S. and Thakur, R. A. 2017. 165–181. Publisher: Springer.
Stelmach, E., Pohl, P. and Szymczycha-Madeja, A. *Food Chem.* 2013;141:1956–1961.
Stelmach, E., Pohl, P. and Szymczycha-Madeja, A. *Food Chem.* 2015;182:302–308.
Stelmach, E., Szymczycha-Madeja, A. and Pohl, P. *Food Chem.* 2016;197:388–394.
Suseela, B., Bhalke, S., Vinod Kumar, A., Tripathi, R. M. and Sastry, V. N. *Food Addit. Contam.* 2001;18:115–120.
Szymczycha-Madeja, A., Welna, M. and Pohl, P. *Food Addit. Contam. Part A.* 2015;32:1488–1500.
Szymczycha-Madeja, A., Pohl, P., Welna, M., Stelmach, E. and Jedryczko, D. *Food Res. Int.* 2016;89:958–966.
Tagliaferro, F. S., de Nadai Fernandes, E. A. and Bacchi, M. A. J. *Radioanal. Nucl. Chem.* 2006;269:371–375.
Tagliaferro, F. S., De Nadai Fernandes, E. A., Bacchi, M. A., Bode, P. and De Franca, J. E. J. *Radioanal. Nucl. Chem.* 2007;271:371–375.
Trindade, A.C., Araújo, S.A., Amorim, F.A.C. Silva, D. S., Alves, J. P. S., Trindade, J. S., Aguiar, R. M. and Bezerra, M. A. *Food Anal. Methods.* 2020;13:203–211.
Trugo, L. C. 2003. 1498–1506. Publisher: Amsterdam: Elsevier Academic Press.
Twyman, R. M. 2005. 146–153. Publisher: Amsterdam: Elsevier Academic Press.
Valentin, J. L. and Watling, R. J. *Food Chem.* 2013;141:98–104.
Van Cuong, T., Hong Ling, L., Kang Quan, G. Jin, S., Shu Jie, S., Le Linh, T. and Duc Tiep, T. *Acta Universitatis Cibiniensis Series E: Food Technol.* 2014;18:19–34.
Vázquez, M. and Vélez, D. 2015. 699–708. Publisher: West Sussex: John Wiley & Sons.
Vega-Carrillo, H. R., Iskander, F. Y. and Manzanares-Acuna, E. J. *Radioanal. Nucl. Chem.* 2002;252:75–80.
Welna, M., Szymczycha-Madeja, A. and Zyrnicki, W. *Anal. Lett.* 2013;46:2927–2940.

Welz, B. and Sperling, M. 1999. Publisher: Weinheim: Wiley-VCH.
Windisch, J., Keppler, B. K., and Jirsa, F. *ACS Omega*. 2020;5:15335–15343.
Worku, M., Ram Upadhayay, H., Latruwe, K., Taylor, A., Blake, W., Vanhaecke, F., Duchateau, L. and Boeckx, P. *Food Chem*. 2019;290:295–307.
Yeung, V., Miller, D. D., and Rutzke, M. A. 2017. 129–150. Publisher: Springer International Publishing.
Zaidi, J., Fatima, I., Arif, M. and Qureshi, I. H. J. *Radioanal. Nucl. Chem*. 2006;267:109–112.
Zhao, Y., Zhang, B., Chen, G., Chen, A., Yang, S. and Ye, Z. *Food Chem*. 2014;145:300–305.

6

Coffee fermentation: New approaches to enhance quality

Rosane Freitas Schwan, Nádia Nara Batista, Silvia Juliana Martinez, Ana Paula Pereira Bressani, and Disney Ribeiro Dias

CONTENTS

6.1 Introduction

Coffee is one of the most popular beverages globally, and changes in production, processing, trading, appreciation, and consumer culture are noticeable (Guimarães et al., 2019). Significant changes in the history of the coffee market were divided into "waves" (Skeie, 2003) and described different moments of coffee consumption/production in the world.

During the Industrial Revolution, the first wave occurred due to advances in processes and significantly increased the beverage's global consumption. Coffee production was high, and the main reason for its consumption was the benefits that caffeine provided (Quintão and Brito, 2016; Guimarães et al., 2019; Ardekani and Rath, 2020). The second coffee wave was marked by the beginning of specialty coffees, bringing a quality standard of big brands. Arabica coffee blends and espresso machines gained prominence, and the number of coffee shops (such as Starbucks) doubled in the United States (Guimarães et al., 2019; Ardekani and Rath, 2020). The third wave started around the 2000s, in which consumers were searching for higher quality and differentiation of aromas and flavors. The unique origin, light or medium roast, and different extraction methods became especially important. Moreover, the third coffee wave paves a way for new knowledge on specialty coffees, which significantly impacted consumers worldwide and made them more aware of its traceability (Guimarães et al., 2016). A fourth wave movement now arises that involves the interaction between producers, coffee processors, micro-roasters, baristas, and consumers.

Coffee microbiota comprise a wide variety of bacteria, yeast, and filamentous fungi (Silva et al., 2008a). Their primary function is to reduce the mucilage layer and produce

DOI: 10.1201/9781003043133-8

pectinolytic enzymes and flavor precursors. Coffee fermentation was previously correlated to low quality because the process was unknown, and there was a lack of monitoring that induced filamentous fungi proliferation. Later, new approaches and findings (the honey process, inoculation of starter cultures, and bioreactors) gave the fermentative process a new view, which farms started adopting (Silva et al., 2013; Martinez et al., 2017; Bressani et al., 2018; Hii and Borém, 2019).

This chapter describes innovations in coffee fermentation linked to quality enhancement, dividing them into four main topics:

1. Microbiota: factors that influence the coffee microbiota, similarities, and differences in each microbial group (bacteria, yeasts, and filamentous fungi).
2. The influence of processing and bioreactors on starter cultures: the choice of starter cultures according to the producing region, altitude, new fermentation processes, novel bioreactors, and microbial co-cultivation.
3. Fermentation and differences in chemical compounds related to coffee quality: chemical composition by different methods and their relationship with the beverage quality.
4. Different sensory methodologies: the preferences and desires of specialty coffee consumers.

6.2 Coffee fermentation

Fermentation is a biochemical process that converts carbohydrates into primary products. The objective of coffee fermentation is to reduce the drying time, boost the flavor precursors, and increase the quality. During fermentation, many biochemical reactions take place outside and inside the beans. Microbial activity is intense in the outer layers (skin, pulp, mucilage) of coffee seeds. The main microbial groups of coffee involve bacteria, yeast, and filamentous fungi (Frank et al., 1965; Silva et al., 2000; Silva, 2015). The variations of microbial genera and species among coffee-producing areas are probably due to environmental conditions, coffee varieties, and processing methods, consequently affecting the beverage quality. The processing methods vary from countries or farms, depending on climate conditions, fruit maturation, capital availability, machinery, and technology (Schwan et al., 2012). Coffee processing comes after harvesting and fruit selection. The coffee processes are known as the natural (or dry), wet, and pulped natural (or semi-dry) processes; a detailed illustration is shown in Figure 6.1.

The natural coffee process, also known as the dry process, is one of the oldest and most straightforward, and is currently used in Brazil, Ecuador, Cameroon, and Ethiopia. It is applied in countries where prolonged sunshine is available, and rainfall is limited. This process uses whole coffee fruits that can either be non-selected or selected, and this selection is made

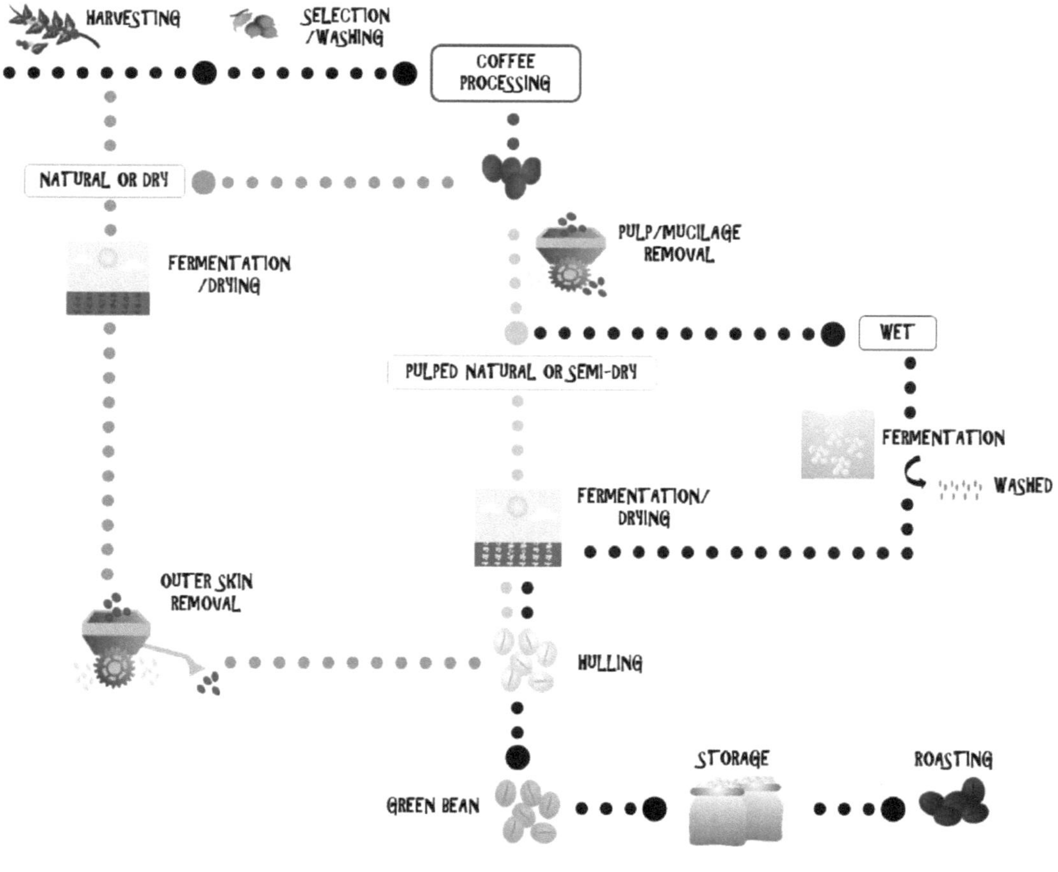

FIGURE 6.1 Steps of dry or natural, wet, and semi-dry coffee processing after harvesting.

manually or by density through washing. Then fruits are directly placed in platforms or on cement patios for fermentation and drying.

During wet coffee processing, after selection, mature fruits are mechanically depulped and sometimes demucilated. Then beans are transferred to tanks with water for fermentation (typically lasting 24–48 h). After fermentation, beans are washed or stay unwashed and then are dried in platforms, on cement patios, or in mechanical dryers. This process is traditionally performed in Colombia, Australia, Kenya, Guatemala, and Tanzania.

The pulped natural coffee process, also known as a semi-dry process, is considered a hybrid between the natural and wet process because fruits are depulped and demucilated as in the wet process, but fermentation is done without water while drying as in the natural process. In this process, coffee is fermented while drying in covered or uncovered platforms or on cement patios.

In Brazil, a new coffee type known as "honey coffee" is derived from the pulped natural process. It requires fruit depulping and demucilation but does not use water for fermentation. Honey coffee emerged as an alternative to change the sensory profile. Different honey coffees are obtained depending on the fermentation hours and drying (yellow, red, and black honey; see Figure 6.2; Hii and Borém, 2019). Yellow honey is depulped and demucilated right before fermentation and dried under the sun with the slightest shade cover, which is the same coffee obtained from the pulped coffee process. At first, for red and black honey, coffee is fermented as fruits (red lasts 48–96 h and black for >96 h), and then they are depulped and demucilated. After fermentation, beans are dried with shade covers until reaching 11%–12% moisture.

Innovative fermentation methods emerged to enhance coffee quality. The first method was tested in Evangelista et al. (2014b), known as direct spraying, (i.e., spraying the microbial solution over the coffee fruits or beans). The second was fermentation with yeasts in open batches for specific hours, as reported by Bressani et al. (2018) and Martinez et al. (2017). This method consists of fermenting coffee in open buckets or bioreactors that favor aerobic conditions and consistency in microbial dynamics. The third innovative method was fermentation in close batches, as published by da Mota et al. (2020) and Martins et al. (2020), which is called self-induced anaerobic fermentation (SIAF). This coffee fermentation method favors anaerobic conditions, yeast activity, and the release of CO_2 due to yeast metabolic activity. All these recently studied methods increased the sensory scores and modified the beverage attributes. Compared to the first method, the sensory scores of the second and third methods are much higher when using wild selected microbial starters.

For all fermentative processes, the length of fermentation varied according to temperature and environmental condition in farms, varying from 8 h to 216 h (Silva et al., 2008a; Vilela et al., 2010; Evangelista et al., 2015; Zhang et al., 2019a, 2019b; Elhalis et al., 2020). The ideal temperature for fermentation oscillates between 25 and 30°C.

The drying process is preferably done under the sun because it maintains quality, mainly in Brazil, Australia, and Ecuador that have prolonged hours of sunshine. Countries that grow coffee in regions with high moisture content (e.g., Colombia, Guatemala) use more mechanical dryers. These processes last 15–20 days until the moisture content reaches 11%–12% (Brando and Brando, 2015). Once beans are dried, parchment from the coffee is removed by hulling to obtain the green beans. Then they are stored and roasted.

6.3 Microbiota

The microbiota involved in fermentation comes from different sources and are considered natural contaminants. Those sources include soil, tools, machinery, plants, natural streams or rivers, rainfall, and workers. Moreover, they are the main motors of fermentation since they produce fruity and floral volatiles and organic acids, such as acetic and lactic acids. Besides producing flavor compounds, microbial activity has other roles, like producing pectinolytic enzymes (Agate et al., 1966; Avallone et al., 2002; Silva et al., 2013; Haile and Kang, 2019) and inhibiting the growth of toxigenic fungi (Masoud et al., 2005; Ramos et al., 2010; Souza et al., 2017).

Apart from the processing methods, other factors cause changes in the microbiota profile, such as coffee variety, fruit

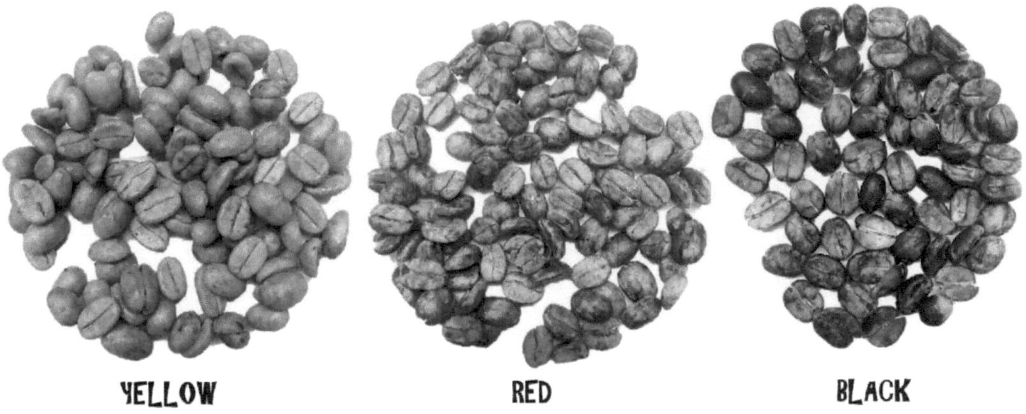

FIGURE 6.2 Types of "honey coffee" after drying.

maturation, season, altitude, moisture, temperature, coffee washing, location/region, water type, and carbohydrate pulp and mucilage.

Coffee microbiota is complex: bacteria is dominant at the initial fermentation stages; then, when water activity reduces, yeast dominates, and later filamentous fungi (Silva et al., 2008a). The microbial population in a spontaneous fermentation from natural processing oscillates from 2.9 to 4.7 Log CFU/g for viable bacteria, from less than 1 to 7.2 Log CFU/g for yeast (Evangelista et al., 2014b), and from 3.2 to 5.2 Log CFU/g for filamentous fungi (Silva et al., 2008b). Spontaneous fermentation in wet processing contains a population of 3.8–8.5 Log CFU/g for mesophilic bacteria and 2–4.9 Log CFU/g for yeast (Evangelista et al., 2015). As for pulped coffee, the population of viable bacteria oscillates from 3.8 to 7 Log CFU/g, yeast from 3.7 to 6.9 Log CFU/g, and filamentous fungi from 2 to 3.7 Log CFU/g (Vilela et al., 2010).

Microbial populations are quantified and then isolated to the knowledge of the microbiota involved during fermentation. Most studies used complex synthetic media to isolate cultivable microorganisms (Table 6.1). However, media containing coffee pulp can study fungal and bacterial populations, as in Gaime-Perraud et al. (1993). As illustrated in Table 6.1, the most common media use to obtain isolates from

TABLE 6.1

Mediums for Isolation and Type of Identification Used in Coffee Fermentations

Media	Type of microorganism			Isolates identification	References
	B	YY	FFF		
Nutrient agar (NA)	■			Bergey's Manual of Determinative Bacteriology, sequencing	Agate and Bath, 1966; Vilela et al., 2010; de Melo Pereira et al., 2015; Martins et al., 2020
Glucose Yeast Extract (GYE) agar		■		Bergey's Manual of Determinative Bacteriology	Agate and Bath, 1966
Wallerstein Laboratory (WL)/Differential (WLD) agar	■			Biochemical tests, Bactray Kits I, II, and III	Silva et al., 2000; Silva et al., 2008a
Eosin Methylene Blue (EMB) agar	■			Biochemical tests, Bactray Kits I, II, and III	Silva et al., 2000; Silva et al., 2008a
De Man, Rogosa, and Sharpe (MRS) agar	■			Biochemical tests, Bactray Kits I, II, and III, API system, sequencing	Silva et al., 2000; Avallone et al., 2001; Silva et al., 2008a; Vilela et al., 2010; Djossou et al., 2011; Leong et al., 2014; Evangelista et al., 2015; Zhang et al., 2019a; Martins et al., 2020
YW agar		■	■	Dichotomous keys, microscope	Silva et al., 2000
Yeast Glucose Chloramphenicol (YGC) agar		■		API system	Avallone et al., 2001
Borras pectin	■			API system	Avallone et al., 2001
Malt Yeast Glucose Peptone (MYGP) agar		■		Sequencing	Masoud et al., 2004
Dicloran Glycerol 18% (DG18) agar		■	■	Dichotomous keys, microscope, sequencing	Silva et al., 2008a; Silva et al., 2008b; Vilela et al., 2010
Malt extract agar		■	■	Sequencing	Vilela et al., 2010
Blood agar	■			Biochemical tests, BBL Crystal kits	Puerta Q et al., 2012
Plate Count agar	■			Biochemical tests, BBL Crystal kits, API system, sequencing	Puerta Q et al., 2012; Avallone et al., 2001; Silva et al., 2008a; Evangelista et al., 2015
Sabouraud Dextrose agar		■	■	Biochemical tests, Minitek kits	Puerta Q et al., 2012
Yeast Extract Peptone Glucose (YEPG) agar		■		Sequencing	de Melo Pereira et al., 2014; Evangelista et al., 2015; de Melo Pereira et al., 2015; de Carvalho Neto et al., 2017a; Martins et al., 2020
Yeast extract Peptone Dextrose (YPD) agar		■		Sequencing	Evangelista et al., 2015; Haile and Kang, 2019
Modified Deoxycholate Mannitol Sorbitol (mDMS) agar	■			Sequencing	Zhang et al., 2019b
Yeast Glucose (YG) agar		■		Sequencing	Zhang et al., 2019b

Notes: B = bacteria; Y = yeast; FF = filamentous fungi.

different bacterial groups are nutrient agar (mesophilic bacteria); De Man, Rogosa, and Sharpe agar (lactic acid bacteria); and Eosin Methylene Blue agar (Enterobacteriaceae). Yeasts are commonly isolated from Yeast Extract Peptone Glucose agar and Dicloran Glycerol 18% agar. This last medium is also commonly used for the isolation of filamentous fungi. Plate Count agar is used to isolate aerobic microbiota and Meat Liver SR agar to isolate anaerobic microbiota (Avallone et al., 2001).

For a while, culture non-dependent techniques have been used to identify microbiota changes during processing, fermentation, and drying (Vilela et al., 2010; Evangelista et al., 2014b; Hamdouche et al., 2016; De Bruyn et al., 2017). These techniques are based on DNA extraction and sequencing and are performed in isolates or whole coffee samples. Among them, the most well-known techniques are polymerase chain reaction denaturing gradient gel electrophoresis (PCR-DGGE) fingerprinting, Sanger sequencing, and high-throughput Illumina sequencing. The groups that have been identified include aerobic spore-forming bacteria, lactic acid bacteria (LAB), Enterobacteriaceae, and other gram-negative bacteria, including acetic acid bacteria (AAB); yeasts; and filamentous fungi (Vilela et al., 2010; Evangelista et al., 2014b, 2015; Hamdouche et al., 2016; Junqueira et al., 2019; Zhang et al., 2019a, 2019b; De Bruyn et al., 2017; Elhalis et al., 2020).

6.3.1 Bacteria

Bacteria are mainly responsible for producing organic acids, such as lactic and acetic acid, which contribute to pH decrease (Silva, 2015; Martinez et al., 2019), encouraging the growth of other microbial groups. Bacteria become dominant when there is high water activity (~0.85) during fermentation and a pH value of 6.5 (Silva, 2015; Silva et al., 2008a). Reports have shown that some bacterial species can excrete pectinolytic enzymes (Avallone et al., 2002).

6.3.1.1 Natural process

The bacterial diversity from natural processing is shown in Table 6.2. Silva (2015) has reported population and aerobic spore-forming bacteria species from natural processing, *Bacillus* being the most representative genus. This genus produces spores and extracellular enzymes that break down cellulose and pectin (Silva et al., 2013), turning this last into a benefit because it can help reduce the mucilage layer. For this process, *Bacillus* species have only been identified in Brazil (Table 6.2). The common species are *B. cereus, B. macerans, B. subtilis, B. anthracis, B. stearothermophilus, B. polymyxa*, and *B. megaterium* (Silva et al., 2000, 2008a).

The intact coffee fruits also have a LAB presence. LAB produces lactic acid, caramel, acidic, and fruity notes and are potential filamentous fungi inhibitors (Arunga, 1992; Djossou et al., 2011; Wang et al., 2019). There are two LAB groups: homofermentative and heterofermentative. Homofermentative LAB converts the energy source (carbohydrate) into lactic acid via the pyruvate pathway; an example is *Lactobacillus plantarum*, already identified in Ecuadorian fermentations (Table 6.2). Heterofermentative LAB, apart from lactic acid, produces ethanol as well as short-chain fatty acids. LAB species *Leuconostoc mesenteroides* and *L. pseudomesenteroides* were dominant in Ecuadorian standard natural process coffees, with abundances from 30% to more than 50% (De Bruyn et al., 2017). Unlike other Brazilian regions, the Caparaó region has abundances over 50% of *Leuconostoc mesenteroides* in coffees from 800, 1,000, 1,200, and 1,400 m altitude and over 25% of *Weissella paramesenteroides* in coffees from 1,000 m altitude (Martins et al., 2020). In addition, LAB population from close batch spontaneous fermentations varied from altitude: 7.1–9.5 Log CFU/g (800 m), 6.2–8.5 Log CFU/g (1,000 m), 4.3–7.8 Log CFU/g (1,200 m), and 2–7.8 Log CFU/g (1,400 m; Martins et al., 2020). Other lactic acid bacteria found in this process are shown in Table 6.2.

Fermentations of 72 h have varying Enterobacteriaceae populations between 5.39 and 8.16 Log CFU/g (Silva, 2015). The species that have been previously identified belong to *Enterobacter dissolvens, E. aerogenes, E. cloacae*, and *Klebsiella* spp. (Pee and Castelein, 1972; Silva et al., 2000, 2008a). Different species from the genus *Pantoea* have only been identified in Brazil; refer to Table 6.2. However, this genus was present in coffees from Cameroon (Hamdouche et al., 2016). SIAF favors *Pantoea agglomerans* growth, especially in coffee from 1,200 m altitude (abundance of 25%–50%; Martins et al., 2020).

According to De Bruyn et al. (2017), the genera *Acetobacter* and *Gluconobacter* take part of the dominant microbiota in this process. AAB's role is not well known, but these bacteria might produce acetic acid that helps the pH value decrease in aerobic fermentation and sometimes remains a flavor note in the beverage. Their growth depends on the degree of oxygen, ethyl alcohol, and reduced sugar content. The only gram-negative genus identified in a Cameroonian process was *Chryseobacterium* (Hamdouche et al., 2016).

6.3.1.2 Wet process

Water temperatures might vary between 16 and 21 °C (Ribeiro et al., 2018). LAB bacteria mainly dominate it because of the high water activity. LAB population oscillates between 2.5 and 7.5 Log CFU/g, varying from a country (Avallone et al., 2001; Evangelista et al., 2015). The LAB species that are frequently found belong to genera *Leuconostoc, Lactococcus, Lactobacillus*, and *Weissella. Leuconostoc citreum, L. mesenteroides, L. pseudomesenteroides, Lactobacillus plantarum*, and *Lactococcus lactis* have been identified in Brazilian, Mexican, Ecuadorian, Australian, Thai, and Indonesian coffee (Avallone et al., 2001; Vilela et al., 2010; Evangelista et al., 2015; Nasanit and Satayawut, 2015; Hamdouche et al., 2016; de Bruyn et al., 2017; Elhalis et al., 2020; Hatiningsih et al., 2018; Ribeiro et al., 2018; Zhang et al., 2019b; Pothakos et al., 2020).

Mesophilic populations are widely evaluated in coffee because they are highly diverse; in wet fermentation, reaching a population of 3.5–8.5 Log CFU/g (Evangelista et al., 2015). The *Enterobacter* genus usually takes part in wet fermentation and performs mixed acid fermentations. In Colombia, this genus dominated up to 60% of the wet fermentations with

TABLE 6.2

Bacterial Diversity Identified in Different Countries and Coffee Processes

Genus/species	Country																																	Reference
	B			M			C			E			A			Ca			I			Th			In			K			Ch			
	D	W	S	D	W	S	D	W	S	D	W	S	D	W	S	D	W	S	D	W	S	D	W	S	D	W	S	D	W	S	D	W	S	
Acetobacter		■									■			■																				l, r, s
A. cibinogensis											■																							r
A. fabarum											■																							r
A. ghanensis											■																							r
A. lambici											■																							r
A. malorum											■																							r
A. okinawensis											■																							r
A. orientalis											■																							r, t
A. persici											■			■																				r, s, t
A. senegalensis											■																							r
A. thaillandicus											■																							r
Acinetobacter		■	■								■			■																				e, l, s, u
Acinetobacter sp.	■																																	b
A. lwoffii														■																				s
A. pittii	■																																	f
A. radioresistens			■																															f, v
A. schindleri		■																																k
A. soli											■																							e
Actinobacterium sp.		■																																k
Aeromonas					■																													g
Agrobacterium		■									■																							l, t
A. tumefaciens	■																																	f
Anabaena														■																				s
Asaia sp.														■																				s
Arthrobacter	■																																	a
Arthrobacter sp.	■																																	b
Bacillus		■																																l
Bacillus sp.			■																		■													u, v
B. subtilis	■		■																				■											a, b, n, u
B. amyloliquefaciens		■																																k
B. anthracis	■																																	a

B. cereus	a, b, u
B. macerans	b, u
B. megaterium	a, u
B. polymyxa	b
B. stearothermophilus	a
Brevibacillus parabrevis	s
Burkholderia	l
Cedecea sp.	a
Cellulomonas	a
Cellulosimicrobium	t
C. cellulans	e, t
C. funkei	t
Chromobacter violaceum	a
Chryseobacterium	d, l, t
C. bovis	k
C. taichungense	k
Citrobacter	l, r
C. freundi	a
C. koseri	s
Clostridium	l, r
Curtobacterium sp.	k
C. flaccumfaciens	f
Dermabacter	a
Devosia	l
Enterobacter	l, o, r, s, t
Enterobacter sp.	e, k
E. aerogenes	a, f
E. asburiae	k
E. agglomerans	b, n, u
E. cloacae	a, f, k
E. gergoviae	a
E. ludwigii	k, s
E. lignolyticus	k
E. sakazakii	a

(*Continued*)

TABLE 6.2 (Continued)

Genus/species	Country																																	Reference
	B			M			C			E			A			Ca			I			Th			In			K			Ch			
	D	W	S	D	W	S	D	W	S	D	W	S	D	W	S	D	W	S	D	W	S	D	W	S	D	W	S	D	W	S	D	W	S	
Enterococcus											■			▒																				s, t
Enterococcus sp.		■																																k
E. casseliflavus																							■			■								m, n
Escherichia hermanni		■																																k
Erwinia		■			■						■			▒																				g, l, r, s
E. dissolvens																							■											n
E. herbicola			■		■																													g, u
E. persicina		■	■																															f, k
E. soli														▒																				s
Fimbriimonas		■																																l
Flavobacterium sp.																							■											v
F. odoratum	■							■																										a, i
Frateuria aurantia											■																							t
Fructobacillus		■									■																							l, r
Gluconobacter		■								■	■			▒																				e, l, r, s, t
G. albidus										■	■																							e, t
G. cerevisiae											■																							r
G. cerinus			■								■			▒																				f, s, t
G. japonicus											■																							r
G. kondonii											■																							r
G. nephelii											■																							r
G. oxydans										■	■																							e, t
Hafnia alvei	■																																	a
Hymenobacter		■																																l
Klebsiella								■			■																							i, r
K. pneumoniae			■		■					■	■			▒									■											e, g, n, s, u
K. oxytoca	■	■			■																													a, g, k
K. ozaenae	■				■																													a, g
Kluyvera																																■		o
Kocuria sp.		■																																k
Kozakia											■																							r

Lactobacillus	a, d, e, i, l, o, p, r, t
Lactobacillus sp.	v
L. brevis	g, m, n, r, t, u
L. coryniformis	o
L. hordei	r
L. plantarum	e, m, n, o, r, t, u
L. fermentum	d, k
L. vaccinostercus	r, t
Lactococcus	e, l, o, p, r, s, t
L. hircilactis	o, r
L. lactis	e, f, k, o, r, s, u
L. lactis subsp. lactis	d, t
Leuconostoc	e, o, p, r, s, t
Leuconostoc sp.	v
L. citreum	e, o
L. fallax	r
L. holzapfelii	o
L. inhae	t
L. lactis	e, t
L. mesenteroides	d, e, f, g, k, m, n, o, s, u
L. pseudomesenteroides	e, k, o, r, t
Methylobacterium	e, l, r, s, t
Microbacterium	a, t
Microbacterium sp.	k
M. foliorum	f
M. laevaniformans	k
M. testaceum	f
Micrococcus latis	f
M. luteus	f
Moraxella osloensis	f

(*Continued*)

TABLE 6.2 (Continued)

Genus/species	Country																																Reference
	B			M			C			E			A			Ca			I			Th			In			K			Ch			
	D	W	S	D	W	S	D	W	S	D	W	S	D	W	S	D	W	S	D	W	S	D	W	S	D	W	S	D	W	S	D	W	S	
Novosphingobium		■																																l
Ochrobactrum pseudogrignonense		■																																k
Oenococcus											■																							t
Paenibacillus amylolyticus		■																																k
Pantoea											■																					■		o, r, t
Pantoea sp.	■															■																		c, d
P. agglomerans	■	■	■																															k, u
P. brenneri	■																																	c
P. dispersa		■																																k
P. eucrina	■																																	c
Pasteurella haemolytica	■																																	a
Pediococcus		■						■			■			■																				l, p, s, t
P. pentosaceus											■																							r
Phenylobacterium		■																																l
Pseudomonas		■						■		■	■										■													e, l, p, t, v
Pseudomonas sp.		■																			■													k, v
P. paucimobilis	■																																	a
P. aeruginosa	■																																	a
P. cepacia	■				■			■																										a, g, i
P. fluorescens	■													■																				a, s
P. pseudoalcaligenes	■																																	a
P. vesicularis	■																																	a
Ralstonia sp.																	■																	d
Raoultella																																■		o
Rhizobium										■																								e
Rhodococcus														■																				s
Rhodoplanes		■																																l
Saccharibacillus														■																				s
Serratia		■												■																				l, s
Serratia sp.			■																															u
S. liquefaciens	■																																	a

S. marcescens	a, f, k, s
S. plymuthica	a
Sphingomonas	l, r
S. desiccabilis	f
Staphylococcus	l, t
S. epidermidis	f
Streptococcus	t
Streptomyces	l
Tatumella	t
Weissella	d, e, l, o, r, t
W. cibaria	k
W. confusa	k
W. paramesenteroides	f, t
W. soli	o
Xanthomonas oryzae	f
Yersinia sp.	b

Countries: A = Australia, B = Brazil, Ca = Cameroon, Ch = China, C = Colombia, E = Ecuador, I = India, In = Indonesia, K = Korea, M = Mexico, Th = Thailand. *Coffee processes*: D = dry, W = wet, S = semidry. a = Silva et al. (2000), b = Silva et al. (2008a), c = Evangelista et al. (2014a), d = Hamdouche et al. (2016), e = De Bruyn et al. (2017), f = Martins et al. (2020), g = Avallone et al. (2001), h = Masoud et al. (2004), i = Puerta Q et al. (2012), j = de Melo Pereira et al. (2014), k = Evangelista et al. (2015), l = de Carvalho Neto et al. (2017b), m = Hatiningsih et al. (2018), n = Nasanit and Satayawut (2015), o = Zhang et al. (2019a), p = Junqueira et al. (2019), r = Zhang et al. (2019b), s = Elhalis et al. (2020), t = Pothakos et al. (2020), u = Vilela et al. (2010), v = Velmourougane (2013).

coffee that was not selected (Puerta Q et al., 2012). Though in Australia, the Enterobacteriaceae was the most abundant in a 36 h fermentation (Elhalis et al., 2020), different results obtained in Ecuador were with 36 h and 64 h fermentation, and the most abundant bacteria were LAB genera *Leuconostoc* and *Lactobacillus* (Pothakos et al., 2020). Moreover, *Klebsiella* was found with frequency in this process and can perform butanediol fermentation. *Pantoea* is also another genus that takes part in this process, and recently, it has been identified in fermentations from Ecuador and Colombia (Junqueira et al., 2019; Pothakos et al., 2020).

Compared to the LAB group, low AAB populations have been reported in fermentations from an experimental coffee farm in China (around 2 Log CFU/g; Zhang et al., 2019a). In the AAB group, genera *Acetobacter, Acinetobacter*, and *Gluconobacter* occur the most, together with *Acetobacter persici* and *Gluconobacter cerinus* (Table 6.2). *Gluconobacter* abundance in Australian fermentation decreased within 24 h (Elhalis et al., 2020), while in Ecuador it decreased after 48 hours (De Bruyn et al., 2017). Other bacteria involved in wet coffee processing can be found in Table 6.2.

6.3.1.3 Pulped coffee process

The bacterial population from this process is always in higher proportions than other microbial groups during 25-h, 48-h, or 96-h fermentation (Vilela et al., 2010; Velmourougane, 2013). The diversity of cultivable microorganisms increases with higher altitudes (Martins et al., 2020). Compared to the natural process, the pulped coffee process has less microbial diversity since the load diminished with skin removal. During a 216 h spontaneous fermentation in Brazil, the bacterial population was around 3.8–7 Log CFU/g (Evangelista et al., 2014a), meaning they were metabolically active until the end of processing. In open batches, spontaneous fermentations have a mesophilic population of 4–4.5 Log CFU/g (Martinez et al., 2017), whereas close batches (SIAF) had a mesophilic population of 6.8–9.3 Log CFU/g (800 m), 6.2–9.4 Log CFU/g (1,000 m), 5.1–7.7 Log CFU/g (1,200 m), and 4.7–7.1 Log CFU/g (1,400 m; Martins et al., 2020). The isolates from Evangelista et al. (2014b) belonged to *Enterobacter agglomerans, Escherichia coli, Bacillus cereus, Bacillus megaterium, Bacillus macerans*, and *Lactobacillus plantarum*. Species that were not isolated were identified through PCR-DGGE belonging to *Enterobacter cowanii, L. plantarum, Pantoea agglomerans*, and *B. macerans* (Vilela et al., 2010). Only *Enterobacter aerogenes* and *Erwinia persicina* were isolated in the Caparaó region of Brazil (Martins et al., 2020).

LAB population in open batches was around 3.8–4.5 Log CFU/g and in close batches was 5.5–7.2 Log CFU/g (800 m), 4.5–6.9 Log CFU/g (1,000 m), 4.6–6.4 Log CFU/g (1,200 m), and 4.6–6.3 Log CFU/g (1,400 m; Martinez et al., 2017; Martins et al., 2020). The Arabica fermentation in India obtained higher total lactic counts, lactose positive bacteria, negative lactic bacteria, and *Pseudomonas* than Robusta coffee (Velmourougane, 2013).

Further bacterial identifications made in different countries and processes are summarized in Table 6.2.

6.3.2 Yeast

Yeasts metabolize sucrose, glucose, fructose, and citric acid at low pH values for their growth during fermentation. They are potential pectinolytic enzyme producers (Haile and Kang, 2019; Silva et al., 2013) and survive the acidic fermentation environment. Moreover, yeast increased the number of volatiles, produced alcohol volatiles, and increased the beverage quality (Evangelista et al., 2014a, 2014b; Martinez et al., 2019; Martins et al., 2019; Bressani et al., 2020a). The yeast diversity in different countries is shown in Table 6.3. *Hanseniaspora uvarum, Pichia kluyveri, Torulaspora delbrueckii*, and *S. cerevisiae* are the most referred species in different countries.

6.3.2.1 Natural process

Yeast starters dominating the fermentation process right after the bacterial population decreases (Silva et al., 2008a). In an early study, 90 yeasts were isolated from 15 farms south of Minas Gerais, Brazil. From that total, 38.9% belonged to *Pichia*, 22.3% to *Candida*, 18.9% to *Arxula*, and 9.7% to *Saccharomycopsis* (Silva et al., 2000). Silva et al. (2008a) identified the genera *Debaryomyces, Pichia, Candida*, and species *D. hansenii* and *P. guilliermondii* as part of the microbial succession. Evangelista et al. (2014b) fermented coffee spontaneously in trays, and results from PCR-DGGE showed a *Saccharomyces cerevisiae* persistence.

In an open batch, spontaneous fermentation *S. cerevisiae* was at a concentration of 3–5 Log cell/g (Bressani et al., 2018). In a SIAF done in the Caparaó region of Brazil, cultivable *S. cerevisiae* colonies were not identified at 800, 1,000, 1,200, and 1,400 m (Martins et al., 2020). Suggesting that the presence of this species changes depending on the fermentation method. Instead, there were high abundances of other yeast such as *Cystofilobasidium ferigula* in more than 50% in coffees from 1,000 and 1,200 m altitude and a total population of 4.51 Log CFU/g, followed by *Meyerozyma caribbica* with a total population of 3.83 Log CFU/g and *Wickerhamomyces anomalus* with a total population of 3.81 Log CFU/g. In coffees from Ecuador, among the fungal community, there were high abundances of *Pichia, Starmerella bacillaris*, and *Candida* spp. (de Bruyn et al., 2017).

6.3.2.2 Wet process

During fermentation, yeast cells adhere to the beans' mucilage layers; a clear view is displayed in Figure 6.3. Total yeast population from spontaneous fermentations oscillate from 2.0 to 4.8 Log CFU/g in Brazil, from 4.6 to 6.0 Log CFU/g in Mexico, from 6.2 to 9.1 Log CFU/g in Indonesia, and from 4 to 5.5 Log CFU/g in Australia (Avallone et al., 2001; Evangelista et al., 2015; Hatiningsih et al., 2018; Elhalis et al., 2020). Evangelista et al. (2015) detected *Hanseniaspora uvarum* and *Torulaspora delbrueckii* in higher numbers from two Brazilian farms. Additionally, *Pichia fermentans* was also detected in both farms and produced volatile compounds that interfere with the coffee beverage quality (Evangelista et al., 2015). Haile and Kang (2019) performed a 48-h fermentation in Korea, and the isolated species comprised *Wickerhamomyces*

TABLE 6.3
Yeast Diversity Identified in Different Countries and Coffee Processes

Genus/species	Country																																Reference
	B			M			C			E			A			Ca			T			I			Th			K			Ch			
	D	W	S	D	W	S	D	W	S	D	W	S	D	W	S	D	W	S	D	W	S	D	W	S	D	W	S	D	W	S	D	W	S	
Arxula sp.			■																															u
Arxula adeninivorans	■																																	a, b
Blastobotrys proliferans	■																																	a
Bensingtonia sp.								■																										p
Candida								■		■	■			■																				e, i, r, s
Candida sp.		■						■																		■								k, n, p
C. auringiensis	■																																	a
C. carpophila			■																															u
C. ernobii			■																															u
C. fermentati	■																																	b
C. glabrata		■																																j
C. glucosophila	■																																	a
C. humilis											■																					■		o, r
C. incommunis	■																																	a
C. membranifaciens	■		■																															b, u
C. paludigena	■																																	a
C. pseudointermedia																				■														h
C. quercitrusa		■								■	■																					■		e, j, o, r
C. railenensis														■																				s
C. saitoana	■																																	b
C. schatarii	■																																	a
C. solani																																■		o
C. tropicalis								■																										i
C. vanderwaltii																																■		p
C. vartiovaarae	■																																	a
C. xylopsoci														■																				s
Citeromyces matritensis	■																																	a
Cordyceps brongniartii																																■		o
Cryptococcus											■																							e, r
C. albidus					■																													g

(*Continued*)

TABLE 6.3 (Continued)

Genus/species	Country																																	Reference
	B			M			C			E			A			Ca			T			I			Th			K			Ch			
	D	W	S	D	W	S	D	W	S	D	W	S	D	W	S	D	W	S	D	W	S	D	W	S	D	W	S	D	W	S	D	W	S	
C. flavescens	■																																	f
C. laurentii					■																													g
C. randhawai			■																															f
C. terreus								■																										i
Cystofilobasidium										■				■																				e, s
Debaryomyces														■																				s
D. polymorphus	■																																	b
D. hansenii	■																																	b, f
Dipodascus tetrasporeus								■																										p
Geotrichum fermentans	■																																	a
Hanaella											■			■																				r, s
Hanseniaspora											■			■																		■		o, s, t
Hanseniaspora opuntiae		■																																j
H. uvarum	■		■								■			■		■				■														d, f, h, o, r, s, t, u
H. vineae																																■		o
Issatchenkia orientalis																				■														h
Kloeckera sp.			■																															u
K. apiculata			■																															u
K. apis apiculata					■																													g
Kluyveromyces sp.			■																							■								n, u
K. marxianus			■																	■														h, u
Lachancea lanzarotensis																																■		o
Malassezia sp.								■																										p
Martiniozyma asiatica								■																										p
Meyerozyma											■			■																				e, s
M. caribbica	■	■	■													■																		d, f, k
Papiliotrema														■																				s
P. flavescens																													■					q
P. terrestris																																■		o
Pichia										■	■																							e, t
Pichia sp.			■																							■								n, u

P. acaciae	a
P. anomala	a, b, h, u
P. burtonii,	b
P. caribbica	j, u
P. ciferii	a
P. fermentans	j, r, s
P. guilliermondii	b, g, j
P. jadinii	a
P. kluyveri	c, e, f, h, j, o, r
P. kudriavzevii	q, s, t
P. lynferdii	a
P. nakasei	p
P. ofunaensis	a
P. sydowiorium	a, b
Rhodotorula	i, r, s
R. mucilaginosa	f, r, u
Rhynchogastrema	r
Saccharomyces	e
Saccharomyces sp.	j, n, u, v
S. bayanus	u
S. cerevisiae	a, b, c, i, o, q, u
Saccharomycopsis crataegensis	r
S. fermentans	a
S. fibuligera	a, q
Schizosaccharomyces sp.	u
S. pombe	a
Sporopachydermia cereana	a
Starmerella	e, r
S. bacillaris	o, r
Torulaspora	s
T. delbrueckii	c, h, k, o, u
Torulopsis pintolopessi	i

(*Continued*)

TABLE 6.3 (Continued)

Genus/species	Country																																	Reference
	B			M			C			E			A			Ca			T			I			Th			K			Ch			
	D	W	S	D	W	S	D	W	S	D	W	S	D	W	S	D	W	S	D	W	S	D	W	S	D	W	S	D	W	S	D	W	S	
Trichosporonoides oedocephales																																		a
Wickerhamomyces																																		s, t
W. anomalus																																		f, o, q, s

Countries: A = Australia, B = Brazil, Ca = Cameroon, Ch = China, C = Colombia, E = Ecuador, I = India, K = Korea, M = Mexico, T = Tanzania, Th = Thailand. *Coffee processes*: D = dry, W = wet, S = semidry.

a = Silva et al. (2000), b = Silva et al. (2008a), c = Evangelista et al. (2014a), d = Hamdouche et al. (2016), e = De Bruyn et al. (2017), f = Martins et al. (2020), g = Avallone et al. (2001), h = Masoud et al. (2004), i = Puerta Q et al. (2012), j = de Melo Pereira et al. (2014), k = Evangelista et al. (2015), n = Nasanit and Satayawut (2015), o = Zhang et al. (2019a), p = Junqueira et al. (2019), q = Haile and Kang (2019), r = Zhang et al. (2019b), s = Elhalis et al. (2020), t = Pothakos et al. (2020), u = Vilela et al. (2010), v = Velmourougane (2013).

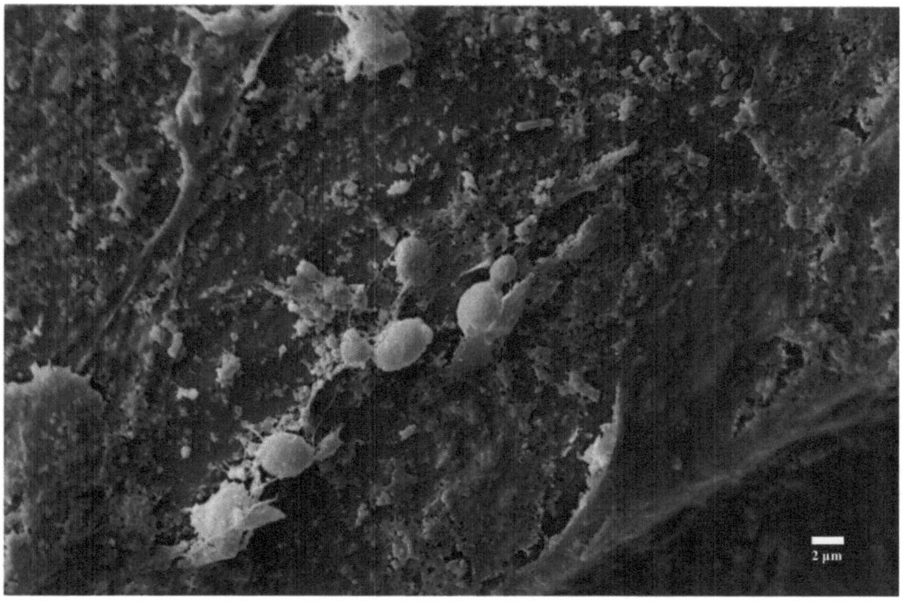

FIGURE 6.3 Yeast cell adherence to the surface of wet process beans.

Source: Martinez et al. (2019).

anomalus, Papiliotrema flavescens, Pichia kudriavzevii, Saccharomycopsis fibuligera, and *S. cerevisiae* (Table 6.3).

Recently, high-throughput analyses showed that operational taxonomic units (OTUs) increased along with fermentation in Ecuador. Those OTUs were from fungal taxon *Pichia, Starmerella*, and *Candida* (de Bruyn et al., 2017). Another fermentation performed in Ecuador showed a dominance of *P. kluyveri* (Zhang et al., 2019b). *P. kluyveri* counts in Tanzania along with fermentation and drying varied from 4.20 to 7.53 Log CFU/g (Masoud et al., 2004). A different kind of *Pichia* dominated the fermentation process in Colombia, *Pichia nakasei*, then *Candida* sp., *Dipodascus tetrasporeus*, and *Malassezia* sp. (Junqueira et al., 2019). At the end of a 36-h fermentation, *Hanseniaspora* and *Meyerozyma* were the dominant genera reported in Australia (Elhalis et al., 2020). Moreover, after isolation, the identification showed that the predominant species belong to *H. uvarum* and *Pichia kudriavzevii* (Elhalis et al., 2020). *Pichia kudriavzevii* has been previously detected in wine, cheese, and cocoa fermentation and was reported in coffee fermentation for the first time. This yeast produces high protease contents and might be considered a quality improver. *H. uvarum* was frequently found in other foods and had high protease and b-glucosidase activity.

6.3.2.3 Pulped coffee process

Yeast is the second group with the highest population after bacteria (Vilela et al., 2010). The yeast biodiversity in Brazil was comprised nine distinct species (*Pichia anomala, P. caribbica, Saccharomyces cerevisiae, S. bayanus, Hanseniaspora uvarum, Candida ernobii,* C. *carpophila, C. membranifaciens*, and *Arxula* sp.) that were detected with traditional and molecular methods (Vilela et al., 2010). Open batch spontaneous fermentation in Brazil demonstrated that the fermentation method favored yeast population and conditions. Yeast was found in high population until the end of processing, particularly for *Torulaspora delbrueckii*; the total population during fermentation was between 4.2 and 7.3 Log CFU/g (Martinez et al., 2017). SIAF batches performed in the southeast of Minas Gerais and Cerrado in Brazil identified higher populations of *S. cerevisiae* than *T. delbrueckii* (da Mota et al., 2020). A 25-h fermentation of Arabica coffee in India had a yeast population above 5.4–5.6 Log CFU/mL, and a 98-h fermentation of Robusta coffee had a population above 5.40–5.95 Log CFU/mL (Velmourougane, 2013). Further, *Saccharomyces* sp. and *Schizosaccharomyces* sp. were recorded in Arabica coffee. Other yeast identification is shown in Table 6.3.

6.3.3 Filamentous fungi

As water activity keeps lowering during processing, filamentous fungi populations start increasing, though their population drops in storage. The colonization of these microorganisms is slow because they compete against yeast and bacteria for nutrients (Silva, 2015). Many filamentous fungi are associated with toxins (ochratoxin A) and enzyme production. Filamentous fungi must be controlled to avoid the production of off-flavors. Different filamentous fungi are shown in Table 6.4.

6.3.3.1 Natural process

Cladosporium, Fusarium, Penicillium, and *Aspergillus* are the primary fungi detected during the natural process. They are protease producers, and their growth is favored with high temperatures and low water activities.

The diversity of filamentous fungi is influenced by each coffee variety's characteristics, which was expected because the coffee pulp and mucilage composition varied concerning

TABLE 6.4
Other Fungal Diversity Identified in Different Countries and Coffee Processes

Genus/specie	Country																		Reference
	B			E			A			Ca			I			Th			
	D	W	S	D	W	S	D	W	S	D	W	S	D	W	S	D	W	S	
Acremonium sp.																	■		n
Arthrobotrys spp.	■																		a
Aspergillus								■											s
Aspergillus sp.	■		■												■		■		a, n, u, v
A. chevalieri			■																u
A. flavus	■																		b
A. foetidius			■																u
A. niger	■		■												■				b, u, v
A. ochraceus	■		■																b, u
A. sydowii	■																		b
A. tamarii	■														■				b, v
A. tubingensis			■																u
A. versicolor			■																u
Beauvaria bassiana	■																		a
Cladosporium								■											s
Cladosporium sp.	■		■												■		■		a, n, u, v
C. cladosporioides	■		■																b, u
C. sphaerospermum											■								d
Colletotrichum								■											s
Cylindrocarpo sp.			■																u
Eurotium chevalieri			■																u
Fusariella sp.			■																u
Fusarium					■														t
Fusarium sp.	■		■												■		■		a, n, u
F. chlamydosporum			■																u
F. lateritium			■																u
F. nivale			■																u
F. solani	■		■																b, u
F. sporotrichioides			■																u
Geotrichum sp.			■																u
Monilia spp.	■																		a

Mucor sp.														■				v
Nectriaceae *Gibberella*								■										s
N. fusarium								■										s
Neodevriesia								■										s
Penicillium								■										s
Penicillium sp.	■		■											■		■		a, n, u, v
P. corylophilum	■																	b
P. chrysogenum	■																	b
P. brevicompactum	■		■															b, u
P. commune			■															u
P. decumbens			■															u
P. fellutanum			■															u
P. implicatum			■															u
P. roqueforti			■															u
Phoma sp.			■															u
Rhizoctonia spp.	■																	a
Rhizopus sp.														■				v
Ulocladium sp.			■															u
Wallemia										■								d

Countries: A = Australia, B = Brazil, Ca = Cameroon, E = Ecuador, I = India, Th = Thailand. *Coffee processes*: D = dry, W = wet, S = semidry.

a = Silva et al. (2000), b = Silva et al. (2008a), d = Hamdouche et al. (2016), n = Nasanit and Satayawut (2015), s = Elhalis et al. (2020), t = Pothakos et al. (2020), u = Vilela et al. (2010), v = Velmourougane (2013).

varieties. Silva et al. (2000) detected *Cladosporium* (39.7%), *Fusarium* (34.2%), *Penicillium* (28.8%), and *Aspergillus* (2.7%) in the Mundo Novo variety. Silva et al. (2008) detected *Cladosporium* (15.9%), *Fusarium* (12.1%), *Penicillium* (27.8%), *Aspergillus* (36%), *Pestalotia* (4.1%) and *Paecelomyces* (3.5%) in the Acaiá variety. *Cladosporium cladosporioides* were the fungi species most found in coffee cherries, and it has been related to good quality coffee beans (Pereira et al., 2005).

Usually, the filamentous fungi population is low (2–3 Log CFU/g) at the beginning of the fermentation process and increases slowly during fermentation and drying (4–5 Log CFU/g). The increase in the filamentous fungi population might occur by external contamination, because the presence of mycelium was not detected during processing (Silva et al., 2008b).

Further, inadequate storage conditions can lead to increased humidity in coffee that favors the proliferation of *Penicillium* and *Aspergillus*. Silva et al. (2008b) detected the predominance of *A. flavus* and *A. niger* during the coffee storage. However, the implementation of quality tools during coffee processing significantly influenced the reduction of microorganism contamination risk.

6.3.3.2 Wet process

Filamentous fungi are the least detected group in this process because the depulped coffee beans are submerged in water. The genus *Penicillium* was recently found in wet process–fermented coffee from Australia, with *Aspergillus, Cladosporium, Pleosporales*, and *Neodevriesia* (Elhalis et al., 2020). The cultivable filamentous fungi from a Thai wet process belong to *Penicillium spp.*, followed by *Acremonium* spp., *Cladosporium* spp., *Aspergillus* sp., and *Fusarium* sp. (Nasanit and Satayawut, 2015). Refer to Table 6.4 to see more species and in other countries.

Though low levels of ochratoxin A (0.2 μg kg^{-1}) by *Aspergillus ochraceus* and *A. niger* have been reported in green coffee from the wet process, fermentation proved essential to decrease the contamination of toxigenic fungi (Silva, 2015; Suarez-Quiroz et al., 2004).

6.3.3.3 Pulped coffee process

According to Evangelista et al. (2014a), the pulped coffee process is less favorable for filamentous fungi development, which in terms of quality is what producers expect. Its population was detected after several days of fermentation and drying and less than bacteria and yeast populations. Vilela et al. (2010) identified *Aspergillus tubingensis, Aspergillus versicolor, Cladosporium cladosporioides, Aspergillus* sp., and *Penicillium decumbens* with counts of 3–5.7 Log CFU/g. The authors also identified other filamentous fungi such as *Aspergillus chevalieri, Cylindrocarpon* sp., *Eurotium chevalieri, Fusariella* sp., *Fusarium solani*, and *Phoma* sp. with population sizes of 1.4 Log CFU/g or lower.

A study performed in India identified the dominance of *Penicillium* sp., *Fusarium* sp., *Rhizopus* sp., *Mucor* sp., *Cladosporium* sp., *Aspergillus terreus, A. nidulans*, and *A. tamarii* in Arabica fermenting mass, and *A. niger* in both Arabica and Robusta fermenting mass (Velmourougane, 2013). A summary of more filamentous fungi in different countries is shown in Table 6.4.

6.3.4 Endophytic microorganisms

Endophytic bacteria and fungi from coffee fruits start microbial activity in the plant and continue after harvesting. According to Vega et al. (2005), endophytic bacteria vary from country to fruit tissue, showing a high percentage of berry seeds with 32.2%. The study identified two bacterial isolates in the crown (*Bacillus subtilis* and *Salmonella typhimurium*), seven in the pulp (*Burkholderia* sp., *Gordona* sp., *Methylobacterium radiotolerans, Pseudomonas putida, Rhodococcus equi, Stenotrophomonas maltophilia*, and unknown), one in the peduncle (*Micrococcus* sp.), and 28 in the seed (*Bacillus* sp., *Burkholderia* sp., *Burkholderia gladioli, Clavibacter michiganense* subsp. *insidiosum, Curtobacterium flaccumfaciens-flaccumfaciens, Curtobacterium flaccumfaciens-poinsettiae, Escherichia vulneris, Micrococcus* sp., *Pantoea agglomerans, Pseudomonas putida, Serratia liquefaciens, Stenotrophomonas maltophilia*, and unknown) from Hawaiian, Colombian, and Mexican samples. Cloned sequences from endophytic bacteria of fruits derived from five cultivars (Catuaí Vermelho, Catuaí Amarelo, Bourbon Vermelho, Bourbon Amarelo, and Catucaí) resulted in four significant genera: *Firmicutes* (≥97%), *Bacillus* (60%), *Staphylococcus* (8%), and *Paenibacillus* (4%) (Oliveira et al., 2013).

Regarding fungi, the identifications by Vega et al. (2010) also varied from fruit tissue and country (Colombia, Mexico, Hawaii, and Puerto Rico). Some isolates were identified as *Aspergillus* sp., *Aspergillus niger, Aspergillus phoenicis, Aspergillus westerdijkiae, Beauveria bassiana, Beauveria brongniartii, Botryosphaeria* sp., *Cryptococcus* sp., *Cladosporium* sp., *Colletotrichum* sp., *Fusarium* sp., *Penicillium* sp., *Paecilomyces* sp., and *Xylariaceae* sp.

6.4 Starter cultures

Implementation of starter cultures (bacteria, yeasts, and filamentous fungi) are well known for several fermented foods and beverages. The selection of an appropriate starter strain is essential to improve the processing conditions and product quality. Quality can be reached through metabolic activities, the improved fermentation process, desirable sensory attributes, safety improvement, and toxicological risks (Holzapfel, 1997; Marsh et al., 2014).

Researchers have proposed technological alternatives for better control over fermentation. It involves a standardized process that guarantees the production of novel coffee flavors and desirable sensory profiles. Among them, the use of starter cultures is a prominent alternative (Table 6.5).

Saccharomyces cerevisiae, Candida parapsilosis, Torulaspora delbrueckii, Yarrowia lipolytica, Pichia kluyveri, Pichia anomala, Hanseniaspora uvarum, Pantoea dispersa, Bacillus subtilis, Arthrobacter koreensis, Lactobacillus rhamnosus, and *Lactococcus lactis* subsp. *cremoris* were used as

TABLE 6.5
Potential Starter Cultures Used in Coffee Processed Via Dry, Semi-dry, and Wet

Starter culture	Variety	Process	Method	Score (SCA)	Time (hour)	Volume/ Weight	Concentration	Location	References
Saccharomyces cerevisiae CCMA 0543	Catuaí Amarelo	Natural	Direct	81.58	16	10 kg	7 Log cells/g	Lavras-MG, Brazil	Bressani et al., 2018
			Open container*	84					
Torulaspora delbruckii CCMA0684			Direct	81.5					
			Open container*	81.5					
Candida parapsilosis CCMA 0544			Direct	80.67					
			Open container*	81.33					
Saccharomyces cerevisiae UFLA YCN727		Natural	Direct			0.8 kg	6 Log cells/mL	Lavras-MG, Brazil	Evangelista et al., 2014b
Saccharomyces. cerevisiae UFLA YCN724									
Candida parapsilosis UFLA YCN448									
Pichia guilliermondii UFLA YCN731									
Saccharomyces cerevisiae CCMA 0543	Catuaí Amarelo	Pulped natural	Direct	81.3	16	10 kg	7 Log cells/g	Lavras-MG, Brazil	Martinez et al., 2017
			Open container*	81.4					
Torulaspora delbruckii CCMA0684			Direct	80.8					
			Open container*	81					
Candida parapsilosis CCMA 0544			Direct	81					
			Open container*	81.3					
Saccharomyces cerevisiae CCMA 0200	Ouro Amarelo	Pulped natural	Direct	83.25		60 kg	5 Log cells/g	Patrocínio-MG, Brazil	Ribeiro et al., 2017a
	Mundo Novo			80.13					
Saccharomyces cerevisiae CCMA 0543	Ouro Amarelo			82.88					
	Mundo Novo			82.63					
Saccharomyces cerevisiae UFLA YCN727	Acaiá	Pulped natural	Direct	81.08		0.6 kg	7 Log cells/mL	Lavras-MG, Brazil	Evangelista et al., 2014a
Saccharomyces cerevisiae UFLA YCN724				79.33					
Candida parapsilosis UFLA YCN448				80					
Pichia guilliermondii UFLA YCN731				81.08					
Saccharomyces cerevisiae CCMA 0543	Ouro Amarelo	Pulped natural	Direct	82.17		60 L	7 Log cells/g	Patrocínio-MG, Brazil	Ribeiro et al., 2017b
	Mundo Novo			75.88					
Saccharomyces cerevisiae CCMA 0200	Ouro Amarelo			79.92					
	Mundo Novo			73.75					
Torulaspora delbrueckii CCMA 0684	Ouro Amarelo			79.96					
	Mundo Novo			75.83					

(*Continued*)

TABLE 6.5 (Continued)

Starter culture	Variety	Process	Method	Score (SCA)	Time (hour)	Volume/ Weight	Concentration	Location	References
Saccharomyces cerevisiae CCMA 0543	Bourbon Amarelo	Pulped natural	Open containers**		40	20 L	6 Log cells/mL	São Sebastião da Grama-MG, Brazil	Bressani et al., 2020a
Candida parapsilosis CCMA 0544	Canário Amarelo								
Torulaspora delbrueckii CCMA 0684									
Saccharomyces cerevisiae CCMA 0543	Bourbon Amarelo	Natural	Closed container*	84.37	70	40 L	7 Log cells/g	Carmo de Minas-MG, Brazil	Da Mota et al., 2020
Torulaspora delbrueckii CCMA0684				87.25					
Saccharomyces cerevisiae CCMA 0543		Pulped natural		86.87					
Torulaspora delbrueckii CCMA0684				87.25					
Saccharomyces cerevisiae CCMA 0543	Catuaí Amarelo	Natural	Closed container*	83	70	40 L	7 Log cells/g	Três Pontas-MG, Brazil	Da Mota et al., 2020
Torulaspora delbrueckii CCMA0684				85					
Saccharomyces cerevisiae CCMA 0543		Pulped natural		84.25					
Torulaspora delbrueckii CCMA0684				85					
Saccharomyces cerevisiae CCMA 0543	Bourbon Amarelo	Natural	Closed container*	82.25	70	40 L	7 Log cells/g	Araxá-MG, Brazil	Da Mota et al., 2020
Torulaspora delbrueckii CCMA0684				82.25					
Saccharomyces cerevisiae CCMA 0543		Pulped natural		84					
Torulaspora delbrueckii CCMA 0684				83					
Saccharomyces cerevisiae CCMA 0543	Rubi	Natural	Closed container*	84	70	40 L	7 Log cells/g	Monte Carmelo-MG, Brazil	Da Mota et al., 2020
Torulaspora delbrueckii CCMA0684				84.37					
Saccharomyces cerevisiae CCMA 0543		Pulped natural		84.6					
Torulaspora delbrueckii CCMA 0684				84.25					
Saccharomyces cerevisiae CCMA 0543	Catuaí Amarelo	Natural	Closed container**	86	70	40 L	7 Log cells/g	Lajinha-MG, Brazil	Da Mota et al., 2020
Torulaspora delbrueckii CCMA 0684				85.5					
Saccharomyces cerevisiae CCMA 0543		Pulped natural		84.75					
Torulaspora delbrueckii CCMA 0684				86.5					
Lactobacillus plantarum LPBR01	Catuaí	Wet	Closed container***	91.5	24	2 kg	8 Log CFU/mL	Patrocínio-MG, Brazil	de Carvalho Neto et al., 2018
Pichia fermentans YC5.2	Mundo Novo	Wet	Closed container***		48	1.5 kg	6 Log CFU/mL	Lavras-MG, Brazil	de Melo Pereira et al., 2014
Saccharomyces sp. YC9.15									
Pichia fermentans YC5.2 + *Saccharomyces* sp. YC9.15									
Pichia fermentans YC5.2	Catuaí	Wet	Closed container***	90	24	2 kg	6 Log CFU/mL	Patrocínio-MG, Brazil	de Carvalho Neto et al., 2020

Saccharomyces cerevisiae CCMA 0200	Catuaí Vermelho	Wet	Concrete tanks	81	40		7 Log cells/g	Patrocínio-MG, Brazil.	Martins et al., 2019
	Mundo Novo			79					
Torulaspora delbrueckii CCMA 0684	Catuaí Vermelho			82					
	Mundo Novo			79.2					
Lactobacillus plantarum LPBR01	Catuí	Wet	Concrete tanks	88.3	24	20 kg	7 Log cells/mL	Cerrado Mineiro-MG, Brazil	Pereira et al., 2016
Yarrowia lipolytica NCYC 2904		Wet			96	0.8 g	5 Log CFU/g	Toraja, Sulawesi, Indonesia	Lee et al., 2017
Lactobacillus rhamnosus HN001		Wet			72	1 kg	7.5 Log CFU/mL	Indonesia	Wang et al., 2019
Saccharomyces cerevisiae CCMA 0543 *Candida parapsilosis* CCMA 0544 *Torulaspora delbruecki* CCMA 0684 *Bacillus subtilis* CCMA 1236 *Arthrobacter koreensis* CCMA 1279	Catuaí Vermelho	Wet			48	0.15 kg	9 Log cells/mL	Lavras-MG, Brazil	Martinez et al., 2019

Notes: * High-density polyethylene; ** stainless steel; *** New Brunswick BioFlo 110 fermenter.

potential starter cultures in coffee due to their ability to produce pectinolytic enzymes and volatile compounds (Masoud et al., 2005; Silva et al., 2013; Lee et al., 2017a; Martinez et al., 2019; Wang et al., 2020a, 2020b, 2019).

Yeasts and filamentous fungi produce pectinase. Pectin lyase, polygalacturonase, and pectin methylesterase are the pectinolytic enzymes involved in the degradation of coffee mucilage. These enzymes catalyzed the hydrolysis of α-1.4 glycosidic bonds into pectic acid, pectin breakage by transelimination, and methoxyl group de-esterification of the pectin forming pectic acid and methanol (Agate et al., 1966; Masoud et al., 2005).

Mucilage reduction helps decrease water activity, turns drying more efficiently, and prevents fungal proliferation. Furthermore, the hexose sugars metabolism is essential to produce a wide range of secondary metabolites that will influence the coffee's sensory characteristics.

Starter culture presence influences some volatile compounds, such as 2-phenylethanol, 4-vinylguaiacol, and 4-vinylphenol by *Y. lipolytica* NCYC 2904 (Lee et al., 2017a). In pulped natural coffee, 2-methyl-2-pentenal is influenced by *T. delbrueckii* CCMA0684 and formic acid, 2-methyl-1,3-butadiene, 3,4-dimethyl-2-pentanone, and pantolactone by *S. cerevisiae* CCMA0543. In natural coffee, *T. delbrueckii* CCMA0684 produced 2-phenyl-2-butenal, 5-amino-1-naphthol, 3,4-dimethyl-2-pentanone, and 1-hydroxy-2-propanone. *S. cerevisiae* CCMA0543 inoculation produced 2-butanone (Da Mota et al., 2020). In the wet process, 4-ethenyl-1,2-dimethoxybenzene by *T. delbrueckii* CCMA0684 (Martinez et al., 2019). The production of specific compounds is due to the different fermentation pathways of starter cultures during the complex fermentation process (Lee et al., 2017b).

During coffee fermentation, the starter culture fermentative performance is influenced by its interaction with the coffee fruit. Two methods were proposed for the natural and pulped natural coffee fermentation process: direct by spraying and close bioreactor (SIAF, previously described in Section 6.2). In the direct method, inoculation is conducted rapidly after harvesting; however, part of the inoculum could be lost to the environment, reducing the starter culture's final concentration. The fermentative performance of starter cultures changed with processing and coffee variety (da Mota et al., 2020).

The direct method's impact on coffee quality was studied. *S. cerevisiae, T. delbrueckii*, and *C. parapsilosis* were inoculated through the direct method during natural and pulped coffee processing. Catuaí Amarelo variety quality was improved using the direct method in natural coffee, resulting in sensory scores of 81.58, 81.5, and 81.33 (Bressani et al., 2018). In studies with pulped natural coffee, *S. cerevisiae* UFLA YCN727 (now designated as CCMA0543) and *S. cerevisiae* CCMA200 stood out among the other starters, generating sensory scores of 81.08 and 83.25 for the Acaiá and Ouro Amarelo varieties (Evangelista et al., 2014a; Ribeiro et al., 2017a).

In the closed bioreactor method (SIAF), the inoculum is added to the coffee fruits (with or without pulp) inside the container where they are fermented. Fermentation monitoring is performed by measuring the temperature and concentration of soluble solids. After fermentation, coffee beans are sun-dried or mechanical dried. This method reduces the loss of inoculum to the environment and allows more contact with the microorganisms with the coffee fruits (Martinez et al., 2017). The interaction between the coffee fruits and the starter *Saccharomyces cerevisiae* (CCMA 0543) was perceptible in the Catuaí Amarelo variety processed via natural, providing an increase of 3 points in the final score of the beverage (Bressani et al., 2018). The sensory profile of Bourbon Amarelo and Canario Amarelo processed via natural and pulped natural changes with inoculation of *T. delbrueckii* CCMA 0684 (Bressani et al., 2020a). In general, the microbial performance was improved by self-induced anaerobiosis conditions in close batches. Besides, it helped by inhibiting the growth of undesirable fungi and bacteria (Hameed et al., 2018).

The sensory quality of fermented coffees under induced anaerobiosis (SIAF) has increased by up to 8 points in the beverage's final score, which has aroused much interest in producers and traders. According to da Mota et al. (2020), the metabolic activity of the strains under SIAF conditions might be different according to the processing type. Also, *T. delbrueckii* CCMA0684 and *S. cerevisiae* CCMA0543 improved coffee quality from five different coffee regions when inoculated in natural and pulped coffee.

Different bioreactors were suggested for the fermentative coffee process (Figure 6.4). Natural and pulped coffee methods have been fermented using open buckets, covered plastic tanks, and open stainless steel tanks (Correa et al., 2014; Martinez et al., 2017; Bressani et al., 2018, 2020a; da Mota et al., 2020). Plastic buckets and tanks are easily found by producers to be flexible, translucent/waxy, with a higher working temperature, tensile strength, and toughness at low temperatures. In addition, they have moisture and oxygen barriers and are easy to clean.

Open-bucket vessels have a 20-L coffee capacity, are made of high-density polyethylene, and favor aerobic conditions (Figure 6.4a). The covered plastic tanks can either have a capacity for 20, 50, and 500 L of coffee, are made of high-density polyethylene, favor anaerobic conditions, and temperature monitoring is done manually or with sensors (Figure 6.4b, 6.4c, and 6.4d). The 20-L and 50-L tanks are cylindrical; their dimensions are Ø = 31 cm and height = 42.5 cm (20 L), and Ø = 40.39 cm and height = 53.79 cm (50 L). The 50-L tank has a seal to close the lid (made of galvanized steel) and an aluminum mesh fixed at 10 cm from the bottom, separating the coffee beans from the exudate produced during the fermentative process. The 500-L tank has a conic shape; its dimensions are bottom Ø = 67 cm, top Ø = 104 cm, and height = 85 cm.

The stainless steel tanks for fermentation can contain 20 L and 316 L of coffee (Figure 6.4e and 6.4f). The dimensions of the 20-L tank are Ø = 30 cm and height = 38 cm; they have a cylindrical shape and are not covered. The 316-L tank has covered bioreactors (with lid) that has dimensions of Ø = 65 cm and height = 90 cm. The bottom of each tank has a conical funnel with an inclination of 25 degrees to allow liquid from the exudation of fruits or beans to slide down without interfering with substrate consumption by microorganisms. The tank also possessed a butterfly valve manufactured in stainless steel (Ø = 4 in.) and two mechanical stirrers (resembling an anchor and an impeller-type stirrer with a 1.5 and 1 CV motor;

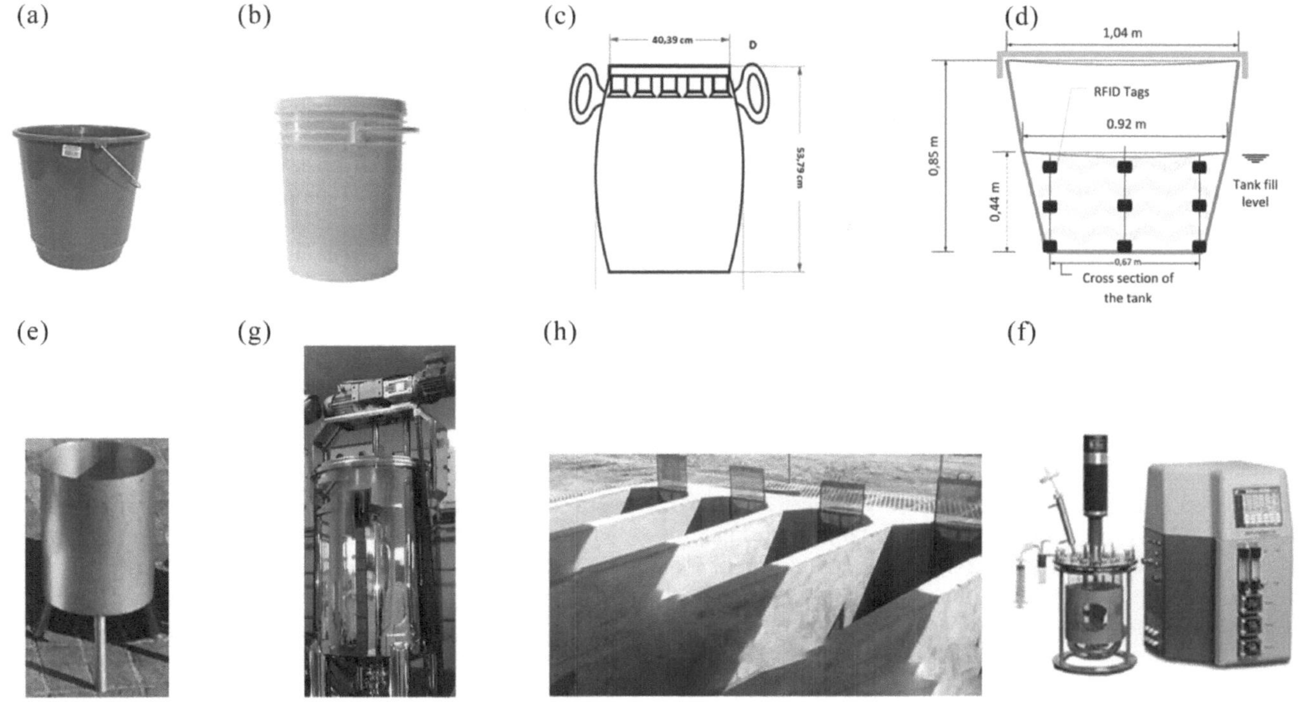

FIGURE 6.4 Bioreactors used to ferment coffee. Dry or natural and semi-dry methods: (a) Open bucket 20 L. Covered plastic tanks: (b) 20 L, (c) 50 L, (d) 500 L. Stainless steel tanks: (e) 20 L, (f) 316 L. (g) Concrete tanks. (h) Stirred-tank bioreactor.

Martinez et al. 2021). All proposed bioreactors allowed effective fermentative performance to improve the coffee beans and beverage quality, whether spontaneous or inoculated fermentation.

Wet fermentation is usually conducted in concrete tanks (Figure 6.4g). Another proposed vessel for fermentation involves using a stirred-tank bioreactor (STR; Figure 6.4h) for laboratory research. It is used for small-scale fermentations and is surrounded by a water jacket with built-in controls to monitor temperature, agitation, pH, and dissolved oxygen. The bioreactor confers controlled conditions and a positive environment that favors starter culture growth (de Carvalho Neto et al., 2018), but it is not suitable for coffee producers.

Starter cultures were selected for the wet process, such as *Pichia fermentans* YC5.2, *Saccharomyces* sp. (YC9.15), *Saccharomyces cerevisiae* CCMA 0200, *Torulaspora delbrueckii* CCMA 0684, *Pichia anomala* (P.aS12, P.aS14, P.aS16), *Pichia kluyveri* (PS7Y1, PkS4Y3, PkS13Y4), *Lc. lactis* subsp. *cremoris* ATCC 19257, *Lactobacillus plantarum, Lactobacillus plantarum* LPBR01, *Leuconostoc/Weissella* (BFE 6997, 6989), homofermentative *Lactobacillus* spp. (BFE 6853, 6823), *Lactobacillus* spp. heterofermentative (BFE 6812), and Enterococcus (BFE 6803; Masoud et al., 2005; Massawe and Lifa, 2010; Pereira et al., 2016; de Carvalho Neto et al., 2018; Martins et al., 2019; de Carvalho Neto et al., 2020; Wang et al., 2019, 2020a). Yeast and bacteria starters used during the wet fermentative process improved the volatile alcohol and acid production (Martinez et al., 2019). However, each strain provided a unique metabolism and sensorial flavor.

Lactic acid bacteria are essential for the wet process, and their inoculation efficiently promotes the consumption of fermentable sugars and, consequently, leads to faster acidification. In addition, LAB help reduces the required fermentation time. De Carvalho Neto et al. (2018) studied the behavior of *L. plantarum* LPBR01 during the wet fermentative process in the STR system and reduced the time of fermentation from 24 to 10 h. *L. plantarum* LPBR01 inoculation in coffee fermentation contributed to the dominance of LAB without interfering with the yeast growth (Pereira et al., 2016).

The yeast has a complementary role associated with coffee quality and contributes to the efficient removal of the coffee bean mucilage layer, promoting the synthesis of yeast-specific volatile constituents (de Carvalho Neto et al., 2020). *P. fermentans* YC5.2 and *Saccharomyces* sp. YC9.15 increased ethanol, acetaldehyde, ethyl acetate, isoamyl acetate, 2,3-butanedione, and hexanal concentration during wet fermentation. These compounds play a role in aroma development during coffee fermentation (de Melo Pereira et al., 2014).

Yeast and bacterial behavior during co-culturing while fermenting may suggest new synergistic flavor-modulating mechanisms that would further contribute to the flavored coffee (Wang et al., 2020b). *S. cerevisiae* and *P. kluyveri* co-culture followed inoculation of *Lc. lactis* subsp. *cremoris* in wet fermentation increased the acidity of green coffee beans, the formation of caramel-smelling furfurals during roasting, and increased fruity and winey acetate esters. Elevated isoamyl acetate and 2-phenylethyl acetate contributed to the fruity and winey aroma of the roasted coffees (Wang et al., 2020b).

The selected starter cultures from dominant microbiota in the coffee substrate ensured a better performance than strains from other substrates and processing because they are already adapted to the processing conditions. *S. cerevisiae* CCMA

0200, previously isolated from sugar cane, was used as a culture starter in pulped coffee and wet process and failed to produce higher quality attributes (Ribeiro et al., 2017a, 2017b; Martins et al., 2019). In pulped coffee processing, *S. cerevisiae* CCMA 0200 was described as having a high astringency that is undesirable for the coffee beverage (Ribeiro et al., 2017b).

The use of starter cultures during fermentation improves the sensory characteristics and produces antifungal properties against the growth and production of mycotoxins by filamentous fungi. *A. ochraceus* growth and its ochratoxin A production was controlled through inoculation of *P. anomala, P. kluyveri*, and LAB during wet processing (Massawe and Lifa, 2010). Other yeast, such as *Pichia, Debaryomyces, Saccharomyces*, and *Rhodotorula*, can also be used as biocontrol agents (Souza et al., 2017).

Food safety problems are usually associated with contamination by pathogenic or toxin-forming bacteria or fungi. During coffee processing, the use of suitable starter culture, yeasts, and LAB, accompanied by proper application of food safety management system, guaranteed coffee safety and improved the beverage's sensory quality.

6.5 Coffee chemistry

The coffee fruit is a drupe composed of six parts (skin, pulp, mucilage, parchment, silver skin, and beans), as shown in Figure 6.5.

The chemical composition of the fruits and beans is complex and varies according to the species, variety, cultivar, type of processing, drying, and roasting. The chemical components of the beans are well known, and they can improve genetic diversity, maintain/improve quality, and expand the export market via competitiveness (Malta et al., 2020). Therefore, they are still studied. Table 6.6 shows the difference in concentration in the chemical composition found in green and roasted coffee beans (Arabica and Robusta coffee).The main constituents of green coffee are carbohydrates – nitrogen compounds such as proteins, trigonelline and caffeine, lipids, organic acids, and water – and most of them are precursors of aroma and flavor (Belitz et al., 2009). They are an essential group of soluble compounds (Belitz et al., 2009). During roasting, those flavor/aroma precursors are achieved through thermal reactions (Worku et al., 2018), such are the breakdown of free sugars, amino acids, trigonelline, and chlorogenic acids.

Fermentation can change the concentration of compounds in green coffee. During fermentation, some microorganisms can generate volatile precursors, improving the beverage. The influence of fermentation on some important quality compounds will be discussed in the next sections.

6.5.1 Carbohydrates

Carbohydrates are generated during coffee fruit maturation and are present in all its parts of the coffee fruit. Glucose, fructose, and pectin are the main components of the mesocarp. The endocarp and silver skin are rich in cellulose, hemicellulose, and monosaccharides. The bean is also rich in soluble carbohydrates (arabinose, fructose, galactose, glucose, sucrose, raffinose, and stachyose) and insoluble (cellulose and hemicellulose), with 59%–61% of dry weight (Esquivel and Jiménez, 2012; Poisson et al., 2018). Differences in fructose and glucose concentrations are related to the different processes. However, other low molecular weight sugars are not significantly affected by processing methods (Knopp et al., 2006).

During fermentation, microorganisms quickly consume sugars that turn into important flavor and aroma metabolites. Depending on the starter culture used, fructose and glucose concentration may be different at the end of the fermentation process (Martinez et al., 2017; Bressani et al., 2018). Some microorganisms produce enzymes, such as polygalacturonases and pectin lyases, necessary to depolymerize and hydrolyze the mucilage pectin (Silva et al., 2013), releasing glucose, rhamnose, and galactose (Garna et al., 2004). Several pectinolytic strains have already been identified, as previously mentioned.

During roasting, sugars present in beans also play an essential role. Sucrose is hydrolyzed by releasing glucose and fructose, involved in caramelization and Maillard reactions. Other free sugars, such as galactose and mannose, are only present in small amounts. Low-molecular-weight carbohydrates contributed to the formation of acids and volatile compounds. Polysaccharides assist in forming volatile compounds, as they are essential constituents for the retention of volatiles (de Melo Pereira et al., 2019), and the soluble fraction favors body perception in the beverage (Campos-Vega et al., 2015).

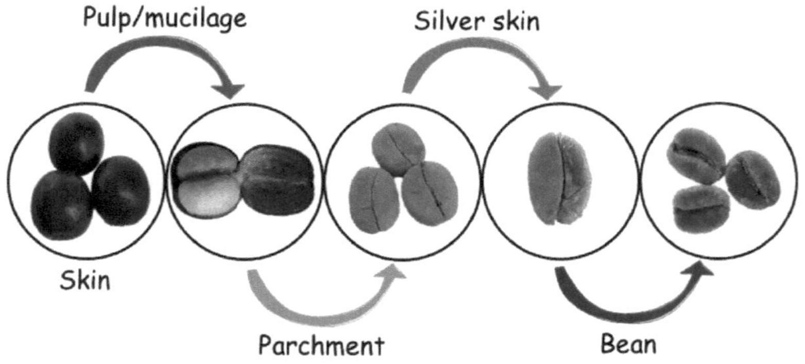

FIGURE 6.5 Parts of coffee fruits.

TABLE 6.6
Difference in the Chemical Composition of Arabica and Robusta Species for Both Green and Roasted Coffee (Dry Weight % w/w)

Compound*	Green bean (Arabica)	Green bean (Robusta)	Roasted** bean (Arabica)	Roasted** bean (Robusta)	References
Caffeine	0.8–1.5	1.7–4	1–1.3	1.3–2.4	Belitz et al., 2009; Esquivel and Jiménez, 2012; Mussatto et al., 2011; Oestreich-Janzen, 2010; Selmar et al., 2014
Chlorogenic acid	4.1–17.4	4.1–12.1	2.5–2.7	2.7–3.8	
Trigonelline	0.6–1.2	0.3–0.9	0.3–1	0.3–0.7	
Polyphenols	5.3–6.5	5.3–6	–	–	
Carbohydrates	5.4–12.5	3–11.5	38	38–42	
Proteins	8.5–12	8.5–12	10–15	10–15	
Lipids	15–18	8–12	14.5–17	11–17	
Cellulose	41–43	32–40	–	–	
Hemicellulose	5–10	3–4	–	–	
Lignin	1–3	1–3	–	–	
Tannins	0.4–0.9	0.4–0.9	–	–	
Volatile aromas	traces	traces	0.1	0.1	
Ash (minerals)	3–5.4	3–5.4	3.5–4.5	4.5–5.0	

Notes: * The change in concentration may be due to a cultivar, agricultural practices, climate, soil composition, and/or methods of analysis; ** considering a medium roast.

6.5.2 Organic acids

Acidity affects beverage quality (Cheng et al., 2016) and is one of the parameters evaluated in specialty coffees (SCA, 2018). For example, citric acid is very appreciated in cupping because it produces fruity and berry flavors (Vandenberghe et al., 2018).

Several aliphatic acids have already been identified in coffee (Avallone et al., 2001; Evangelista et al., 2014a; Koshiro et al., 2015; Peñuela-Martínez et al., 2018; Martinez et al., 2019; Bressani et al., 2020a); however, citric, malic, acetic, lactic, and formic acids are the most important. Acids can represent 11%–12% of the dry weight in green coffees. Besides, *Coffea arabica* has more excellent acidity than *Coffea canephora* (Farah and Lima, 2019).

The concentration and type of acid found can be affected by genetic aspects, growth conditions (altitude, agricultural methods, degree of fruit maturity), choice of post-harvest method, and roast degree (Koshiro et al., 2015; Martinez et al., 2017; Diviš et al., 2019). There is not yet concise data about the concentration of these acids during maturation. Therefore, it can oscillate by increasing or decreasing. Moreover, reports have shown that malic and oxalic acids accumulate in more significant amounts in the bean's pericarp and citric acid. The type of extraction influences the results during the analysis of organic acids (whole fruit or only beans) (Evangelista et al., 2014a; Martinez et al., 2017).

Most organic acids are generated because of an incomplete Krebs cycle, occurring under anaerobic conditions. Through the glycolytic pathway, pyruvate is produced and converted into other organic acids through the Krebs cycle (Yoshida and Yokoyama, 2012). However, some yeasts and bacteria may use different metabolic routes as a survival strategy based on the substrate and medium conditions. Some microorganisms in coffee metabolize different organic acids. Differences in concentration will depend on the species or strain (Silva, 2015).

Some yeasts may use organic acids as a carbon source during aerobic respiration or through the glyoxylate cycle in gluconeogenesis. This fact was reported by Lee et al. (2017a), where coffee was fermented using *Y. lipolytica* as a starter culture. However, yeasts can produce succinic acid under different conditions through the isocitrate's oxidation via the glyoxylate cycle and the citric acid–reducing cycle (Jayaram et al., 2014). Bressani et al. (2020a) stated that the increase in succinic acid during fermentation is a good indicator, especially when using *M. caribbica* (CCMA 0198), *S. cerevisiae* (CCMA 0543), and *C. parapsilosis* (CCMA 0544) as starter cultures. In this sense, the production of desirable acids can be a criterion for selecting starter cultures.

6.5.3 Antioxidants

Coffee has high antioxidant activity. Chlorogenic acid, caffeic acid, caffeine, trigonelline, and other phenolic compounds are the group's main constituents. However, coumaric, ferulic, and melanoidins (in roasted coffee) also have this activity (Aguiarb et al., 2016). These compounds can quickly donate a hydrogen atom to an oxidized molecule or one at risk of transformation (Arellano-González et al., 2011).

Several methodologies might be used to quantify the antioxidant capacity in food. Because there are different types of antioxidants, concentrations, and compositions, it is necessary to perform two different methodologies to have a better estimate of these properties (Górnaś et al., 2016).

The total polyphenol content is analyzed using the Folin-Ciocalteu reagent. Many studies have already evaluated the antioxidant activity of different coffees through the combination of different methodologies, such as DPPH (2,2-difenil-1-

picril-hidrazil), ABTS (2,2′-azinobis (3-ethylbenzothiazoline-6-sulfonic acid), FRAP (ferric reducing ability), ORAC (oxygen radical absorbance capacity), and CUPRAC (cupric ion reducing antioxidant capacity; Arellano-González et al., 2011; Kwak et al., 2018; Zhang et al., 2018). However, few studies have evaluated the influence of fermentation on these compounds. Arellano-González et al. (2011) compared the antioxidant activity of fermented and unfermented coffees. Despite a low total polyphenol content, there was no significant difference between treatments. The authors also found that the antioxidant activity of ABTS assays was also lower in fermented coffee. A recent study by Kwak et al. (2018) showed that the total polyphenol content and antioxidant activity (measured by the ORAC method) were lower in wet fermented coffee compared to unfermented coffee. Despite this, the inoculated coffees with three different species of *Saccharomyces cerevisiae* showed similar and higher values. The same authors also claimed that during fermentation, phenolic compounds firmly bind and weaken the cell wall. Thus, the extraction of these compounds is more painless after roasting (Kwak et al., 2018).

Caffeine is not affected by the fermentation process and roasting (Duarte et al., 2010; Bressani et al., 2018). However, the content of chlorogenic acids can be modified. The content of chlorogenic acids increases during natural and pulped coffee fermentation and decreases after roasting (Martinez et al., 2017; Bressani et al., 2018). This behavior was expected because these compounds are degraded during roasting to form different volatile compounds among the chlorogenic acids (CGA) of some coffee hybrids and varieties studied by Duarte et al. (2010), the most relevant 5-CGA content. Also, almost all wet fermented coffee samples showed a higher content of this compound than pulped fermented coffee.

These results suggested that starter cultures' inoculation may influence the content of total polyphenols, antioxidant activity, and other bioactive compounds. However, the interaction of these compounds with different fermentation processes needs further study.

6.5.4 Protein content

The free amino acids, peptides, and proteins in green coffee are potential nitrogen precursors of coffee flavor and aroma. Storage of 11S protein is one of the most important and stores amino acids and peptides (Montavon et al., 2003). Montavon et al. (2003) suggested that besides the Maillard reaction, storage protein subunits undergo oxidative degradation and contribute to coffee flavors.

The late embryogenesis abundant protein (LEA protein) was identified in different post-harvesting coffee processes (Livramento et al., 2017). These authors suggest that it was related to the drying tolerance that the fruit had. The LEA protein is in high abundance when natural coffees are dried on terraces. This protein is present in plants and microorganisms and is strongly induced in conditions where the body needs to lower an oxidative process's adverse effects. The protein synergistically acts with sucrose and ensures an organism's cell stability to dryness (Livramento et al., 2017).

When high proteins remain after roasting, they produce a lighter color and increase the pyrazine group that generates desirable sensory attributes in the beverage, such as nuts, almonds, and sweet flavors (Hwang et al., 2012; Bressanello et al., 2017). A recent study by Bressani et al. (2020b) evaluated the protein profile of fermented coffees with and without starter cultures using matrix-assisted laser desorption ionization-time of flight mass spectrometry (MALDI-TOF MS). The results suggested that starter cultures, inoculation, and processing methods influenced coffee's protein profile during drying. The treatment inoculated with *S. cerevisiae* CCMA 0543 showed an increase in protein synthesis and less clustering with the other treatments without starter culture and inoculated with *Candida parapsilosis* (CCMA 0544), resulting in the highest sensory score (Bressani et al., 2018). Continuous studies on protein interaction must be carried out to elucidate the biochemical changes in coffee.

6.5.5 Volatile compounds

Specialty coffees with unique origins, high scores (according to the Specialty Coffee Association), and different flavors and aromas are highly valued by the influence of the so-called third coffee wave. In this scenario, consumers have turned the production chain's preparation methods, and sustainability is important (Guimarães et al., 2019). The origin, fruit maturation, post-harvest processes, roasting, grinding, and preparation methods influence the beverage's aromatic profile. In addition, each person may have a different perception when tasting the same beverage (Farah, 2019).

Biochemical changes that occur in green coffee influence the generation of volatile compounds during roasting. Most volatile compounds are generated during this process, explaining the differences in volatile profiles between green and roasted beans. More than 1000 compounds from different chemical groups have been identified in coffee (Lee and Shibamoto, 2002; Farah, 2019). The most abundant chemical classes in green beans are alcohols, aldehydes, and ketones (Yeretzian et al., 2002; Martinez et al., 2017; Bressani et al., 2020a). Hexanal, 1-hexanol, 2-heptanol, benzyl alcohol, and benzaldehyde found in coffee beans can be lost during thermal processing (Caporaso et al., 2018; Bressani et al., 2020a). However, many compounds are considered standard in green and roasted beans, such as phenylethyl alcohol, linalool, vanillin, limonene, and γ-butyrolactone (Bressani et al., 2020a; Lee et al., 2017a).

The increase in the beverage quality, flavor, and aroma of coffee promoted by the fermentative process has been studied in the last decade (Velmourougane, 2013; Evangelista et al., 2014b; Puerta Q and Molina, 2015; De Bruyn et al., 2017; de Carvalho Neto et al., 2017a; Bressani et al., 2018; Peñuela-Martínez et al., 2018; Haile and Kang, 2019). However, the microbiota in each region might influence the aromas. Evangelista et al. (2015) showed that the microbiota and flavor compound present during the fermentation of wet-processed coffee was different in the two regions studied (south of Minas Gerais and Cerrado regions). The proton-transfer-reaction mass spectrometry (PTR-MS) technique found that the volatile profile of wet fermented coffee samples can change according to the fermentation time. The fermentation process affects the development of flavor precursors; excessive fermentation time

(>72 h) might decrease the number of volatile compounds in coffee (Bodner et al., 2019).

The use of microorganisms as starter cultures in coffee fermentation is being widespread. Different species produce different metabolites that will turn into desirable volatiles during roasting, like esters, furans, pyrazines (Bressani et al., 2020a; de Melo Pereira et al., 2015; Evangelista et al., 2014b, 2014a; Gonzalez-Rios et al., 2007; Lee et al., 2017a; Martinez et al., 2019; Silva et al., 2013; Wang et al., 2019). Some microorganisms can use various metabolic pathways (Lee et al., 2017a; Martinez et al., 2019; Yoshida and Yokoyama, 2012). Therefore, the same microorganism might behave differently based on the type of processing (natural/pulped coffee/wet, aerobic/anaerobic, and others; Bressani et al., 2018, 2020a; Evangelista et al., 2014a, 2014b; Martinez et al., 2017).

The compounds mentioned above influenced the fermentation process and, consequently, the formation of aromatic compounds. Yeasts can catabolize amino acids through the Ehrlich pathway or use the acetyl-CoA produced in glycolysis to generate esters. In the same way, higher alcohols, aldehydes, ketone, and terpenoids can also be produced, and different molecules may be generated through the metabolism of carbon and nitrogen (Dzialo et al., 2017; Ribeiro et al., 2017a; Silva et al., 2013).

Martinez et al. (2019) evaluated the production of aromatic compounds in fermented coffees using yeast and bacteria. They identified several compounds with a pleasant aroma and selected potential flavor marker precursors. Heptadecanol and 4-hydroxy-2-methyl acetophenone can be used as markers by yeasts and guaiacol for *B. subtilis* species.

Despite the influence of numerous variables and processing, there is no doubt that fermentation improves the formation of flavor compounds after roasting, either by intensifying any descriptor or by making the beverage more complex. However, there is still no starter culture that is sensorially efficient for all coffee processing and types. Therefore, further studies on the relationship between fermentation and aromatic/flavor compounds should be investigated.

6.6 Quality, sensory analysis methods, and perspectives

Quality is defined as a degree that fulfills adequacy, conformity, and requirements in a product, system, or process for customers and other interested parties (International Organization for Standardization, 2015). However, the term is also associated with the needs and expectations, directly influencing its definition.

In food studies, quality embraces attributes easily perceived by human senses (e.g., taste, aroma, color, texture) and those that are not (e.g., food security, nutrition). In both circumstances, instrumentation is needed to measure a series of selected attributes with accuracy (Shewfelt, 1999).

Despite its comprehensive meaning, the quality of a product must be clearly defined and understood, mainly in coffee, because the attributes which defined quality have changed over the centuries. For example, ISO 9116 (International Organization for Standardization, 2004) established a standard to evaluate green coffee quality that takes into consideration the geographic origin, botanical origin, crop year, moisture content or loss of mass, total defect and foreign matter, damaged bean content by insects, bulk density, and bean size (Batista and Chalfoun, 2014). Due to the arrival of the third coffee wave, the term "quality" became more frequent and reinforced in the production chain (Samoggia and Riedel, 2018). Coffee quality was classified through protocols defined by the Specialty Coffee Association (SCA). Statistics showed that between 2012 and 2016, there was a 20.6% annual increase in the demand for specialty coffees (Guimarães et al., 2019).

Pre-harvest and post-harvest practices affect the physical and chemical characteristics of the beans. Laboratory protocols were implemented to access those characteristics. Other protocols involving tasting allowed attributes to be perceived through sensations/expectations using trained or untrained individuals, resulting in sensory responses. In the end, they all contribute to coffee quality, as shown in Figure 6.6.

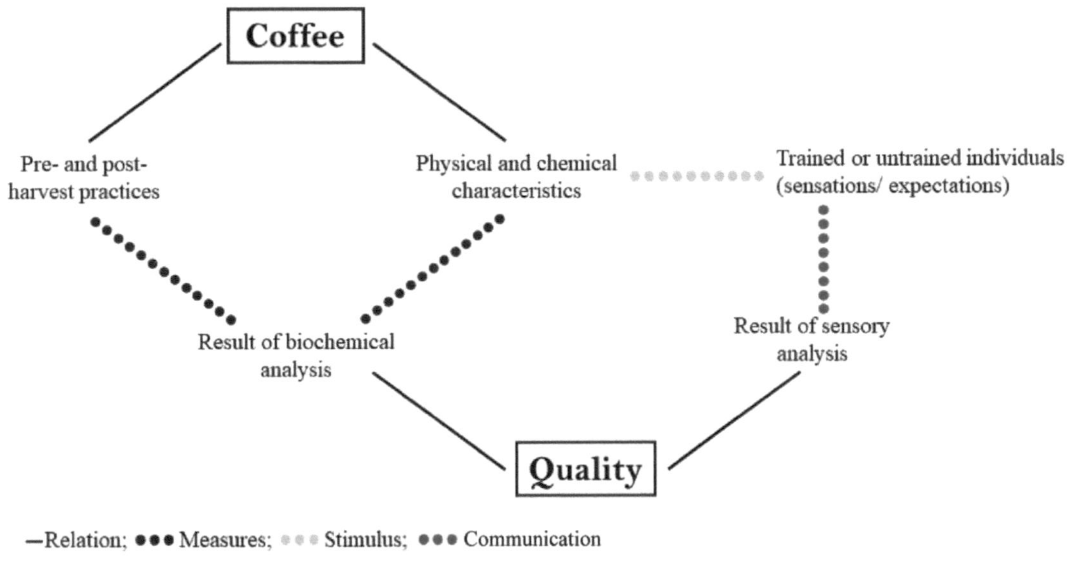

FIGURE 6.6 Simplified block diagram showing the combination of occurrences to obtain quality coffee.

Sensory analysis is the science that uses human senses (sight, smell, touch, taste, and hearing) to evaluate the characteristics or attributes. It is crucial to determine product acceptance, quality, and shelf life (Sharif et al., 2017; Mbela et al., 2018).

Besides the quality analysis of green beans, specialty coffees are evaluated sensorially for classification. The classification protocol is defined by the SCA and used worldwide (Gutiérrez-Guzmán et al., 2018). A panel of trained tasters carries out the analysis for Arabica coffees with Q-Grader certification. The attributes evaluated are aroma/fragrance, flavor, aftertaste, acidity, body, uniformity, balance, clean cup, and sweetness. Each attribute received a score on a scale from 6.0 to 10.0 points, with intervals of 0.25. The final score corresponded to the total sum of the evaluated attributes. Samples with a total score equal to or greater than 80 points are considered "specialty coffees" and can be subdivided into very good (80–84.99), excellent (85–89.99), and outstanding (>90). Tasters may also include sensory descriptors to provide additional information on each coffee (SCA, 2018).

The standardization of the sensory methodology is beneficial for producers, buyers, and tasters because it helps the purchase and sell of coffee. However, it is modified and adapted to different work styles (Gutiérrez-Guzmán et al., 2018). The Q-Graders' opinion is not always correlated with consumer preferences (Di Donfrancesco et al., 2014). Furthermore, consumers can be influenced by extrinsic factors, such as packaging, brands, and information (Li et al., 2019; Samoggia and Riedel, 2018; Bressani et al., 2021; Spence and Carvalho, 2020). Depending on the objective, different sensory methodologies can expand information about coffee consumers' choices and desires. The results from trained tasters and consumers can effectively reach different market segments.

Analyses carried out with consumers must use many participants to facilitate statistical analyzes, considering tasters' variability (Palermo, 2015). They are usually held in individual booths so that there is no interference between participants. However, some analyses can be carried out in coffee shops (Carvalho and Spence, 2019). Samples are handed out randomly, standardly, and balanced with three-digit encodings to minimize errors (Palermo, 2015).

Acceptance tests are used frequently and aim to know how much tasters like or dislike a sample. It can evaluate the acceptance globally from different processing, roasting levels, preparation methods, among others, or a specific attribute (Varela et al., 2014; Nguyen et al., 2016; Kwak et al., 2018). The structured hedonic scale of "n" points is used in applying this test, with the 9-points scale being the most common. The maximum and minimum impressions stand for liked highly (9) and disliked extremely (1) (Sharif et al., 2017). The unstructured and facial hedonic scale can also perform acceptance tests (Palermo, 2015).

Preference tests can express a preference between two (paired comparison) or more samples (ordination test). In this case, the evaluator will know the most/least preferred sample and determine if there is a significant difference between them (Varela et al., 2014; Li et al., 2019). However, the intensity of like or dislike for individual samples cannot be adequately determined by this method. Additionally, they can be applied in conjunction with other descriptive sensory tests to obtain complete results.

The descriptive sensory analysis provides detailed, accurate, reliable, and consistent results. Information on food and beverages' sensory characteristics is critical for the successful development and commercialization of new products (Ares and Jaeger, 2013).

"Check all that apply" (CATA) is a descriptive sensory analysis method used to discriminate samples and optimize food products (Valentin et al., 2012). This analysis consists of a list of attributes (words or phrases) from which the tasters must select all the attributes they consider appropriate to describe a product/food. The evaluators can choose the attributes using a focus group or, in the case of specialty coffees, using descriptors provided by Q-Graders from the "cupping." It is considered a simple methodology for both the evaluator and the taster. Its limitation is that it does not classify the samples and does not provide intensity. Also, CATA requires a large number of evaluators (Valentin et al., 2012). Several recent studies have used CATA as a sensory evaluation of coffees with consumers. Espitia-López et al. (2019) used the CATA task to describe samples of specialty coffees in Mexico. The authors obtained an efficient evaluation of the differences in the samples obtained from different preparation methods (espresso and French press). Bressani et al. (2021) found impressive results when using the CATA methodology. The participants were not able to perceive all the attributes described by trained tasters. Besides, they related fermented coffee to the citrus flavor, being influenced by the information.

On the other hand, the CATA test was used by Giacalone et al. (2019) to assess consumers' perception of different roasting defects. The methodology was also applied to know consumers' perceptions of different iced coffees (Heo et al., 2019). CATA was also used to check the influence of temperature on coffee and tea beverages (Pramudya and Seo, 2018).

Due to the complexity of coffee, a dynamic sensory analysis, such as temporal dominance of sensations (TDS), could help understand the flavors perceived over time (seconds). This analysis allows the evaluation of attributes perceived as dominant (the most striking perception). It is widely used with trained tasters, but it does not have to take a long time. In this step, the list of attributes that describe the product to be evaluated is already defined. Almost all sensory analysis software provides a TDS module (Di Monaco et al., 2014). Sensomaker is an example of free software that has TDS (Pinheiro et al., 2013). All selected attributes are shown on the same computer screen, and the tester identifies the sensations perceived during the time established for analysis (Pinheiro et al., 2013).

It should be noted that a chosen attribute is registered as dominant until another attribute is chosen. During a test, tasters are free to choose the same attribute several times, but they may also not use all the attributes provided (Pineau et al., 2009). At the end of the analysis, the TDS curves show the attributes' dominance rates against the time of each sample (Bruzzone et al., 2013; Meillon et al., 2010).

TDS is a valuable tool for evaluating different food products and is widely used in coffee samples. The method was applied to discriminate and characterize fermented coffee samples (Evangelista et al., 2014a, 2014b; Ribeiro et al., 2017b) and find a perception in the different degrees of roasting (Barbosa et al., 2019).

Due to sensory fatigue, the number of samples can be limiting for the use of some methodologies. The return of participants in several sessions can also make it challenging to perform sensory analysis. When the sample set is relatively large (8–20) (Hamilton and Lahne, 2020), the Sorting task can be used as a quick descriptive method. It can be associated with CATA, facilitating the grouping of the samples. In this methodology, the tasters evaluate the samples and systematically group them into categories based on their criteria. It can be performed by trained tasters or by consumers (with variation in the number of participants; Fleming et al., 2015; de Alcantara and Freitas-Sá, 2018).

The Sorting task can be a helpful tool in quality control, marketing research to select products/samples before another test, and even as a substitute for triangular tests (Chollet et al., 2011). Although the methodology is widely used, there are few studies carried out with coffee. Spencer et al. (2016) applied the Sorting task to develop a new wheel of flavors to simplify and standardize the industry's coffee description process. More recently, Hamilton and Lahne (2020) carried out a study to classify cold coffee samples from different origins and toast at different levels using sorting.

The choice of a sensory test(s) depends on the sensory measurement's objective and quality. Tests provide useful information about human perception of a product's characteristics, ingredients, processing, packaging, and shelf-life changes. They are essential in quality control and marketing research. Sensory analysis can help understand the preferences and desires of consumers of specialty coffees, creating better communication between the market and the final consumer. An appropriate sensory program is convenient for the company to meet consumer expectations and guarantee a higher chance of success.

6.7 Conclusions

Environmental conditions and pre- and post-harvest coffee processes affect both the coffee fruit and microbial diversity composition. The coffee processes are known as the natural, pulped coffee, and wet process. However, the variations process – such as fermentation hours, fruit fermentation followed by pulped and drying under shade covers (honey, red and black honey) – is an alternative to change and intensifies the sensory profile. Therefore, more and more, the specialty coffee market and consumers have been interested in fermented coffee.

Starter cultures are essential to control better and standardize the fermentative process of coffee to guarantee coffee production with unique and desirable sensory profiles. Three methods of inoculation of starter cultures have been proposed to natural and pulped coffee processing: (1) the direct method, which consists of spraying the microbial solution over the coffee fruits or beans, (2) an open batches bioreactor that consists of fermenting coffee in open containers, and (3) a close batches bioreactor that consists of fermenting coffee in closed containers. All methods increased the sensory scores and modified the sensory profile of the beverage. However, the second and third methods are more suitable when using starter cultures. Starter cultures in processing in close batch bioreactors reduced fermentation time and improved beverage quality in wet processing. The starter culture's fermentative performance must be evaluated for each type of processing, region, and variety of coffee.

The chemical modifications that occur during the post-harvest coffee process are essential for forming flavor and aroma precursors. Analysis of temperature, Brix, and pH value are simple parameters to assess coffee fermentation on the farms (in loco). Complementary sugar concentration analysis, volatile compounds is also essential to verify more specific changes during this stage. Furthermore, the changes in each microorganism's metabolic routes, the influence of amino acids, and protein concentration involved in the aroma and flavor of the coffee after roasting must be studied. More than 1000 compounds from different chemical groups have been identified in green and roasted coffee, and there is an evident change in the predominant chemical classes before and after fermentation and after roasting, influencing the flavor and aroma of the beverage. However, the elucidation of how the changes occur in the coffee seed during the fermentation process with and without starter cultures is a great challenge for coffee researchers.

Coffee quality is assessed by a panel of trained tasters with Q-Grader certification, following the methodology defined by the Specialty Coffee Association (SCA). Furthermore, the Q-Graders' opinions did not always correlated with consumer preferences. Thereby, the validation of other sensory methodologies – acceptance test, CATA, TDS, and sorting test – is an alternative to strengthening data to describe and evaluate coffee quality.

REFERENCES

Agate AD et al., 1966. Applied Microbio. 14:256–260.

Aguiarb J et al., 2016. Trends in Food Sci Tech. 58: 21–39. https://doi.org/10.1016/j.tifs.2016.10.012

Ardekani RS., Rath J., 2020. J Consumer Culture. 20:122–140. https://doi.org/10.1177/1469540517736557

Arellano-González MA et al. 2011. Food Tech Biotech. 49: 374–378.

Ares G., Jaeger SR. 2013. Food Qual Prefer. 28:141–153. https://doi.org/10.1016/j.foodqual.2012.08.016

Arunga RO. 1992. In: Wood, B. J. B. (Ed.), *The lactic acid bacteria*. Springer US, pp. 409–429. https://doi.org/10.1007/978-1-4615-3522-5_16

Avallone S et al. 2001. Current Microbio. 42:252–256.

Avallone S et al. 2002. Int. l J Food Sci Tech. 37:191–198.

Barbosa MSG et al. 2019. J. Culinary Science & Tech. 17:453–464.

Batista LR., Chalfoun M., 2014. *Cocoa and Coffee Fermentations*. CRC Press, pp. 477–508.

Belitz HD et al. 2009. *Coffee, Tea, Cocoa*. Springer, Berlin, Heidelberg. https://doi.org/10.1007/978-3-662-07279-0_22

Bodner M et al 2019. European Food Res. Tech. 245:1499–1506.

Brando CHJ, Brando MF. 2015. *Cocoa and Coffee Fermentations*. CRC Press, Boca Raton, FL, pp. 367–396.

Bressanello D et al. 2017. Food Chem. 214:218–226. http://dx.doi.org/10.1016/j.foodchem.2016.07.088

Bressani APP et al. 2018. LWT – Food Sci. Tech. 92:212–219. https://doi.org/10.1016/j.lwt.2018.02.029

Bressani APP et al. 2020a. Int. Food Res. J. 128:108773. https://doi.org/10.1016/j.foodres.2019.108773

Bressani APP et al. 2020b. Pesqui. Agropecu. Bras. 55:e01159. https://doi.org/10.1590/S1678-3921.pab2020.v55.01159

Bressani APP et al. 2021. Food Sci. Tech. https://doi.org/10.1590/fst.30720

Bruzzone F et al. 2013. Int. Dairy J. 29:124–134. https://doi.org/10.1016/j.idairyj.2012.10.012

Campos-Vega R et al. 2015. Trends in Food Sci. Tech. 45:24–36.

Caporaso N et al. 2018. Food Res. Int. 108:628–640.

Carvalho FM., Spence C., 2019. Food Qual. Prefer. 75:157–169.

Cheng B et al. 2016. Trends in Food Sci. Tech. 57:20–30.

Chollet S et al. 2011. Food Qual. Prefer. 22:507–520.

Correa EC et al. 2014. Food Bioprocess Technol. 7:3166–3174.

da Mota MCB et al. 2020. Food Res. International. 136:109482.

de Alcantara M., Freitas-Sá DGC., 2018. Braz. J Food Tech. 21:1–12.

de Bruyn F et al. 2017. Applied and Environ. Microbiol. 83:1–16.

de Carvalho Neto DP et al. 2017a. Fermentation. 3:2–11. https://doi.org/10.3390/fermentation3010011

de Carvalho Neto DP et al. 2017b. Food Technol. Biotechnol. 56: 90–95. https://doi.org/10.17113/ftb.56.01.18.5441

de Carvalho Neto DP et al. 2018. Food Biosci. 26:80–87. https://doi.org/10.1016/j.fbio.2018.10.005

de Carvalho Neto DP et al. 2020. Food Biotech. 34:172–192. https://doi.org/10.1080/08905436.2020.1746666

de Melo Pereira GV et al. 2014. Int. J. Food Microbiol. 188:60–66. https://doi.org/10.1016/j.ijfoodmicro.2014.07.008

de Melo Pereira GV et al. 2015. Food Res. Int. 75:348–356. https://doi.org/10.1016/j.foodres.2015.06.027

de Melo Pereira GV et al. 2019. Food Chem. 272:441–452. https://doi.org/10.1016/j.foodchem.2018.08.061

Di Donfrancesco B et al. 2014. J. Sens. Stud. 29:301–311. https://doi.org/10.1111/joss.12104

Di Monaco R et al. 2014. Trends Food Sci Technol. 2:104–112. https://doi.org/10.1016/j.tifs.2014.04.007

Diviš P et al. 2019. Potravinarstvo Slovak J. Food Sci. 13:344–350. https://doi.org/10.5219/1062

Djossou O et al. 2011. Anaerobe. 17:267–272. https://doi.org/10.1016/j.anaerobe.2011.03.006

Duarte GS et al. 2010. Food Chem. 118:851–855. https://doi.org/10.1016/j.foodchem.2009.05.042

Dzialo MC et al. 2017. FEMS Microbiology Rev. 41:S95 – S128. https://doi.org/10.1093/femsre/fux031

Elhalis H et al. 2020. Int. J Food Microbiol. 321:108544. https://doi.org/10.1016/j.ijfoodmicro.2020.108544

Espitia-López J et al. 2019. Int. J. Food Prop. 22:967–973. https://doi.org/10.1080/10942912.2019.1619577

Esquivel P., Jiménez VM., 2012. Food Res. Int. 46:488–495. https://doi.org/10.1016/j.foodres.2011.05.028

Evangelista SR et al. 2014a. Food Microbio. 44:87–95. https://doi.org/10.1016/j.fm.2014.05.013

Evangelista SR et al. 2014b. Food Res Int. 61:183–195. https://doi.org/10.1016/j.foodres.2013.11.033

Evangelista SR et al. 2015. Int. J Food Microbiol. 210:102–112. https://doi.org/10.1016/j.ijfoodmicro.2015.06.008

Farah A., 2019. *Coffee Production, Quality and Chemistry*. Royal Society of Chemistry.

Farah A., Lima ÂG., 2019. In: Farah, Adriana (Ed.), *Coffee: Chemistry, Quality and Health*. The Royal Society of Chemistry, pp. 517–536. https://doi.org/10.1039/9781782622437

Fleming EE et al 2015. Food Qual. Prefer. 45:41–49. https://doi.org/10.1016/j.foodqual.2015.05.004

Frank HA et al. 1965. Applied Microbiol. 13:201–207.

Gaime-Perraud I et al. 1993. Micología Neotropical Aplicada. 6:95–103.

Garna H et al. 2004. J Agric. Food Chem. 52:4652–4659. https://doi.org/10.1021/jf049647j

Giacalone D et al. 2019. Food Qual. Prefer. 71:463–474. https://doi.org/10.1016/j.foodqual.2018.03.009

Gonzalez-Rios O et al. 2007. J Food Comp. Analysis. 20:297–307. https://doi.org/10.1016/j.jfca.2006.12.004

Górnaś P et al. 2016. Europ. Food Res. Tech. 242: 641–653. https://doi.org/10.1007/s00217-015-2572-1

Guimarães ER et al. 2016. Organizações Rurais e Agroindustriais. 18:214–227.

Guimarães ER et al. 2019. J. Food Products Marketing. 25:49–71. https://doi.org/10.1080/10454446.2018.1478757

Gutiérrez-Guzmán N et al. 2018. Coffee Sci. 13:401–409.

Haile M., Kang WH., 2019. J. Food Quality. 2019:1–6. https://doi.org/10.1155/2019/4836709

Hamdouche Y et al. 2016. Control. 65:112–120. https://doi.org/10.1016/j.foodcont.2016.01.022

Hameed A et al. 2018. Compr. Rev. Food Sci. Food Saf.17:1184–1237. https://doi.org/10.1111/1541-4337.12365

Hamilton LM, Lahne J. 2020. Food Qual. Prefer. 83:103889. https://doi.org/10.1016/j.foodqual.2020.103889

Hatiningsih S et al. 2018. J. Food Sci. Tech. 5:123–138.

Heo J et al. 2019. Foods. 8:1–14. https://doi.org/10.3390/foods8080344

Hii CL., Borém FM., 2019. *Drying and roasting of cocoa and coffee*. CRC Press, Taylor & Francis Group.

Holzapfel W. 1997. Food Control. 8:241–258. https://doi.org/10.1016/S0956-7135(97)00017-0

Hwang CF et al. 2012. Int. J. Food Sci. Tech. 47:2117–2126. https://doi.org/10.1111/j.1365-2621.2012.03078.x

International Organization for Standardization, 2004. ISO 9116: 2004. Green coffee – Guidelines on methods of specification. https://doi.org/10.1163/ej.9789004163300.i-1081.154

International Organization for Standardization, 2015. ISO 9000:2015-Quality management systems -Fundamentals and vocabulary. https://doi.org/10.1007/978-3-662-48986-4_1603

Jayaram VB et al. 2014. Food Chem. 151:421–428. https://doi.org/10.1016/j.foodchem.2013.11.025

Junqueira ACO et al. 2019. Sci. Rep. 9:1–10. https://doi.org/10.1038/s41598-019-45002-8

Knopp S et al. 2006. Eur. Food Res. Technol. 223:195–201. https://doi.org/10.1007/s00217-005-0172-1

Koshiro Y et al. 2015. European Chemical Bulletin. 4:378–383. https://doi.org/10.17628/ECB.2015.4.378

Kwak HS et al. 2018. J. Food Quality. 2018:1–8. https://doi.org/10.1155/2018/5967130

Lee KG., Shibamoto T., 2002. J. Agri. Food Chem. 50:4947–4952. https://doi.org/10.1021/jf0255681

Lee LW et al. 2017a. LWT – Food Sci. Technol. 80:32–42. https://doi.org/10.1016/j.lwt.2017.01.070

Lee LW et al. 2017b. LWT – Food Sci. Technol. 77:225–232. https://doi.org/10.1016/j.lwt.2016.11.047

Leong K et al. 2014. Curr. Microbiol. 68:440-447. https://doi.org/10.1007/s00284-013-0495-2

Li J et al. 2019. Food Qual. Prefer. 71:447–451. https://doi.org/10.1016/j.foodqual.2018.08.006

Livramento KG et al. 2017. Food Chem. 221:1874–1882. https://doi.org/10.1016/j.foodchem.2016.10.069

Malta MR et al. 2020. Food Res. Int. 134:109219. https://doi.org/10.1016/j.foodres.2020.109219

Marsh AJ et al. 2014. Trends in Food Sci. Tech. 38:113–124. https://doi.org/10.1016/j.tifs.2014.05.002

Martinez SJ et al. 2017. Food Res. Int. 102:333–340. https://doi.org/10.1016/j.foodres.2017.09.096

Martinez SJ et al. 2019. Frontiers in Microbiol. 10:1287. https://doi.org/10.3389/fmicb.2019.01287

Martinez SJ et al. 2021. Food Res. Int. 139: 109921-10. https://doi.org/10.1016/j.foodres.2020.109921

Martins PMM et al. 2019. J. Sci. Food Agri. 99:5638–5645. https://doi.org/10.1002/jsfa.9820

Martins PMM et al. 2020. Food Res. Int. 129(2020):108872. https://doi.org/10.1016/j.foodres.2019.108872

Masoud W et al. 2004. Yeast. 21:549–556. https://doi.org/10.1002/yea.1124

Masoud W et al. 2005. Yeast. 22:1133–1142. https://doi.org/10.1002/yea.1304

Massawe GA., Lifa SJ., 2010. Int. J. Postharvest Tech. Innovation. 2:41–81. https://doi.org/ 10.1504/IJPTI.2010.038187

Mbela DEN et al. 2018. African J. Food Sci. 12:297–308. https://doi.org/10.5897/ajfs2018.1738

Meillon S et al. 2010. Food Qual. Prefer. 21:732–740. https://doi.org/10.1016/j.foodqual.2010.06.005

Montavon P et al. 2003. J. Agri. Food Chem. 51:2328–2334. https://doi.org/10.1021/jf020831j

Mussato SI et al. 2011. Food Bioprocess Technol. 4, 661. https://doi.org/10.1007/s11947-011-0565-z

Nasanit R., Satayawut K., 2015. Kasetsart Journal – Natural Science. 49:32–41.

Nguyen T et al. 2016. J. Food Sci. 81:S2817–S2823. https://doi.org/10.1111/1750-3841.13521

Oliveira NMV et al. 2013. Can. J. Micro. 59:221–230. https://doi.org/10.1139/cjm-2012-0674

Palermo JR. 2015. Métodos para avaliação sensorial, in: Palermo, J. R. (Ed.), *Análise Sensorial – Fundamentos e Métodos*. Atheneu, Rio de Janeiro, Brazil, pp. 57–111.

Pee W., Castelein JM., 1972. J. Food Sci. 37:171–174. https://doi.org/10.1111/j.1365-2621.1972.tb03412.x

Peñuela-Martínez AE et al. 2018. Coffee Sci. 13:465–476. https://doi.org/10.25186/cs.v13i4.1486

Pereira RTG et al. 2005. Ciênc. Agrotec. 29(6):1112–1116. https://doi.org/10.1590/s1413-70542005000600002

Pereira GVM et al. 2016. Int. J. Food Sci. Tech. 51:1689–1695. https://doi.org/10.1111/ijfs.13142

Pineau N et al. 2009. Food Qual. Prefer. 20:450–455. https://doi.org/10.1016/j.foodqual.2009.04.005

Pinheiro ACM et al. 2013. Agrotecnologia. 37(3):199–201. https://doi.org/10.1590/S1413-70542013000300001

Poisson L et al. 2018. J. Agri. Food Chem. 66:2422–2431. https://doi.org/10.1021/acs.jafc.6b04849

Pothakos V et al. 2020. Curr. Res. Biotech. 2:1–15. https://doi.org/10.1016/j.crbiot.2020.02.001

Pramudya RC., Seo HS., 2018. Fron. Psychology. 8:2264. https://doi.org/10.3389/fpsyg.2017.02264

Puerta Q GIP., Molina JGE., 2015. Gerencia Técnica/Programa Investigación Científica Fondo Nacional del Café. 1–12.

Puerta Q GIP et al. 2012. Cenicafé, 63(2):58–78.

Quintão RT., Brito EPZ., 2016. Revista Brasileira de Marketing. 15:01–15. https://doi.org/10.5585/remark.v15i1.3042

Ramos DMB et al. 2010. Acta Scientiarum Agronomy. 32(3):397–402. https://doi.org/10.4025/actasciagron.v32i3.3361

Ribeiro LS et al. 2017a. Food Res. Int. 92:26–32. https://doi.org/10.1016/j.foodres.2016.12.011

Ribeiro LS et al. 2017b. LWT – Food Sci. Technol. 82:32–38. https://doi.org/10.1016/j.lwt.2017.04.008

Ribeiro LS et al. 2018. Ann. Micro. 68:705–716. https://doi.org/10.1007/s13213-018-1377-4

Samoggia A., Riedel B., 2018. Appetite. 129:70–81 https://doi.org/10.1016/j.appet.2018.07.002

SCA, 2018. Coffee Standards Table of Contents (p. 14). https://static1.squarespace.com/static/584f6bbef5e23149e5522201/t/5bd985c1352f53cb4cc1be48/1540982325719/Coffee+Standards-Digital.pdf

Schwan RF et al. 2012. In: Hui, Y. H. and Evranuz, E. O. (Eds.), *Handbook of Plant-Based Fermented Food and Beverage Technology*. CRC Press, Boca Raton, FL, pp. 677–690.

Sharif MK et al. 2017. In: Jeantet, R. et al. (Eds.) *Handbook of Food Science and Technology*. pp. 362–386. https://doi.org/10.1002/9781119296225

Shewfelt RL., 1999. Postharvest Bio. Tech. 15:197–200. https://doi.org/10.1016/S0925-5214(98)00084-2

Silva CF., 2015. In: Schwan, R. F. and Fleet, G. H. (Eds.), *Cocoa and Coffee Fermentations*. CRC Press, pp. 397–423.

Silva CF et al. 2000. Int. J. Food Microbiol. 60:251–260. https://doi.org/10.1016/S0168-1605(00)00315-9

Silva CF et al. 2008a. Food Microbiol. 25:951–957. https://doi.org/10.1016/j.fm.2008.07.003

Silva CF et al. 2008b. Braz. J. Microbio. 39:521–526. https://doi.org/10.1590/S1517-83822008000300022

Silva CF et al. 2013. World J. Microbiol. Biotech. 29:235–247. https://doi.org/10.1007/s11274-012-1175-2

Skeie TR., 2003. Norway and Coffee. Newsletter of the Roasters Guild. URL https://web.archive.org/web/20030824171335/www.roastersguild.org/052003_norway3.shtml (accessed 8.16.20).

Souza ML et al. 2017. Acta Sci. Agron. 393:349–358. https://doi.org/10.4025/actasciagron.v39i3.32659

Spence C., Carvalho FM., 2020. Food Qual. Prefer. 80:103802. https://doi.org/10.1016/j.foodqual.2019.103802

Spencer M et al. 2016. J. Food Sci. 81: S2997–S3005. https://doi.org/10.1111/1750-3841.13555

Suarez-Quiroz M et al. 2004. Int. J Food Sci. Tech. 39:501–507. https://doi.org/10.1111/j.1365-2621.2004.00810.x

Valentin D et al. 2012. Int. J Food Sci. Tech. 47:1563–1578. https://doi.org/10.1111/j.1365-2621.2012.03022.x

Vandenberghe LPS et al. 2018. Current Dev. Biotech. Bioeng. pp. 415–434. https://doi.org/10.1016/B978-0-444-63990-5.00018-9

Varela P et al. 2014. Food Qual. Prefer. 32:152–159. https://doi.org/10.1016/j.foodqual.2013.03.004

Vega FE et al. 2005. J Basic Microbio. 45:371–380.

Vega FE et al. 2010. Fungal Ecol. 3:122–138.

Velmourougane, K. 2013. Proc. Natl. Acad. Sci. India Sect. B Biol. Sci. 83:233–239. https://doi.org/10.1007/s40011-012-0130-1

Vilela DM et al. 2010. Food Microbio. 27:1128–1135.
Wang et al. 2019. J. Sci. Food and Agri. 99:409–420. https://doi.org/10.1002/jsfa.9202
Wang C et al. 2020a. LWT – Food Sci. Tech. 120:108930 https://doi.org/10.1016/j.lwt.2019.108930
Wang C et al. 2020b. Food Res. Int. 136:109452. https://doi.org/10.1016/j.foodres.2020.109452
Worku M et al. 2018. Food Res. Int. 105:278–285. https://doi.org/10.1016/j.foodres.2017.11.016
Yeretzian C et al. 2002. Eur. Food Res. Tech. 214:92–104. https://doi.org/10.1007/s00217-001-0424-7
Yoshida S., and Yokoyama, A., 2012. J. Biosci. Bio. 113:556–561. https://doi.org/10.1016/j.jbiosc.2011.12.017.
Zhang O et al. 2018. J. Food Sci. Tech. 55:964–976. https://doi.org/10.1007/s13197-017-3004-y
Zhang SJ et al. 2019a. Frontiers in Micro. 10:2621. https://doi.org/10.3389/fmicb.2019.02621
Zhang SJ et al. 2019b. Applied and Env. Microbio. 85:02635–18. https://doi.org/10.1128/AEM

7

Coffee polyphenols: Biochemical, processing, and health insights

Wenny Bekti Sunarharum, Tunjung Mahatmanto, Dego Yusa Ali, Yuniar Ponco Prananto, and Paulus Immanuel Nugroho

CONTENTS

7.1 Introduction

Coffee is a multibillion-dollar industry with an increasing trend in total global production (ICO, 2020a). The global trade of coffee has been supported by two major species, *Coffea arabica* (Arabica) and *C. canephora* (Robusta), which account for around 62% and 38% of total traded coffee, respectively (ICO, 2020b). The growing popularity of coffee has been attributed to its social utility, economic value, and political importance, boosted by the global economic development and connectedness (Tucker, 2017). In addition, the health benefits associated with coffee consumption appear to have gained more interest, particularly with regard to the polyphenol content in coffee brews. In the following, we will discuss the biochemistry of coffee polyphenols and the effect of processing on these molecules. In addition, we will also briefly discuss about the potential health effects of coffee polyphenols.

7.2 Phenolic compounds in coffee

Phenolic compounds are a large class of secondary metabolites widely found in higher plants and plant foods such as fruits, vegetables, cereals, and legumes (Cheynier, 2005; Manach et al., 2004; Vuolo et al., 2018). In plants, phenolic compounds play important roles in defense mechanisms (protection against UV, microbial infection, and pests), pollination, and seed dispersal (Crozier et al., 2006; Jaganath and Crozier, 2010). Beyond their native functions in plants, phenolic compounds in tea, wine, coffee, juice, beer, and cocoa are known to have various beneficial health effects, including protection against chronic and degenerative diseases (Manach et al., 2004; Del Rio et al., 2013).

The structures of phenolic compounds are diverse, from simple phenolic acids to complex polyphenols, characterized based on the number of phenol rings, bearing at least one benzene ring with one or more hydroxyl groups (Manach et al., 2004; Crozier et al., 2006; Del Rio et al., 2013). Based on the number of phenol rings that act as the basic structure and what bind the rings together, phenolic compounds can be classified into two major classes: the flavonoids and the non-flavonoids. Further grouping into subclasses is based on what substitutions apply to the main structure (Manach et al., 2004; Del Rio et al., 2013)

Unlike the non-flavonoid structures, flavonoids consist of two aromatic rings with three-carbon bridge connections. Flavonoids can be divided into flavones, flavonols, flavan-3-ols, isoflavones, flavanones, and anthocyanidins, while non-flavonoids can be divided into phenolic acids, stilbenes, and lignans (Manach et al., 2004; Del Rio et al. 2013). Phenolic acids can further be divided into derivatives of benzoic acid, such as gallic acid and hydroxybenzoic acids, and derivatives of cinnamic acid or hydroxycinnamic acids (Manach et al., 2004; Maru et al., 2013). The latter, hydroxycinnamic acids, mainly occur as conjugates and are rarely found in their free form (Manach et al., 2004; Del Rio et al., 2013). Examples of hydroxycinnamic acids include caffeic, ferulic, *p*-coumaric, and sinapic acids (Figure 7.1). Certain *trans*-hydroxycinnamic acids (e.g., caffeic acid, can form esters with quinic acid and create chlorogenic acids (CGAs; Figure 7.1). CGAs constitute

DOI: 10.1201/9781003043133-9

the most abundant polyphenols in plants and plant-derived foods and beverages, including coffee (Clifford, 1985, 1999, 2000).

In coffee, flavonoids and non-flavonoids are found in every part of the cherry, including the pulp and seeds, with varying compositions. The pulp contains more tannins, anthocyanins, and lignans, whereas the seeds contain more chlorogenic acids with minor amounts of other phenolic compounds (Farah and Donangelo, 2006). The compositional variation of coffee polyphenols is affected by a number of factors including plant growth, fruit maturity, and post-harvest processing. These factors are important to consider because they will also affect the flavor and thus the quality of coffee, which is characterized by the unique compositions of the volatile and non-volatile components of coffee formed during their processing (Sunarharum et al., 2014).

In comparison with other popular beverages, coffee is regarded as the major source of polyphenols. Filtered coffee provides polyphenols in approximately twice the concentration of black tea, green tea, and red wine (Figure 7.2). The polyphenol content of filtered coffee is ten times higher than that of beer (dark and black beer). Remarkably, among the many different types of coffee, filtered coffee contains the lowest total phenolic compounds, the highest being instant coffee followed by espresso and Turkish coffee (Gorjanović et al., 2017).

Differences in the composition and concentration of polyphenols in coffee may arise from the different techniques applied to brew the ground coffee beans. For example, brewing time and method (e.g., espresso and filtering) are known to affect the extraction efficiency of polyphenols from coffee (Ludwig et al., 2012). In addition, coffee-to-water ratio also affects the extraction of polyphenols and other flavor compounds from coffee, thus creating coffee with different sensory attributes (Andueza et al., 2007; Sunarharum et al., 2020; Fibrianto et al., 2018).

From a compositional perspective, coffee contains CGAs, CGA isomers, and caffeic acids as its major polyphenols (Tresserra-Rimbau et al., 2015). In comparison to coffee, beer contains gallic acid (Habschied et al., 2020), ferulic acid, and sinapic acid (Socha et al., 2017), while black and green tea infusions contain catechins as the major phenolic compounds (Zhao et al., 2019). In apple juice (da Silva et al., 2019) and wine (Georgiev et al., 2014), flavonoids and flavanols are the dominant polyphenols. The phenolic compounds contained in the aforementioned beverages have been associated with many health outcomes related to their antioxidant properties. The antioxidant activity of some popular beverages is given in Table 7.1.

Table 7.1 shows that different beverages have different antioxidant activity. It should be noted that the values listed in

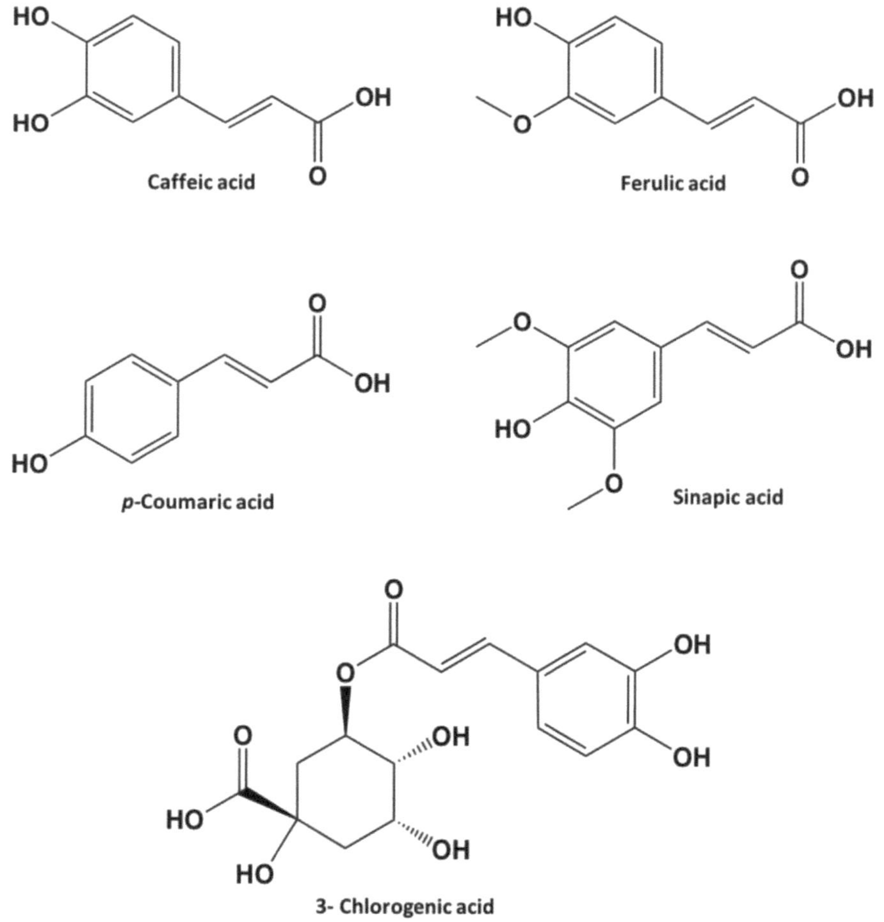

FIGURE 7.1 Structures of caffeic, ferulic, *p*-coumaric, sinapic acids, and 3-chlorogenic acid.

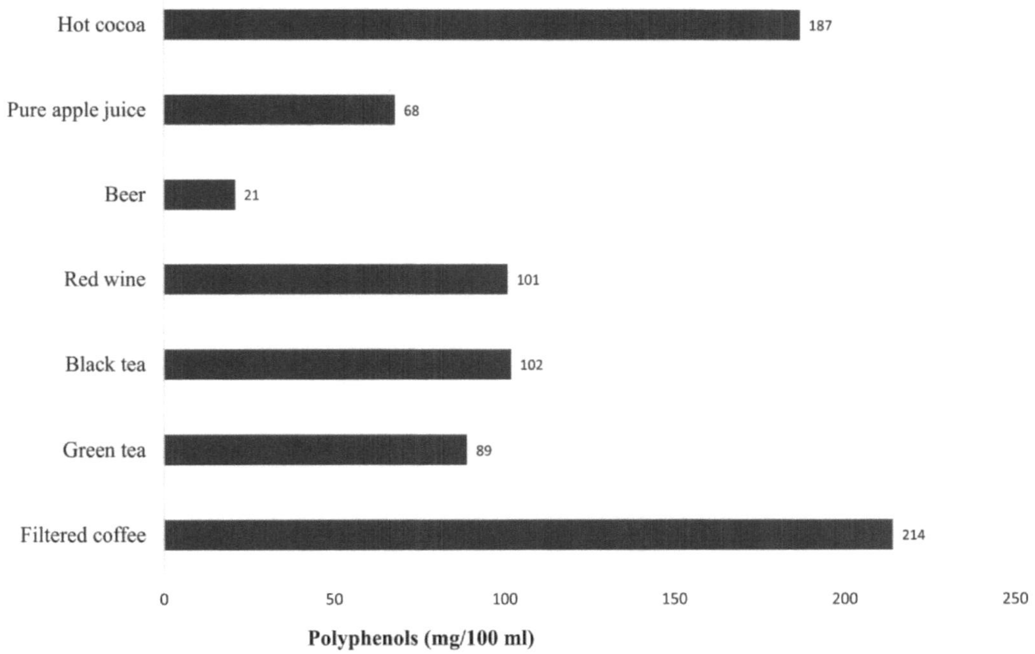

FIGURE 7.2 Polyphenol content of some popular beverages (polyphenol content expressed as the sum of individual polyphenols as determined by chromatography).

Sources: Concentration in filtered coffee, green tea, black tea, red wine, and pure apple juice were obtained from Pérez-Jiménez et al. (2010); concentration in beer was averaged from polyphenols identified in dark beer and black beer by Habschied et al. (2020); concentration in hot cocoa was reported as flavanols in Wiswedel et al. (2004).

TABLE 7.1
Antioxidant Activity of Some Popular Beverages

Beverages	Antioxidant activity		Method/ standard*	Reference
Filtered coffee	3.99–6.89	mmol TE/L	DPPH	Gorjanović et al., 2017
	267	mg/100 mL	Folin	Pérez-Jiménez et al., 2010
Green tea	1.498.51–2.532.41	mmol TE/kg	TEAC	Zhao et al., 2019
	62	mg/100 mL	Folin	Pérez-Jiménez et al., 2010
Black tea	471.17–1034.29	mmol TE/kg	TEAC	Zhao et al., 2019
	104	mg/100 mL	Folin	Pérez-Jiménez et al. 2010
Wine	270–581	mmol TE/L	ABTS	Di Lorenzo et al., 2017
	215	mg/100 mL	Folin	Pérez-Jiménez et al. 2010
Beer	5.9	mmol TE/L	DPPH	Habschied et al., 2020
	27.9[a]	mg/100 g	Folin	Pérez-Jiménez et al., 2010
Apple juice	0.358–0.908	mmol TE/L	FRAP	da Silva et al., 2019
	34	mg/100 mL	Folin	Pérez-Jiménez et al., 2010
Hot cocoa	43	mmol TE/kg	ORAC	Stahl et al., 2009

Notes: [a] Calculated based on per serving data of 160 mg in 574 g beer from Pérez-Jiménez et al. (2010), presented as mg/100 g; * DPPH = 2,2-diphenyl -1-picryl-hydrazyl-hydrate; TEAC = Trolox equivalent antioxidant capacity (mentioned as TEAC assay); ABTS = 2,2 -azino-bis(3-ethylbenzothi azoline-6-sulphonic acid); FRAP = ferric reducing antioxidant power; ORAC = oxygen radical absorbance capacity; Folin = Folin assay.

Table 7.1 are calculated as Trolox (vitamin E analog) equivalent, and it is difficult to directly compare the data because of the different methods used for the analysis. Furthermore, other factors from the nature of samples to the methods of preparation (e.g., water ratio and temperature) may also influence the results.

The consumption of polyphenol-rich beverages such as coffee in considerable amounts has been associated with a number of health benefits because polyphenols exert many functional properties including protection against antioxidative stress and degenerative diseases (Miranda et al., 2017; Vamanu, 2019; Silva and Pogačnik, 2020). One of the most important polyphenols in coffee that has been extensively studied with regard to their associated health effects is CGAs. As one of the richest sources of CGAs in the human diet (Clifford, 1999, 2000), coffee has contributed to the increase in the daily intake of polyphenols in the population of coffee consumers. A cup of Arabica coffee brew (200 mL) typically contains 70 to 200 mg of CGAs, whereas Robusta coffee typically contains a higher concentration, between 70 and 350 mg (Clifford, 1999).

In green coffee beans, the major CGA groups are caffeoylquinic acids (CQA) (3-,4-,5-CQA), dicaffeoylquinic acids (diCQA) (3-,4-,5-diCQA), feruloylquinic acids (FQA) (3-,4-,5-FQA), *p*coumaroylquinic acids (*p*CoQA), and six mixed diesters of caffeoylferuloyl-quinic acids (CFAQ; Clifford et al., 2003; Farah and Donangelo, 2006). The abundance of the CGA isomers from the most to the least abundant is CQA, FQA, and diCQA, followed by other minor CGAs such as diferuloylquinic acids, dimethoxycinamoylquinic acids, caffeoyldimethoxycinamoylquinic acids, and feruloyldimethoxycinamoylquinic (Clifford et al., 2006; Farah and Donangelo, 2006; Farah and de Paula Lima, 2019). Among the 30 known CGA isomers, 5-caffeoylquinic acid (5-CQA) is considered the most important as it accounts for at least 50% of the total CGA in coffee (Farah and Donangelo, 2006; Farah and de Paula Lima, 2019) and therefore measurement of 5-CQA often represents coffee chlorogenic acid. The structures of some important CGAs in coffee are presented in Figure 7.3.

There are more than 124 coffee species worldwide, but only a few are being cultivated such as *C. canephora, C. arabica*, and *C. liberica* (Davis et al., 2019; Sunarharum et al., 2019**a**). In terms of CGA content, Robusta coffee has approximately twice the concentration of CGAs as Arabica coffee (Farah and Donangelo, 2006), depending on the cultivars or varieties. Robusta shows higher levels of FQA and can be characterized

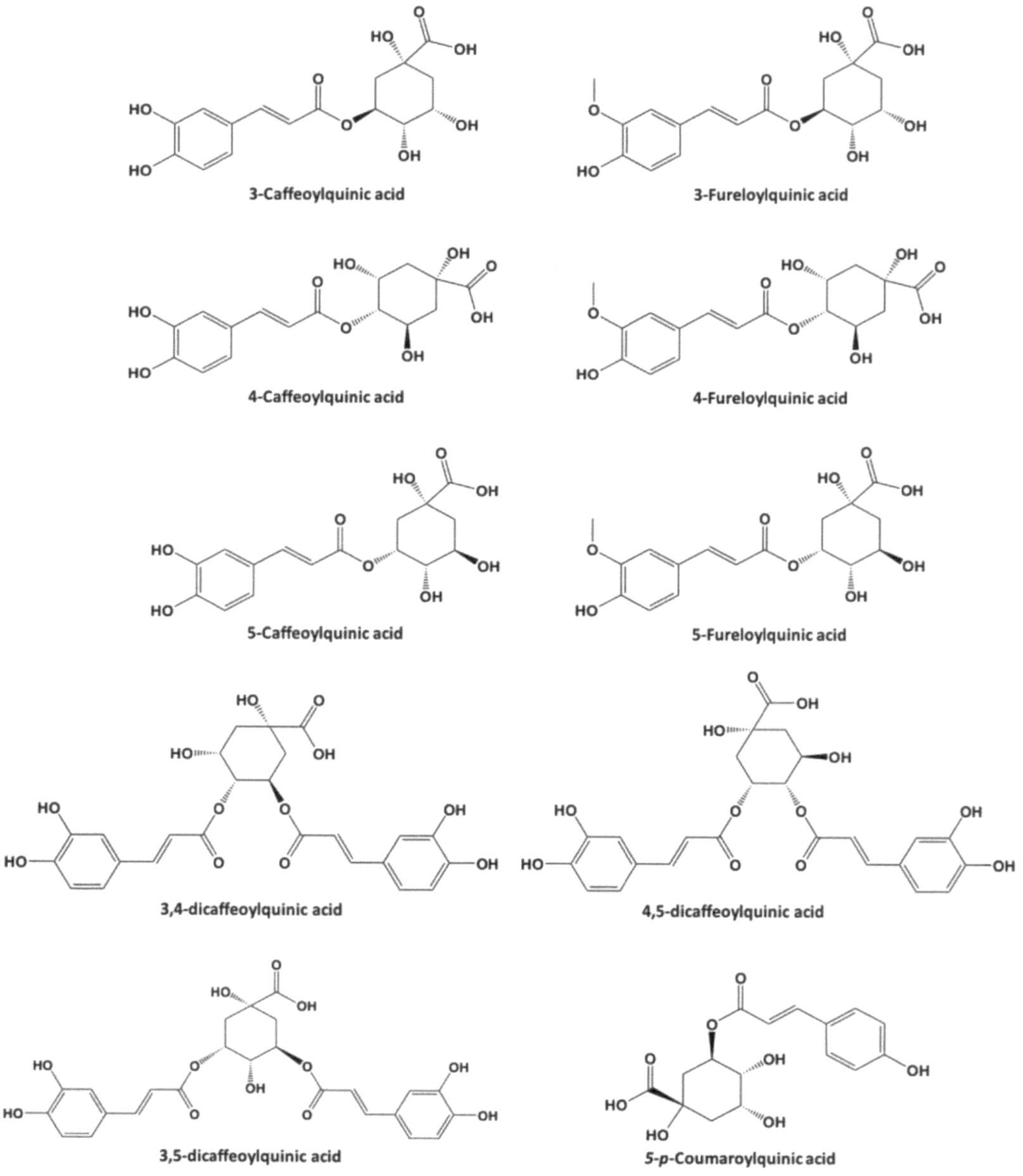

FIGURE 7.3 Important chlorogenic acid isomers in coffee.

by its diCQA content (Sunarharum, 2016). The CGA profiles of green coffee beans are not necessarily similar to that of the corresponding roasted coffees and brews. Subsequent processing of the green coffee beans, most importantly roasting process, results in an approximate 75% loss in CGAs (Sunarharum, 2016) and up to 93% for dark roast (Farah et al., 2005) while further brewing may reduce this CGA level to another 12%–13% from the content in the dark roast. In addition, different brewing techniques also influence the CGA content in coffee extracts. Espresso coffee provides significantly higher CGA than dripped or filtered coffee and other methods (Farah and de Paula Lima, 2019). The composition and concentration of phenolic compounds in green coffee beans, the interaction and changes of the phenolic compounds during processing, and the preparation methods of coffee brews all have a great effect on the quality of coffee and will be discussed next.

7.3 Polyphenols and coffee quality

Apart from quality in terms of nutritional function and importance to health, polyphenols also affect the sensory quality of beverage products. The association between polyphenol content and astringency (the dry, rough, and puckery sensation in the mouth) as well as bitterness in beverages has been studied by Dinnella et al. (2010); Higgins et al. (2020); Farah et al. (2005); Ginz and Engelhardt (2001); and Ong et al. (2018), but the detailed mechanisms of how polyphenols contribute to astringency and bitterness are not yet fully understood. For example, the association between tannins and flavonoids with the astringency and bitterness of tea, coffee, cocoa, wine, and beer has been well documented (Noble, 2002; Vidal et al., 2018; Soares et al., 2020). Certain polyphenols such as flavanols or flavan-3-ols (an important flavonoid in cocoa) can form complexes with salivary proteins, and thus are thought to be responsible for cocoa bitterness and astringency (Manach et al., 2004; Katz et al., 2011). However, the biochemical reactions between polyphenols and saliva or other compounds contributing astringency and bitterness, the products of these reactions, and the mechanisms that underlie the human brain perception of astringency and bitterness from these compounds are still poorly understood.

In coffee, bitterness and astringency have been studied extensively because both properties are important to quality. Polyphenols, mainly CGAs, contribute to the astringency (Buffo and Cardelli-Freire, 2004) and bitterness (Oestreich-Janzen, 2010) of coffee brews. As noted earlier, the perceived bitter-astringent character of coffee presumably arises from a series of complex mechanisms that involve the interaction of chemical compounds with other constituents, the transformation of each contributing compound, the interaction of the compounds with the saliva, and the oral processing mechanisms during coffee consumption (Noble, 2002; Dinnella et al., 2010; Ong et al., 2018).

Similar to astringency, bitterness in coffee is influenced by multiple compounds. During roasting, coffee polyphenols such as CGAs may undergo isomerization and hydrolysis to form quinic acid and cinnamic acids. CGAs can also form chlorogenic lactones (CGLs), such as 1,5-g-quinolactone from 3-CQA, where the quinic acid moiety loses one water molecule and forms an intramolecular ester bond (Farah et al., 2005; Farah and Donangelo, 2006; Farah, 2012; Farah and de Paula Lima, 2019). This CGL (Figure 7.4) may contribute to coffee bitterness (Farah et al., 2005; Ginz and Engelhardt, 2001). In addition, CGAs and other phenolic compounds can also be incorporated into melanoidins (Bekedam et al., 2008), brownish-colored polymers formed through Maillard reaction between sugars and amino acids at high temperature.

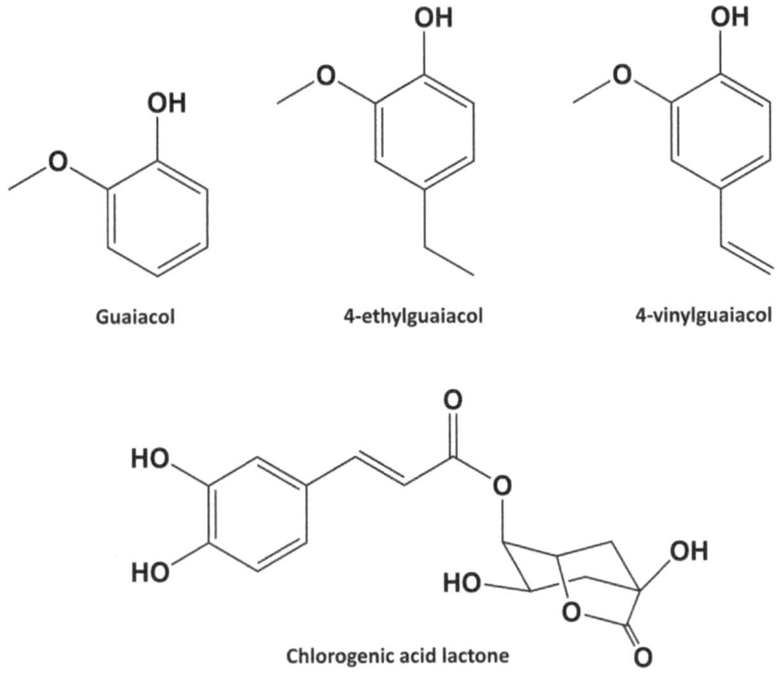

FIGURE 7.4 Structures of simple phenols (guaiacol, 4-ethylguaiaol, and 4-vinylguaiacol) and CGL.

Melanoidins are also thought to contribute to coffee bitterness, along with other bitter tasting compounds such as caffeine and trigonelline (Clarke and Vitzthum, 2001) as well as bengalensol, cafestol, and kahweol (Lang et al., 2020). It should be noted that the perception of bitterness is not only determined by the molecules contained in coffee but also by the consumption behavior of the consumer (Ong et al., 2018).

The correlation between coffee polyphenols and quality is somewhat controversial. The presence of polyphenols with different composition and concentration in beverages may be desirable or undesirable. In coffee, higher levels of CGAs may indicate a lower quality of coffee (Silva, 1999). For example, the immature and immature-black defective coffee beans contain higher CGA isomers, mainly CQA and FQA, compared with normal coffee beans (Mazzafera, 1999; Farah et al., 2005).

Apart from the aforementioned sensory quality related to taste, CGAs also play an important role in coffee flavor. Apparently, this flavor contribution comes from the indirect reaction or precursor formation of CGAs. The higher the CGA concentration, the more derivatives and oxidation products formed during coffee processing, preparation, and storage. The resulting compounds are responsible for the aroma or flavor that distinctly characterizes coffee. Robusta coffee contains higher CGA levels than Arabica and is perceived to be harsher with a stronger spicy and smoky aroma. This aroma is thought to be contributed by volatile phenols such as guaicol, 4-ethylguaiacol, and 4-vinylguaiacol, produced from the thermal degradation of CGAs (Toci and Farah, 2014; Sunarharum, 2016). In addition, there is a strong association between the levels of CQA, FQA, and their oxidation products with poor cup quality and the Rio off-flavor (negative phenolic note) of coffee (Amorim et al., 1975; Farah et al., 2006). In this context, polyphenols, and importantly CGAs, could be used as an indicator of coffee quality. Thus, attempts to improve coffee quality may be done through modifications and control of certain coffee quality indicators that include the composition and concentration of CGAs. In the next section, we will discuss about how phenolic compounds are biosynthesized in coffee.

7.4 The biosynthesis of phenolic compounds in coffee beans

As mentioned earlier, phenolic compounds can differ in the number of phenol ring(s) that they contain (can be mono-, di-, or polyphenols) and in the structural elements that bind the rings together (Manach et al., 2004). Despite their differences, the majority of phenolic compounds are mainly synthesized through the phenylpropanoid pathway from phenylalanine (Yu and Jez, 2008), which derives from chorismate, the end product of the shikimate pathway (Herrmann and Weaver, 1999). Both the shikimate and phenylpropanoid pathways that produce phenolic compounds have been extensively reviewed (e.g., Bentley and Haslam, 1990; Cheynier et al., 2013; Deng and Lu, 2017; Duke, 1985; Herrmann and Weaver, 1999; Tohge et al., 2013; Tzin and Galili, 2010; Vogt, 2010; Yu and Jez, 2008) and thus will only be discussed briefly here.

The shikimate pathway is a seven-step biosynthetic pathway that produces chorismate from two metabolic precursors: phosphoenolpyruvate (PEP, from the glycolysis pathway) and erythrose-4-phosphate (E4P, from the non-oxidative branch of the pentose phosphate pathway). The seven reactions start with the condensation of PEP and E4P into 3-deoxy-d-arabino -heptulosonate-7-phosphate (DAHP) by DAHP synthase. The second reaction of the pathway is the elimination of a phosphate group from DAHP to yield 3-dehydroquinate (DHQ) by DHQ synthase. The third and fourth reactions are the dehydration of DHQ into 3-dehydro shikimate (DHS) and its subsequent reduction into shikimate by a bi-functional enzyme DHQ dehydratase-shikimate dehydrogenase. The fifth reaction of the pathway is the phosphorylation of shikimate to yield shikimate 3-phosphate (S3P) by shikimate kinase. The sixth reaction is the condensation of a second PEP and S3P into 5-enolpyruvylshikimate 3-phosphate (EPSP) by EPSP synthase. Finally, the seventh reaction of the pathway is the *trans*-1,4 elimination of phosphate from EPSP by chorismate synthase. This final step yields chorismate, the precursor for phenylalanine, which is formed through the sequential actions of chorismate mutase, prephenate aminotransferase, and arogenate dehydratase.

Subsequently, the synthesis of phenolic compounds from phenylalanine occurs through the phenylpropanoid pathway and is mediated by many enzymes. The first two to three steps of the pathway are known as the general phenylpropanoid pathway (Ferrer et al., 2008; Winkel, 2004; Vogt, 2010) and provide the precursors for specific branch pathways to produce various phenolic compounds. The first reaction of the general phenylpropanoid pathway is the deamination of phenylalanine to yield cinnamic acid by phenylalanine ammonia lyase. The second reaction is the hydroxylation of cinnamic acid into *p*-coumaric acid by cinnamate 4-hydroxylase. In the third step, *p*-coumaric acid is converted into *p*-coumaroyl-CoA by 4-coumaroyl CoA ligase. The intermediates of this general phenylpropanoid pathway serve as the precursors for the synthesis of flavonoids, monolignols, phenolic acids, coumarins, and stilbenes through their respective branch pathways (Tohge et al., 2013; Deng and Lu, 2017). Once produced, these phenolic compounds serve important roles in plant development and stress response (Bhattacharya et al., 2010).

As mentioned earlier, the majority of the phenolic compounds in coffee beans are CGAs – esters of certain *trans*-hydroxycinnamic acids (mainly *p*-coumaric, caffeic, ferulic, and sinapic acids) and quinic acid (Clifford et al., 2017). CGAs are synthesized from *p*-coumaric acid or *p*-coumaroyl-CoA through the actions of many enzymes: 4-coumaric acid 3-hydroxylase/4-coumaroylester 3′-hydrolase, CoA ligase, caffeoyl-CoA:quinic acid caffeoyl transferase, and *O*-methyltransferase (Koshiro et al., 2007). The final composition of CGAs in coffee is directly related to their synthesis, accumulation, and storage within the beans and is affected by many factors, including genetics, physiological, and environmental conditions (Farah and Donangelo, 2006). For example, different levels of CGAs are known to be affected by the species and geographical origins of the coffee plant (Campa et al., 2008; De Rosso et al., 2018; Jeszka-Skowron et al., 2016); the maturation stage of the cherries (Smrke et al., 2015; Koshiro et al., 2007); and the conditions (elevation, rainfall, irrigation, fertilization, temperature, sunlight, and shade) under which the coffee plant is cultivated (Hameed et al., 2018). As the main group of phenolic compounds in coffee, CGAs play an important role in determining the cup quality of coffee. Albeit some uncertainties, it appears that good quality coffee might, among other attributes, have a CGA concentration threshold that should not be exceeded (Farah and Donangelo, 2006). Although the factors that affect the final composition of CGAs in coffee beans (and thus their quality) before harvest are well documented, the standardization of coffee bean quality remains a major challenge, partly because many of the aforementioned factors are beyond our ability to control.

7.5 The effect of fermentation on the phenolic compounds of coffee

After harvest, coffee cherries are typically processed with one of three fermentation methods – wet, dry, or semi-dry – depending on the maturation level of the coffee cherries, availability of water, local weather conditions, and commonly held traditional practice (Haile and Kang, 2019; Schwan et al., 2012; Sunarharum et al., 2019**a**). In practice, coffee fermentation typically refers to the process of mucilage removal from coffee beans after washing and pulping. However, in theory, coffee fermentation – when broadly defined as the biochemical degradation of complex molecules into simpler ones – occurs during fruit ripening (Bouzayen et al., 2010; Prasanna et al., 2007) and probably up to drying, albeit at a limited rate. In addition, coffee fermentation also refers to the fermentation of coffee beans after drying, through the addition of microbial culture. For the following discussion, we refer to the practical definition where post-harvest processing has been applied to ferment coffee to improve its quality.

The effect of fermentation process on the final composition of phenolic compounds in coffee beans is somewhat contradictory. Some studies found an increase in the phenolic compounds of coffee beans following fermentation, whereas others a decrease or no change (Table 7.2). This inconsistency may arise from differences in the biochemical composition of the coffee beans, the processing methods applied to the coffee beans, the fermentation culture, and the dynamic interplay between these three factors (Selmar et al., 2014; de Oliveira Junqueira et al. 2019; Zhang et al., 2019**a,** 2019b). The reason is that, first, the biochemical composition of coffee, including its phenolic contents, is affected by the metabolic processes that occur in the coffee beans. For example, different CGA compositions have been observed in coffee cherries during their maturation (Koshiro et al., 2007; Smrke et al., 2015). Second, the metabolic processes that convert the phenolic compounds in the coffee beans may further be substantiated by the processing conditions applied to them (Selmar et al., 2014). Indeed, different fermentation methods have been reported to yield different phenolic contents in coffee beans (Balyaya and Clifford, 1995; Joët et al., 2010). Third, the microbial communities that perform the fermentation itself are affected by both the biochemical composition of the coffee beans and the method used to ferment them (de Oliveira Junqueira et al., 2019; Zhang et al., 2019a). Thus, the dynamics of these three factors are important to consider, as they affect the composition of the phenolic compounds in coffee beans before roasting

TABLE 7.2

Effect of Fermentation on Coffee Phenolic Compounds from Recent Studies

Sample	Phenolic compounds*		Reference
	Control	Fermented	
Coffee cherries treated with different yeasts (*S. cerevisiae* CCMA 0543, *C. parapsilosis* CCMA 0544, and *Torulospora delbrueckii* CCMA 0684) suspension; fermented using dry method for 400 h	17.40 mg kg^{-1}	27.32 mg kg^{-1}	Bressani et al., 2018
Coffee cherries (washed and non-washed) treated with different yeasts (*Saccharomyces cerevisiae* UFLA YCN 727, *S. cerevisiae* UFLA YCN724, *Candida parapsilosis* UFLA YCN448, and *Pichia guilliermondii* UFLA YCN731); fermented under the sun until reaching 11%–12% moisture	ND (green coffee washed and non-washed), 0.21% (roasted washed), and 0.39% (roasted non-washed)	ND (green coffee washed and non-washed), 0.19–0.21% (roasted washed), and 0.20–0.39% (roasted non-washed)	Evangelista et al., 2014b
Semi-dried pulped coffee treated with different yeasts and fermented as above	ND (green coffee), 0.79% (roasted coffee)	ND (green coffee), 0.26–0.79% (roasted coffee)	Evangelista et al., 2014a
Hydrated and heat-treated green coffee beans; fermented (in solid state) with *Rhizopus oligosporus* for 5 days at 30°C	23.33 mg g^{-1}	19.61 mg g^{-1}	Lee et al., 2016
Green coffee beans; fermented (submerged) with *Lactobacillus rhamnosus* for 72 h at 37°C	32.2×10^3 mg kg^{-1}	27×10^3 mg kg^{-1}	Wang et al., 2019
Green coffee beans; fermented (submerged) with commercial *Saccharomyces* species for 24 h at 30°C	14.11 µM mL^{-1}	34.95 µM mL^{-1}	Kwak et al., 2018

Notes: * For brevity, representative values are selected from the samples and details regarding the type of phenolic compounds and their measurements are not provided. Please refer to the respective reference for details; ND = not detectable.

and brewing. In addition, it should be noted that thermal processing such as roasting also affects the formation of specific phenolic compounds in coffee after fermentation (e.g., guaiacol from lignin pyrolysis; Evangelista et al., 2014a, 2014b).

7.6 Thermal processing effect on coffee polyphenols

After fermentation, coffee cherries or beans are dried until reaching 11%–12% moisture, then hulled and stored as green beans before roasting (Schwan et al., 2012; Sunarharum et al., 2019**a**). The process of drying and roasting in coffee production typically involves the use of heat (at high temperatures). Because polyphenols are heat-sensitive, drying and roasting will affect the final contents of phenolic compounds in coffee beans. In addition, before being consumed, coffee is prepared by brewing at different conditions, and this will affect the composition of the extracted phenolic compounds.

7.6.1 Coffee bean drying

In coffee production, there are several drying methods that can be used (e.g., room-temperature drying, sun drying, heat-pump drying, hot-air drying, freeze drying, and microwave-vacuum drying). Each method has advantages and disadvantages, and the method of choice will depend on the availability of the equipment and the practicality of its usage, as many coffee plantations are still traditionally managed and operated. For example, room-temperature drying is simple and cheap but time-consuming (on the order of weeks), whereas sun drying is faster (on the order of days) but dependent on the availability of the sun. Heat-pump and hot-air drying are faster (on the order of days) than sun drying but require equipment and energy input. Both heat-pump and hot-air dryers are typically run at around 50°C. Freeze drying and microwave-vacuum drying require equipment and energy input but are faster (on the order of hours to minutes for microwave-vacuum drying) than heat-pump and hot-air dryers as they operate under vacuum. Freeze drying is typically run between −50 and −80°C, whereas microwave-vacuum drying is run at around 60°C. Traditionally, a combination of room-temperature and sun drying is used by farmers to dry their coffee cherries or beans.

Because different drying methods have different operating conditions, the method of choice will affect the polyphenol contents and the quality of the dried coffee beans. For example, Dong et al. (2017) showed that quinic acid – an important source of phenolic compounds (Wang et al., 2013) – was preserved when coffee cherries were dried using freeze drying but was degraded when dried using room-temperature drying, sun drying, heat-pump drying, and hot-air drying. In addition, freeze drying resulted in the highest volatile contents compared with the other drying methods. More recently, Cheng et al. (2019) showed that microwave-vacuum drying was the most suitable method for preserving the phenolic compounds of green coffee beans compared with room-temperature drying, heat-pump drying, and freeze drying. In addition, microwave-vacuum drying was the least time- and energy-consuming method. As an emerging technology, it would be interesting to see how this method will be developed into a scalable and affordable technology for field applications. For example, Monteiro et al. (2015) developed a low-cost microwave-vacuum drying method using a domestic microwave oven with turntable. The adoption of such technology could enable a better drying process to produce high-quality coffee products.

7.6.2 Coffee bean roasting

Roasting of green coffee beans is a fundamental step in coffee production, as it determines the color, taste, and aroma of coffee. The process of roasting can roughly be divided into three phases: (1) the initial drying phase, where the moisture content is removed from the beans (from around 12% to 2%) by increasing the temperature to 180°C; (2) the roasting phase, where complex chemical reactions occur in the beans at around 200–300°C; and (3) the cooling phase, where the roasting is brought to a stop by using air or water (Buffo and Cardelli-Freire, 2004; de Melo Pereira et al., 2019). The chemical changes that occur to the phenolic compounds in the beans during roasting will depend on the degree (whether light, medium, or dark) and on the duration of the roasting. For example, the higher the roasting degree and the longer the roasting time, the more CGAs and other polyphenols will be degraded (Bobková et al., 2020; Duarte et al., 2005; Król et al., 2020; Wei and Tanokura, 2015).

With regard to CGAs, roasting is known to cause progressive degradation of CGAs. In the early stage of roasting, CGAs undergo isomerization and partial hydrolysis releasing various cinnamic acids (Clifford, 2000). This event appears to be temperature dependent and has been observed in a recent study where the total phenolic content of Arabica and Robusta coffee slightly increased during initial drying phase at 95°C. At elevated temperatures (125 and 165°C), the total phenolic content decreased below the no-roasting control (Sunarharum et al., 2019b). Consistent with other studies, the decrease in the polyphenol contents tended to decrease the antioxidant activities of the brewed coffee (Bobková et al., 2020; Duarte et al., 2005; Sunarharum et al., 2019**b**).

7.6.3 Coffee brewing

Coffee brewing is the process of preparing coffee from ground (green or roasted) coffee beans using water. There are various methods to brew coffee, from the simple direct infusion and stirring of ground coffee beans in boiling water (the Javanese "tubruk" method) to the more sophisticated temperature-controlled and pressurized coffee brewing using an espresso machine. During the brewing process, the biochemical contents (including polyphenols) that are contained in coffee beans will be extracted. The amount of polyphenols extracted from particular ground coffee beans will vary depending on the brewing method used (Farah and de Paula Lima, 2019) but will generally increase following the brewing temperature and time. Thus, as with other processing conditions, coffee brewing should be taken into consideration and performed with care because, as pointed out previously, polyphenols not only affect the sensory attributes of coffee but also its potential health effects.

7.7 Health potential of coffee polyphenols

Coffee consumption has been associated with many health benefits. A recent evaluation of existing evidence on the associations between coffee consumption and its health outcomes found that coffee consumption lowers the incidence and risk of various diseases (Poole et al., 2017). The benefits of coffee on degenerative diseases such as cardiovascular disease, cancer, and diabetes have been, to some extent, attributed to the antioxidant and anti-inflammatory activities of CGAs in coffee (Liang and Kitts, 2015; Tajik et al., 2017). To exert their antioxidant and anti-inflammatory activities, CGAs (and other polyphenols) must be absorbed through the digestive system and remain bioavailable in the circulation. On average, it is estimated that one-third of the amount of CGA consumed is absorbed, although inter-individual variation is high (Farah and de Paula Lima, 2019). Following absorption, polyphenols exert their antioxidant activity by suppressing the generation of free radicals through the inhibition of enzymes involved in free radical formation, through the scavenging of reactive species and their precursors and through the chelation of metal ions (Tsao, 2010). On the other hand, polyphenols exert their anti-inflammatory activity by inhibiting pro-inflammatory enzymes and molecules as well as by modulating signaling pathways and key transcription factors of genes involved in inflammatory responses (Hussain et al., 2016).

In addition to the antioxidant and anti-inflammatory activities of polyphenols, another route to the beneficial health effects of coffee may stem from the modulation of the gut microbial communities by the unabsorbed polyphenols (Cardona et al., 2013; Nishitsuji et al., 2018). The reason is that although polyphenols can be absorbed through the digestive system, their plasma concentrations are low, on the order of nmol L^{-1} – a range probably too low for exerting significant physiological outcomes (van Duynhoven et al., 2011; Rechner et al., 2002). In the colon, the polyphenols are metabolized by the gut microbiota into various compounds. These microbial metabolic derivatives may then be absorbed and further metabolized in the liver (where they undergo first-pass metabolism) before being circulated to other organs. On the other hand, the microbial metabolic derivatives may also function as bioactive compounds that affect the ecology of the intestine, influencing the composition of the gut microbial communities and consequently our health (Cardona et al., 2013). Previous studies have shown that coffee consumption appears to alter specific members of the colonic bacteria, principally *Bacteroides* and *Bifidobacterium*, in which polyphenols might play a role (González et al., 2020; Jaquet et al., 2009).

7.8 Conclusion

Coffee is a natural source of polyphenols. From coffee cherries on the farm to a cup of coffee on the table, coffee undergoes major transformations. Changes in the physical, chemical, and biological properties of coffee occur because of the nature of the processing conditions, from harvesting, sorting, fermentation, drying, hulling, roasting, grinding, packaging, and storage to brewing. Accordingly, major chemical changes also occur to the polyphenols throughout coffee processing. Because polyphenols affect the sensory attributes and the potential health effects of coffee, the conditions of coffee processing should ideally be monitored, controlled, and evaluated at each step. In the majority of regions where coffee plantation is managed by local farmers, there is a huge gap between the ideal process and the reality in the field, partly because of the socioeconomic pressure and the limited resources. To help close this gap, the government, industry, farmers, and universities must collaborate and work to tackle not only the social and economic issues but also the technical problems related with coffee processing. These challenges should be addressed in a concerted manner by supporting and empowering local communities through continuous partnership programs with a sustainability framework. Because coffee quality is inherently tied to its growers and processors, the investments that we put into such programs will be fruitful in the long run.

REFERENCES

Amorim HV, Teixeira AA, Melo M, Cruz VF, Malavolta E. Turrialba 1975; 25:18–24.

Andueza S, Vila M, De Peña M, Cid C. J Sci Food Agr. 2007;586–592.

Balyaya K, Clifford MN. J Food Sci Technol. 1995; 32:104–108.

Bekedam, Koen E, Schols HA, Van Boekel MAJS, Smit G. J Agric Food Chem. 2008;56(6):2055–2063.

Bentley R, Haslam E. Crit Rev Biochem Mol Biol. 1990;25:307–384.

Bhattacharya A, Sood P, Citovsky V. Mol. Plant Pathol. 2010;11(5):705–719.

Bobková A, Hudáček M, Jakabová S, Belej L, Capcarová M, Čurlej J, Bobko M. J Environ Sci Heal B. 2020;55(5):495–500.

Bouzayen M, Latché A, Nath P, Pech JC. In: Pua E, Davey M, editors. Plant Developmental Biology. Springer, Berlin, Heidelberg. 2010.

Bressani APP, Martinez SJ, Evangelista SR, Dias DR, Schwan RF. LWT. 2018; 92:212–219.

Buffo RA, Cardelli- Freire C. Flavour Fragr J. 2004;19(2): 99–104.

Campa C, Jean-Jacques R, de Kochko A, Hamon S. Adv in Plant Physiology 2008; 10:421–437.

Cardona F, Andrés-Lacueva C, Tulipani S, Tinahones FJ, Queipo-Ortuño MI. J Nutr Biochem. 2013;24(8):1415–1422.

Cheng K, Dong W, Long Y, Zhao J, Hu R, Zhang Y, Zhu K. Food Sci Nutr. 2019;7(3):1084–1095.

Cheynier V. Am J Clin Nutr. 2005;81(1 Suppl):223–229.

Cheynier V, Comte G, Davies KM, Lattanzio V, Martens S. Plant Physiol Biochem. 2013; 72:1–20.

Clarke RJ, Vitzthum OG. Coffee: Recent developments. Blackwell Science. 2001.

Clifford MN. In Clarke RJ, Macrae R, editors. Coffee. 1985.

Clifford MN. J Sci Food Agric. 1999; 79:362–372.

Clifford MN. J Sci Food Agric. 2000; 80:1033–1043.

Clifford MN, Johnston KL, Knight S, Kuhnert N. J Agric Food Chem. 2003;51(10):2900–2911.
Clifford MN, Knight S, Surucu B, Kuhnert N. J Agric Food Chem. 2006;54(6):1957–1969.
Clifford MN, Jaganath IB, Ludwig IA, Crozier A. Nat Prod Rep. 2017;34(12):1391–1421.
Crozier A, Clifford MN, Ashihara H. In Plant secondary metabolites: Occurrence, structure and role in the human diet. Blackwell Pub. 2006.
da Silva KM, Zielinski AAF, Benvenutti L, Bortolini DG, Zardo DM, Beltrame FL, Nogueira A, Alberti A. Food Sci Tech. 2019;39(2):294–300.
Davis AP, Chadburn H, Moat J, O'Sullivan R, Hargreaves S, Lughadha EN. Sci Adv. 2019;5(1):1–10. de Melo Pereira et al. Food Chem. 2019; 272:441–452.
Del Rio D, Rodriguez-Mateos A, Spencer JPE, Tognolini M, Borges G, Crozier A. Antioxid Redox Signal. 2013;18(14):1818–1892.
de Melo Pereira GV, Neto DPC, Júnior AIM, Vásquez ZS, Medeiros ABP, Vandenberghe LPS, Soccol CR. Food Chem. 2019;272:441–452.
Deng Y, Lu S. Crit Rev Plant Sci. 2017;36(4):257–290.
de Oliveira Junqueira AC, de Melo Pereira GV, Coral Medina JD, Alvear MCR, Rosero R, de Neto DPC, Enríquez HG, Soccol CR. Sci Rep. 2019;9(1):8794.
De Rosso M, Colomban S, Flamini R, Navarini L. J Mass Spectrom. 2018;53(9):763–71.
Dinnella C, Recchia A, Vincenzi S, Tuorila H, Monteleone E. Chem Senses. 2010;35(1):75–85.
Dong W, Hu R, Chu Z, Zhao J, Tan L. Food Chem. 2017;234:121–130.
Duarte SMS, de Abreu CMP, de Menezes HC, dos Santos MH, Gouvêa CMCP. Food Sci Tech. 2005; 25:387–393.
Duke SO. SE – 8. ACS Symposium Series. American Chemical Society. 1985; 268:113–131.
Duynhoven J, Vaughan EE, Jacobs DM, Kemperman RA, van Velzen EJJ, Gross G, Roger LC. Proceedings of the National Academy of Sciences. 2011;108 (Supplement 1): 4531 LP – 4538.
Evangelista SR, Miguel MGCP, Cordeiro CS, Silva CF, Pinheiro ACM, Schwan RF. Food Microbiol. 2014a;44:87–95.
Evangelista SR, Silva CF, Miguel MGCP, Cordeiro CS, Pinheiro ACM, Duarte WF, Schwan RF. Food Res Int. 2014b;61:183–95.
Farah A. In *Coffee: Emerging health effects and disease prevention*, ed. Y. Chu. Iowa: IFT Press, Wiley- Blackwell. 2012;21–58.
Farah A, de Paula Lima J. Beverages. 2019;5(1):11.
Farah L, Donangelo CM. Braz J Plant Physiol. 2006;18(1):23–36.
Farah A, De Paulis T, Trugo LC, Martin PR.. J Agric Food Chem. 2005;53(5):1505–13.
Farah A, Monteiro MC, Calado V, Franca AS, Trugo LC. Food Chem. 2006;98(2):373–80.
Ferrer JL, Austin MB, Stewart C, Noel JP. Plant Physiol Biochem. 2008;46(3):356–70.
Fibrianto K, Febryana YR, Wulandari ES. IOP Conf Ser Earth Environ Sci. 2018;131(1).
Georgiev V, Ananga A, Tsolova V. Nutrients. 2014;6 (1):391–415.
Ginz M, Engelhardt UH. Eur Food Res Tech. 2001;213(1):8–11.
González S, Salazar N, Ruiz-Saavedra S, Gómez-Martín M, de los Reyes-Gavilán CG, Gueimonde M. Nutrients. 2020;12(5):1287.
Gorjanović S, Komes D, Laličić-Petronijević J, Pastor FT, Belščak-Cvitanović A, Veljović M, Pezo L, Sužnjević D. J Food Sci Tech. 2017;54(8):2324–31.
Habschied K, Loncarić A, Mastanjević K. Foods. 2020;9(2).
Haile M, Kang WH. J Food Qual. 2019:4836709.
Hameed A, Hussain SA, Suleria HAR. In *Co- evolution of secondary metabolites*. 2018;1–67.
Herrmann KM, Weaver LM. Annu Rev Plant Physiol Plant Mol Biol. 1999;50(1):473–503.
Higgins MJ, Hayes JE. Nutrients. 2020;12(6):1560.
Hussain T, Tan B, Yin Y, Blachier F, Tossou MCB, Rahu N. Oxid Med Cell Longev. 2016:7432797.
ICO. Historical data on the global coffee trade. 2020a.
ICO. Trade Statistics Tables. 2020b.
Jaganath IB, Crozier A. John Wiley & Sons, Inc. 2010.
Jaquet M, Rochat I, Moulin J, Cavin C, Bibiloni R. Int J Food Microbiol. 2009;130(2):117–21.
Jeszka-Skowron M, Sentkowska A, Pyrzyńska K, De Peña MP. Eur Food Res Tech. 2016;242 (8):1403–9.
Joët T, Laffargue A, Descroix F, Doulbeau S, Bertrand B, de kochko A, Dussert S. Food Chem. 2010;118(3):693–701.
Katz DL, Doughty K, Ali A. Antioxid Redox Signal. 2011;15(10): 2779–2811.
Koshiro Y, Jackson MC, Katahira R, Wang ML, Nagai C, Ashihara H. J. Biosci. 2007;62:731–42.
Król K, Gantner M, Tatarak A, Hallmann E. Eur Food Res Tech. 2020;246(1):33–39.
Kwak HS, Jeong Y, Kim M. J Food Quality. 2018:5967130.
Lang T, Lang R, Di Pizio A, Mittermeier VK, Schlagbauer V, Hofmann T, Behrens M. J Agric Food Chem. 2020, 68, 24, 6692–6700.
Lee LW, Cheong MW, Curran P, Yu B, Liu SQ. Food Chem. 2016;211:916–24.
Liang N, Kitts DD. Nutrients. 2015;8(1):16.
Lorenzo CD, Badea M, Colombo F, Orgiu F, Frigerio G, Pastor RF, Restani P. BIO Web Conf. 2017;9 04008.
Ludwig I, Sanchez L, Cämmerer B, Kroh L, De Peña MP, Cid C. Food Res Int. 2012;48(1):57–64.
Manach C, Scalbert A, Morand C, Rémésy C, Jiménez L. Am J Clin Nutr. 2004;79(5):727–747.
Maru GB, Kumar G, Ghantasala S, Tajpara P. Elsevier Inc. 2013;2:1141–1179.
Mazzafera P. Food Chem. 1999;65:547–54.
Miranda AM, Steluti J, Fisberg RM, Marchioni DM. Nutrients. 2017;9(3):276.
Monteiro RL, Carciofi BAM, Marsaioli A, Laurindo JB. J Food Eng. 2015;166:276–84.
Nishitsuji K, Watanabe, Xiao J, Nagatomo R, Ogawa H, Tsunematsu T, Umemoto H. Sci Rep. 2018;8 (1):16173.
Noble AC. ACS Symposium Series. 2002;825:192–201.
Oestreich- Janzen S. In Comprehensive natural products II. editors. H- W Liu and L Mander. Oxford: Elsevier. 2010
Ong JS, Hwang DLD, Zhong VW, An J, Gharahkhani P, Breslin PAS, Wright MJ. Sci Rep. 2018;8:16414.
Pérez-Jiménez J, Neveu V, Vos F, Scalbert A. Eur J Clin Nutr. 2010;64:S112–20.

Poole R, Kennedy OJ, Roderick P, Fallowfield JA, Hayes PC, Parkes J. BMJ. 2017;359:j5024.

Prasanna V, Prabha TN, Tharanathan RN.Crit Rev Food Sci Nutr. 2007;47(1):1–19.

Rechner AR, Kuhnle G, Bremner P, Hubbard GP, Moore KP, Rice-Evans CA. Free Radic Biol Med. 2002;33(2):220–35.

Schwan R, Silva C, Batista L. Coffee fermentation. In Handbook of plant- based fermented food and beverage technology editors. YH Hui and EÖ Evranuz, CRC Press. 2012;677–690

Selmar D, Kleinwächter M, Bytof G. Cocoa and coffee fermentations. 2014:431–476.

Silva EB. *Fontes e doses de potássio na produção e qualidade de café provenientes de plantas cultivadas em condições edafoclimáticas*. Universidade Federal de Lavras, Brazil. 1999.

Silva and Pogačnik, 2020. Antioxidants. MDPI AG. 2020.

Smrke S, Kroslakova I, Gloess AN, Yeretzian C. Food Chem. 2015;174:637–42.

Soares S, Brandão E, Guerreiro C, Soares S, Mateus N, de Freitas V. Molecules. 2020;25(11):2590

Socha R, Pająk P, Fortuna T, Buksa K. Int J Food Prop. 2017;20(1):S595–609.

Stahl L, Miller KB, Apgar J, Sweigart DS, Stuart DA, Mchale N, Ou B, Kondo M, Hurst WJ. J Food Sci. 2009;74(6):456–61.

Sunarharum WB. 2016. *The compositional basis of coffee flavour*. The University of Queensland, Australia.

Sunarharum WB, Williams DJ, Smyth HE. Food Res Int. 2014;62:315–325.

Sunarharum WB, Fibrianto K, Yuwono SS, Nur M. 2019a. Sains Kopi Indonesia, Malang, UB Press.

Sunarharum WB, Yuwono SS, Aziza OF. IOP Conf Ser Earth Environ Sci. 2019b; 230:12045.

Sunarharum WB, Ahmad R, Primadiani E. IOP Conf Ser Earth Environ Sci. 2020;475:012020.

Tajik N, Tajik M, Mack I, Enck P. Eur J Nutri. 2017;56(7):2215–44.

Toci AT, Farah A. Food Chem. 2014;153:298–314.

Tohge T, Watanabe M, Hoefgen R, Fernie AR. Front Plant Sci. 2013;4:62.

Tresserra-Rimbau A, Medina-Remón A, Estruch R, Lamuela-Raventós RM. In Preedy VR (Ed.). *Coffee in health and disease prevention*. Elsevier Inc. 2015;387–394

Tsao R. Nutrients. 2010;2(12):1231–46.

Tucker CM. *Coffee culture: Local experiences, global connections*. New York, Routledge, Taylor & Francis Group. 2017.

Tzin V, Galili G. In *Arabidopsis book*. American Society of Plant Biologists, PMC, National Center for Biotechnology Information, U.S. National Library of Medicine. 2010; 8:e0132.

Vamanu E. Front Pharmacol. 2019:10:492.

Vidal L, Antúnez L, Rodríguez-Haralambides A, Giménez A, Medina K, Boido E, Ares G. Food Res Int. 2018;112:25–37.

Vogt T. Phenylpropanoid biosynthesis. Molecular Plant. 2010;3(1): 2–20.

Vuolo MM, Lima VS, Junior MRM. *Bioactive compounds: Health bene ts and potential applications*. Elsevier. 2018;33–50.

Wang Z, Li X, Zhen S, Li X, Wang C, Wang Y. J Therm Anal Calorim. 2013;114(3):1231–38.

Wang C, Sun J, Lassabliere B, Yu B, Zhao F, Zhao F, Chen Y, Liu SQ. J Sci Food Agric. 2019;99(1):409–20.

Wei F, Tanokura M. *Coffee in health and disease prevention*. San Diego: Academic Press. 2015;83–91.

Winkel BSJ. Annu Rev Plant Biol. 2004;55: 85–107.

Wiswedel I, Hirsch D, Kropf S, Gruening M, Pfister E, Schewe T, Sies H. Free Radic Biol Medic. 2004;37(3):411–21.

Yu O, Jez JM. Plant J. 2008;54(4):750–62.

Zhang SJ, De Bruyn F, Pothakos V, Contreras GF, Cai Z, Moccand C, Weckx S, De Vuyst L. Front Microbiol. 2019;10:2621.

Zhang SJ, De Bruyn F, Pothakos V, Torres J, Falconi C, Moccand C, Weckx S, De Vuyst L. Appl Environ Microbiol. 2019;85(6):e02635–18.

Zhao CN, Tang GY, Cao SY, Xu XY, Gan RY, Liu Q, Mao QQ, Shang A, Li HB. Antioxidants. 2019;8(7):25.

8

*Effect of solid-liquid extraction in the art of preparing fine coffee drinks (*Coffea arabica *L. var.* Castillo*) on the antioxidant activity in 28 preparation methods*

Angela M. Ormaza-Zapata, Félix O. Díaz-Arango, and Benjamín A. Rojano

CONTENTS

DOI: 10.1201/9781003043133-10

8.1 Introduction

Coffee is among the most desired agricultural products worldwide (Liang et al., 2016). It forms an important part of the Colombian economy as a characteristic agricultural chain in the Nariño, Norte de Santander, Antioquia, Valle del Cauca, Cundinamarca, Huila, Cauca, Tolima, Caldas, Quindío, and Risaralda departments, the last three of which are known as the Coffee Axis (Perfect Daily Grind, 2017). Similarly, coffee production has promoted coffee-themed and coffee cultural landscape–related tourism, represented by the mountainous Coffee Axis zone, as well as other regions of Colombia. The industrialization of coffee constitutes a symbol both of identity and national pride. Coffee provides flavor, stimulants, and biofunctional metabolites. These are generated by way of the operation parameters applied during its processing, as well as by the benefits, fermentation, roasting and grind parameters, climatic conditions such as altitude, temperature, relative humidity, solar energy received, and intrinsic factors such as the origin, variety, and genomics of the processed coffee (FNCC, 2017; Arango et al., 2012). Coffee's chemical compounds contribute a great deal to its cup profile, and may define its quality.

In accordance with the QDA methodology (CTG 4883, 2000) applied in the coffee sensorial analysis, the attributes evaluated in coffee include fragrance, aroma, acidity, bitterness, body, and overall impression. Other descriptors were considered as well, for each beverage's cup test, to indicate flavor defects such as acrid, vinegar-like, fermented, astringent, stagnant, rancid, phenolic, earthy, or chemical notes (CTG 3566, 2011). The sensorial analysis requires training and discipline on the part of tasters, who guarantee the quality and standardization of coffee to be consumed nationally and internationally (Cheng et al., 2016; Díaz-Arango et al., 2018).

Various authors have reported the functional aspects of different coffee varieties, recommending their continuous consumption for the attainment of health benefits (Aguiar et al., 2016; Cano-Marquina et al., 2013). Numerous investigations on the topic have indicated the role of foods, such as coffee, fruit, grains, and vegetables, as functional products which can diminish cellular oxidative stress. This is an initiator of many illnesses, including diabetes, cancer, and Parkinson's disease (Aguiar et al., 2016; Oroian and Escriche, 2015; Ramakrishna et al., 2011; Ramakrishna et al., 2012; Ravishankar and Ramakrishna, 2017; Yashin et al., 2013). Further, coffee roasting, via the Maillard reaction, generates antioxidant compounds, and with antiradical activity, given its principally phenolic components (Liu and Kitts, 2011; Moreira et al., 2017).

Phenolic compounds act as antioxidant agents, by way of single electron transfer (SET) and hydrogen atom transfer (HAT). Additionally, flavonoids, together with the phenolic acids and tannins present in the coffee, create a compositional complex that provides greater functionality, such as antioxidants and flavors, to the coffee beverage (Ross et al., 2011). Hydroxycinnamic acids, such as chlorogenic and ferulic acids, add acidity and bitterness to the beverage, while tannins confer astringency to the cup profile (Cheng et al., 2016).

In terms of commercial coffee varieties and sensorial attributes, the Arabica variety is characterized by its sweet and caramel notes, while Robusta coffee provides spicy, coarse, earthy notes to the coffee beverage. In this sense, the antioxidant compounds in the coffee also serve to highlight the most characteristic flavors in the final infusion (Sunarharum et al., 2014). It should be noted that ferulic acid, as one of coffee's hydroxycinnamic acids, participates in decarboxylation reactions during roasting and generates phenolic compounds as products of said reaction (Perrone et al., 2012). These compounds increase the concentration of those with antioxidant properties in coffee and simultaneously confer a bitter flavor and acidic notes (Lee et al., 2015).

U.G.Q coffee calls green. Coffee on circular 14/64-inch mesh, with a diameter that retains 1.5% of the beans on 12/64-inch mesh (Ormaza-Zapata et al., 2019a). The phenolic compound content is determined via Folin-Ciocalteu's colorimetric method. Total flavonoids and condensed tannins are quantified by way of UV-VIS spectrophotometry (Zapata et al., 2013). The content of phenolic acids as chlorogenic acid, p-coumaric acid, caffeic acid, and ferulic acid was determined by HPLC analysis, according to the protocol of Kelebeck et al. (2009) with slight modifications indicated by Rosales-Delgado et al. (2019). ABTS (2,2′-azino-bis (3-ethylbenzothiazoline-6 -sulfonic acid)) antioxidant assay was determined by UV-VIS spectrophotometry (Zapata et al., 2013), while ORAC (oxygen radical absorbance capacity) antioxidant assay was quantified by fluorescence spectrophotometry (Rosales-Delgado et al., 2019).

Barism is an experimental discipline, in which different coffee methods, types of milk, flavorings, and liquors are added to coffee to obtain unique and distinctive flavors in each cup, with an artistic or decorative presentation (Díaz-Arango et al., 2018). Barista culture generated an emerging market for diverse coffee preparations, supported by different preparations, creativity, and the ingenious principles of connoisseurs and coffee lovers. Each preparation is characterized by its sensorial profile and composition, in terms of antioxidant compounds, in accordance with the preparation employed (Illy, 2017; Ormaza-Zapata et al., 2019a; Ormaza-Zapata et al., 2019b). Barism also stimulates the consumption of specialty coffees, called origin coffees, as well as the development of emerging methods for the preparation of new and exotic beverages, so as to add value to coffee-based products (OIC, 2015).

The relationship between the preparation method, cup profile, and antioxidant compound composition has neither been established nor used as an opportunity to offer coffee varieties with defined sensorial or functional qualities to the market. It is important to analytically establish the relationship between coffee preparation methods and their biofunctional potential, from composition and antioxidant activity, so as to obtain specific scientific information for healthy alternatives for the consumption of coffee-based beverages. These can improve our health and prevent chronic and neurodegenerative illness, all with an impeccable, aromatic sensorial profile (Santini et al., 2011; Çelik and Gökmen, 2018). In this chapter, the principle of coffee preparation is related to the composition and antiradical activity of coffee's most characteristic antioxidant compounds (total phenols, flavonoids, condensed tannins, hydroxycinnamic phenolic acids (chlorogenic acid, caffeic acid, p-coumaric acid, and ferulic acid) and its ABTS and ORAC antioxidant activity, and the sensorial profiles of the ten best coffee preparations with the greatest antioxidant biofunctionality are included.

8.2 Commercial coffee preparation method classification

Coffee preparation methods are classified into five large groups, in accordance with the mechanism by which the leaching process is carried out, for the obtention of the hot or cold coffee beverage. Parameters such as method material, coffee/water contact time, the temperature reached during preparation, the grind, and other operation variables during the extraction (Liang et al., 2016) were considered, for the establishment of the following classification.

8.2.1 Direct contact preparation methods

Direct contact methods are characterized by leaving water, at 90°C, in contact with coffee for a period of time so as to carry out a mass transference phenomenon, in which coffee solids pass to the water via a solid-liquid extraction mechanism. Contact time varies by method, as does the coffee/water proportion and the total preparation time. These methods are represented by Ibrik coffee makers, French presses, and electric percolators. The French press method uses a piston in its preparation. The beverage is extracted (water/coffee) before pressing the piston to the base of the glass container. The function of the piston, or press, in this case, is to separate the bean grounds from the beverage following extraction. For this reason, this method forms part of the direct contact category.

8.2.2 Pressure filtration methods

Pressure filtration methods are characterized by the application of force to a mass of ground, compacted coffee, before the extraction, via leaching. By compacting the coffee, a fixed bed is created, through which the approximately 90°C pressurized water passes. Consequently, its extraction capacity increases. The application of pressure may vary with the use of a piston (Staresso and Aeropress), lever (Presso), small holes through which the water/coffee mix is forced to recirculate (Mocca), or simply pressurized water (Expresso). Generally, with these types of methods, contact time is low (seconds or minutes). However, despite the application of pressure for the obtention of a coffee beverage, a mass transference phenomenon is carried out, where the coffee solids pass to the water through a leaching process. With these methods, the following variables are determinant: contact time, which is always short (90 s), the coffee/water ratio, and the element that generates pressure (Ormaza-Zapata et al., 2019a).

8.2.3 Cold coffee preparation methods

Cold methods for coffee-based beverage preparation are based on direct contact between the coffee and cold water, or in some applications, ice. Coffee makers used for said purpose use one of two main systems: the slow, or prolonged, drip, which may last up to 24 continuous hours (Cold Tower and Cold Brew); and immersion, where the beverage is created via coffee diffusion, by way of cold water or packed ice in a reservoir, within a pitcher or container (Toddy, Fretta, and Midzudashi). Some baristas agree that cold coffee–based beverages are characterized by their low acidity and bitterness (Baumann et al., 2012).

There is limited information available regarding the relationships between cold coffee preparation methods, sensorial profiles, antioxidant composition, and their expression, for the provision of a commercial coffee-based product alternative with varied sensorial profiles and the functional contribution of this complex, stimulating beverage (Hamilton and Lahne, 2020).

8.2.4 Drip gravity filtration methods

Drip gravity filtration methods are characterized by involving a leaching process, which occurs by way of a filter. Said filter may be made of thermoresistant plastic (RS 16), paper with different pore sizes (V60 and Clever), ceramic (V60 Kalita), or a perforated metallic surface (Kinto, Yama). Should it employ the drip method, contact and extraction times are prolonged (minutes). The size of the cone (Clever, Kinto, Yama, V60 Kalita, and V60), glass (Vietnamese, RS 16), or filtration unit determines volume of the beverage obtained. The presence of rough, smooth, or oblique surfaces may yield different cup profiles. All gravity filtration methods that involve extraction via drip occur with 90°C water, which serves to improve the performance of the extraction process. The mass transfer velocity during leaching may depend upon the pore diameter of the paper filter, or if paper is not used, the size, number of orifices, and material of the filter employed (Ormaza-Zapata et al., 2019b).

8.2.5 Gravity filtration methods

Gravity filtration methods are characterized by passing coffee through semi-permeable structures (cloth, paper, or wire mesh) so as to measure out the coffee extraction that occurs with gravity, while the coffee and 90°C water are in contact, resulting in a coffee beverage via the leaching-filtration method. The filter type influences the retention of antioxidant compounds, and antiradical activity in the use of cloth (Cloth filter, Japanese siphon), wire mesh (Fellow Duo, Silvertone,

Neapolitan) and paper (Chemex, Kalita, Drip), given that these metabolites may remain trapped in the insoluble coffee fraction and modify antioxidant content and, consequently, its activity. With these methods, contact time is short, except with the Silvertone method, which has a very small pore size in the wire mesh filter (Díaz-Arango et al., 2018).

8.3 Mode of preparation of the 28 most commercially popular coffee preparation methods

8.3.1 French Press

The coffee utilized must be coarsely ground (701–900 μm). Contact time between the coffee and 90°C water totals 4 minutes. To prepare, 14 g of coffee is mixed with 15 mL of water. This works by way of a direct contact filtration system, also known as a piston coffee maker, pump, or French press.

8.3.2 Ibrik

This method uses a metallic recipient with a reduced volume, which is placed on low flame until boiled three consecutive times. Between each boil, the mixture is removed from the fire and left to rest for 1 minute. With this method, 14 g of medium-ground coffee (501–700 μm) is mixed with 150 mL of 90°C water. The total water/coffee contact time is approximately 4 minutes. This is also known as the Turkish method.

8.3.3 Percolator

In this method, 56 g of medium-ground coffee (501–700 μm) is mixed with 600 mL of water, for 4 minutes, via a recirculation system. In this case, a significant volume of water passes through a wide-pore filter containing coffee. In Figure 8.1, the direct contact methods evaluated in the present study are listed.

8.3.4 Toddy

Exactly 100 g of medium-ground coffee (501–700 μm) is mixed with 500 mL of cold water and 500 g of ice. The three elements are mixed in a plastic container placed over the glass jar in which the prepared beverage is collected. This content passes, via gravity, through a porous filter and into a glass pitcher secured to a plastic container. The water/coffee contact time is 24 hours, during which time it is stored in the refrigerator.

8.3.5 Cold Tower

Roasted, medium-ground coffee (501–700 μm) is used. Contact time with water, at 4°C, is 12 hours. Exactly 24 g of coffee is mixed with 240 mL of water. This method uses the slow drip system until the glass container is full. The lower part of the tower consists of a vertical series with three sections (two funnels, the second with a coil and container, all three made of glass). It is also known as a "cold brew coffee tower."

8.3.6 Midzudashi

With this coffee maker, roasted, medium-ground coffee (501–700 μm) is used. This method has a filtration system below, and a glass container with plastic netting. The coffee is filtered, cold, for 24 hours, and stored in the refrigerator (4°C). Exactly 500 mL of cold water is mixed with 500 g of ice and 100 g of coffee. The ice and coffee are placed in the plastic netting. The water comes into direct contact with the ice and coffee when the pitcher is put together, as the netting is submerged in the pitcher with water, and there the cold extraction process begins.

8.3.7 Cold Brew

This method has a slow drip system, in which the water, at 4°C, is left in contact with the coffee for 4 hours. Exactly 21

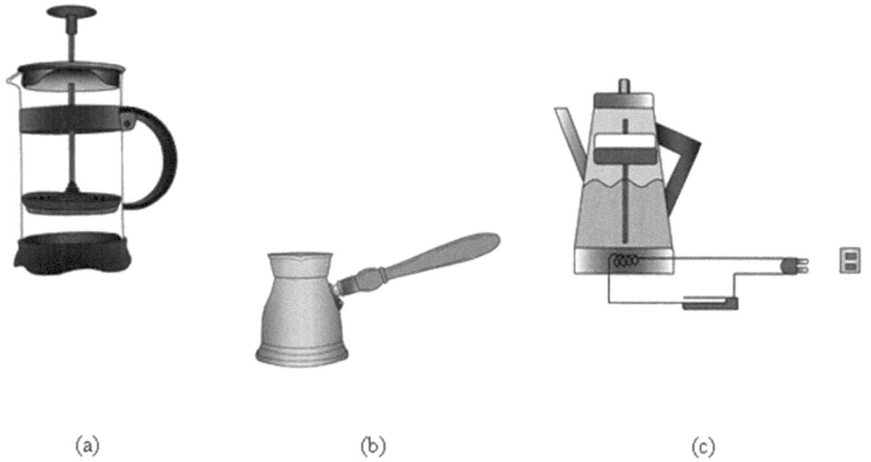

FIGURE 8.1 Methods of direct-contact coffee preparation: (a) French press, (b) Ibrik, (c) percolator.

Source: Authors.

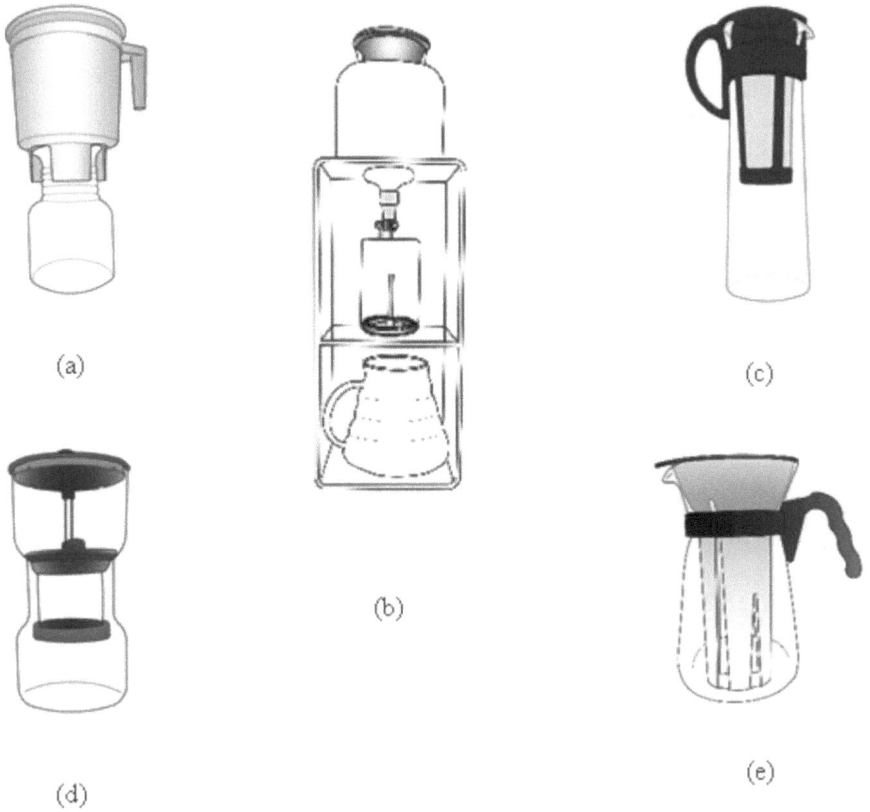

FIGURE 8.2 Cold methods for coffee preparation: (a) Toddy, (b) Cold tower, (c) Midzudashi, (d) Cold brew, and (e) Fretta.

Source: Authors.

g of medium-ground coffee (501–700 μm) is mixed with 210 mL of cold water.

8.3.8 Fretta

This method has a reservoir in which to place ice. The coffee is in contact with cold water, and falls, drop by drop, onto the ice. Exactly 14 g of medium-ground coffee (701–900 μm) and 150 mL of water, at 4°C, were mixed. Total contact time between the water and coffee was 4 hours. This is also known as an "ice-coffee maker." Figure 8.2 shows the cold coffee preparation methods evaluated in the present study.

8.3.9 Aeropress

Roasted, medium-ground coffee (501–700 μm) is used. Contact time with the 90°C water totals 4 minutes. Exactly 14 g of coffee is mixed with 150 mL of water. This functions with a piston or pump system, which permits the filtration. This method consists of two plastic cylinders, which together function as a syringe that introduces pressurized air to the mixture of water and ground coffee, forcing said mixture through a paper or cylindrical metal filter. The objective of the Aeropress is to diminish coffee infusion time and thus reduce its bitterness.

8.3.10 Presso

Finely ground (350–500 μm) roasted coffee is used. This is known as the manual action Presso coffee maker. This method uses a lever with two arms on a pump, which applies pressure to the 90°C water. This is placed in the container, in the upper part of the manual pump. For beverage preparation, 14 g of roasted, ground coffee is placed in 60 mL of hot water. This method does not use electric energy. Further, it is 100% designed in reclaimed aluminum.

8.3.11 Staresso

For beverage preparation with this coffee maker, 14 g of medium-ground coffee is mixed (501–700 μm) with 150 mL of hot water (90°C). The coffee is placed in the section indicated and pressed with the spoon with which the coffee is measured. The container with the roasted, ground coffee is fastened to the elongated cylinder with a piston. The piston is removed, and hot water is added in the available space. The piston is secured once again. Pressure is applied manually to the piston, and one waits until total beverage extraction is achieved. The beverage passes to the glass container, which serves as a cup from which to drink the hot coffee.

8.3.12 Expresso

With this method, hot water at high pressure (5–18 bars, depending on the machine) is forced through previously compacted, roasted, ground coffee. Extraction time is 20 seconds. Exactly 14 g of finely ground coffee (350–500 μm) is mixed with 75 mL of hot water, at 90°C.

8.3.13 Mocca or domestic expresso

Here, 14 g of finely ground coffee (350–500 μm) is combined with 150 mL of water, for one minute, via a pressure filtration system opposite gravity. The coffee is placed in a hermetically sealed space in the middle of the two containers. The water that ascends is below, and this causes the coffee to pass through several orifices. The beverage remains in the upper container. In Figure 8.3, the pressure filtration methods evaluated in the present study are listed.

8.3.14 Vietnamese

With this coffee maker, roasted, coarsely ground (701–900 μm) coffee is used. The contact time with the 90°C water totals 1 minute. Exactly 14 g of coffee is mixed with 150 mL of water. The Vietnamese (or Phin drip) coffee maker, is made of stainless steel and has a perforated base with a total of 80 orifices. The coffee is added directly to the container and later combined with hot water. The weight of the water causes the coffee to infuse and pass directly to the cup, owing to the effect of gravity. The coffee maker is small, can only prepare one cup at a time, and is placed directly over the cup.

8.3.15 Clever

With this method, roasted, medium-ground coffee (501–700 μm) is used. This coffee maker consists of a thermoresistant plastic cone with a small orifice at the tip. A paper filter is placed in the cone, and the coffee on the filter. The cone has a closure valve that traps the coffee until a contact time of approximately 4 minutes with the hot water has concluded. When the cone is placed over a coffee cup, the valve opens and the drip process begins. On lifting the cone to the lip of the cup, the beverage flow automatically stops. With this preparation, 14 g of roasted, ground coffee is mixed with 150 mL of hot water.

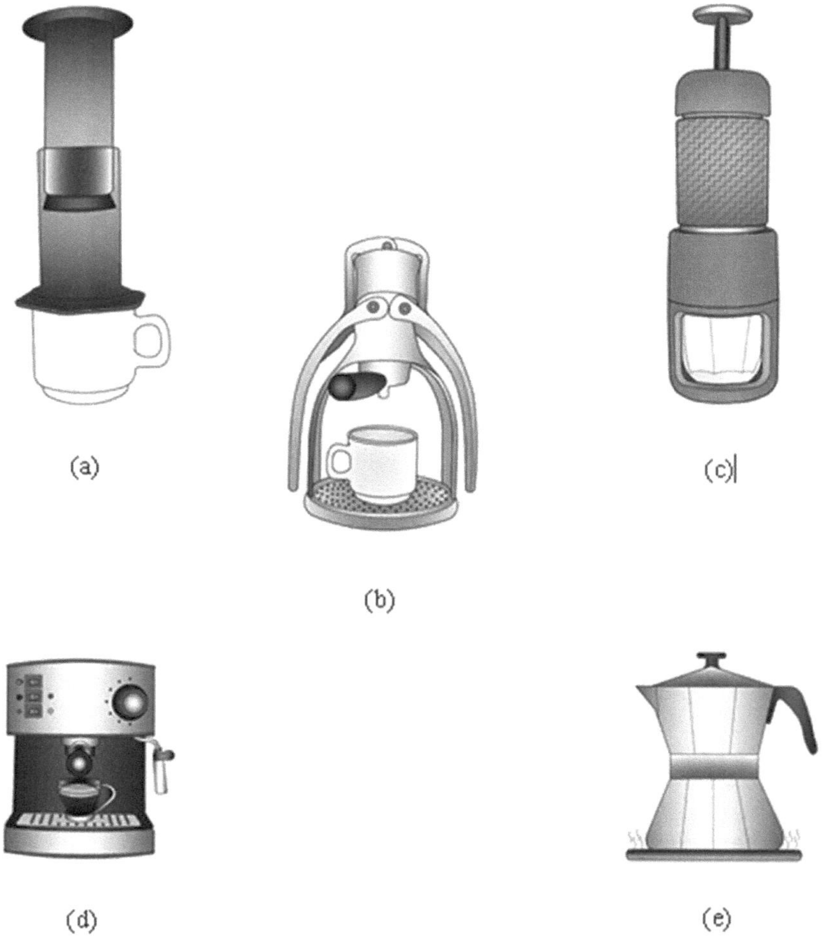

FIGURE 8.3 Pressure filtration coffee preparation methods: (a) Aeropress, (b) Presso, (c) Staresso, (d) Expresso, and (e) Mocca.

Source: Authors.

8.3.16 Kinto

This beverage is prepared with 21 g of medium-ground coffee (501–700 μm) and 225 mL of hot water (90°C). The coffee is placed in a stainless steel Kinto cone, with an ultrafine pore size, which is also located at the tip of the cone. Later, the hot water is added on top of the coffee. The water/coffee contact time is 4 minutes. The cone is placed over the mouth of a glass pitcher designed especially for this preparation. Once the water passes through the filter, the pitcher's content may be poured into a glass.

8.3.17 Yama

This method consists of a stainless-steel cone with numerous orifices, placed over a glass support. The 90°C water is added on top of the coffee until all of the water filters into a glass container. The coffee/water contact time is 1 minute. With this coffee maker, 21 g of finely ground coffee (350–500 μm) is placed in contact with 225 mL of hot water.

8.3.18 RS 16

In the beverage preparation, 18 g of coarsely ground coffee (701–900 μm) is combined with 250 mL of water, with 5 minutes of contact time, via a filtration system which consists of 48 small orifices, distributed uniformly in the lower part of the glass cup. These permit the beverage to pass via drip and gravity. The coffee is placed inside the glass on top of the coffee cup. Hot water is added, at 90°C, to the coffee, and extraction commences immediately. The beverage is deposited into the cup, ready for consumption. With this method, paper filters are not employed.

8.3.19 V60 Kalita

Roasted, medium-ground coffee (501–700 μm) is used. Contact time with the 90°C water totals four minutes. Exactly 44 g of coffee is mixed with 150 mL of water. A Kalita paper filter is placed inside the porcelain cone. The inside of the cone has three small orifices, with a triangular distribution. The waves of the V60 Kalita coffee maker cone are in horizontal position.

8.3.20 V60

For the preparation of this beverage, 14 g of medium-ground coffee (501–700 μm) is mixed with 150 mL of hot water. A V60 paper filter (Hario brand) is placed on the methacrylate cone, with oblique waves that form a 60-degree angle. The coffee is added on the paper, and the water is then added (90°C). The water/coffee contact time is 4 minutes. The beverage is deposited directly into the cup, and extraction occurs until the water passes completely through the paper filter on the cone. The cone has a single orifice at the tip. In Figure 8.4, the gravity drip filtration methods evaluated in the present study are listed.

8.3.21 Kalita

Roasted, medium-ground coffee (501–700 μm) is used. Contact time with the 90°C water totals 4 minutes. For beverage preparation, 14 g of coffee is mixed with 150 mL of water. The paper Kalita filter (Wave #185-white) is placed inside the glass cone. In the upper part of the container, where the coffee goes, hot water is added onto the coffee, and a filtration process occurs.

8.3.22 Drip

With this coffee maker, roasted, medium-ground coffee (501–700 μm) is used. For the extraction, a paper filter is used, in accordance with the dimensions of the drip coffee maker. Exactly 21 g of roasted, ground coffee are used in 225 mL of hot water (90°C). The contact time between the roasted, ground coffee and water is 3 minutes.

8.3.23 Cloth filter

This beverage is prepared with 21 g of medium-ground coffee (501–700 μm) and 225 mL of hot water (90°C). The coffee is placed inside a cloth filter over the cup or container, hot water is added to the inside of the cloth filter with the coffee, and extraction occurs.

8.3.24 Japanese siphon

The water is deposited into the round-bottom glass container over the alcohol burner. The coffee is placed in a glass funnel, with a cloth filter placed therein, and over the round-bottom glass container. The burner is turned on. The water boils in the round-bottom container, rises to the funnel, and enters into contact with the roasted, ground coffee. It is mixed manually, with a coffee spoon, and the water rises to the funnel for 1 minute. At that point, the coffee is completely prepared but still contains ground coffee. The round-bottom glass jar is removed from the burner. This is left to cool, and the prepared coffee falls, via gravity, through the cloth filter in the funnel, toward the round-bottom glass container. The coffee is ready for consumption. This equipment is also known as the empty brew coffee maker. Extraction time is 1 minute. This beverage is prepared with 21 g of medium-ground coffee (501–700 μm) and 225 mL of boiling water.

8.3.25 Fellow Duo

The metallic container in the upper part houses the filter or metal filtration grate, into which the roasted, ground coffee and hot water (90°C) are placed. With this preparation method, the upper container may be turned (it only has two graduated positions), in accordance with the degree of extraction desired, for the obtention of a concentrated or weak beverage. In this study, a concentrated beverage was prepared. Finally, after a 4-minute contact time, the coffee is filtered with gravity, and remains in the lower part's glass container, ready for consumption. This beverage is prepared with 21 g of medium-ground coffee (501–700 μm) and 225 mL of water.

8.3.26 Chemex

With this coffee maker, roasted, medium-ground coffee (501–700 μm) is used. Exactly 14 g of coffee is mixed with 150 mL of water. Contact time with the 90°C water totals 4 minutes.

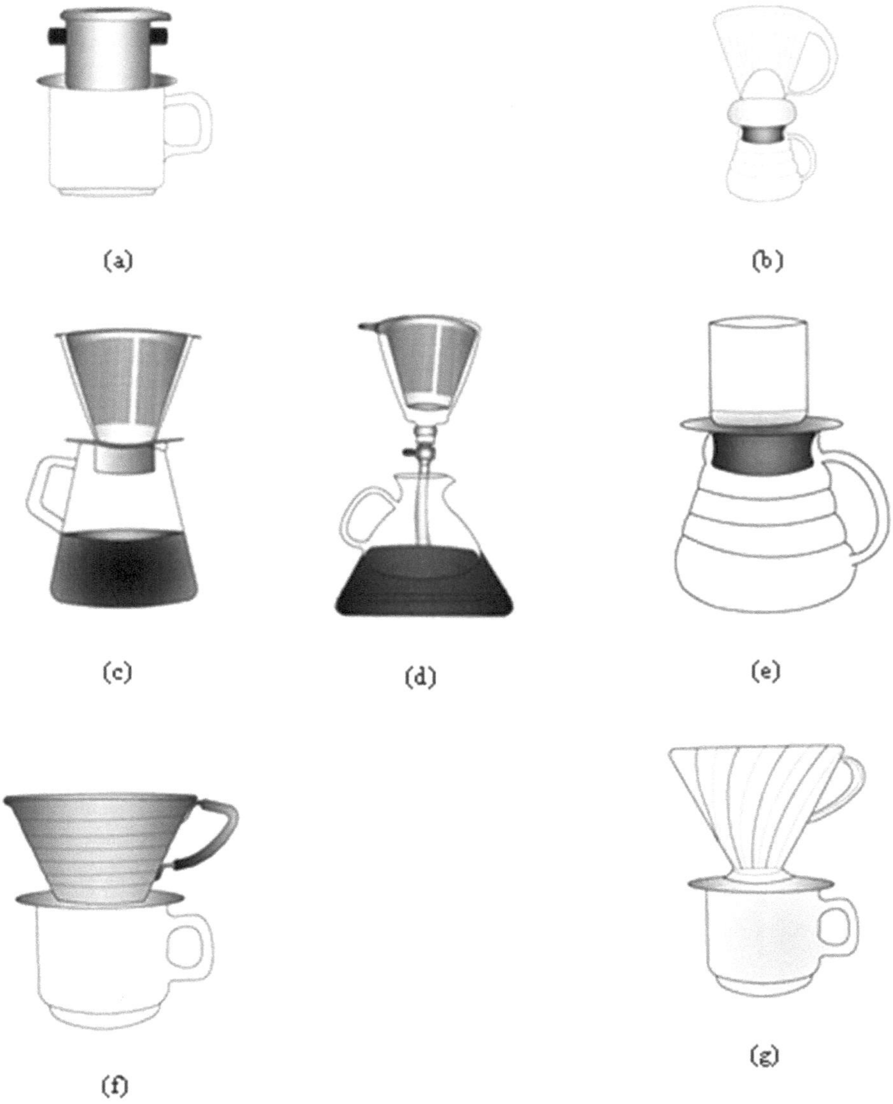

FIGURE 8.4 Drip gravity filtration coffee preparation methods: (a) Vietnamese, (b) Clever, (c) Kinto, (d) Yama, (e) RS 16, (f) V60 Kalita, and (g) V60.

Source: Authors.

The paper Chemex filter (FS-100) is placed inside the glass cone, in the upper part of the container, hot water is added, and one awaits the conclusion of the filtration process.

8.3.27 Neapolitan

This method consists of a coffee maker with two reservoirs, one on top of the other, with a screw mechanism to keep them fastened together as a single unit. Each reservoir has a handle with which to move the container. Once the coffee maker has been dismantled, the water is placed into the lower container, which has the spout from which to pour. This is filled with water, only up to the valve. The filter is placed on top of the lower recipient, and the roasted, ground coffee is added. The upper part is secured by screwing it securely to the lower part. The coffee maker is placed on the stove, with the part containing water below. When the water begins to boil, the coffee maker is flipped. One of the two handles is gripped and quickly turned, such that the water falls through the filter and begins to fill the lower container. After 2 minutes, the water filters through the ground coffee, and the beverage is in the lower container, ready to be served, with the help of the spout. This beverage is prepared with 14 g of medium-ground coffee (501–700 μm) and 150 mL of water.

8.3.28 Silvertone

This method uses a metallic filter with numerous holes, into which roasted, ground coffee is placed. The metallic filter goes on top of a fixed support, so as to facilitate the coffee infusion process. The coffee/water contact time is 45 minutes. This beverage is prepared with 14 g of medium-ground coffee (501–700 μm) and 150 mL of hot water (90°C). In Figure 8.5,

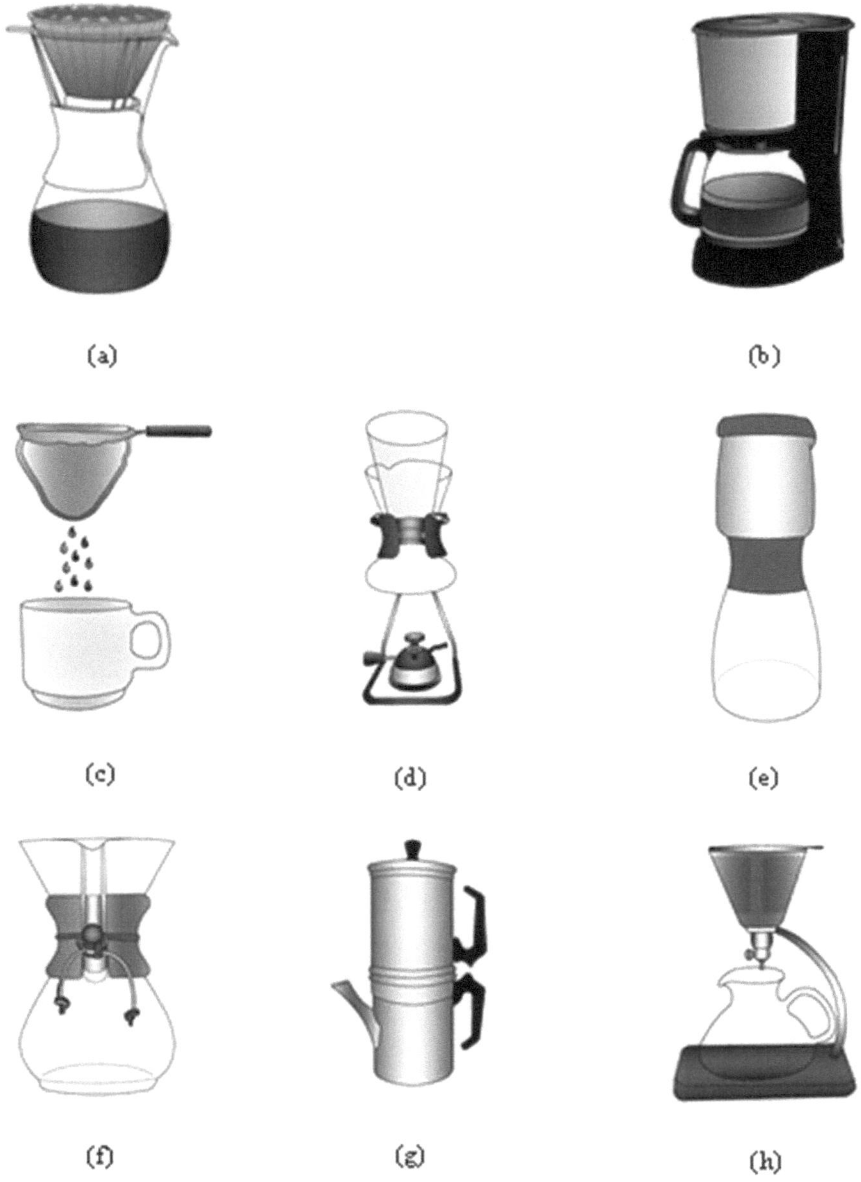

FIGURE 8.5 Gravity filtration coffee preparation methods: (a) Kalita, (b) Goteo, (c) coth filter, (d) Japanese siphon, (e) Fellow Duo, (f) Chemex, (g) Neapolitan, and (h) Silvertone.

Source: Authors.

the gravity filtration methods evaluated in the present study are listed.

8.4 Antioxidant compound content in 28 coffee preparation methods

Antioxidant compound content (e.g., total phenols, flavonoids, and condensed tannins) were evaluated in coffee beverages, based on 28 coffee preparations or devices for solid-liquid coffee extraction. Results are shown in Table 8.1. In accordance with the data indicated in Table 8.1, observe that phenol content is greatest, followed by flavonoid concentration, while condensed tannins are present in minor quantity.

8.5 Antioxidant capacity measured in 28 coffee preparation methods

Antioxidant activity is measured by the ORAC and ABTS assays, as the expression of certain metabolites to scavenger free radicals and reduce other structures, respectively. Table 8.2 shows the expression of the antioxidant activity of compounds presented in the coffee beverages.

TABLE 8.1

Total Phenol, Flavonoid, and Condensed Tannin Content in 28 Coffee Preparation Methods

Preparation method	Total phenols (mg AG eq./100 g sample)	Flavonoids (mg C eq./100 g sample)	Condensed tannins (mg C eq./100 g sample)
Aeropress	1505.4 (±0.1)	1864.3 (±0.3)	90.6 (±0.4)
Chemex	8545.9 (±0.1)	13,223.9 (±0.1)	55.1 (±0.4)
Presso	474.1 (±0.2)	401.7 (±0.3)	118.1 (±0.3)
Ibrik	3024.0 (±0.3)	3089.7 (±0.2)	162.6 (±0.2)
Expresso	5306.2 (±0.4)	10,705.2 (±0.4)	197.4 (±0.1)
Staresso	2170.3 (±0.3)	2352.6 (±0.1)	129.5 (±0.1)
V60 Kalita	922.7 (±0.4)	1414.9 (±0.3)	50.5 (±0.1)
French Press	2112.9 (±0.4)	2467.9 (±0.4)	95.9 (±0.2)
Coth filter	1126.3 (±0.1)	1634.9 (±0.3)	65.8 (±0.4)
Fretta	493.2 (±0.2)	998.9 (±0.4)	39.4 (±0.4)
Kalita	837.8 (±0.2)	1319.4 (±0.2)	74.3 (±0.3)
V60	1040.2 (±0.3)	1552.3 (±0.1)	72.7 (±0.3)
Percolator	1230.9 (±0.3)	1587.5 (±0.2)	51.3 (±0.2)
Mocca	2929.3 (±0.1)	3003.1 (±0.2)	114.7 (±0.1)
RS 16	1862.4 (±0.1)	2359.4 (±0.3)	120.5 (±0.4)
Yama	1144.2 (±0.1)	1537.7 (±0.2)	81.3 (±0.4)
Toddy	1968.8 (±0.4)	2263.1 (±0.1)	123.5 (±0.3)
Neapolitan	1062.1 (±0.4)	1486.8 (±0.1)	88.2 (±0.3)
Silvertone	2423.7 (±0.2)	2794.6 (±0.2)	147.9 (±0.4)
Midzudashi	988.2 (±0.3)	1481.1 (±0.2)	107.3 (±0.2)
Cold Brew	1512.7 (±0.3)	1855.8 (±0.1)	109.4 (±0.2)
Vietnamese	1425.9 (±0.1)	1972.5 (±0.2)	146.1 (±0.3)
Kinto	1240.6 (±0.2)	1678.7 (±0.2)	82.3 (±0.2)
Drip	560.3 (±0.2)	1161.3 (±0.3)	53.7 (±0.2)
Japanese siphon	2449.1 (±0.2)	2578.2 (±0.2)	140.6 (±0.1)
Cold Tower	2146.1 (±0.3)	2082.0 (±0.1)	126.3 (±0.3)
Clever	1991.4 (±0.3)	2170.7 (±0.2)	119.4 (±0.2)
Fellow Duo	1164.3 (±0.3)	1447.8 (±0.2)	66.9 (±0.1)

Notes: GA = gallic acid; C = catechin; average values (n = 3).

TABLE 8.2

ABTS and ORAC Antioxidant Capacity Measured in 28 Coffee Preparation Methods

Preparation method	ABTS (μmol Trolox eq./100 g sample)	ORAC (μmol Trolox eq./100 g sample)
Aeropress	28,745.7 (±0.3)	50,172.4 (±0.1)
Chemex	14,852.2 (±0.4)	37,099.6 (±0.2)
Presso	34,713.5 (±0.3)	49,733.5 (±0.3)
Ibrik	49,664.0 (±0.4)	74,801.8 (±0.1)
Expresso	106,493.5 (±0.2)	159,781.9 (±0.3)
Staresso	28,606.0 (±0.3)	59,370.8 (±0.3)
V60 Kalita	17,239.0 (±0.4)	24,867.6 (±0.1)
French Press	33,739.7 (±0.4)	57,106.7 (±0.2)
Coth Filter	22,325.5 (±0.4)	30,325.8 (±0.3)
Fretta	10,412.2 (±0.3)	18,120.4 (±0.3)

Preparation method	ABTS (μmol Trolox eq./100 g sample)	ORAC (μmol Trolox eq./100 g sample)
Kalita	11,182.6 (±0.2)	23,801.9 (±0.1)
V60	19,021.5 (±0.4)	30,901.4 (±0.2)
Percolator	22,156.3 (±0.4)	16,667.1 (±0.3)
Mocca	46,014.7 (±0.4)	85,655.1 (±0.3)
RS 16	33,851.4 (±0.3)	54,898.8 (±0.2)
Yama	21,892.7 (±0.2)	28,432.2 (±0.4)
Toddy	30,658.2 (±0.4)	60,406.1 (±0.3)
Neapolitan	18,948.5 (±0.4)	42,498.1 (±0.1)
Silvertone	38,767.9 (±0.4)	70,688.5 (±0.2)
Midzudashi	16,485.5 (±0.3)	31,650.6 (±0.4)
Cold Brew	24,269.9 (±0.3)	43,690.3 (±0.4)
Vietnamese	21,739.7 (±0.3)	43,700.1 (±0.3)
Kinto	21,331.6 (±0.1)	41,432.7 (±0.2)
Drip	9153.4 (±0.5)	22,889.3 (±0.3)
Japanese siphon	38,057.3 (±0.5)	67,325.3 (±0.1)
Cold Tower	31,823.1 (±0.3)	46,938.5 (±0.2)
Clever	32,291.9 (±0.3)	52,339.0 (±0.2)
Fellow Duo	17,499.2 (±0.3)	24,982.9 (±0.2)

Note: Average values ($n = 3$).

8.6 Hydroxycinnamic phenolic acid content measured in 28 coffee preparation methods

Hydroxycinnamic acids of phenolic nature were measured in 28 different preparations (Table 8.3). These acids included chlorogenic acid, caffeic acid, p-coumaric acid, and ferulic acid. In relation to the concentrations indicated in Table 8.3, chlorogenic acid is the predominant acid, followed by ferulic, p-coumaric, and caffeic acid, from the phenolic fraction evaluated.

8.7 The ten coffee preparation methods that provide greatest antioxidant compound content and activity

In accordance with the data reported in Tables 8.1, 8.2, and 8.3, which correspond to antioxidant content and the antioxidant capacity of coffee beverages, a classification was performed, and the ten coffee preparations with the greatest functional compound contribution of antioxidant nature were obtained. The selection criteria used was total phenol content. The ten methods with greatest total phenol concentration exhibit principles of pressure filtration (3), followed by direct contact (3), cold methods (2), gravity filtration (1), and drip (1). The ten methods selected with greatest antioxidant functionality, in descending order, were Expresso, Ibrik, Mocca, Silvertone, Japanese siphon, Cold Tower, Staresso, Clever, French Press, and Toddy.

Figure 8.6 indicates total phenol content in the ten coffee preparations with the highest amounts of this antioxidant compound. Total phenol concentration was greatest with the Expresso method, followed by the Ibrik and Mocca methods. The greatest concentration of total phenols occurred in the coffee prepared with the Expresso method, with a concentration of 5306.2 mg of gallic acid eq./100 g of sample, a pressure filtration method. The seven remaining methods had total phenol concentrations between 1991.4 and 2449.1 mg of gallic acid eq./100 g of sample. The preparation method had a significant effect ($p < 0.05$) on total phenol retention.

Excelso-quality coffee contains elevated concentrations of phenolic acids compared to other types of Colombian coffee. As such, phenolic compounds are responsible for antioxidant activity in coffee (Naranjo et al., 2011). However, in the present study, it was identified that the antioxidant fraction composed of flavonoids was superior in concentration to total phenol content. Roasting liberates caffeine, owing to phenol thermal degradation, and generates an increase in body and the bitter flavor characteristic of the coffee beverage (Cheng et al., 2016; Swak et al., 2017). Additionally, phenols participate in the synthesis of products secondary to the Maillard reaction, owing to the roasting temperature (Moreira et al., 2017). This is also the case of melanoidins, which support antioxidant activity, such as beverage flavor and color (Lee et al., 2017).

Phenolic compounds influence the caffeine and chlorogenic and caffeic acid concentrations of the prepared beverage. Further, on the chemical stability level, they are sensitive to pH changes and participate in transesterification reactions (Hečimović et al., 2011). These pH variations may occur on dissolving coffee in large quantities of water. Ingesting phenolic compounds in one's diet protects the body from illnesses such as diabetes and those of a cardiovascular nature and regulates oxidative stress. Further, they possess proven

TABLE 8.3

Hydroxycinnamic Phenolic Acid Content Measured in 28 Coffee Preparation Methods

Preparation method	Chlorogenic acid (mg/100 g sample)	Caffeic acid (mg/100 g sample)	P-coumaric acid (mg/100 g sample)	Ferulic acid (mg/100 g sample)
Aeropress	11.8 (±0.1)	N.D.	23.0 (±0.3)	24.5 (±0.4)
Chemex	66.8 (±0.3)	N.D.	35.3 (±0.3)	37.8 (±0.2)
Presso	33.4 (±0.5)	N.D.	13.5 (±0.3)	5.7 (±0.5)
Ibrix	94.9 (±0.2)	N.D.	38.2 (±0.2)	40.7 (±0.3)
Expresso	96.2 (±0.4)	N.D.	36.7 (±0.5)	39.1 (±0.1)
Staresso	129.6 (±0.3)	98.5 (±0.1)	27.2 (±0.2)	29.0 (±0.3)
V60 Kalita	35.1 (±0.3)	32.3 (±0.4)	26.1 (±0.4)	27.8 (±0.3)
French Press	127.2 (±0.3)	100.2 (±0.3)	27.2 (±0.4)	28.9 (±0.5)
Coth Filter	15.7 (±0.1)	N.D.	12.9 (±0.2)	13.8 (±0.5)
Fretta	38.4 (±0.4)	N.D.	25.9 (±0.8)	27.4 (±0.1)
Kalita	40.8 (±0.2)	N.D.	26.2 (±0.4)	27.9 (±0.3)
V60	72.1 (±0.6)	N.D.	36.3 (±0.5)	34.0 (±0.7)
Percolator	87.3 (±0.5)	N.D.	39.8 (±0.1)	42.6 (±0.6)
Mocca	81.6 (±0.2)	N.D.	34.0 (±0.7)	36.3 (±0.1)
RS 16	13.2 (±0.3)	N.D.	5.9 (±0.3)	8.1 (±0.5)
Yama	87.7 (±0.1)	N.D.	38.1 (±0.3)	40.7 (±0.3)
Toddy	65.1 (±0.5)	N.D.	25.4 (±0.4)	25.4 (±0.3)
Neapolitan	114.5 (±0.5)	N.D.	41.8 (±0.1)	44.6 (±0.2)
Silvertone	73.9 (±0.3)	N.D.	34.0 (±0.7)	31.4 (±0.3)
Midzudashi	49.8 (±0.7)	27.4 (±0.4)	34.0 (±0.7)	28.4 (±0.1)
Cold Brew	71.4 (±0.1)	N.D.	34.0 (±0.7)	26.7 (±0.3)
Vietnamese	72.9 (±0.3)	N.D.	30.5 (±0.3)	32.6 (±0.4)
Kinto	8.6 (±0.2)	N.D.	7.0 (±0.1)	7.4 (±0.3)
Drip	48.9 (±0.4)	45.2 (±0.3)	32.0 (±0.3)	34.2 (±0.2)
Japanese siphon	200.9 (±0.5)	N.D.	66.8 (±0.4)	34.0 (±0.7)
Cold Tower	N.D.	N.D.	N.D.	N.D.
Clever	130.6 (±0.3)	N.D.	45.2 (±0.5)	48.2 (±0.2)
Fellow Duo	103.4 (±0.2)	N.D.	44.5 (±0.4)	47.7 (±0.4)

Notes: Average values ($n = 3$); N.D. = no data.

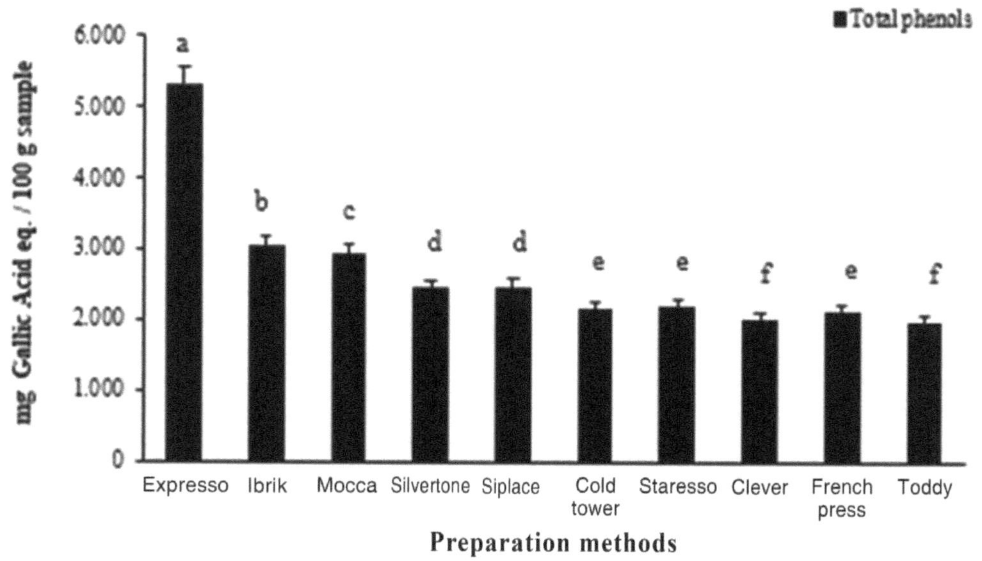

FIGURE 8.6 The ten coffee preparations with greatest total phenol content. Average values ($n = 3$) with different letters in the same variable indicate statistical differences at the 5% significance level ($p > 0.05$).

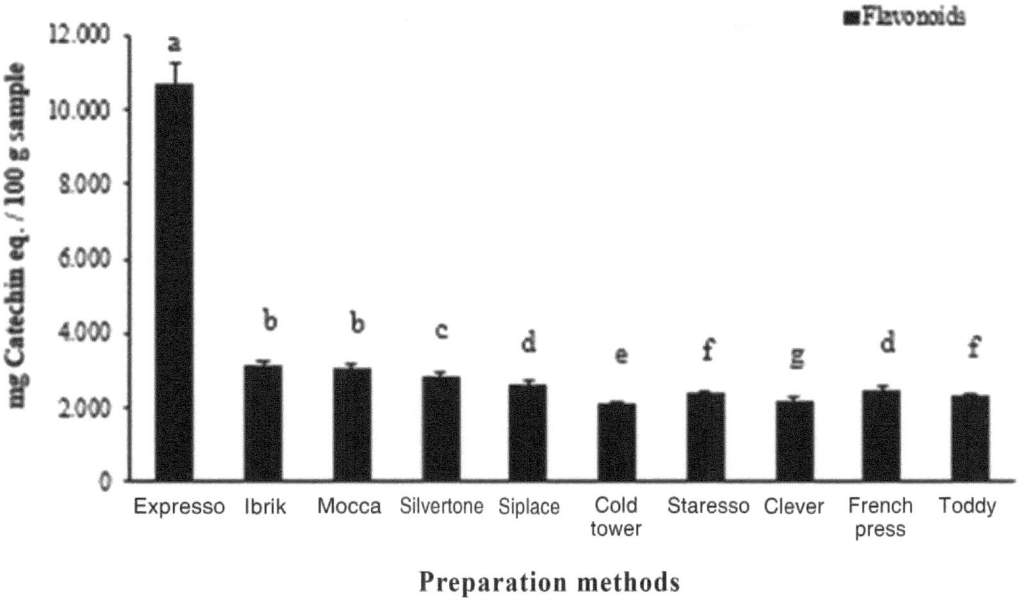

FIGURE 8.7 The ten coffee preparations with greatest flavonoid concentration. Average values ($n = 3$) with different letters in the same variable indicate statistical differences at the 5% significance level ($p > 0.05$).

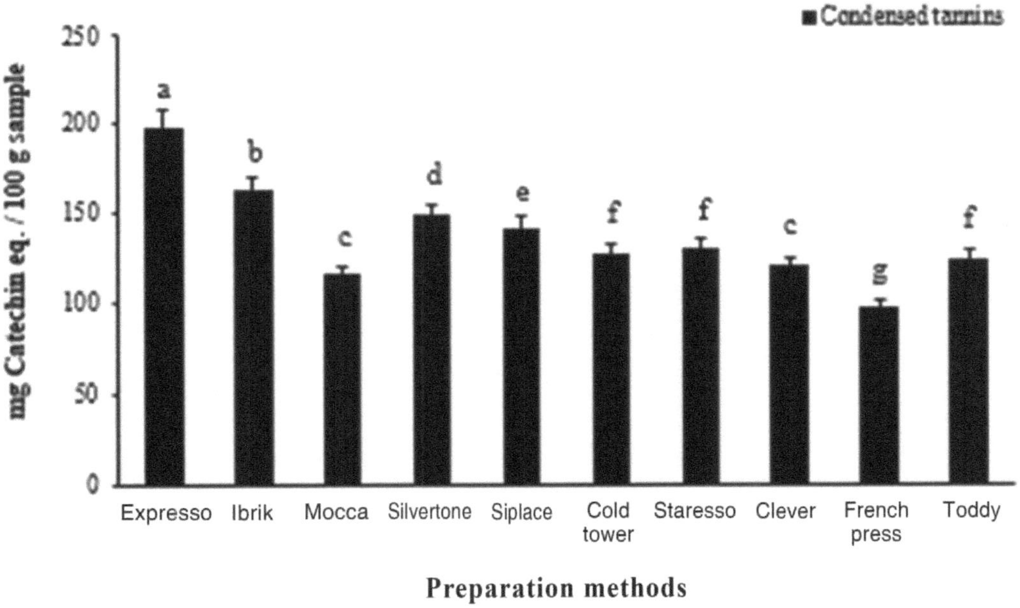

FIGURE 8.8 The ten coffee preparations with greatest condensed tannin content. Average values ($n = 3$) with different letters in the same variable indicate statistical differences at the 5% significance level ($p > 0.05$).

antiallergenic, antimicrobial, and anti-inflammatory properties (Mussatto, 2015).

Figure 8.7 shows the flavonoid content of those ten coffee preparations with the highest content of this antioxidant component. The maximum flavonoid content was determined to be in the Expresso method beverage, followed by that of Ibrik and Mocca. Expresso coffee presented a concentration of 10,705.2 mg of catechin/100 g of sample, significantly higher than the other methods evaluated. The remaining methods presented flavonoid concentrations between 3089.7 and 2082.0 mg of catechin/100 g of sample. The preparation method had a significant effect ($p < 0.05$) on total flavonoid retention in the selected preparations.

Flavonoids have anti-atherosclerotic, anti-cancer, and anti-inflammatory properties. They also prevent illnesses such as Alzheimer's and Parkinson's diseases (Kurzawa-Zegota et al., 2012). They further contribute to the reduction in prevalence of coronary illnesses and to intestinal microbial equilibrium (Cardona et al., 2013). Flavonoids contribute sensoriality to coffee with the bitter flavor.

Condensed tannin content is represented graphically, in Figure 8.8, in the ten coffee preparations with the highest content of this antioxidant metabolite. Tannin content is quite low, compared to the concentration indicated for total phenols and flavonoids. Coffee preparation methods with the greatest condensed tannin concentrations were Expresso, followed by Ibrik and Silvertone, which are pressure filtration, direct contact, and gravity filtration methods, respectively. The Expresso method reported a concentration of 197.4 mg of catechin/100 g of sample. The remaining methods presented concentrations between 162.6 and 95.9 mg of catechin/100 g of sample. The preparation method had a significant effect ($p < 0.05$) on total tannin retention in the selected preparations from the initial group of 28 methods.

Tannins are substances with high nutritional value, and their presence in the cup test is closely related to astringent notes in the coffee-based beverage (Caetano et al., 2017). In accordance with the obtained cup profile, astringent notes in the evaluated beverages were within an optimal range, given that the tannin concentration was not representative in the QDA profile score, on presentation of a score similar to that of the Excelso pattern. On the biofunctional level, tannins possess antihemorrhagic and anti-inflammatory properties, and also help to prevent premature aging and degenerative disease (Zapata et al., 2013). These substances provide an undesirable tactile sensation to coffee, characterized by constriction, puckering, and dryness in the oral cavity. Further, they influence the sensation of bitterness, owing to the degree of polymerization of condensed tanning molecules. Tannins also present synergy with saliva proteins, and impart an astringent sensation in the palate (Poncet-Legrand et al., 2010; Chira and Teissedre, 2015; Watrelot et al., 2018).

Figure 8.9 shows the hydroxycinnamic acid concentration of those ten coffee preparations with the highest content of said compounds, differentiated into chlorogenic acid, caffeic acid, p-coumaric acid, and ferulic acid content, respectively. The predominant organic acid in all preparations was chlorogenic acid, indicating that the Japanese siphon method had the greatest concentration thereof (200 mg chlorogenic acid/100 g of sample). Caffeic acid was not detected in the Expresso, Ibrik, Mocca, Silvertone, Japanese siphon, Cold Tower, Clever, or Toddy methods, as this compound was not retained therein. Caffeic acid was detected in the Staresso and French Press beverages, at 98.5 and 100.2 mg of this acid per 100 g of sample, respectively. P-coumaric acid was detected in all selected preparations except in the Cold Tower method beverage. The Japanese siphon method had the greatest concentration thereof. Ferulic acid also had a maximum concentration with the Japanese siphon method, and was not detected with the Cold Tower method. The latter did not register content of any of the four hydroxycinnamic acids evaluated. The preparation method had a significant effect ($p < 0.05$) on retention of the four hydroxycinnamic acids evaluated, in the ten preparations with greatest antioxidant retention.

Hydroxycinnamic acids influence the sensorial perception of the beverage and its antioxidant activity. The chemical breakage and transformation of chlorogenic acids during roasting generates phenolic compounds (Ross et al., 2011). Chlorogenic acids provide acidity, astringency, and bitter notes to the beverage (Cheng et al., 2016), as well as flavonoids and caffeine. One study reported that the predominant hydroxycinnamic acid was chlorogenic acid, whereas caffeic acid and p-coumaric acid were found in the lowest concentrations in samples of *Coffea arabica* (Cheong et al., 2013). These results are similar to the findings of the present study.

Melanoidins and chlorogenic, caffeic, coumaric, and ferulic acids help to promote antioxidant capacity in coffee (Hečimović et al., 2011; Vignoli et al., 2011). One study

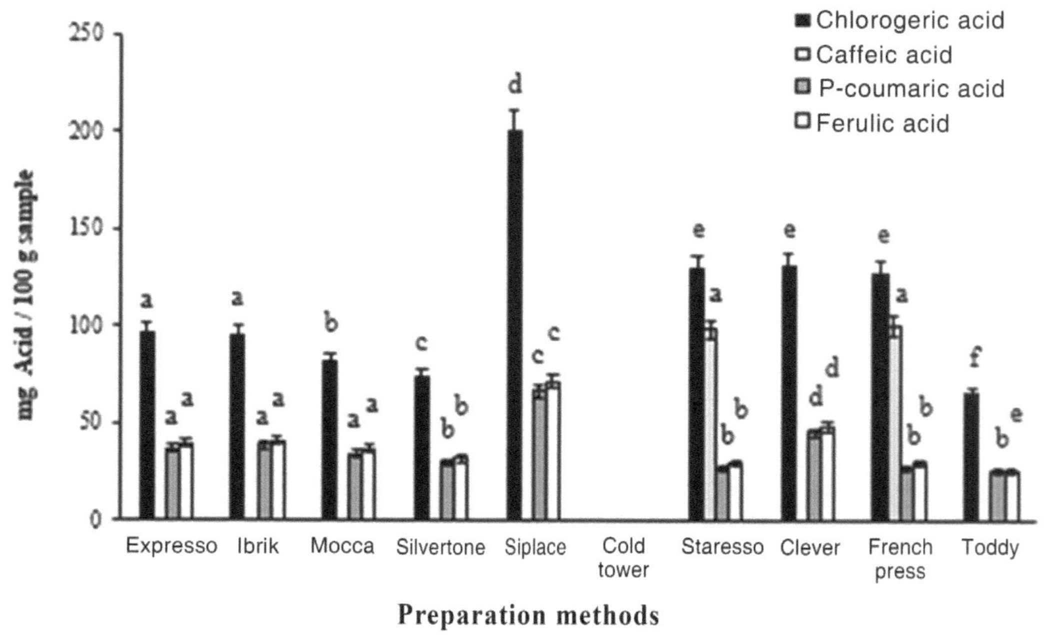

FIGURE 8.9 The ten coffee preparations with greatest hydrocinnamic phenolic acid concentration. Average values ($n = 3$) with different letters in the same variable indicate statistical differences at the 5% significance level ($p > 0.05$).

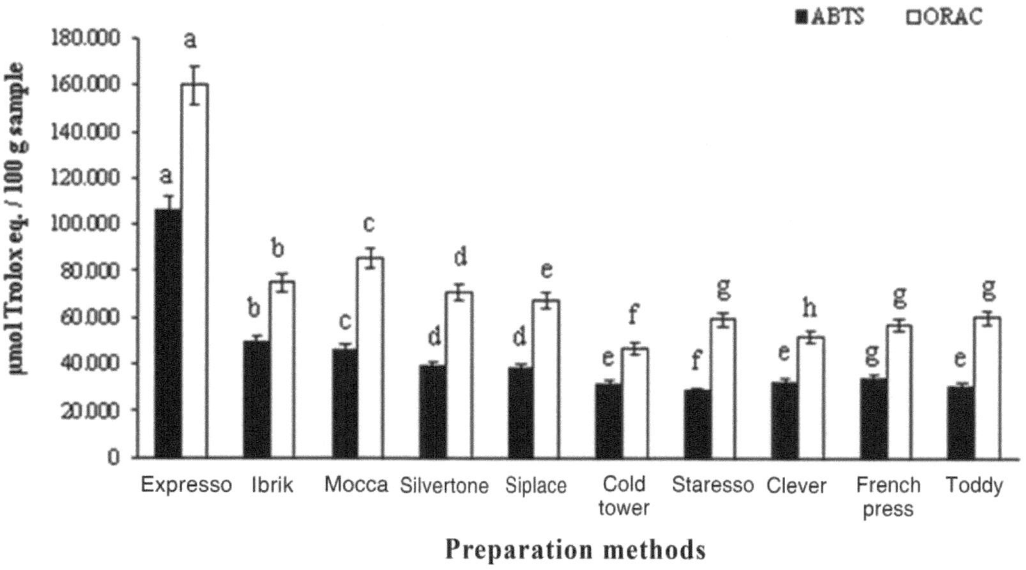

FIGURE 8.10 The ten coffee preparations with greatest ABTS and ORAC antioxidant capacity. Average values ($n = 3$) with different letters in the same variable indicate statistical differences at the 5% significance level ($p > 0.05$).

indicated a direct relationship between non-enzymatic browning and the antioxidant capacity of Colombian roasted coffee (Contreras-Calderón et al., 2016).

Figure 8.10 graphically represents ABTS and ORAC antioxidant activity in the ten coffee preparations with greatest concentration of antioxidant compounds (total phenols, flavonoids, condensed tannins, and hydroxycinnamic phenolic acids). This parameter is more important than the concentration of antioxidant compounds, given that it is a direct measurement of the chemical radical inhibition capacity and represent the efficiency of each compound as an antioxidant (Zapata et al., 2013). All ten preparations presented ABTS and ORAC antioxidant activity. The greatest ABTS antioxidant activity was equivalent to the content obtained, with greatest activity in the Expresso beverage, followed by the Ibrik and Mocca beverages, with concentrations of 106,493.5, 49,664.0, and 46,014.7 µmol Trolox equivalent per 100 g sample, respectively. ORAC activity presented greatest antioxidant activity with the Expresso method, followed by the Mocca and Ibrik methods, with contents of 159,781.9, 85,655.1, and 74,801.8 µmol Trolox equivalent per 100 g sample respectively. The Cold Tower method was among the three coffee preparations with the least antioxidant activity for both methodologies studied.

8.8 Cup test of the ten coffee preparation methods that contribute the greatest content of biofunctional compounds with antioxidant capacity, compared to a UGQ-quality coffee beverage

Figure 8.11 shows, in radial graphic form, the sensorial evaluations, or cup tests, performed on the ten coffee preparations that presented the greatest antioxidant content and activity, respectively, compared to an Excelso pattern (UGQ). The preparations selected to be compared to the UGQ pattern presented different cup profiles. Expresso coffee presented greater bitterness, body, and less acidity than the pattern. The Ibrik method showed less acidity and greater bitterness than the UGQ pattern. The Mocca preparation had slightly greater body, acidity, and bitter notes than the reference pattern. The Silvertone method presented a higher overall impression, body, and bitterness, with less acidity than the pattern. The Japanese siphon method displayed greater acidity than the UGQ pattern. The Cold Tower method had higher acidity and body, and less acidity than the reference pattern. The Staresso method presented the same scores as the pattern, except for the overall impression, which was slightly higher than that of Excelso coffee. The Clever method showed slightly higher scores in the bitterness, body, and overall impression parameters than the pattern. The coffee prepared with the French Press registered greater bitter flavor, lower acidity, and slightly more body than Excelso coffee. The coffee prepared with the Toddy coffee maker scored slightly higher in body and overall impression than the Excelso pattern. The sensorial descriptors, which were not previously named, presented the same score as the UGQ pattern in all of the abovementioned methods. Lopane (2018) obtained cold coffee with sweeter and less bitter notes, a conclusion which does not coincide with the results obtained in the present study.

In terms of antioxidant activity and content in coffee-based beverages, this expression responds to factors including the grind, coffee/water ratio, water temperature, and extraction time in the method employed (Niseteo et al., 2012).

8.9 Conclusion

Of the 28 methods of preparation evaluated for antioxidant content and activity, the ten best coffee preparation methods

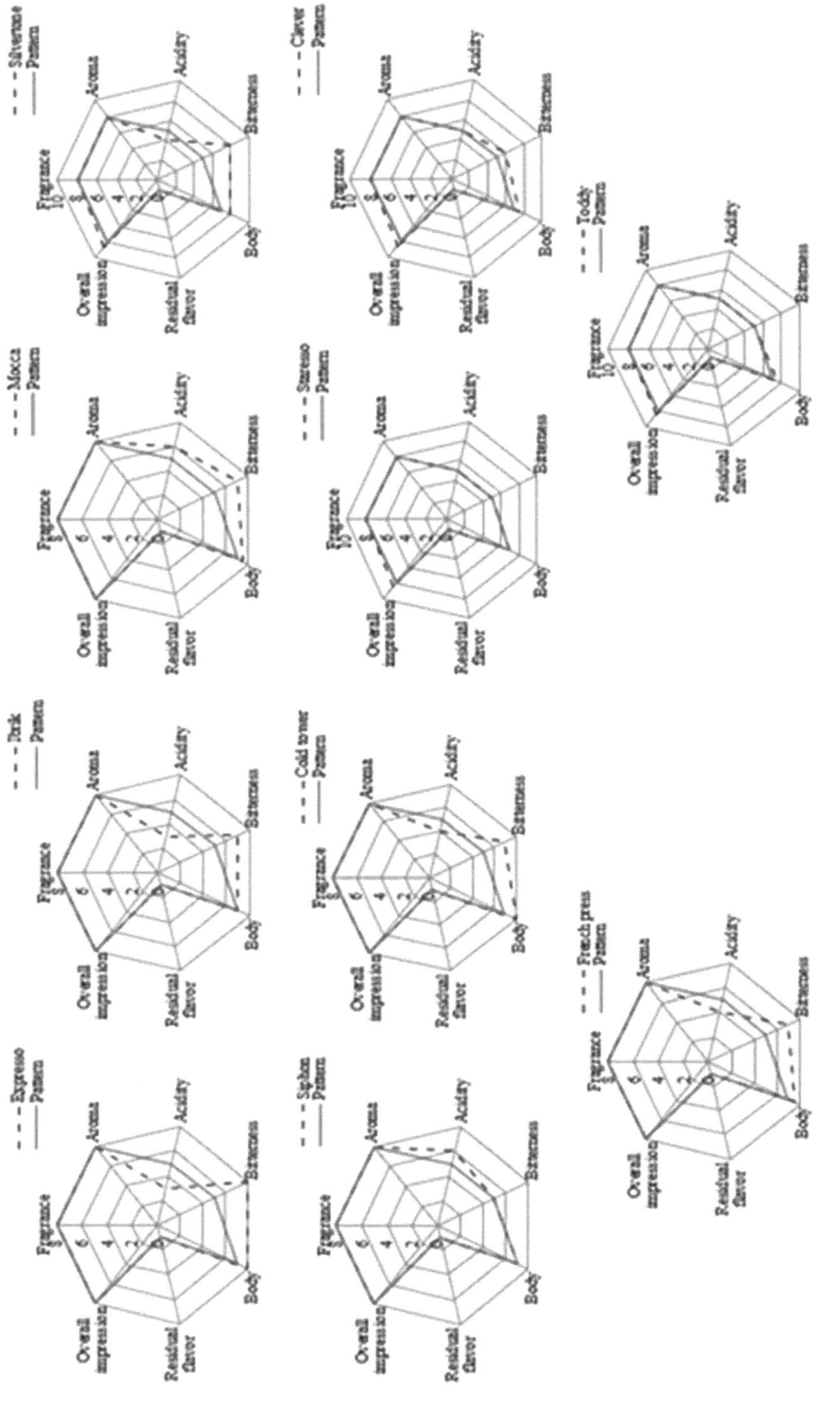

FIGURE 8.11 Cup test of the ten coffee preparations with greatest concentrations of antioxidant chemical types, compared to the UGQ average values ($n = 3$).

were selected based on the phenolic concentrations obtained. It is inferred, from this selection, that the Expresso, Ibrik, and Mocca methods are the coffee preparations that contribute the greatest antioxidant content and activity (ABTS and ORAC) to the hot beverage. There was a direct relationship between antioxidant varieties and antioxidant activity. The Expresso, Ibrik, and Mocca methods are classified as pressure filtration, direct contact, and pressure filtration methods, respectively, in accordance with their extraction methods. Of the cold method group, the Toddy and Cold Tower methods were selected to be among the ten best methods. The method with greatest hydroxycinnamic phenolic acid content was the Japanese siphon method, followed by the Staresso method. The predominant hydroxycinnamic acid in the ten best preparations was chlorogenic acid. In the cup profile of the ten best methods selected, Silvertone, Staresso, Clever, and Toddy presented the highest overall impression scores. The remaining preparations presented the same score as the pattern. Of the ten best methods, those with least acidity were Expresso, Silvertone, Ibrik, Cold Tower, and French Press. Of the ten methods, those with greatest bitter flavor were Expresso, Ibrik, Mocca, Silvertone, Cold Tower, and French Press. Finally, coffee-based beverage preparation is recommended with the Expresso, Ibrik, and Mocca preparations so as to guarantee high consumption of antioxidant compounds with the highest antioxidant capacity. These methods all use finely ground coffee and short extraction times. If the consumer desires a marked bitter flavor, they may choose Expresso or Ibrik preparations. In contrast, if they seek a coffee with a balance of acidic and bitter flavors, Mocca is the best preparation method.

8.10 Acknowledgments

The authors are grateful for the support provided by the Sensorial analysis and coffee industrialization laboratory at the Universidad de Caldas and the food science laboratory at the Universidad Nacional de Colombia, Medellin branch, during the planning and implementation of the present study. They also appreciate the contribution of Diana Paola Cardona Tangarife in the creation of the illustrations found in the present text.

REFERENCES

Aguiar J, Estevinho BN, Santos L. Trends Food Sci Tech. 2016;58:21–39.

Arango M.R. El beneficio ecológico del café en Colombia. 2012. Available online: www.revistas.javeriana.edu.co/index.php/desarrolloRural/article/. . ./1671. (Accessed 11 March 2020).

Baumann et al. Patent US20120021108A1. 2012. Available online: https://patents.google.com/patent/US20120021108A1/en. (Accessed 7 March 2020).

Caetano et al. Waste Biomass Valori. 2017;8:1981–1994.

Cano-Marquina et al. Maturitas. 2013;75:7–21.

Cardona et al. J. Nutr. Biochem. 2013;24:1415–1422.

Çelik EE, Gökmen V. Food Chem. 2018;255:8–14.

Cheng et al. Trends Food Sci. Technol. 2016;57:20–30.

Cheong et al. Food Res. Int. 2013;51:388–396.

Chira K, Teissedre P, Technol. 2015;240:533–547.

Contreras-Calderón et al. Food Chem. 2016;209:162–170.

CTG 3566:2011. *Café verde. Preparación de muestras para uso en análisis sensorial.* ICONTEC, Bogotá D.C., Colombia, 2011:1–9.

CTG 4883. *Análisis Sensorial café. Metodología para análisis sensorial cuantitativo descriptivo del café.* ICONTEC, Bogotá, Colombia, 2000:1–13.

Díaz-Arango FO, Mejía-Gutiérrez LF, León LA. 2018. *In text: (Capítulo 8. Preparación de la bebida y sus beneficios para la salud 143).* Editorial Universidad de Caldas.

FNCC. Federación nacional de cafeteros de Colombia. Descripción del proceso productivo y del beneficio del café. Guía tecnológica del cultivo. 2017. Available online: www.federaciondecafeteros.org/static/files/8Capitulo6.pdf. (Accessed 10 March 2020).

Hamilton LM, Lahne J. Food Qual Prefer. 2020;83:1–11.

Hečimović et al. Food Chem. 2011;129:991–1000.

Illy. El café filtro, o café americano: cómo se prepara. 2017. Available online: www.illy.com/wps/wcm/connect/es/cafe/cafe-filtro. (Accessed 3 April 2020)

Kelebeck et al. Microchemical J. 2009;91:187–192.

Kurzawa-Zegota et al. Food Chem. Toxicol. 2012;50:124–129.

Lee et al. Food Chem. 2015;185:182–191.

Lee et al. LWT-Food Sci. Technol. 2017;80:32–42.

Liang et al. Food Chem. 2016;213:251–259.

Liu Y, Kitts DD. Food Res Int. 2011;44:2418–2424.

Lopane SN. Clemson University. 2018. Available online: https://tigerprints.clemson.edu/all_theses (Accessed 2 March 2020)

Moreira et al. Food Chem. 2017;227:422–431.

Mussatto SI. *Coffee in Health and Disease Prevention.* Elsevier. 2015.

Naranjo et al. Rev. Cubana Plant Med. 2011;16:164–173.

Niseteo et al. Food Chem. 2012;134:1870–1877.

OIC. Organización Internacional del café. Informe del mercado de café. 2015. Available online: www.ico.org/documents/cy2014-15/cmr-0315-c.pdf. (Accessed 3 March 2020).

Ormaza-Zapata AM, Díaz-Arango FO, Rojano BA. The effect of gravity-drip filtration methods on the chemical and sensorial properties of coffee (Coffea arabica L. var. Castillo). Coffee Sci. 2019a;14:415–426.

Ormaza-Zapata et al. Rev. DYNA. 2019b;86:270–279.

Oroian M, Escriche I. Food Res Int. 2015;74:10–36.

Perfect Daily Grind. Una Guía De las Regiones de Café de Colombia. 2017. Available online: https://perfectdailygrind.com/es/2017/10/31/una-guia-de-las-regiones-de-cafe-de-colombia/(Accessed 3 March 2020).

Perrone et al. J. Agric. Food Chem. 2012;60:4265–4275.

Poncet-Legrand et al. Biomacromolecules. 2010;11:2376–2386.

Ramakrishna et al. Plant Signaling & Behavior. 2011;6:800–809.

Ramakrishna et al. J. Pineal Res. 2012;52:470–476.

Ravishankar GA, Ramakrishna A. *Serotonin and Melatonin: Their Functional Role in Plants, Food, Phytomedicine, and Human Health.* Taylor & Francis, CRC Press. 2017.

Rosales-Delgado et al. Acta Sci. Pol. Technol. Aliment., 2019;18:235–248.

Ross et al. J. Food Sc. 2011;76:C884–C890.

Santini et al. Food Control. 2011;22:1240–1245.

Sunarharum et al. Food Res. Int. 2014;62:315–325.

Swak et al. Food Control. 2017;71:210–216.

Vignoli et al. Food Chem. 2011;124:863–868.

Watrelot et al. LWT – Food Sci. Technol. 2018;91:330–338.

Yashin et al. Antioxidants. 2013;2:230–245.

Zapata et al. CIT. 2013;24:103–112.

9

Phenolic compounds in coffee seeds and by-products

Juliana de Paula and Adriana Farah

CONTENTS

9.1 Introduction

Phenolic compounds are secondary metabolites generally involved in plant adaptation to environmental stress conditions found in plant-based foods and beverages. Many epidemiological studies, intervention studies, and mechanistic *in vitro* data support a role for phenolic compounds against the development of chronic diseases related mainly to their potent antioxidant and anti-inflammatory activities. Chlorogenic acids (CGA) and related compounds are the main components of the phenolic fraction of all parts of the coffee plant, although at different proportions. Additional phenolic compounds in different parts of the plant include flavonoids and xanthones. These topics are approached in this chapter.

9.2 Phenolic compounds in nature

Phenolic compounds are ubiquitous constituents of higher plants found in a wide range of commonly consumed plant foods such as fruits, vegetables, cereals, and legumes, and in beverages of plant origin, such as wine, *Camellia sinensis* teas, mate tea, and coffee (Manach et al., 2004; Cheynier, 2005; Farah and Donangelo, 2006; Lima et al., 2016). These compounds are secondary metabolites of plants generally involved in response to biotic and abiotic stresses, for example, defense against ultraviolet radiation or aggression by pathogens (Cheynier et al., 2013). In this way, they may play a role in adaptation to environmental changes and in co-evolution with pests and diseases (Cheynier et al., 2013; Caretto et al., 2015). Several thousands of phenolic compounds have been described in plant foods and can be grouped into different classes according to their basic chemical structure (such as type and number of phenol rings), and into different subclasses according to specific substitutions in the basic structure, association with carbohydrates and polymerized forms (Table 9.1; Zeb, 2020). Currently, various compositional databases containing detailed information on the classes and distribution of phenolic compounds are available (www.phenol-explorer.eu; Neveu et al., 2010).

A large number of phenolic compounds have received considerable attention as potentially protective factors against human chronic disease development, such as type 2 diabetes, cancer, and neurodegenerative and cardiovascular diseases in general (Scalbert et al., 2005; Williamson, 2017; Grosso et al., 2017). These potential benefits result mostly from their antioxidant and anti-inflammatory effects. In coffee, CGA and derivatives play a major role in producing such effects. In other parts of coffee plants (and very discretely in the seeds), anthocyanins, isoflavones, rutin, quercetin, lignans and other minor phenolic compounds play a complementary role.

9.3 The coffee plant

The coffee tree is a shrub belonging to the botanical family of Rubiaceae, genus *Coffea*, of which several species are known. Among them, *Coffea arabica* and *Coffea canephora* are the most economically relevant species (Davis et al., 2011; Farah, 2019). The coffee plant is a perennial dicotyledon that, in nature, can reach a height of 4–6 m in the *C. arabica* bushes and 8–10 m in the *C. canephora* bushes. However, the bushes in the crop are pruned to reach a maximum height of 3 m in order to facilitate the harvest and maintain the ideal shape of

DOI: 10.1201/9781003043133-11

TABLE 9.1
Main Classes of Phenolic Compounds in Higher Plants

Classes and subclasses		Examples of specific compounds
Non-flavonoid compounds		
Phenolic acids	Hydroxybenzoic acids	Benzoic acid, ellagic acid, gallic acid, hippuric acid, protocatechuic acid, syringic acid, vanillic acid, 2,4-dihydroxybenzoic acid, *p*-hydroxybenzoic acid
	Hydroxycinnamic acids	Cinnamic acid, coumaric acid, caffeic acid, ferulic acid, isoferulic acid, sinapic acid
	Hydroxyphenylacetic acids	3,4-dihydroxyphenylacetic acid, 4-hydroxyphenylacetic acid, phenylacetic acid
	Hydroxyphenylpropanoic acids	Dihydrocaffeic acid, dihydroferulic acid
	Elagitans	Pentagalloylglucose
	Stilbenes	Resveratrol
	Lignins and lignans	Secoisolariciresinol, matairesinol, lariciresinol, pinoresinol
	Xanthones	Mangiferin, isomangiferin
Flavonoid compounds		
	Anthocyanins	Cyanidin, delphinidin, peonidin, pelargonidin, malvidin
	Flavanols	Catechins, epicatechins, gallocatechins, procyanidin, theaflavin
	Flavanones	Hesperetin, hesperidin, naringenin, naringin, narirutin
	Flavones	Apigenin, luteolin, tangeritin
	Flavonols	Rutin, quercetin, kaempferol, myricetin
	Isoflavonoids	Daidzein, genistein, glycitein
	Hydrolysable tannins	Ellagitannins, gallotannins
	Condensed tannins or proanthocyanidins	Trimeric procyanidin, prodelphinidins

Note: Adapted from Farah and Donangelo (2006); www.phenol-explorer.eu.

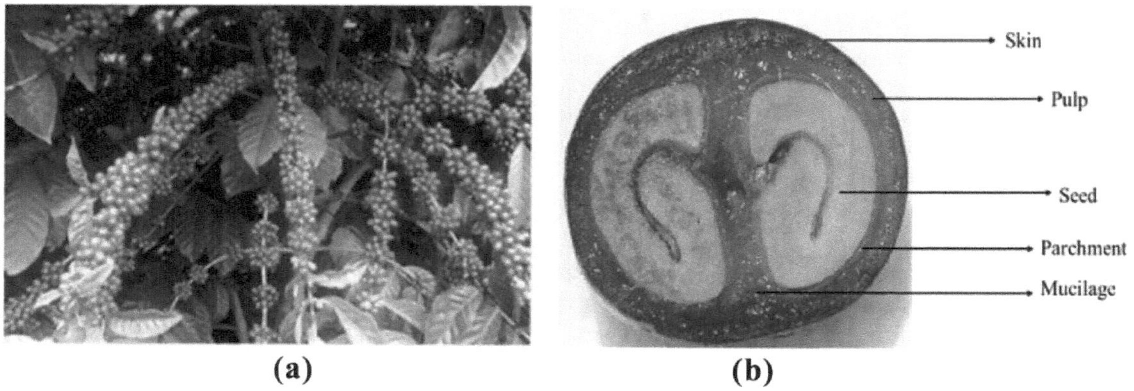

FIGURE 9.1 (a) *Coffea canephora* branches with ripe fruits. (b) Transverse cut of a coffee fruit.

Source: (A) Reproduced from Ferreira et al. (2019), with permission from the Royal Society of Chemistry.

the plant (Ferreira et al., 2019). The first flowers are produced when the plant is between 3 and 4 years old. After the flowers fall, the ovaries slowly develop into oval fruits up to 18 mm in length and 10–15 mm in diameter. Initially, the fruits have an intense green color and, on ripening, a bright red, pink, or yellow color (Ferreira et al., 2019).

The coffee fruit consists of skin, pulp, mucilage, and seeds, which are usually found in two units in each fruit, covered by a thin silver film (silver skin) around which a yellow film (parchment) is found (Smith, 1985). In relation to the thin and shiny leaves, they grow in pairs, opposite to each other, on the branches. The *Coffea* fruit branches and a structure of a coffee fruit are shown in Figure 9.1.

9.4 Harvest and postharvest

Traditionally, coffee fruit harvesting methods can be classified as manual, semi-mechanized, and mechanized. In manual harvesting, the fruits can be selectively collected (finger picking, ideally from ripe fruits) or stripped from the branches. Although picking ripe fruits tends to favor quality over mechanical methods, it is not usually chosen by large producers, as it is a costly and time-consuming practice. It is usually performed in small farms that are dedicated to the production of specialty coffees. A manual harvesting alternative is stripping. In this practice, fruits of different degrees of ripeness and leaves are stripped simultaneously from the branches. Afterward shaking, or sieving, must be carried out in order to separate the fruits from leaves and impurities in general (Duarte et al., 2010; Guimarães et al., 2019). The mechanized harvesting is carried out by harvesting machines or by a device pulled by a tractor as a viable alternative to reduce harvest time and labor demand. However, it depends on flat surfaces and, like stripping, harvests fruits of different degrees of ripeness. Variations of these methods do exist.

After harvesting the coffee fruits, there are different ways to produce green coffee seeds. Traditionally, there are two main methods of postharvest processing, dry and wet, which generate differences in the chemical composition of the seeds and, consequently, in their sensory characteristics after roasting (Farah, 2019). In the traditional dry method, also known as "natural processing," the harvested fruits are exposed to the sun and/or air dryers, until the moisture content reaches 10%–12% (Farah, 2019). Then, the fruits are stored, sanitized, and peeled, with complete removal of the skin, pulp, mucilage, and parchment (Esquivel and Jimenez, 2012). This waste is called husk. Only the silver skin, rich in polysaccharides, remains attached to the seeds. The dry processing method has traditionally been used to dry coffee fruits that are harvested by manual stripping or mechanically. Wet processing is more complex than dry processing, although it requires less time (about a third compared to dry) and less area for drying the fruits, which are first selected in floating tanks (Farah, 2019). With the aid of a pulper, the fruits are partially pulped (the mucilage remains on the seed, in addition to the silver skin), and fermented, a step that increases the acidity and develops flavor. During fermentation – of variable duration – enzymes, selected microorganisms, or the local natural microbiota can be used to remove mucilage and most of the silverskin, while flavor is developed. *Saccharomyces cerevisiae, Pichia kluyveri, Pichia anomala, Hansenia sporauvarum, Debaryomyces hansenii*, and *Torulaspora delbrueckii* are some of the species of yeasts that have been used during coffee processing. Afterward the seeds are carefully washed, polished, and sun-dried or air-dried. Coffees treated by this method are known as "washed or pulped" (Farah, 2019). Also, for post-harvesting methods, there are several variations, like the semi-dry and hybrid methods, which become more and more sophisticated in the search for flavor development.

During roasting of green coffee seeds, the remaining silver skin still adhering to the unfermented seeds is released and accrues as waste (Narita and Inouye, 2014). Spent coffee grounds are another type of waste produced either by coffee consumers or by the industrial production of instant coffee (Murthy and Naidu, 2012a).

Significant amounts of by-products are generated from coffee fruit processing to obtain seeds. Approximately 90% of the fruit is discarded (del Castillo et al., 2019). Coffee pulp alone corresponds to approximately 28% of the coffee fruit on a dry weight basis. Coffee husk corresponds to about 12%, whereas the coffee seed corresponds to 50%–55% (Heuzé and Tran, 2015). It was estimated in the 1990s that for every 1 million 60-kg bags of dried coffee seeds, some 218,400 tons of pulp and mucilage wastes were generated (Veenstra, 1995). This sheer mass of coffee processing by-products is largely untapped at present and, in addition to causing environmental problems, represents a large resource that can be used for value-added products (Alves, 2017; Dorsey and Jones, 2017; del Castillo et al., 2019). Leaves are classified as coffee by-products as well, because they are also discarded during harvesting and tree pruning (Klingel et al., 2020). The coffee production–related by-products are presented in Figure 9.2.

9.5 Phenolic compounds in coffee seeds

9.5.1 Green coffee

CGA are the major phenolic compounds in coffee seeds. Despite the large distribution of some CGA in the plant kingdom, green coffee is known as one of the main food sources of these compounds, with contents being equaled or exceeded only by the green leaves of *Ilex paraguariensis* (yerba mate; Marques and Farah, 2009). This family of compounds includes different molecules formed by the esterification of one molecule of (–)-quinic acid and one to three molecules of *trans*-hydroxycinnamic acids (*trans*-phenyl-3-propenoic acids, with different substitutions in the aromatic ring). *Trans*-hydroxycinnamic acids are the most relevant secondary plant metabolites in the human diet (Williamson and Clifford, 2017). Caffeic acid (3,4-dihydroxy-cinnamic acid) is the most abundant in coffee, followed by ferulic acid (3-metoxy-4-hydroxy-cinnamic acid) and *p*-coumaric acid (4-hydroxy-cinnamic acid). After esterification, the main classes of CGA found in *C. arabica* and *C. canephora*seeds are caffeoylquinic acids (CQA), dicaffeoylquinic acids (diCQA), feruloylquinic acids (FQA), and *p*-coumaroylquinic acids (*p*-CoQA), with at least three major isomers each, and caffeolyl-feruloyl-quinic acids (CFQA), with six isomers (Farah and Paula Lima, 2019; Figure 9.3).

Total CGA content in green coffee seeds may vary according to genetics – species and cultivar, degree of maturation, and, less importantly, agricultural practices, climate, and soil (Farah and Donangelo, 2006; Farah, 2012). Literature values for CGA in regular green coffee seeds on dry matter (dm) vary from about 2100 to 7800 mg/100 g for *C. arabica*, with 4000–6000 mg/100 g being a more common range. Although reported values in *C. canephora* have reached 14,400 mg/100 g (dm), 7000–8000 mg/100 g is the more usual range (Farah and Paula Lima, 2019).

Green coffee is safe for human consumption. Because of its high antioxidant property and thermogenic effects among others, its dried aqueous extract has been used as a nutraceutical (23–40 g CGA 100 g^{-1}; Farah et al., 2008; Farah and Paula Lima, 2019) or to formulate products like green and roasted soluble coffee blend (4.6–6.7 g CGA 100 g^{-1}; Farah and Duarte, 2015; Farah and Paula Lima, 2019).

CQA isomers account for approximately 70%–85% of total CGA content, followed by diCQA (10%–20%) and FQA (5%–10%) isomers. Particularly, 5-caffeoylquinic acid (5-CQA) alone accounts for about 60% of total CGA in green coffee seeds; 3-caffeoylquinic acid (3-CQA) and 4-caffeoylquinic acid (4-CQA) each make up approximately 10% (Farah and Paula Lima, 2019). The same tendency is observed for FQA

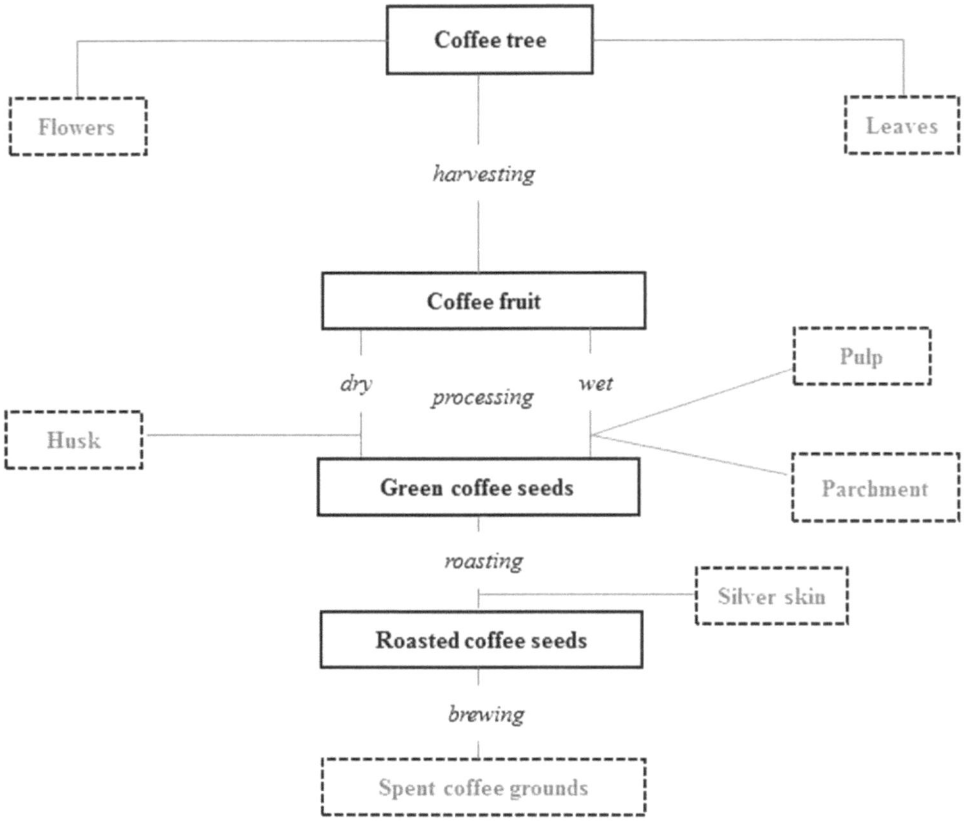

FIGURE 9.2 Coffee production–related by-products.

Source: Adapted from Klingel et al. (2020).

isomers: 5-feruloylquinic acid (5-FQA) is the major FQA and accounts for about 5% of total CGA; 3-feruloylquinic acid (3-FQA) and 4- feruloylquinic acid (4-FQA) together constitute approximately 2%. Among diCQA, 3,5-dicaffeoylquinic acid (3,5-diCQA) accounts for about 7% of total CGA, while 3,4- and 4,5- dicaffeoylquinic acids together make up about 6% of CGA composition (Farah and Paula Lima, 2019). A number of minor CGA and related compounds have been identified in green *C. arabica* and *C. canephora* seeds. Together they correspond to less than 1% of total phenolic compound content (Farah and Paula Lima, 2019). They include acyl dicaffeoylquinic acids, dimethoxycinamoylquinic acids, caffeoyl-dimethoxycinamoylquinic acids, diferuloylquinic acids, feruloyl-dimethoxycinamoylquinic acids, sinapoylquinic acids, sinapoyl-caffeoylquinic acids, sinapoyl-feruloylquinic acids, feruloyl-sinapoylquinic acids and new minor *p*-coumaric acid-containing compounds (Clifford et al., 2005, 2006; Jaiswal et al., 2010).

CGA composition of coffee seeds varies considerably during fruit maturation. Its content increases in coffee seeds until fruits are completely formed and then decreases as they ripe until senescence (Clifford and Ramirez-Martinez, 1991; Farah and Donangelo, 2006). When coffee fruits are harvested by stripping the branches or mechanically, the different fruit maturation stages are reflected on the chemical composition of raw coffee seeds, and unripe or overripened seeds are called intrinsic defects. They are usually marketed for lower prices and may integrate lower-quality commercial blends. Allowing for variations between cultivars, average total CGA content of 7.3–8.7 g/100 g (dm) has been reported for raw *C. arabica* immature seeds (originated mainly from immature fruits) and 7.7 g/100 g (dm) for black-immature seeds (originated from use of high temperature during drying of regular or immature seeds; Farah et al., 2006a); therefore the amount of green defective seeds in a coffee blend may increase the total CGA content, depending on the amount in the blend. Sour defective seeds (generated by water paucity during fruit development or by abnormal fermentation of immature or ripe seeds) may also contain significant levels of total CGA (about 7.0 g/100 g); lower contents have been reported for black defective seeds, originated from overripened cherries (1.3 mg/100 g; Farah et al., 2006a). The distribution of CGA classes in coffee seeds does not tend to change along fruit maturation. However, from immature to black seeds, the content of 5-acyl isomers tends to decrease, while those of 3-acyl and 4-acyl isomers tend to increase suggesting acyl migration (Farah and Donangelo, 2006).

It is well known that postharvest processing has pronounced effects on the chemical composition of coffee seeds, especially in water-soluble components like sugars, caffeine, trigonelline, and CGA when fruits are soaked in water and/or fermented. Studies have reported higher CGA content in wet-processed *C. canephora* seeds compared to dry-processed seeds (Leloup et al., 2004; Duarte et al., 2010). This difference in CGA

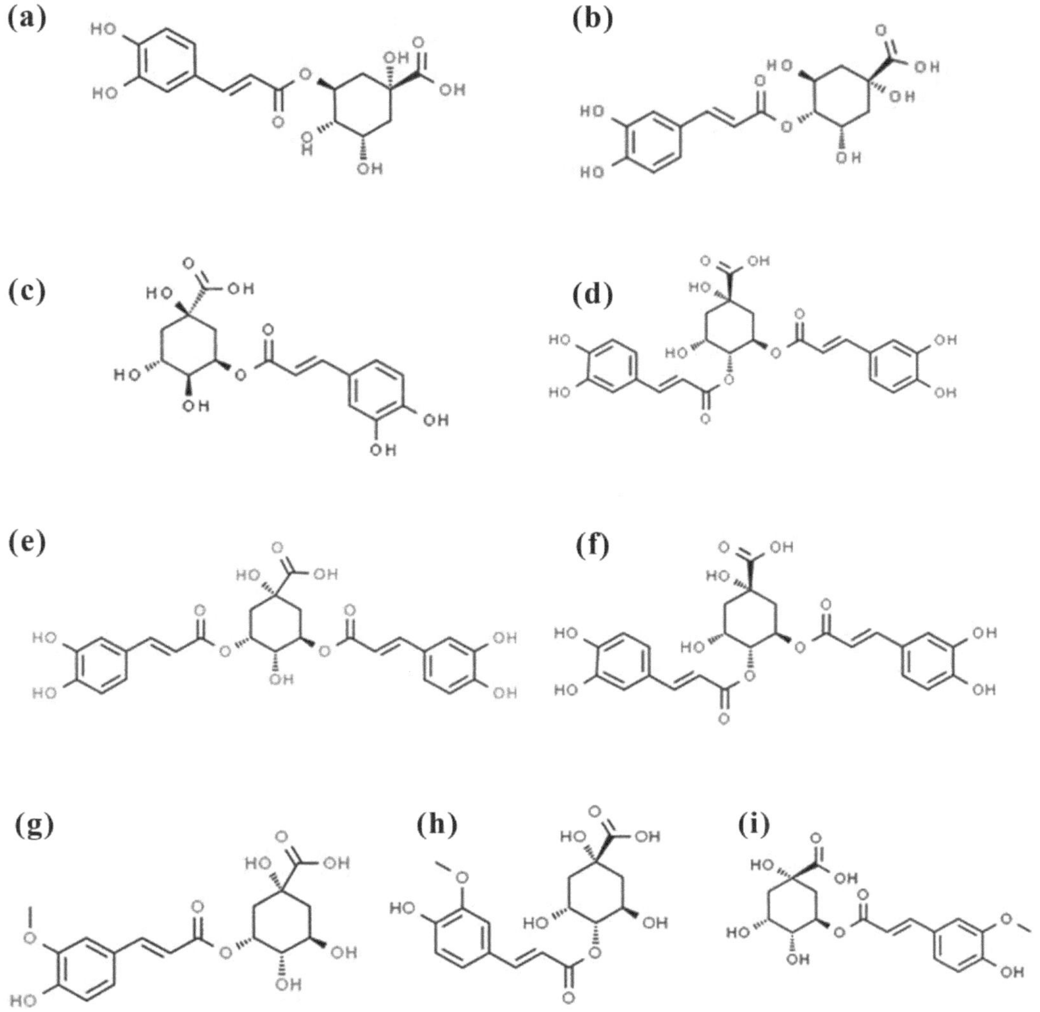

FIGURE 9.3 Major chlorogenic acids in coffee. (a) 3-caffeoylquinic acid; (b) 4-caffeoylquinic acid; (c) 5-caffeoylquinic acid; (d) 3,4-dicaffeoylquinic acid; (e) 3,5-dicaffeoylquinic acid; (f) 4,5-dicaffeoylquinic acid; (g) 3-feruloylquinic acid; (h) 4-feruloylquinic acid; (i) 5-feruloylquinic acid.

contents has been attributed to losses of other water-soluble components by lixiviation and fermentation in the wet method (Wootton, 1974) in addition to slight degradation of CGA in dry processed seeds due to exposure to sun heat (Duarte et al., 2010). About animal depulped coffees, the contents of Kopi Luwak–depulped and Jacu bird–depulped coffees have not changed compared to the original seeds, neither significant changes in CGA isomers distribution have been observed (Farah and Paula Lima, 2019).

CGA contents reported for a few green *C. arabica* and *C. canephora* seeds samples are presented in Table 9.2.

Small amounts of free hydroxycinnamic acids, anthocyanins, isoflavones, and lignans have been identified in green coffee seeds, with about 1% of these phenolics being glycosides (Clifford, 1985, 1987; Mazza and Miniati, 1993; Alves et al., 2014; Milder et al., 2005). The anthocyanins cyanidin, pelargonidin, and peonidin have been identified in *C. arabica* seeds as residues of the red skin of fruits (Mazza and Miniati, 1993). The main isoflavones identified in *C. arabica* and *C. canephora* were formononetin, daidzein, genistein and glycitein. Biochanin A has been identified only in vestigial amounts (Alves et al., 2014). Regarding lignans, secoisolariciresinol, lariciresinol, matairesinol and pinoresinol, in order of abundance, have been identified in green coffee (Milder et al., 2005). Free volatile phenols have also been identified (Moreira et al., 2000; Toci et al., 2020).

9.5.2 Roasted coffee

CGA take part in the generation of color, flavor, and aroma of coffee during roasting (Farah and Donangelo, 2006). Given their thermal instability and vulnerability to the intense heat conditions (200–310°C, commonly 210–230°C), they undergo many changes, including those occurring in Maillard and pyrolysis reactions. In the beginning of roasting, acyl migration is observed (the levels of substitutes in the 5-position of the quinic acid decrease substantially, whereas levels of substitutes in the 3- and 4- positions increase in some cases to almost double their original levels) in all main CGA classes (Farah and Donangelo, 2006; Farah and Duarte, 2015). As roasting

TABLE 9.2

Contents of Major Chlorogenic Acids (CGA) in Green *C. arabica* and *C. canephora* Cultivars from Different Origins

Origin	Cultivar	Total CQA	Total FQA	Total diCQA	Total CGA	References
		mg/100 g			g/100 g	
Coffea arabica						
Brazil	Mundo Novo, Red Catuaí, Bourbon, Rubi, Topazio	4153–5086	193.3–286.5	511.8–928.8	4.0–6.1	Farah et al. (2005), Perrone et al. (2008), Duarte et al. (2010)
Venezuela	Red Caturra, Red Bourbon	4660–4760	330.0–340.0	550.0–650.0	5.6	Clifford and Ramirez-Martinez, (1991)
Honduras	Nr	1830–2320	90.2–110.1	170.0–280.0	2.1–2.7	Babova et al. (2016)
Peru	Nr	2070.0–2250	100.3–110.0	240.0–242.1	2.4–2.6	Babova et al. (2016)
Ethiopia	Longberry, Heirloom	4558–4529	292.3–367.1	840.4–709.8	5.6–5.8	Farah et al. (2005, 2006b)
Kenya	Nr	1495–2476	140.1–640.0	190.1–290.2	1.8–3.4	Ky et al. (2001)
Indonesia	Kopi Luwak depulped	3909	241.1	583.6	4.7	Farah and Paula Lima (2019)
Brazil	Jacu bird depulped	3809	375.1	499.1	4.6	Farah and Paula Lima (2019)
Coffea canephora						
Brazil	Conilon	5860–6457	610.0–720.0	1160.0–1299.0	7.7–8.3	Toci et al. (2006), Perrone et al. (2008)
Vietnam	Robusta	1800–1950	150.0–160.4	300.0–370.1	2.2–2.5	Babova et al. (2016)
Africa	Robusta	2870–4872	527.6–861.9	611.4–1137.8	4.0–6.9	Alonso-Salces et al. (2009)
Asia	Robusta	3134–5523	602.9–1057.3	668.4–1216.3	4.4–7.8	Alonso-Salces et al. (2009)
Colombia	Robusta	1158–2741	243.6–476.3	620.8–989.9	2.0–4.2	Guerrero et al. (2001)
Venezuela	Robusta	5350	790.3	1050.1	7.2	Clifford and Ramirez-Martinez, (1991)

progresses, CGA can be lactonized or epimerized. CGA lactones (1,5-γ-quinolactones or quinides [CGL]), which contribute to the characteristic mild bitterness of coffee, are formed by dehydration of the quinic acid moiety. Both CGA and CGL are then degraded to form free organic acids and low-molecular-mass volatile derivatives, including pyrogallol, hydroxyhydroquinone, catechols, methylcatechols, ethylcatechols, vinylcathecols, and guaiacols among a series of di- and trihydroxybenzenes (Dorfner et al., 2003; Yaylayan and Haffenden, 2003; Farah et al., 2005, 2006b; Farah, 2019; Figure 9.4). It has also been reported that as roasting proceeds, 4-vinylcatechol is oligomerized to produce a family of polyhydroxylated phenylindans, which exhibit a harsh and lingering bitter taste profile (Frank et al., 2007; Blumberg et al., 2010). Moreover, di- and trihydroxybenzenes, catechol, pyrogallol, 3-methylcatechol, and 4-methylcatechol were reported to react with furfuryl alcohol during roasting to give (furan-2-yl) methylated benzene diols and triols, which are also bitter-tasting compounds derived from coffee CGA (Kreppenhofer et al., 2011). CGA and derivatives participate in the formation of polymeric materials like melanoidins as well (Bekedam et al., 2008; Perrone et al., 2010), which explains at least part of the antioxidant activity of melanoidins in coffee brews (Fuster et al., 2000; Delgado-Andrade and Morales, 2005; Farah, 2019).

Total CGA content in ground roasted coffees ranges considerably for the same species, depending mainly on the roast degree. In general, values reported for total CGA (including lactones) in regular roasted *C. arabica* seeds (dm) are 2.9–5.1 g/100 g for light roasts, 1.7–3.5 g/100 g for medium roasts, and 0.4–2.1 g/100 g for dark roasts (Farah and Paula Lima, 2019). In regular *C. canephora* seeds, the contents (dm) are usually in the range of 5.9–7.1 g/100 g in light roasts, 1.0–4.3 g/100 g in medium roasts, and 0.4–1.7 g/100 g in dark roasts (Farah and Paula Lima, 2019). It is noted that as roasts get darker, the difference between species disappears. In commercial ground roasted coffees reported contents are in the range of 0.2–3.1 g/100 g (Farah and Paula Lima, 2019). Contents of CGA reported for roasted *C. arabica* and *C. canephora* seeds are presented in Table 9.3.

In roasted coffees, CQA account for approximately 78% of total CGA content, followed by diCQA (5%–8%) and FQA (5%–7%). Breaking into CGA groups of isomers, 5-CQA alone accounts for about 40% of total CGA in seeds (dm), and 3-CQA and 4-CQA account for approximately 16% and 20%, respectively. The same tendency is observed for FQA isomers: 5-FQA is the major FQA isomer in roasted coffee seeds and accounts for about 4% of total CGA; 3-FQA and 4-FQA together constitute approximately 3%. Among diCQA, 3,4-diCQA constitutes, on average, 3% of total CGA; and 3,5- diCQA, and 4,5-diCQA about 2% and 2.5%, respectively (Farah and Paula Lima, 2019).

About 7%–10% of CGA in *C. arabica* and 6%–9% of *C. canephora* seeds were reported to be converted to CGL. 3-caffeoyl-1,5-quinide (3-CQL) and 4-caffeoyl-1,5-quinide (4-CQL) are the main CGA lactones in coffee (Farah et al., 2005; Frank et al., 2007; 2008), although lactones from FQA, diCQA, and *p*-CoQA and other minor quinides, in order of relevance, are present in roasted coffee as well (Scholz-Böttcher et al., 1991; Farah et al., 2005). CGL levels peak most commonly from light medium to dark medium roast degree, depending on the roaster and roasting profile. Average maximum levels vary from 380 to 430 mg/100 g (dm) (Farah and Paula Lima, 2019).

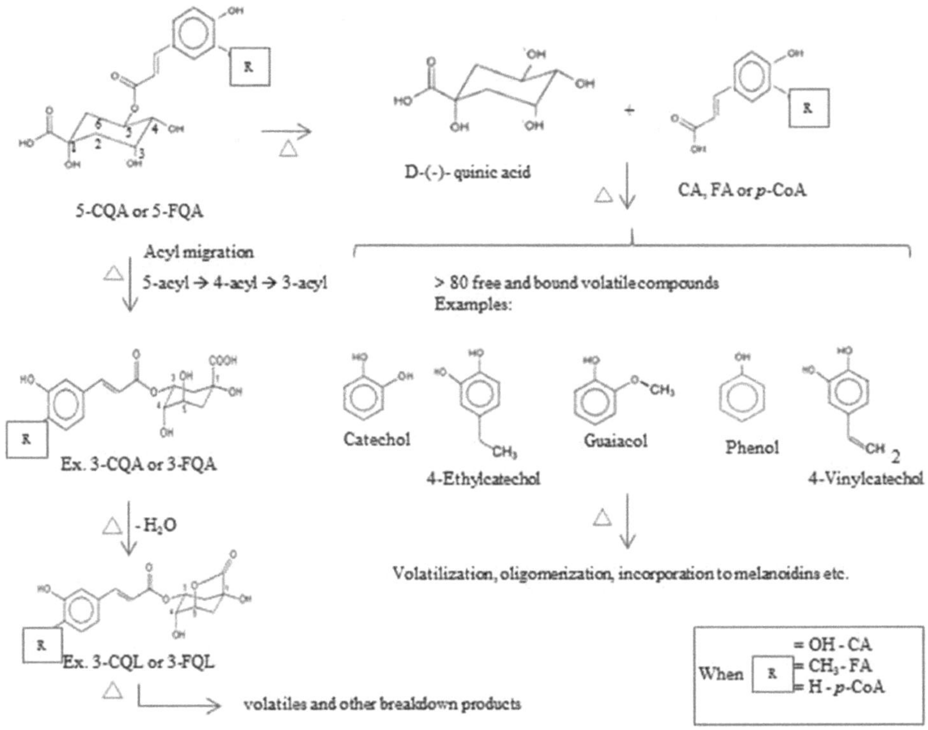

FIGURE 9.4 Acyl migration and formation of major 1,5-γ-quinolactones (exemplified by 3-caffeoylquinic-1,5-γ-lactone) and volatiles from chlorogenic acid (5-caffeoylquinic acid) or 5-feruloylquinic acid during roasting. Although under IUPAC rules the numbering system for the lactones is different from that for the acids, in order to avoid confusion, most reports in the literature have used for lactones the same numbering of the carbons as for the acid precursors.

Notes: CQA= caffeoylquinic acid; CQL = caffeoylquinide; FQA= feruloylquinic acid; FQL = feruloylquinide; *p*-CoA.
Source: Reproduced from Farah and Paula Lima, 2019, with permission from the Royal Society of Chemistry.

TABLE 9.3
Contents of Major Chlorogenic Acids (CGA) in Roasted *C. arabica* and *C. canephora* Cultivars

Roast degree	Total CQA	Total FQA	Total diCQA	Total CGL	Total CGA and CGL	References
	mg/100 g				g/100 g	
Coffea arabica						
Very light	5178	280.5	585.6	89.2	6.1	Perrone et al. (2008)
Light	2812–4427	151.7–232.0	213.0–471.0	252.3–294.4	2.9–5.1	Farah et al. (2005), Perrone et al. (2008), Moon et al. (2009), Farah et al. (2011), Bicho et al. (2011)
Light-medium	1654–5040	130.0–165.4	131.3–440.2	326.5–349.2	2.2–5.4	Farah et al. (2005), Toci et al. (2006), Perrone et al. (2008)
Medium	1490–2688	90.1–162.3	90.0–189.5	214.1–328.0	1.6–3.0	Toci et al. (2006), Moon et al. (2009), Farah et al. (2011), Bicho et al. (2011)
Dark	289.3–1483	40.7–156.1	14.6–76.1	142.4–279.7	0.4–1.9	Farah et al. (2005), Toci et al. (2006), Perrone et al. (2008), Moon et al. (2009), Farah et al. (2011), Bicho et al. (2011)
Very dark	270.8	25.2	5.6	74.3	0.4	Perrone et al. (2008)
Coffea canephora						
Very light	6150–6541	508.5–665.4	951.1–968.9	83.5–264.6	8.0–8.2	Farah et al. (2005), Perrone et al. (2008)
Light	2014–4950	149.7–596.1	112.0–793.0	308.8–357.6	4.5–6.5	Farah et al. (2005), Perrone et al. (2008), Farah et al. (2011), Bicho et al. (2011)
Light-medium	3140–5690	335.1–579.1	235.1–640.2	392.4–419.2	3.7–6.7	Farah et al. (2005), Toci et al.2006), Perrone et al. (2008)
Medium	1090–3387	145.4–447.1	60.1–456.0	256.4–291.3	1.6–4.3	Farah et al. (2005), Toci et al. (2006), Farah et al. (2011), Bicho et al. (2011)
Dark	611.9–2415	101.5–468.3	19.4–273.1	132.6–365.1	0.8–3.5	Farah et al. (2005), Toci et al. 2006), Perrone et al. (2008), Farah et al. (2011), Bicho et al. (2011)
Very dark	318.9–377.7	56.3–125.6	7.6–36.2	80.3–86.3	0.4–0.6	Farah et al. (2005), Perrone et al. (2008)

The speed at which coffee is roasted, related to time and temperature conditions (roasting profile), affects the content of CGA in the coffee seeds. The higher the roasting speed to obtain one specific color degree, the lower the CGA loss tends to be (Farah and Duarte, 2015).

9.6 Phenolic compounds in coffee by-products

While the chemical composition and bioactive properties of *C. arabica* and *C. canephora* seeds have been extensively investigated due to the economic importance and quality attributes associated with the coffee beverage, little research has been conducted to study other parts of the coffee tree considered as by-products, such as the flowers, leaves, and pulp.

Flowers: The flower of the coffee tree is crucial for the quantity and quality of the fruits because it helps in pollination and, consequently, in production. The flower also plays a key role in preventing coffee plant dehydration in addition to protecting it from insects by releasing substances (exudates) during its development (Mayer et al., 2013). Every year, approximately 30,000–40,000 flowers of adult coffee naturally fall to the ground and represent abundant residues from the coffee industry (Nguyen et al., 2019; Figure 9.5).

FIGURE 9.5 *C. canephora* inflorescence.

Source: Reproduced from Ferreira et al., 2019, with permission from the Royal Society of Chemistry.

It is worth mentioning that harvesting the flowers does not result in low productivity of the seeds because they are only harvested after pollination, when the flowers become light brown, before the formation of the "pinhead fruits" (small infructescence that, after development, gives rise to the coffee seed; Barros et al., 1999; DaMatta et al., 2007; de Abreu Pinheiro et al., 2021).

The main components of flowers can be divided into pollen, nectar, and petals. Pollen is rich in flavonoids, in addition to proteins, carbohydrates, lipids, and carotenoids. Petals and other parts contain a significant amount of phenolic compounds such as gallic acid, kaempferol, quercetin, and 5-CQA (Murayama et al., 2012; Kumar et al., 2009), as well as vitamins and minerals (Anesini and Perez, 1993).

Few studies on the bioactive composition of coffee flowers have been conducted so far. A preliminary study provided a brief investigation of phytochemical constituents of *C. canephora* flowers obtained from a coffee farm in Surat Thani province, Thailand, and extracted with methanol (Chairgulprasert and Kongsuwankeeree, 2017). The extract contained an amount of total phenolic compounds estimated in 139.9 mg gallic acid equivalent (GAE)/g extract.

Nguyen et al. (2019) identified minor amounts of phenolic acids such as protocatechuic acid (77.5 ± 0.1 mg/100 g) and gallic acid (28.7 ± 0.0 mg/100 g) in an extract of flower collected from *C. canephora* in a coffee farm in Tay Nguyen, Vietnam. The compound 5-CQA was investigated but not detected. Total concentration of phenolic compounds was estimated in 72.4 mg GAE/g extract.

More recently, 5-CQA (74.3 ± 3.4 mg/100 g dm) and gallic acid (3.9 ± 1.3 mg/100 g dm) were identified in coffee flowers of *C. arabica* and *C. canephora* species harvested in Brazil and dried by different techniques and extraction methods. Vanillic acid, caffeic acid, *p*-coumaric acid, and ferulic acid were investigated but not detected (de Abreu Pinheiro et al., 2021).

Eight major CGA compounds (3-CQA, 4-CQA, 5-CQA, 4-FQA, 5-FQA, 3,4-diCQA, 3,5-diCQA, and 4,5-diCQA) were recently identified in coffee flowers of *C. canephora* cv. Conilon harvested in Espírito Santo, Brazil. Total CQA contents were in the range of 0.6–1.0 g/100 g (dm); diCQA 90.4–223.1 mg/100 g (dm), and FQA 10.4–30.2 mg/100 g dm (dePaula, 2020).

(a) (b)

FIGURE 9.6 Leaves of (a) *C. arabica* and (b) *C. canephora*.

Source: Reproduced from Ferreira et al. (2019), with permission from the Royal Society of Chemistry.

Leaves: As the productivity of coffee seeds increases in the world, so does the production of leaves. Fast and efficient methods for harvesting coffee fruits on a large scale, such as the use of strippers and mechanical harvesters, frequently used in Brazil, for example, select the fruits, and dispense the leaves, which are returned to the soil for decomposition (Trevisan et al., 2019). It is estimated that 3.3 tons of leaves per hectare are discarded during coffee harvest (Matiello et al., 2010) (Figure 9.6).

The main phenolic compounds present in the leaves are also CGA (3-CQA, 4-CQA, 5-CQA, 3,4-di-CQA, 3,5-di-CQA, and 4,5-di-CQA; Chen, 2018; Monteiro et al., 2019; Trevisan et al., 2019; dePaula, 2020). Campa et al. (2012) reported content (dm) of 3.5 g/100 g for total CQA, 260 mg/100 g for total diCQA, and 120 mg/100 g for total FQA in leaves of *C. arabica* cv. Laurina from Ethiopia. Mondolot et al. (2006) identified five CGA compounds (5-CQA, 3,4-diCQA, 3,5-diCQA, 4,5-diCQA, and 5-FQA) in leaves of *C. canephora Pierre.* The oldest leaves presented lower mean CGA contents (dm) (4.1 mg/100 g) when compared to the youngest ones (7.8 mg/100 g). Recently, Monteiro et al. (2019) evaluated the major CGA compounds (3-CQA, 4-CQA, 5-CQA, 3,4-diCQA, 3,5-diCQA, 4,5-diCQA) in a variety of coffee leaf species at different development stages. Total CGA content, calculated as the sum of CQA and diCQA, varied from 3.5 to 8.0 g/100 g (dm) in young leaves and from 1.9 to 4.4 g/100 g (dm) in mature leaves; *C. arabica* and *C. eugenioides* showed the highest content, while *C. canephora* and *C. racemosa* showed the lowest, which is consistent with the trend reported by Campa et al. (2012).

The predominance of 5-CQA and 3,5-diCQA isomers and absence of the 3-FQA isomer in leaves of the genus *Coffea* has been consistently reported in the literature (Mondolot et al., 2006; Campa et al., 2012; Chen et al., 2018; de Almeida et al., 2019; Monteiro et al., 2019).

Major additional compounds in coffee leaves include flavonoids (rutin glucoside, quercetin-3-*O*-glucoside, and kaempferol-3-*O*-rhamnoglucoside; Figure 9.7) and xanthones (mangiferin and isomangiferin; de Almeida et al., 2019;

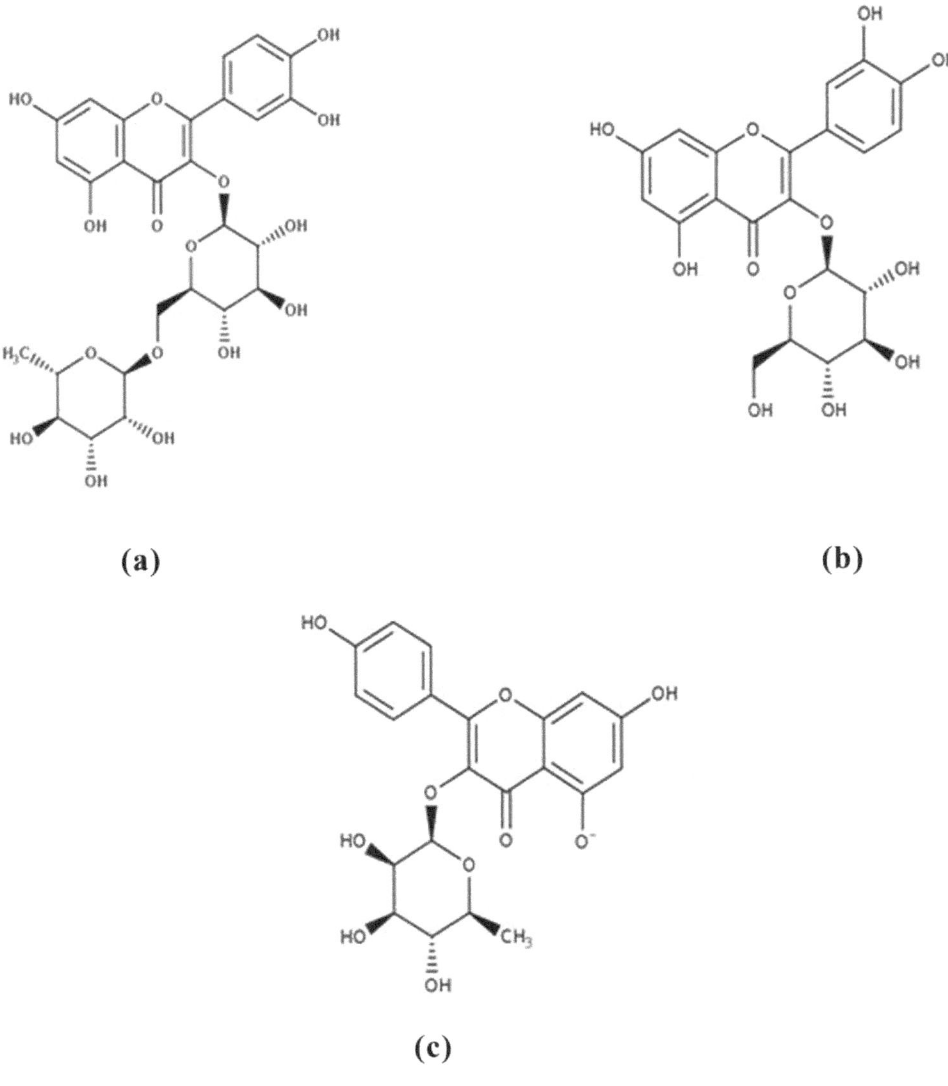

FIGURE 9.7 (a) Rutin glucoside, (b) quercetin-3-*O*-glucoside, and (c) kaempferol 3-*O*-rhamnoglucoside.

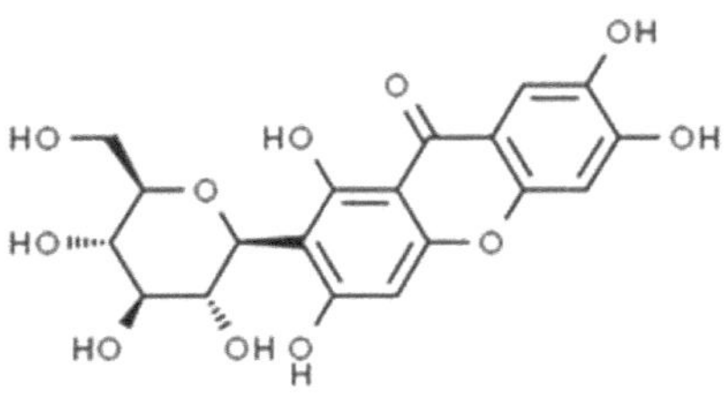

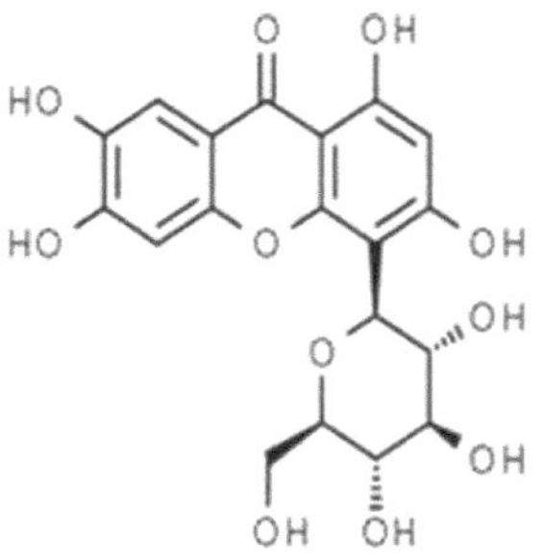

FIGURE 9.8 (a) Mangiferin and (b) isomangiferin.

Figure 9.8). The main flavonoid compound in coffee leaves is rutin. The content depends on many factors, but 500–800 mg/100 g (dm) have been reported for leaves of *C. arabica* (de Almeida et al., 2019; Trevisan et al., 2019). These contents of rutin are much higher than those found in *Camellia sinensis* black tea, for example (80 mg/100 g; Jiang et al., 2015). The green leaves and stems of *Ilex paraguariensis* used for chimarrão preparation also contain a considerable amount of rutin (559 mg/100 g dm) (Mateos et al., 2018). Mangiferin (1,3 ,6,7-tetrahydroxyoxanthone-C2-β-glucoside) is a glycosylated xanthone, widely distributed in the plant kingdom (Trevisan et al., 2019). Isomangiferin is a structural isomer of mangiferin, differing only in the position of the glycosidic fraction (de Beer and Jouber, 2010). Mangiferin is the major xanthone in coffee leaves (Trevisan et al., 2016,2019). These compounds are found in many plant species, in particular *Cyclopia genistoides* and *Cyclopia subternata* (de Beer and Jouber, 2010), popular for the preparation of a medicinal infusion known as "honey bush," although it was first observed in fruits and mango peels in 1922. In coffee leaves, mangiferin was recently discovered, having been isolated for the first time in 2008 (Talamond et al., 2008). This bioactive compound is identified in leaves of different species of the genus *Coffea* (except in *C. canephora* species; Campa et al., 2012). Literature contents of mangiferin and isomangiferin are 48–1570 mg/100 g and 19–138 mg/100 g, respectively, in young *Coffea* leaves and 35 to 788 mg/100 g and about 5 to 100 mg/100 g, respectively, in adult leaves (Monteiro et al., 2019; Ngamsuk et al., 2019).

Minor amounts of different groups of flavonoids, including flavanols (catechin, epicatechin, epigallocatechin gallate, epicatechin gallate), anthocyanins (delphinidin 3,5-diglucoside and delphinidin 3-(600-malonyl-glucoside)), flavonols (myricetin), and flavones (luteolin, apigenin) have been reported in *Coffea* leaves (Chen, 2018). The presence of free hydroxycinnamic acids (benzoic, 3-hydroxybenzoic acid, caffeic, *p*-coumaric, ferulic, sinapic, and protocatechuic acids) have been observed in methanolic and ethanolic extracts of *Coffea* leaves, especially in young leaves (Patay et al., 2016; Chen, 2018).

Since the early nineteenth century, the consumption of coffee leaf tea (like green *C. sinensis* tea) as an "ethnomedicinal" infusion is traditional in countries where coffee grows naturally, like Indonesia (as *kopidaon*), Yemen (as *geyser*) and Ethiopia (as *jenuai*), for the treatment of intestinal disorders, AIDS complications, and tuberculosis (Patay et al., 2016; Chen, 2018; Trevisan et al., 2019). In the West, nevertheless, this type of tea infusion is not usual because of its strong "green" taste. However, since 2014, *C. arabica* leaves processed similar to black or oolong tea have been marketed in Canada and the United States, with reasonable acceptance among consumers (Chen, 2018). Despite the partial loss of bioactive compounds, the slow drying process at controlled temperatures allows oxidation, fermentation, and development of different nuances of aroma and flavor, just like with the *C. sinensis* oolong and black teas, which increases acceptance. If monitored for pesticides, properly processed, and safe from the microbiological point of view, *C. arabica* leaves can produce quality infusions.

Skin and pulp: In recent years, the number of studies on the chemical composition of coffee skin and pulp has been increasing due to their potential use in animal feeding, especially in coffee producer countries, where the disposal of wastes represents an important pollution problem (del Castillo et al., 2019; Figure 9.9).

Three major classes of phenolic compounds have been described in the fruit pulp of *C. arabica* and *C. canephora*: tannins, anthocyanins, and hydroxycinnamic acids (Figure 9.10). Condensed tannins, which are the main phenolic compounds in these parts of the coffee fruit, have received special attention because they are considered as anti-nutrients for ruminants (Barcelos et al., 2001, 2002; Ulloa Rojas et al., 2002, 2003). Soluble (hydrolysable) tannins may account for 0.8%–2.8% of *C. arabica* and *C. canephora* skin and pulp, with higher contents observed in *C. canephora*, and with prodelphinidins exceeding trimeric procyanidins (Clifford and Ramirez-Martinez, 1991; Barcelos et al., 2001; Ulloa Rojas et al., 2003; Farah and Donangelo, 2006). Small amounts of insoluble condensed tannins, especially in the yellow varieties, may also be found in the pulp. Treatment with alkali solutions, ensilation with molasses, and inoculation with selected microorganisms may destroy or reduce the content of tannins (Ulloa Rojas et al., 2002, 2003). Storage of dehydrated *C. arabica* skin and pulp has also produced a linear decrease in tannins content (approx. 39%/year; Barcelos et al., 2001).

Anthocyanins are flavonoid compounds that act as plant pigments widely distributed in colored fruits, such as the coffee

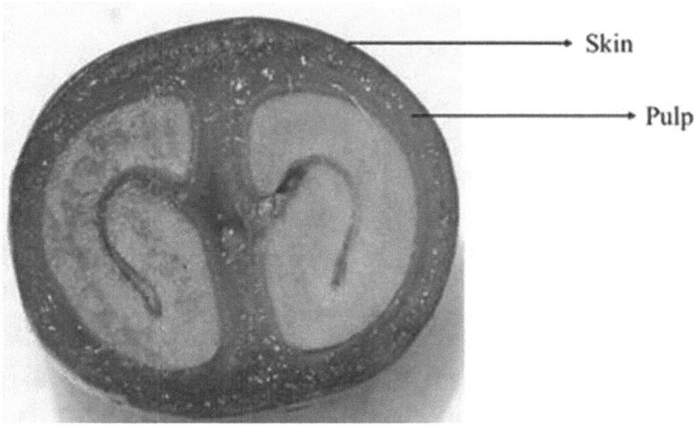

FIGURE 9.9 Coffee pulp and skin.

Source: Reproduced from Ferreira et al. (2019), with permission from the Royal Society of Chemistry.

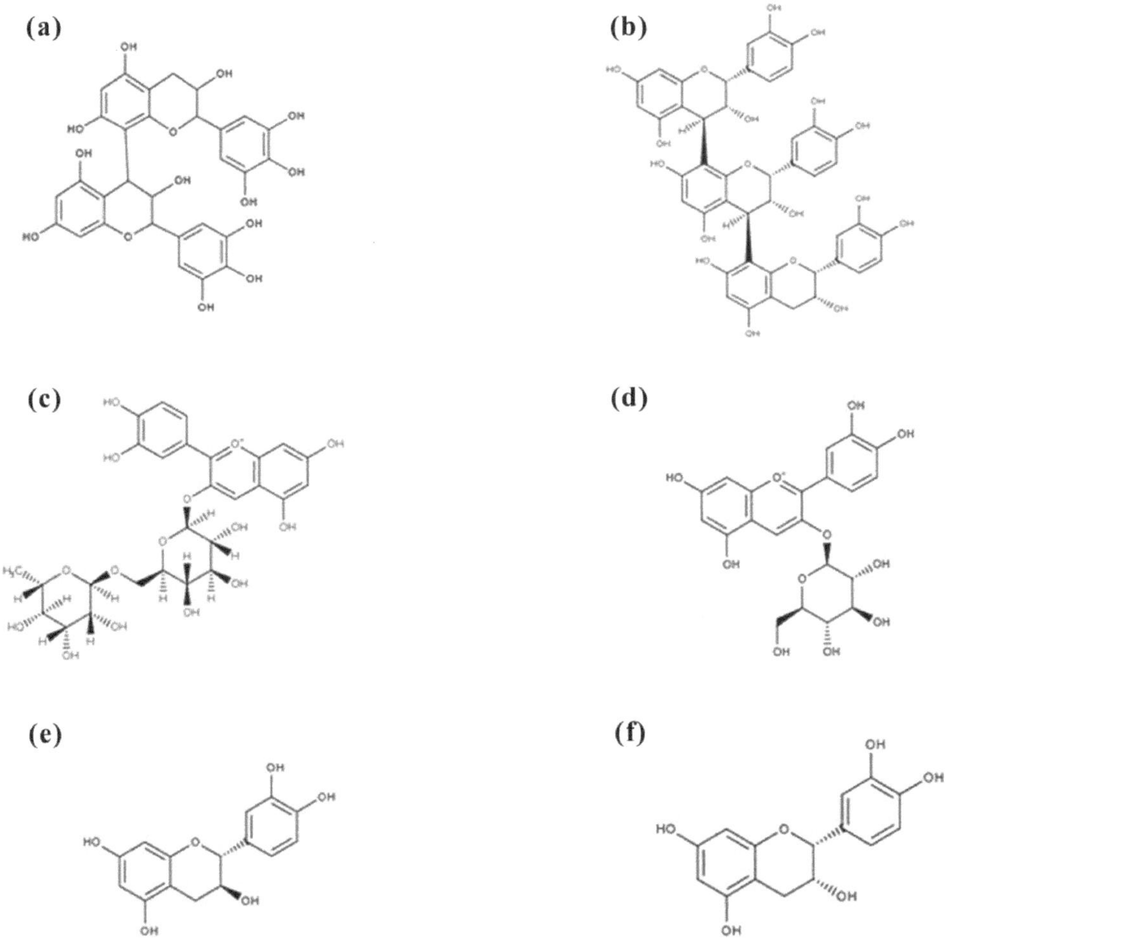

FIGURE 9.10 (a) Prodelphinidin dimer B3, (b) procyanidin trimer C1, (c) cyanidin-3-rutinoside, (d) cyanidin-3-glucoside, (e) (+)-catechin, and (f) (–)-epicatechin.

cherry. These phenolic compounds are commonly found in the skin material of red fruit cultivars (Mazza and Miniati, 1993), being responsible for the red color of the coffee cherry. The major anthocyanins present in the pulp derived from wet-processed fruits are cyanidin-3-rutinoside and cyanidin-3-glucoside (Esquivel and Jimenez, 2012).

Small amounts of the main groups of CGA found in coffee seeds (CQA, FQA, and diCQA) have been identified in the pulp (Clifford and Ramirez-Martinez, 1991). Griffin and Stonier (1975, cited by Clifford, 1985) have reported that coffee pulp contains CGA-protein complexes that are able to inhibit the enzyme indole acetic acid oxidase. Ramirez-Martinez

(1988) identified four CGA compounds in pulps from eight coffee cultivars: 5-CQA, 3,4-diCQA, 3,5-diCQA, and 4,5-diCQA. As in the seeds, 5-CQA was the major CGA (42%). 3,4-diCQA, 3,5-diCQA and 4,5-diCQA corresponded to 5.7%, 19.3%, and 4.4%, respectively (Farah and Donangelo, 2006). Later, Clifford and Ramirez-Martinez (1991) identified and quantified the same four CGA isomers in addition to 3-CQA, 4-CQA, and 5-FQA, in pulps of two *C. arabica* (Red Caturra and Red Bourbon) and one *C. canephora* cultivars. Total CGA contents (dm) were similar in the pulps of *C. arabica* cultivars, 1.01 g/100 g in Red Caturra and 1.16 g/100 g in Red Bourbon; and higher in *C. canephora* pulp, about and 1.61 g/100 g (dm).

Additional phenolic compounds such as catechin (0.07–0.6 mg/100 g dm), epicatechin (1.5–4.8 mg/100 g dm), rutin (0.09–0.7 mg/100 g dm), and protocatechuic acid (0.09–0.31 mg/100 g dm) have also been reported in fresh coffee pulp (Ramirez-Martinez, 1988).

Husk: Coffee husks, which comprise the outer skin, pulp, and parchment, are the main residues obtained in the dry post-harvest processing of the coffee berry. Husks are large sources of potentially useful molecules of economic value in coffee-growing countries (del Castillo et al., 2019; Figure 9.11). They are particularly of interest because the dry method does not use high temperatures, and thus the antioxidant capacities tend to remain high (Dorsey and Jones, 2017).

Coffee fruits husk have been traditionally used for infusions for centuries in several countries, being known as *sultana* in Bolivia, *hashara* in Ethiopia, and *qishr* in Yemen (del Castillo et al., 2019). In these places, different types of spices (e.g., cinnamon and ginger) and fruits are added to make the drink more aromatic. The name *cascara* originated in El Salvador, where a coffee producer named it after the word "husk" in Spanish. Worldwide, this drink is known as "Coffee cherry tea." Recently, controlled fermentations of varying intensities using or not using yeasts, associated or not with other microorganisms have been developed during seed processing. Husk and pulps of these fermented coffees are valued for their wealth of sensory attributes (Farah, 2019). Such fermentations tend to degrade phenolic compounds and other components present in coffee mucilage, providing positive changes in flavor development (de Carvalho Neto et al., 2017).

Polyphenol profile and total content in coffee husks vary with species, origin, and agricultural practices (Mullen et al., 2013). The total polyphenols content has been reported to be about 1200 mg GAE/g (Murthy and Naidu, 2012b). About CGA, eight major compounds (3-CQA, 4-CQA, 5-CQA, 4-FQA, 5-FQA, 3,4-diCQA, 3,5-diCQA, and 4,5-diCQA) have been identified in *C. arabica* husks from Brazil, Nicaragua, El Salvador, Bolivia, Costa Rica, Mexico, and India, with average levels varying considerably from 173.1 to 665.5 mg/100 g (dm) (Mullen et al., 2013; Farah and Paula Lima, 2019; dePaula, 2020). CGA compounds have also been identified in *C. canephora* (110–320 mg/100 g) husks from Mexico and India (Mullen et al., 2013). In*C. arabica* husks, higher and more variable values were observed when the fruits were fermented (310.3 ± 210.8 mg/100 g dm), compared to unfermented ones (212.7 ± 16.2 mg/100 g dm; dePaula, 2020). In relation to distribution of CGA subgroups, CQAs were the most abundant compounds (51%–82%), followed by diCQA (14–43%) and FQA (8%–12%). Regarding individual isomer distribution, 5-CQA predominated, corresponding, on average to 55%–77% of total CGA; 3-CQA and 4-CQA together comprised 10%–24%; among diCQA, 3,5-diCQA predominated, corresponding, on average, to 11%–36% of total CGA, while 3,4-diCQA and 3,5-diCQA together comprised 6%–15%; 5-FQA and 4-FQA corresponded to about 4%–6% and 2%–4% of total CGA content, respectively. CGA percent distribution was similar in unfermented and fermented husks. As observed in *C. arabica* leaves, the isomer 3-FQA was not identified in the husks (dePaula, 2020).

Hydroxybenzoic acids (gallic, syringic, and vanillic acids, 0.3–4.4 mg/100 g) and hydroxycinnamic acids (caffeic, *p*-coumaric, and ferulic acids, 0.09–53.8 mg/100 g) have been identified in aqueous extracts of *C. arabica* husk from Colombia (Rebollo-Hernanz et al., 2019). Procyanidins (1.1 to 5.3 mg/100 g), flavonols (rutin, quercetin and kaempferol,

FIGURE 9.11 *C. arabica* husk.

1.9–26.1 mg/100 g), and flavanols (catechin and epicatechin, 0.17–5.0 mg/100 g) have also been identified in coffee husks (Mullen et al., 2013; Rebollo-Hernanz et al., 2019.

Caffeic, ferulic, *p*-coumaric, gallic, and 3,4-dihydroxybenzoic acids were identified in all unfermented and fermented husks (dePaula, 2020). Individual contents ranged from 1.2 to 10.2 mg/100 g (dm). Higher average content of caffeic acid was observed in fermented samples (8.6 ± 0.8 mg/100 g, dm), indicating degradation of CGA during fermentation. Higher contents of gallic acid were also observed in fermented samples (6.3 ± 1.0 mg/100 g dm), indicating hydrolysis of flavonoid compounds such as catechins and epicatechins during fermentation or metabolism of CGA by the fermentation agents. Vanillic and dihydrocaffeic acids were identified in three commercial samples of husk, with contents ranging from 7.1 to 11.5 mg/100 g (dm). Their presence suggests that these samples were subjected to fermentation of greater intensity compared to the remaining samples. No difference was observed in the contents of ferulic, *p*-coumaric, and 3,4-dihydroxybenzoic acid contents when fermented and unfermented husks were compared.

Silver skin: Coffee silver skin (which is very lightweight; Figure 9.12) is considered to be a good source of bioactive compounds, especially CGA (Narita and Inouye, 2014; Murthy and Naidu, 2012a; Bresciani et al., 2014; Jiménez-Zamora et al., 2015). Eight CGA compounds have been identified: three CQA, two FQA, and three di-CQA. 5-CQA and 3-CQA are more relevant, totaling 130–430 mg/100 g (dm). The content of FQA ranged between 142 and 274 mg/100 g (dm), while that of diCQAranged between 44 and 56.8 mg/100 g (dm) (Bresciani et al., 2014; Jiménez-Zamora et al., 2015; Regazzoni et al., 2016). Two CGL, which contributed only 3% of total CGA compounds, have also been reported in *C. arabica* silver skin in consequence of heat exposure (Bresciani et al., 2014).

Free hydroxybenzoic acids (gallic, protocatechuic, 4-hydroxybenzoic, and vanillic acids, 0.33–4.40 mg/100 g), hydroxycinnamic acids (caffeic, *p*-coumaric, and ferulic acids, 0.09–53.80 mg/100 g), and flavanols such as catechin (1.02 mg/100 g) have been identified in aqueous extracts from *C. arabica* silver skin from Colombia (Rebollo-Hernanz et al., 2019).

Regarding total phenolics content, Costa et al. (2014) reported 30.25 mg GAE/100 mL of an extract prepared using released silverskin during roasting of a commercial coffee blend (about 40% of *C. arabica* and 60% of *C. canephora*). The authors also found tannins (about 43 mg of tannic acid equivalents–TAE/100 mL) and flavonoids (about 8.3 mg of epicatechin equivalents–ECE/100 mL). In another study, Toschi et al. (2014) reported 390–730 mg GAE/100 g of coffee silverskin.

The presence of melanoidins has also been reported in coffee silver skin from roasted coffee (Borrelli et al., 2004; Ballesteros et al., 2014; Costa et al., 2014) as a product of the Maillard reaction mainly after degradation of polysaccharides, proteins, and CGA (Moreira et al., 2012).

Spent coffee grounds: In addition to by-products from coffee production, more than six million tons of spent coffee grounds (Figure 9.13) are produced after brewing the beverage in soluble coffee industries, commercial establishments, and homes (Mussatto et al., 2011). For every ton of green coffee, about 650 kg of spent coffee grounds are produced (Murthy and Naidu, 2012b).

Regarding CGA compounds, 3-CQA, 4-CQA, 5-CQA, 3,4-diCQA, 3,5-diCQA, and 4,5-diCQA have been identified in spent coffee grounds obtained from different brewing processes and coffee species (Bravo et al., 2012; Jiménez-Zamora et al., 2015). CQA (530–770 mg/100 g dm). DiCQA (490–580 mg/100 g dm) have been identified in *C. arabica* spent grounds (Bravo et al., 2012). The presence of melanoidins has also been reported in this material (Monente et al., 2015).

Free hydroxybenzoic acids (gallic, syringic, and vanillic acids, 0.001–0.2 mg/100 g dm) and hydroxycinnamic acids (caffeic, *p*-coumaric, and ferulic acids, 0.02 to 0.8 mg/100 g dm) have been recently reported in spent *C. arabica* grounds from Ethiopia, Brazil, India, Colombia, and Costa Rica (Angeloni et al., 2020). Traces of anthocyanins (cyanidin 3-glucoside) and flavonols (rutin and quercetin) have also been identified. Kaempferol, catechins, and epicatechins have been investigated but not detected. Additional compounds, such as stilbenes (resveratrol) and flavanones (naringin) have also been identified in spent *C. arabica* grounds (Ramón-Gonçalves et al., 2019).

Total contents of phenolic compounds in spent coffee ground varies in the literature from 900 to 3500 mg GAE/100 g (Bravo et al., 2012; Panusa et al., 2013; Ramón-Gonçalves

FIGURE 9.12 *C. arabica* silver skin.

FIGURE 9.13 Spent coffee grounds.

Source: Reproduced from del Castillo et al. (2019), with permission from the Royal Society of Chemistry.

TABLE 9.4

Contents (mg/100 g) of Major Phenolic Compounds in Coffee By-Products

Phenolic compound	Flower	Leaf	Skin and pulp	Husk	Silver skin	Spent coffee grounds
3-CQA	–	7.0–300.2	10.0–130.4	10.2–22.0	130.4–178.0	39.6–1502.0
4-CQA	–	20.2–77.3	10.2–140.4	10.0–24.3	84.9–100.2	20.2–1623.1
5-CQA	7.0–74.3	78.3–2139.5	60.4–920.4	20.3–442.0	190.3–3000.0	25.3–1870.3
3,4-diCQA	–	5.4–1200.3	10.4–113.5	9.8–12.3	160.1–180.2	20.4–289.1
3,5-diCQA	–	5.0–2312.1	5.3–445.6	10.2–24.5	110.0–150.0	10.3–126.0
4,5-diCQA	–	6.1–240.4	6.0–490.0	9.7	120.3–150.7	14.6–207.7
Gallic acid	3.9–74.7	–	–	9.2–86.9	17.0	0.02–0.04
Protocatechuic acid	77.5	–	9.0–31.2	44.8–488.6	44.0	–
Mangiferin	–	35.2–788.7	–	–	–	–
Isomangiferin	–	5.2–138.5	–	–	–	–
Anthocyanins	–	–	3.5–50.5	17.2–203.4	–	0.001–0.005
Catechins	–	5.0–18.2	6.2–58.0	1.7–103.4	10.2	–
Epicatechins	–	26.2–550.5	17.1–436.0	9.7–606.3	–	–
Procyanidins	–	124.1–148.1	–	9.8–306.4	–	–
Rutin	–	19.1–800.0	9.1–70.6	–	–	0.002–0.009
Quercetin	–	9.2–113.1	–	8.9–153.6	–	0.0004–0.0008
Kaempferol	–	0.2–29.7	–	122.6	–	–
Tannins	–	–	13.6–1500.4	500.1–9300.4	–	–
Total phenolic compounds (mg GAE/100 g)	72.4–1911.2	1340.1–5910.2	254.6–4532.2	302.5–730.0	44.8–1900.0	1700.2–8000.0
References	Nguyen et al. (2019), De Abreu Pinheiro et al. (2021)	Campa et al. (2012), Chen et al. (2018), Ratanamarno and Surbkar (2017), de Almeida et al. (2019), Trevisan et al. (2019), Monteiro et al. (2019), Ngamsuk et al. 2019	Ramirez-Martinez (1988), Clifford and Ramirez-Martinez (1991), Ramirez-Coronel et al. (2004), Rosales-Delgado et al. (2019), Murthy and Naidu (2012b), da Silveira et al. (2019)	Brand et al. (2001), Prata and Oliveira (2007), Franca and Oliveira (2009), Gouvea et al. (2009), Mullen et al. (2013), Rodríguez-Durán et al. (2014), Heeger et al. (2017), Iriondo-Dehond et al. (2019), Rebollo-Hernanz et al. (2019)	Murthy and Naidu (2012b), Bresciani et al. (2014), Jimenez-Zamora et al. (2015), Regazzoni et al. (2016), Iriondo-Dehond et al. (2019), Rebollo-Hernanz et al. (2019)	Bravo et al. (2012), Panusa et al. (2013), Monente et al. (2015), Jimenez-Zamora et al. (2015), Regazzoni et al. (2016), Ramón-Gonçalves et al. (2019), Angeloni et al. (2020)

et al., 2019). The contents of the major phenolic compounds in coffee by-products are summarized in Table 9.4.

9.7 Contribution of polyphenols from coffee and by-products to health

The coffee seeds and other parts of the coffee plant tend to contain about the same classes of phenolic compounds, although at different proportions, with hydroxycinnamates being more abundant. In *Coffea* leaves, additional compounds include xanthones and flavonoids. The latter are also identified in the pulp/skin and in the silver skin.

The beneficial effects of coffee consumption can be attributed to the joint action of many bioactive compounds (Farah, 2019). It is likely that most contribution to decrease the risk of certain diseases is caused by synergistic or additive effects of the various compounds present in coffee (Farah, 2017). However, *in vitro* and animal studies have linked certain compounds to specific mechanisms. Because CGA are the main phenolic compounds in coffee plant (Farah and Donangelo, 2006; Trevisan et al., 2019; Iriondo-DeHond et al., 2019) they are responsible for most of coffee's and by-products' antioxidant capacity observed in *in vitro* and *in vivo* studies (Liang and Kitts, 2016; Tajik et al., 2017). They are also referred to as powerful anti-inflammatory compounds (Torres and Farah, 2016; Naveed et al., 2018), acting through mechanisms that involve signal transduction (Li et al., 2012; Han et al., 2017). As a consequence of these two key properties, CGA play important roles in decreasing the relative risk of development of a number of diseases, for example, type 2 diabetes mellitus (Yan et al., 2020), liver diseases (Shi et al., 2016), some types

TABLE 9.5

Potential Applications in Foods and Health Properties of Coffee By-Products

Coffee by-product	Use in foods	Potential health properties
Coffee flower	Tea-like drinks (Nguyen et al., 2019; de Abreu Pinheiro et al. 2021) Tisane or infusions, pure or in blends with other flowers or medicinal herbs (Klingel et al. 2020)	Antioxidant (Nguyen et al. 2019; de Abreu Pinheiro et al. 2021)
Coffee leaf	Tea-like drinks (Ratanamarno and Surbkar, 2017; Rilma et al. 2018)	Antioxidant, anti-aging (Takagaki and Tsuzaki, 2000)
Coffee husk	Source of anthocyanins to produce food coloring (Prata and Oliveira, 2007) Dietary fiber (Murthy and Naidu, 2012a) Tea-like drinks (Heeger et al. 2017) Ingredient for the preparation of flours (Iriondo-DeHond et al. 2019)	Antioxidant (Heeger et al. 2017) Anti-inflammatory (Rebollo-Hernanz et al. 2019) Anti-diabetic (Iriondo-DeHond et al. 2020)
Coffee silver skin	Dietary fiber (Pourfarzad et al. 2013) Bakery products (Garcia-Serna et al. 2014) Tea-like beverages, antioxidant beverages (Martinez-Saez et al. 2014)	Prebiotic (Borrelli et al., 2004; Jiménez-Zamora et al. 2015) Anti-diabetic (del Castillo et al. 2016) Antioxidant (Iriondo-DeHond et al. 2019)
Spent coffee grounds leaves	Dietary fiber (Campos-Vega, 2016) Distilled beverage (Sampaio et al. 2013)	Prebiotic (Jiménez-Zamora et al. 2015) Anti-diabetic (Vázquez-Sánchez et al. 2018) Antioxidant (Jiménez-Zamora et al. 2015; Martinez-Saez et al. 2017) Anti-inflammatory (López-Barrera et al. 2016)

of cancer (Bułdak et al., 2018), neurodegenerative diseases (Socała et al., 2020), and other chronic degenerative diseases, as well as the risk of death (Grosso et al., 2017).

Coffee by-products have attracted great attention in the last few years due to the large amounts generated during coffee production. Some value-added applications have been proposed to reuse these coffee wastes, but they are only in their early stage, while others have already been patented and even used industrially (Alves et al., 2017; del Castillo et al., 2019). Some of these proposals are related to health. Anthocyanin-rich coffee pulp, for example, is a powerful inhibitor of glucosidase and amylase enzymes. As these enzymes play an important role in the management of glucose metabolism, the use of this extract has been proposed to improve postprandial blood glucose metabolism (del Castillo et al., 2019); coffee silver skin have exhibited a powerful antioxidant effect in vitro (Mesías et al., 2014; Martinez-Saez et al., 2014), decreasing the risk of chronic diseases such as cardiovascular diseases, cancer, type 2 diabetes, and Alzheimer's and Parkinson's diseases. Coffee silver skin also showed relevant prebiotic activity in vitro, increasing the count of healthy bacteria such as *Lactobacillus* spp. and *Bifidobacterium* spp. (Jimenéz-Zamora et al., 2015). Spent coffee grounds may serve as natural sources of insoluble dietary antioxidant fiber (del Castillo et al., 2014). As insoluble dietary fiber material is indigestible by the enzymes of the gastrointestinal tract, it affects gut motility and slows down gastric emptying. It also has a significant antioxidant capacity, mainly ascribed to the remaining presence of phenolic molecules after brewing (del Castillo et al., 2019).

The bioactive potential of mangiferin and isomangiferin present in the leaves of *C. arabica* has been widely studied and continues to attract attention, especially with regard to its antioxidant potential (Gold-Smith et al., 2016). Even though they seem to be poorly absorbed, they can act as antioxidants in the digestive tract, protecting it against free radicals. *In vitro*, mangiferin has been shown to exert anti-inflammatory, immunomodulatory, and anti-apoptotic activities, resulting in antitumor activity (Gold-Smith et al., 2016). Additional potential properties associated with mangiferin and isomangiferin antioxidant effect are pending confirmation of their bioavailability (Martínez et al., 2001; Halliwell, 2006; Vithana et al., 2019).

Regarding flavonoids, like the CGA, there are evidences suggesting that quercetin and rutin are endowed with promising antioxidant (La Casa et al., 2000), anti-inflammatory (Negahdari et al., 2020), anticarcinogenic (Araújo et al., 2011), antibacterial (Luca et al., 2019), neuroprotective (Enogieru et al., 2018), and cardioprotective properties (Ugusman et al., 2014), especially rutin (Gullón et al., 2017). Anthocyanins have been shown to reduce biomarkers (Kalt et al., 2020) and risk of cardiovascular disease (Cassidy et al., 2013), type 2 diabetes mellitus (Wedick et al., 2012), and neurological decline (Devore et al., 2012). Also, there is much evidence from cell culture and animal studies that catechins and epicatechins have beneficial effects against several pathological diseases, including cancer, diabetes, and cardiovascular diseases (Khan and Mukhtar, 2018). Proanthocyanidins have a wide range of health beneficial properties that include antioxidant, antitumor, and immunostimulating activities (Rauf et al., 2019).

Table 9.5 summarizes possible applications in foods and potential health properties of coffee by-products related to their polyphenolic composition.

9.8 Conclusion

Coffee and by-products are among the richest food sources of polyphenols, especially CGA. Additional phenolic compounds in different parts of the coffee plant include flavonoids like rutin, quercetin, isoflavones, and anthocyanins, as well as other classes of phenolic compounds like lignans and xanthones. Despite the existence of substantial information on major CGA, data on minor CGA and other phenolic compounds in seeds and other parts of the plant, including their contribution to flavor and consumer health, are still missing.

While coffee is one of the most consumed food products on Earth, coffee by-products that are generated in large quantities are undervalued. The concern to create and promote sustainable practices and conscious consumption in a world where food deprivation is a growing reality has strengthened the concept of sustainability worldwide, for benefit to the environment, rural producers, and consumers. The adoption of sustainable practices favors better marketing conditions and opens space for new markets, including the production of organic coffee. We hope the knowledge of the potential health benefits of coffee by-products derived from their polyphenol composition will stimulate their consumption and upgrade their status in the near future as healthy food products.

REFERENCES

Alonso-Salces RM, Serra F, Reniero F, Rolyheberger A. *J Agric Food Chem.* 2009; 57(10): 4224–4235.

Alves RC, Soares CMD, Oliveira B. In: Preedy VR, editor. *Coffee in Health and Disease Prevention.* London: Elsevier, 2014.

Alves RC, Rodrigues F, Nunes MA, Vinha AF, Oliveira MBPP. In: Galanakis CM, editor. *Handbook of Coffee Processing By-Products.* United States: Academic Press, 2017.

Anesini, E. and Perez, C. *J Ethnopharmacol.* 1993; 39: 119–128.

Angeloni S, Nzekoue FK, Navarini L, Sagratini G, Torregiani E, Vittori S. *J Mass Spectrom.* 2020; 55(11): e4519. doi:10.1002/jms.4519.

Araújo JR, Gonçalves P, Martel F. *Nutr Res.* 2011; 31(2): 77–87.

Babova O, Occhipinti A, Maffei ME. *Phytochem.* 2016; 123: 33–39.

Ballesteros LF, Teixeira JA, Mussatto SI. *Food Bioprocess Technol.* 2014; 7: 3493–3503.

Barcelos AF, Paiva PCA, Pérez JRO, Santos VB, Cardoso RM. *R Bras Zootec.* 2001; 30(4): 1325–1331.

Barcelos AF, Paiva PCA, Pérez JRO, Santos VB, Cardoso RM. In: *Anais do II Simpósio de Pesquisa dos cafés do Brasil.* Consórcio Brasileiro de Pesquisa e Desenvolvimento do café, Brazil, 2002.

Barros RS, Maestri M, Rena AB. *J Coffee Res.* 1999; 27: 1–54.

Bekedam EK, Schols HA, Van Boekel MAJS, Smit GJ. *J Agric Food Chem.* 2008; 56(6): 2055–2063.

Bicho NC, Leitão AE, Ramalho JC, Lidon FC. *Eur Food Res Technol.* 2011; 233: 303–311.

Blumberg S, Frank O, Hofmann T. *J Agric Food Chem.* 2010; 58(6): 3720–3728.

Borrelli RC, Esposito F, Napolitano A, Ritieni A, Fogliano V. *J Agric Food Chem.* 2004; 52(5): 1338–1343.

Brand D, Pandey A, Rodriguez-Leon JA, Roussos S, Brand I, Soccol CR. *Biotechnol Prog.* 2001; 17(6): 1065–1070.

Bravo J, Juániz I, Monente C, Caemmerer B, Kroh LW. *J Agric Food Chem.* 2012; 60(51): 12565–12573.

Bresciani L, Calani R, Bruni F, Brighenti, D. del Rio. *Food Res Int.* 2014; 61: 196–201.

Bułdak RJ, Hejmo T, Osowski M, Bułdak Ł, Kukla M, Polaniak R, Birkner E. *Molecules.* 2018; 23(12): 3309. doi: 10.3390/molecules23123309.

Campa C, Mondolot L, Rakotondravao A, Bidel LP, Gargadennec A, Couturon E. *Ann Bot.* 2012; 110(3): 595–613.

Campos-Vega R. Patent No. MX/a/2016008578, 2016.

Caretto S, Linsalata V, Colella G, Mita G, Lattanzio V. *Int J Mol Sci.* 2015; 16(11): 26378–26394.

Cassidy A, Mukamal KJ, Liu L, Franz M, Eliassen AH, Rimm EB. *Circulation.* 2013; 127(2): 188–196.

Chairgulprasert V, Kongsuwankeeree K. *Sci Technol Asia.* 2017; 22(1): 1–8.

Chen X. *Crit Rev Food Sci Nutr.* 2018; 59(6): 1008–1025.

Chen X, Ma Z, Kitts DD. *Food Chem.* 2018; 249: 143–153.

Cheynier V. *Amer J Clin Nutr.* 2005; 81(1): 223–229.

Cheynier V, Comte G, Davies KM, Lattanzio V, Martens S. *Plant Physiol Biochem.* 2013; 72: 1–20.

Clifford MN. In: Clarke RJ, Macrae R, editors. *Coffee.* London: Elsevier Applied Science Publications, 1985.

Clifford MN. In: Clifford MN and Willsom KC, editors. *Coffee: Botany, Biochemistry and Production of Beans and Beverage.* Sydney, Australia: Chroom Helm, 1987

Clifford MN, Ramirez-Martinez JR. *Food Chem.* 1991; 40(1): 35–42.

Clifford MN, Knight S, Kuhnert N. *J Agric Food Chem.* 2005; 53(10): 3821–3832.

Clifford MN, Knight S, Surucu B, Kuhnert N. *J Agric Food Chem.* 2006; 54(6): 1957–1969.

Costa ASG, Alves RC, Vinha AF, Barreira SVP, Nunes MA, Cunha LM. et al. *Industr Crops Prod.* 2014; 53: 350–357.

DaMatta FM, Ronchi CP, Maestri M, Barros RS. *Braz J Plant Physiol.* 2007; 19(4): 485–510.

da Silveira JS, Durand N, Lacour S, Belleville M-P, Perez A, Loiseau G et al. *Food Bioprod Process.* 2019; 115: 175–184.

Davis AP, Tosh J, Ruch N, Fay MF. *Bot J Linn Soc.* 2011; 167(4): 357–377.

de Abreu Pinheiro F, Ferreira Elias L, de Jesus Filho M, Modolo MU, Rocha JCG, Lemos MF. *Food Chem.* 2021; 336: 127701.

de Almeida RF, Trevisan MTS, Thomaziello RA et al. *Food Res Int.* 2019; 115: 493–503.

de Beer D, Joubert E. *J Food Compos Anal.* 2010; 23(3): 289–297.

de Carvalho Neto DP, De Melo Pereira GV, Tanobe VOA, Soccol VT, da Silva BJG, Rodrigues C. *Fermentation.* 2017; 3(1): doi 10.3390/fermentation3010011

DePaula, J. (PhD Thesis) – Federal University of Rio de Janeiro, Institute of Nutrition Josué de Castro, 2020.

del Castillo MD, Martinez-Saez N, Ullate M. WO 2014128320 A1, 2014.

del Castillo MD, Fernandez-Gomez B, Ullate M, Mesa MD. International Patent Application No. PCT/ES2015/070915, 23 June 2016.

del Castillo MD, Fernandez-Gomez B, Martinez-Saez N, Iriondo-DeHond A, Mesa MD. In: Farah A, editor. *Coffee: Production, Quality and Chemistry*, London: Royal Society of Chemistry, 2019.

Delgado-Andrade A, Morales FJ. *J Agric Food Chem.* 2005; 53(5): 1403–1407.

Devore E, Kang HJ, Breteler MM, Grodstein F. *Ann Neurol.* 2012; 72(1): 135–143.

Dorfner R, Ferge T, Kettrup A, Zimmermann R, Yeretzian C. *J Agric Food Chem.* 2003; 51(19): 5768–5773.

Dorsey BM, Jones MA. In: Galanakis CM, editor, *Handbook of Coffee Processing By-Products*, United States: Academic Press, 2017.
Duarte GS, Pereira AA, Farah A. *Food Chem.* 2010; 118(3): 851–855.
Enogieru AB, Haylett W, Hiss DC, Bardien S, Ekpo OE. *Oxid Med Cell Longev*. 2018; 6241017. doi: 10.1155/2018/6241017.
Esquivel P, Jimenez VM. *Food Res Int.* 2012; 46(2): 488–495.
Farah A. In: Chu YF, editor. *Coffee: Emerging Health Effects and Disease Prevention.* New York: IFT Press, 2012.
Farah A. In: Lashermes P, editor. *Achieving Sustainable Cultivation of Coffee.* London: Burleigh Dodds Science Publishing, 2017.
Farah A. In: Lik HC, Borém F, editors. *Drying and Roasting of Cocoa and Coffee.* Nova Iorque: CRC Press, 2019.
Farah A, Donangelo CM. *Braz J Plant Physiol.* 2006; 18(1): 23–36.
Farah A, Duarte G. In: Preedy VR, editor. *Coffee and Health Disease Prevention.* Elsevier: Cambridge, 2015.
Farah A, Paula Lima J. In: Farah A, editor. *Coffee: Production, Quality and Chemistry.* London: Royal Society of Chemistry, 2019.
Farah A, De Paulis T, Trugo LC, Martin PR. *J Agric Food Chem.* 2005; 53(5): 1105–1113.
Farah A, Monteiro MC, Trugo LC. In: *Proceedings of the International Conference on Coffee Science* (ASIC), Italy, 2006a.
Farah A, De Paulis T, Moreira DP, Trugo LC, Martin PR. *J Agric Food Chem.* 2006b; 54(2): 374–381.
Farah A, Monteiro M, Donangelo CM, Lafay S. *J Nutr.* 2008; 138(12): 2309–2315.
Farah A, Perrone D, Fernandes JP, Kenny JS. In: *Proceedings of the International Conference on Coffee Science* (ASIC), Paris, 2011.
Ferreira T, Shuler J, Guimarães R, Farah A. In Farah, A, editor. *Coffee: Production, Quality and Chemistry.* London: Royal Society of Chemistry, 2019.
Franca AS, Oliveira LS. In: Ashworth GS, Azevedo P, editors. *Agricultural Wastes.* Nova Science Publishers, New York, 2009.
Frank O, Blumberg S, Krümpel G, Hofmann T. *J Agric Food Chem.* 2008; 56(20): 9581–9585.
Frank O, Blumberg S, Kunert C, Zehentbauer G, Hofmann TJ. Structure determination and sensory analysis of bitter-tasting 4-vinylcatechol oligomers and their identification in roasted coffee by means of *LC-MS/MS. J Agric Food Chem.* 2007; 55(5): 1945–1954.
Fuster MD, Mitchell AE, Ochi H, Shibamoto T. *J Agric Food Chem.* 2000; 48(11): 5600–5603.
Garcia-Serna E, Martinez-Saez N, Mesias M, Morales F, de Castillo MD. *Pol J Food and Nutr Sci.* 2014; 64(4): 243–251.
Gold-Smith F, Fernandez A, Bishop K. *Nutrients.* 2016; 8(7): 396. doi: 10.3390/nu8070396.
Gouvea BM, Torres C, Franca AS, Oliveira LS, Oliveira ES. *Biotechnol Letters.* 2009; 31(9): 1315–1319.
Griffin B, Stonier T. *Physiol Plant.* 1975; 33(2): 157–160.
Grosso G, Stepaniak U, Micek A, Stefler D, Bobak M, Pająk A. *Eur J Nutr.* 2017; 56(4): 1409–1420.
Guerrero G, Suárez M, Moreno G. *J Agric Food Chem.* 2001; 49(5): 2454–2458.
Guimarães RJ, Borém FM, Shuler J. In: Farah, A, editor. *Coffee: Production, Quality and Chemistry.* London: Royal Society of Chemistry, 2019.
Gullón B, Lú-Chau TA, Moreira MT, Lema JM, Eibes G. *Trends Food Sci Technol.* 2017; 67: 220–235.
Halliwell B. *J Neurochem.* 2006; 97(6): 1634–1658.
Han D, Chen W, Gu X, Shan R, Zou J, Liu G, Shahid M, Gao J, Han B. *Oncotarget,* 2017; 8(9): 14680–14692.
Heeger A, Kosińska-Cagnazzo A, Cantergiani E, Andlauer W. *Food Chem.* 2017; 221: 969–975.
Heuzé V, Tran G. 2015. *Coffee hulls, fruit pulp and by-products. Feediédir, a programme by INRAE, CIRAD. AFZ and FAO.* https://www.feedipedia.org/node/549. Accessed on 01 april 2021
Iriondo-DeHond A, García NA, Fernandez-Gomez, Guisantes-Batan E, Escobar FV, Blanch GP. et al. *Innov Food Sci Emerg Technol.* 2019; 51: 194–204.
Iriondo-DeHond M, Iriondo-DeHond A, Herrera T, Fernández-Fernandéz AM, Sorzano COS, Miguel E. et al. *Nutrients.* 2020; 12(3): 627. doi: 10.3390/nu12030627.
Jaiswal R, Patras MA, Eravuchira PJ, Kuhnert N. *J Agric Food Chem.* 2010; 58(15): 8722–8737.
Jiang H, Engelhardt UH, Thräne C, Maiwald B, Stark J. *Food Chem.* 2015; 183: 30–35.
Jiménez-Zamora A, Pastoriza S, Rufián-Henares JA. *LWT Food Sci Technol.* 2015; 61(1): 12–18.
Kalt W, Cassidy A, Howard LR, Krikorian R, Stull AJ, Tremblay F et al. *Adv Nutr.* 2020; 11(2): 224–236.
Khan N, Mukhtar H. *Nutrients.* 2018; 11(1): 39, doi: 10.3390/nu11010039.
Klingel T, Kremer JI, Gottstein V, Rajcic de Rezende T, Schwarz S, Lachenmeier DW. *Foods.* 2020; 9(5): 665, doi: https://doi.org/10.3390/foods9050665.
Kreppenhofer S, Frank O, Hofmann T. *Food Chem.* 2011; 126(2): 441–449.
Kumar N, Bhandari P, Singh B, Bari SS. *Food Chem Toxicol.* 2009; 47(2): 361–367.
Ky CL, Louarn J, Guyot B, Dussert S, Hamon S, Noirot M. *Food Chem.* 2001; 75(2): 223–239.
La Casa C, Villegas I, Alarcón de la Lastra C, Motilva V, Martín Calero MJ. *J Ethnopharmacol.* 2000; 71(1–2): 45–53.
Leloup V, Gancel C, Liardon R, Rytz A, Pithon A. In: *Proceedings of the International Conference on Coffee Science* (ASIC), India, 2004.
Li S, Bian H, Liu Z, Wang Y, Dai J, He W. et al. *Eur J Pharmacol.* 2012; 674(2–3): 65–72.
Liang N, Kitts DD. *Nutrients.* 2016; 8(1): 16. doi: 10.3390/nu8010016
Lima JP, Farah A, King B, De Paulis T, Martin Peter R. *J Agric Food Chem.* 2016; 64(11): 2361–2370.
López-Barrera DM, Vázquez-Sánchez K, Loarca-Piña MGF, Campos-Vega R. *Food Chem.* 2016; 212: 282–290.
Luca SV, Macovei I, Bujor A, Miron A, Skalicka-Woźniak K, Aprotosoaie AC. et al. *Crit Rev Food Sci Nutr.* 2019; 60(4): 626–659.
Manach C, Scalbert A, Morand C, Rémésy C, Jimenez L. *Amer J Clin Nutr.* 2004; 79(5): 727–747.
Marques VX, Farah A. *Food Chem.* 2009; 113(4): 1370–1376.
Martínez G, Giuliani A, Leon OS, Perez G, Núñez Selles AJ. *Phytother Res.* 2001; 15(7): 581–585.
Martinez-Saez N, Ullate M, Martin-Cabrejas MA, Martorell P, Genovés S, Ramon D. et al. *Food Chem.* 2014; 150: 227–234.
Martinez-Saez N, García TA, Pérez ID, Rebollo-Hernanz M, Mesías M, Morales FJ. et al. *Food Chem.* 2017; 216: 114–122.

Mateos R, Baeza G, Sarriá B, Bravo L. *Food Chem.* 2018; 241: 232–241.

Matiello J, Almeida SR, Aguiar EC, Araújo RA, Josino V. et al. In: *36° Congresso Brasileiro de Pesquisas Cafeeiras*. Fundação Procafé, Brazil, 2010.

Mayer JLS, Carmello-Guerreiro SM, Mazzafera P. *Aob Plants.* 2013; 5: plt029.

Mazza G, Miniati E. In: Mazza, G and Miniati, editors. *Anthocyanins in Fruits, Vegetables and Grain*. CRC Press, Boca Raton, 1993.

Mesías M, Navarro M, Martínez-Saez N, Ullate M, del Castillo MD, Morales FJ. *Food Res Int.* 2014; 62: 1120–1126.

Milder IEJ, Feskens EJM, Arts ICW, Mesquita HBB, Hollman PCH, Kromhout D. *J Nutr.* 2005; 135(5): 1201–1207.

Mondolot L, La Fisca P, Buatois B, Talansier E, De Kochko A, Campa C. *Ann Bot.* 2006; 98(1): 33–40.

Monente C, Ludwig IA, Irigoyen A, De Peña MP, Cid C. *J Agric Food Chem.* 2015; 63(17): 4327–4334.

Monteiro Â, Colomban S, Azinheira HG, Guerra-Guimarães L, Silva MC, Navarini L. et al. *Antioxidants (Basel).* 2019; 9(1): 6. doi: 10.3390/antiox9010006.

Moon JK, Yoo HS, Shibamoto T. *J Agric Food Chem.* 2009; 57(12): 5365–5369.

Moreira RFA, Trugo LC, Maria CAB. *Quim Nova.* 2000; 23(2): 195–201.

Moreira SP, Nunes FM, Domingues MR, Coimbra MA. *Food Func.* 2012; 3(9): 903–915.

Mullen W, Nemzer B, Stalmach A, Ali S, Combet E. *J Agric Food Chem.* 2013; 61(22): 5298–5309.

Murayama T, Yada H, Kobori M, Shinmoto H, Tsushida T. *J Jpn Soc Hortic Sci.* 2012; 71(2): 236–242.

Murthy PS, Naidu MM. *Resour Conserv Recy.* 2012a; 66: 45–58.

Murthy PS, Naidu MM. *Food Bioprocess Technol.* 2012b; 5: 897–903.

Mussatto SI, Ercília MS, Martins S, Teixeira JA. *Food Bioprocess Technol.* 2011; 4: 661–672.

Narita Y, Inouye K. *Food Res Int.* 2014; 61: 16–22.

Naveed M, Hejazi V, Abbas M, Kamboh AA, Khan GJ, Shumzaid M, Ahmad F, Babazadeh D, FangFang X, Modarresi-Ghazani F, WenHua L, XiaoHui Z. *Biomed Pharmacother.* 2018, 97:67–74. doi: 10.1016/j.biopha.2017.10.064.

Negahdari R, Bohlouli S, Sharifi S, Dizaj SM, Saadat YR, Khezri K. et al. *Phytother Res.* 2020; 35(4): 1719–1738.

Neveu V, Perez-Jimenez J, Vos F, Crespy V, du Chaffaut L, Mennen L et al. *Database (Oxford)*, 2010: bap024 doi: 10.1093/database/bap024. Epub 2010 Jan 8.

Ngamsuk S, Huang TC, Hsu JL. *Foods.* 2019; 8(9): 389. doi: 10.3390/foods8090389.

Nguyen TMT, Cho EJ, Song Y, Oh CH, Funada R, Bae HJ. *Food Chem.* 2019; 299: 125120. doi: 10.1016/j.foodchem.2019.125120. Epub 2019 Jul 2.

Panusa A, Zuorro A, Lavecchia R, Marrosu G, Petrucci R. *J Agric Food Chem.* 2013; 61(17): 4162–4168.

Patay EB, Németh T, Németh TS, Filep R, Vlase L, Papp N. *Farmacia.* 2016; 64(1): 125–130.

Perrone D, Farah A, Donangelo CM, de Paulis T, Martin PR. *Food Chem.* 2008; 106(2): 859–867.

Perrone D, Donangelo R, Donangelo CM, Farah A. *J Agric Food Chem.* 2010; 58(23): 12238–12243.

Pourfarzad A, Mahdavian-Mehr H, Sedaghat N. *LWT Food Sci Technol.* 2013; 50(2): 599–606.

Prata ERBA, Oliveira LS. *LWT Food Sci Technol.* 2007; 40(9): 1555–1560.

Ramirez-Coronel MA, Marnet N, Kolli VS, Roussos S, Guyot S, Augur C. *J Agric Food Chem.* 2004; 52(5): 1344–1349.

Ramirez-Martinez JR. *J Sci Food Agric.* 1988; 43(2): 135–144.

Ramón-Gonçalves M, Gómez-Mejía E, Rosales-Conrado N, León González ME, Madrid Y. *Waste Managem.* 2019; 96(1): 15–24.

Ratanamarno S, Surbkar S. *Maejo Int J Sci Technol.* 2017; 11(3): 211–218.

Rauf A, Imran M, Abu-Izneid T, Iahtisham-Ul-Haq, Patel S, Pan X et al. *Biomed Pharmacother.* 2019; 116:108999 doi: 10.1016/j.biopha.2019.108999.

Rebollo-Hernanz M, Zhang Q, Aguilera Y, Martín-Cabrejas MA, Gonzalez de Mejia E. *Food Chem Toxicol.* 2019; 132: 110672. Parte inferior do formulário

Regazzoni L, Saligari F, Marinello C, Rossoni G, Aldini G, Carini M. et al. *J Func Foods.* 2016; 20: 472–485.

Rilma N, Anwar K, Tuty A, Deddi PP. *J Ethnic Foods.* 2018; 5(4): 286–291.

Rodríguez-Durán LV, Ramírez-Coronel MA, Aranda-Delgado E, Nampoothiri KM, Favela-Torres E, Aguilar CN et al. *J Agric Food Chem.* 2014; 62(31): 7869–7876.

Rosales-Delgado S, Alzate Arbelaez AF, Rojano B. *Acta Sci Pol Technol Aliment.* 2019; 18(3): 235–248.

Sampaio A, Dragone G, Vilanova M, Oliveira JM, Teixeira JÁ, Mussatto SI. *LWT Food Sci Technol.* 2013; 54(2): 557–563.

Scalbert A, Johnson IT, Saltmarsh M. *Amer J Clin Nutr.* 2005; 81(1): 215–217.

Scholz- Böttcher BM, Ludges E, Maier HG. *Liebigs Ann Chem.* 1991; 10: 1029–1036.

Shi H, Shi A, Dong L, Lu X, Wang Y, Zhao J et al. *Clin Nutr.* 2016; 35(6): 1366–1373.

Smith RF. In: Clifford MN, Wilson KC, editors. *Coffee: Botany, Biochemistry and Production of Beans and Beverage.* London: Avi Pub Co, 1985.

Socała K, Szopa A, Serefko A, Poleszak E, Wlaź P. *Int J Mol Sci.* 2020; 22(1): 107 doi: 10.3390/ijms22010107.

Tajik N, Tajik M, Mack I, Enk P. *Eur J Nutr.* 2017; 56(7): 2215–2244.

Takagaki K, Tsuzaki, S. Japan Patent JP-2002065227-A, filed August 31, 2000, and issued March 5, 2002.

Talamond P, Mondolot L, Gargadennec A, de Kochko A, Hamon S, Fruchier A. et al. *Acta Bot Gall.* 2008; 155: 513–519.

Toci A, Farah A, Trugo LC. *Quim Nova.* 2006; 29(5): 965–971.

Toci A, Azevedo D, Farah A. *Food Res Int.* 2020; 137: 109546 https://doi.org/10.1016/j.foodres.2020.109546

Torres T, Farah A. *Eur J Nutr.* 2016; 56(4): 1523–1533.

Toschi TG, Cardenia V, Bonaga G, Mandrioli M, Rodriguez-Estrada MT. *J Agric Food Chem.* 2014; 62(44): 10836–10844.

Trevisan MTS, de Almeida RF, Soto G, Filho EDMV, Ulrich CM, Owen RW. *Food Anal Method.* 2016; 9(9): 2649–2655.

Trevisan MTS, Almeida RF, Breuer A, Owen RW. In: Farah A, editor. *Coffee: Production, Quality and Chemistry*. London: Royal Society of Chemistry, 2019.

Ugusman A, Zakaria Z, Chua KH, Nordin MM, Anita N, Abdullah Mahdy Z. *Sci World J.* 2014; 169370. doi: 10.1155/2014/169370. Epub 2014 Jun 24.

Ulloa Rojas JB, Verreth JAJ, van Weerd JH, Huisman EA. *Anim Feed Sci Technol.* 2002; 99(1–4): 195–204.

Ulloa Rojas JB, Verreth JAJ, Amato S, Huisman EA. *Bioresour Technol.* 2003; 89(3): 267–274.

Vázquez-Sánchez K, Martinez-Saez N, Rebollo-Hernanz M, del Castillo MD, Gaytán-Martínez M, Campos-Vega R. *Food Chem.* 2018; 261: 253–259.

Veenstra S. In: *Curso taller Internacional sobre tratamiento anaerobio de aguas residuals.* Santiago de Cali, Colombia, 1995.

Vithana MDK, Singh Z, Johnson SK. *J Sci Food Agric.* 2019; 99(8): 3740–3751.

Wedick NM, Pan A, Cassidy A, Rimm EB, Sampson L, Rosner B. et al. *Am J Clin Nutr.* 2012; 95(4): 925–933.

Williamson G. *Nutr Bull.* 2017; 42(3): 226–235.

Williamson G, Clifford MN. *Biochem Pharmacol.* 2017; 139: 24–39.

Wootton AE. In: *ASIC Proceedings of the 16th Colloque Scientifique International sur la Chimie des Cafés Verts.* Paris, 1974.

Yan Y, Zhou X, Guo K, Zhou F, Yang H. *J Immunol Res.* 2020; 28: 9680508.

Yaylayan VA, Haffenden LJK. Food Res Int. 2003; 36(6): 571–577.

Zeb A. *J Food Biochem.* 2020; 44(9): e13394. https://doi.org/10.1111/jfbc.13394

Section III

Advances in post-harvest processing technologies pertaining to coffee

10

*Influence of coffee bean processing on the phenolic acid and flavonoid content in coffee brews (*Coffea arabica *and* Coffea robusta*)*

Magdalena Jeszka-Skowron, Robert Frankowski, and Agnieszka Zgoła-Grześkowiak

CONTENTS

10.1 Introduction

Coffee belongs to the Rubiaceae family and it grows natively in the mountains of southwest Ethiopia, and the first mention of the properties and usage were in the 9th century A.D. in Abyssinia and Arabia (Ukers, 1922). *Coffea arabica* L. and *Coffea canephora* L. (commonly known as *C. robusta*), the main coffees in the global market, grow in tropical and subtropical areas, mainly in the equatorial regions (Figure 10.1). In 2020–2021, world coffee production is predicted to rise by 1.9% to 171.9 million bags. That increase is driven by the high growth of Arabica production, estimated at 5.2% (to 101.9 million bags), which reflects the increase in Brazil's Arabica production, as well as expected growth from other major Arabica producers, such as Colombia. On the other hand, Robusta production is projected to decline by 2.6% (to 70.02 million bags), due to the large decrease in Vietnam's output (ICO, 2021).

About two billion cups of this stimulating beverage are consumed daily worldwide (ICO, 2021). Regular and moderate coffee consumption may have positive health effects in reducing the risk of type 2 diabetes mellitus, Alzheimer's, Parkinson's, cardiovascular diseases, and also the risk of mortality (Abe et al., 2019; Akash et al., 2014; Gökcen and Şanlier, 2019; Ludwig et al., 2014a; Socała et al., 2021). The latest review of epidemiological, in vivo, and in vitro studies of coffee showed that the brew can also protect against gastrointestinal and liver carcinogenesis (Romualdo et al., 2019).

The coffee brew is a source of biologically active compounds such as methylxanthines (caffeine, theobromine, and theophylline), trigonelline, chlorogenic acids, and other phenolic compounds, diterpenes, and organic acids (Angelino et al., 2018; Farah et al., 2005; Farah and Paula Lima, 2019; Jaiswal et al., 2010; Jeszka-Skowron et al., 2015, 2020, 2021; Jeszka-Skowron and Zgoła-Grześkowiak, 2017; Perrone et al., 2008). Chlorogenic acids in coffees are mainly composed of quinic acid and at least one group of caffeic, ferulic, or coumaric acid, and the main chlorogenic acids in green and roasted coffees are 5-caffeoylquinic acids (5-CQA), 4-caffeoylquinic acids (4-CQA), and 3-caffeoylquinic acids (3-CQA). Recently, secoisolariciresinol and lariciresinol, the phenolic compounds from the lignan group, were determined in green and roasted coffee samples (Angeloni et al., 2018, 2020b). Isoflavones (genistein and daidzein), their glucosides (genistin and daidzin), and biochanin A and formononetin were quantified in the green coffees (Angeloni et al., 2020a). Comprehensive characterization of hydroxycinnamoyl derivatives in green and roasted coffee beans was recently described, and a new group of methyl hydroxycinnamoyl quinate was found (Asamenew et al., 2019).

The roasting process, even the light roasting process, the decaffeination, the steaming, and fermentation processes of coffee beans show a noticeable influence on its chemical composition. It was previously found that chlorogenic acids were drastically decreased by the roasting process, even 70% for dark roasted coffees, and chlorogenic acid lactones appeared (Farah et al., 2005; Ludwig et al., 2014b). The Vietnamese Robusta steamed coffee brew contains a 50% lower level of 5-CQA but approximately 35% higher levels of 3-CQA and 4-CQA in comparison with the unroasted coffee brew (Jeszka-Skowron et al., 2016b). On the other hand, the Vietnamese Robusta decaffeinated coffee brew contains a similar level of 5-CQA and almost 50% higher levels of 3- and 4-CQA in comparison with the same unroasted coffee brew (Jeszka-Skowron et al., 2016a, 2016b). The organic acids such as tartaric,

FIGURE 10.1 Coffee seeds on a branch.

Source: Photo by Piotr Werner.

formic, citric, malic, lactic, and quinic acids were determined in Arabica and Robusta coffee brews with the use of isotachophoresis (Jeszka-Skowron and Zgoła-Grześkowiak, 2017). After roasting, the content of organic acids was increased (except for citric and malic acids in Brazilian Arabica coffee brew), and the pH of the brews was decreased. However, there is little information about other phenolic compounds in unprocessed and processed coffee beans.

Therefore, the aim of the study was to determine phenolic acids, flavonols, catechins, and succinic acid in green and then roasted Arabica and Robusta coffee samples with the use of LC-MS/MS. Apart from the roasting process, the influence of decaffeination and steaming processes of Vietnamese Robusta coffee beans on these compounds was also analyzed.

10.2 Coffee samples and the determination of phenolic compounds

Coffee bean samples from four different regions of the world were provided by Strauss Café Poland (Swadzim/Poznań, Poland). Robusta beans were from Vietnam (Grade 2), and Arabica coffee beans originated from Brazil (Type N.Y. Good beans), Guatemala (SHG – Strictly High Grown fermented beans), and Honduras (HG – High Grown). The Vietnamese coffee was processed using different methods: decaffeination, steaming, and roasting. The decaffeination was performed with the use of dichloromethane. The light roasting process of Arabica and Robusta coffee beans was performed at temperatures of 170–190°C in 8 minutes (Jeszka-Skowron et al., 2016a).

Briefly, 1 g of milled beans of coffee were extracted by 100 mL of distilled water at 95°C. Then the solution was cooled to room temperature, filtered, and finally diluted to the proper volume. Succinic acid, gallic acid, protocatechuic acid (dihydroxybenzoic acid), quinic acid, chlorogenic acids (the total of 5-, 4- and 3-caffeoylquinic acids), caffeic acid, *p*-coumaric acid, ferulic acid, and sinapic acid, myricetin, quercetin, kaempferol, rutin, (+)-catechin, (−)-epicatechin, (−)-catechin-3-O-gallate, and (−)-epicatechin-3-O-gallate were investigated in all coffee samples in one analysis with the use of tandem liquid chromatography-mass spectrometry (LC-MS/MS; Jeszka-Skowron et al., 2018). The UltiMate 3000 RSLC chromatographic system (Dionex, Sunnyvale, CA, USA) was used, and 5 μL of samples were injected into a Kinetex Evo C18 column (150 mm × 2.1 mm I.D.; 2.6 μm; Phenomenex, Torrance, CA, USA) maintained at 35°C. The mobile phase employed in the analysis consisted of 0.1% formic acid in water and acetonitrile at a flow rate of 0.3 mL min^{-1} in a gradient. The LC column effluent was directed to the electrospray ionization source (the Turbo Ion Spray) of the API 4000 QTRAP triple quadrupole mass spectrometer (AB Sciex, Foster City, CA, USA). The Turbo Ion Spray source operated in the negative ion mode.

10.3 Influence of coffee bean processing on the phenolic acids and flavonoids

Studies published so far showed a significant variation in the content of polyphenols and their profile in coffee, depending on the variety, cultivation, climatic conditions (altitude, soil, fertilization, sunlight, temperature, precipitation), agricultural practices (traditional or organic), harvesting and post-harvest conditions (fermentation, washing, drying, storage), and technological processing (Alonso-Salces et al. 2009; Duarte et al., 2010; Jeszka-Skowron et al., 2016b; Król et al., 2020; Mehari et al., 2016). The influence of the origin and coffee processing has also been shown in the present study.

Based on the obtained results, it was found that the main constituents from analyzed phenolic compounds were chlorogenic

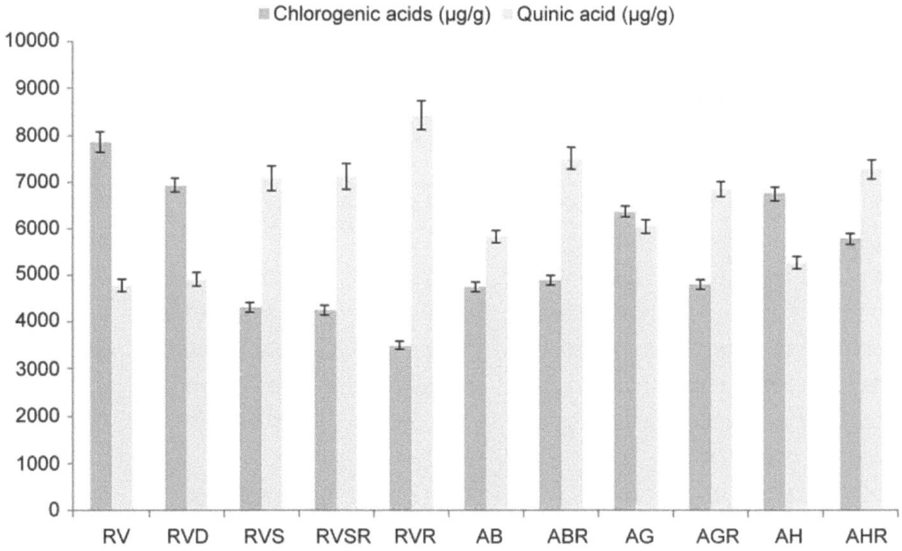

FIGURE 10.2 Concentration of a total of 5-CQA, 4-CQA, and 3-CQA (chlorogenic acids) and quinic acid in *Coffea arabica* and *Coffea robusta* (μg g^{-1}). See Table 10.1 for abbreviation legend.

TABLE 10.1
Content of Phenolic Acids and Succinic Acid in *Coffea robusta* and *Coffea arabica* Brews (μg g^{-1})

Coffee	Phenolic and succinic acids				
	Caffeic acid	**Protocatechuic acid**	**Succinic acid**	**Gallic acid**	**Sinapic acid**
RV	370.1 ± 12.2	16.4 ± 1.0	51.6 ± 0.2	0.042 ± 0.005	0.050 ± 0.002
RVD	274.0 ± 10.1	26.2 ± 1.2	85.4 ± 1.2	1.83 ± 0.11	0.096 ± 0.003
RVS	128.4 ± 7.3	51.3 ± 1.4	117.0 ± 3.2	2.73 ± 0.12	0.343 ± 0.009
RVSR	39.3 ± 3.4	56.7 ± 1.1	89.3 ± 0.9	2.04 ± 0.09	0.338 ± 0.008
RVR	20.6 ± 2.3	36.6 ± 0.8	168.3 ± 3.3	1.06 ± 0.012	0.128 ± 0.004
AB	388.8 ± 11.7	7.4 ± 0.5	93.0 ± 0.9	0.031 ± 0.002	0.022 ± 0.001
ABR	21.5 ± 1.9	29.8 ± 1.1	111.2 ± 2.1	1.14 ± 0.01	0.068 ± 0.002
AG	57.2 ± 2.2	13.2 ± 0.5	<LOD*	0.047 ± 0.007	0.027 ± 0.001
AGR	23.5 ± 1.2	23.0 ± 0.7	91.8 ± 0.8	2.27 ± 0.02	0.164 ± 0.003
AH	90.1 ± 3.3	8.3 ± 0.2	42.7 ± 0.6	0.071 ± 0.002	0.036 ± 0.001
AHR	31.2 ± 1.4	25.4 ± 0.5	98.8 ± 0.9	1.37 ± 0.02	0.146 ± 0.002

Notes: * <LOD = below the limit of detection. AB = Arabica coffee from Brazil, green; ABR = Arabica coffee from Brazil, roasted; AG = Arabica coffee from Guatemala, green; AGR = Arabica coffee from Guatemala, roasted; AH = Arabica coffee from Honduras, green; AHR = Arabica coffee from Honduras, roasted; RV = Robusta coffee from Vietnam, green; RVD = Robusta coffee from Vietnam, green and decaffeinated; RVR = Robusta coffee from Vietnam, roasted; RVS = Robusta coffee from Vietnam, steamed; RVSR = Robusta coffee from Vietnam, steamed and roasted.

acids (the total of 5-, 4- and 3-caffeoylquinic acids) and quinic acid determined in all coffee samples (Figure 10.2). The third important compound was caffeic acid, which dominated in Vietnamese Robusta and Brazilian Arabica unprocessed coffees (Table 10.1). Decaffeination, steaming, and roasting processes affect the concentration of phenolic compounds.

10.3.1 Influence of coffee bean processing on the phenolic acids and flavonoids in Robusta coffees

The highest level of chlorogenic acids, 7845 μg g^{-1}, was found in green coffee from Vietnam (*Coffea robusta*), and all the used processes decreased these compounds to a different level (Figure 10.2). Decaffeination with dichloromethane is the most popular process of caffeine removal from coffee beans. These processes had a different impact on the phenolic acid, flavonols, and flavan-3-ols in the Vietnamese coffee brews (Figures 10.2–10.4, Tables 10.1 and 10.2).

Chlorogenic acids and ferulic acid were decreased, but quinic and *p*-coumaric acid contents were on a similar level as before decaffeination process. Rutin and quercetin were decreased on a similar level, but kaempferol and myricetin concentrations were increased, twofold for kaempferol content (Figure 10.4, Table 10.2).

In the flavan-3-ol group of compounds, such treatment decreased the level of all analyzed compounds except for (–)-epicatechin-3-O-gallate, for which the concentration

TABLE 10.2
Content of Flavonoids in *Coffea robusta* and *Coffea arabica* Brews (μg g[-1])

Coffee	Flavonols			Flavan-3-ols			
	Rutin	Myricetin	Kaempferol	(+)-Catechin	(–)-Epicatechin	(–)-catechin-3-O-gallate	(–)-epicatechin-3-O-gallate
RV	0.656 ± 0.010	<LOD*	0.296 ± 0.002	0.379 ± 0.006	0.347 ± 0.012	0.032 ± 0.006	0.029 ± 0.002
RVD	0.405 ± 0.011	0.381 ± 0.009	0.489 ± 0.003	0.022 ± 0.001	0.075 ± 0.004	0.006 ± 0.003	0.104 ± 0.005
RVS	0.091 ± 0.002	0.191 ± 0.004	1.538 ± 0.018	0.713 ± 0.007	<LOD	0.039 ± 0.002	0.035 ± 0.003
RVSR	0.149 ± 0.007	<LOD	0.016 ± 0.002	0.045 ± 0.004	0.185 ± 0.007	0.123 ± 0.009	0.006 ± 0.001
RVR	0.192 ± 0.008	0.437 ± 0.012	1.065 ± 0.016	0.632 ± 0.015	0.220 ± 0.010	0.026 ± 0.005	0.017 ± 0.001
AB	0.213 ± 0.003	0.169 ± 0.004	<LOD	0.356 ± 0.012	0.035 ± 0.003	0.007 ± 0.001	0.023 ± 0.001
ABR	0.127 ± 0.002	0.880 ± 0.012	<LOD	0.586 ± 0.014	0.277 ± 0.008	0.046 ± 0.003	0.081 ± 0.002
AG	0.181 ± 0.006	0.311 ± 0.009	<LOD	0.207 ± 0.011	<LOD	0.013 ± 0.001	0.012 ± 0.001
AGR	0.181 ± 0.005	0.430 ± 0.011	<LOD	0.109 ± 0.012	0.300 ± 0.009	0.006 ± 0.001	0.029 ± 0.001
AH	0.373 ± 0.003	<LOD	<LOD	0.092 ± 0.009	0.103 ± 0.006	0.019 ± 0.002	0.041 ± 0.001
AHR	0.091 ± 0.001	<LOD	<LOD	0.759 ± 0.027	0.318 ± 0.011	0.052 ± 0.001	0.006 ± 0.001

* *<LOD = below the limit of detection. See Table 10.1 for abbreviation legend.*

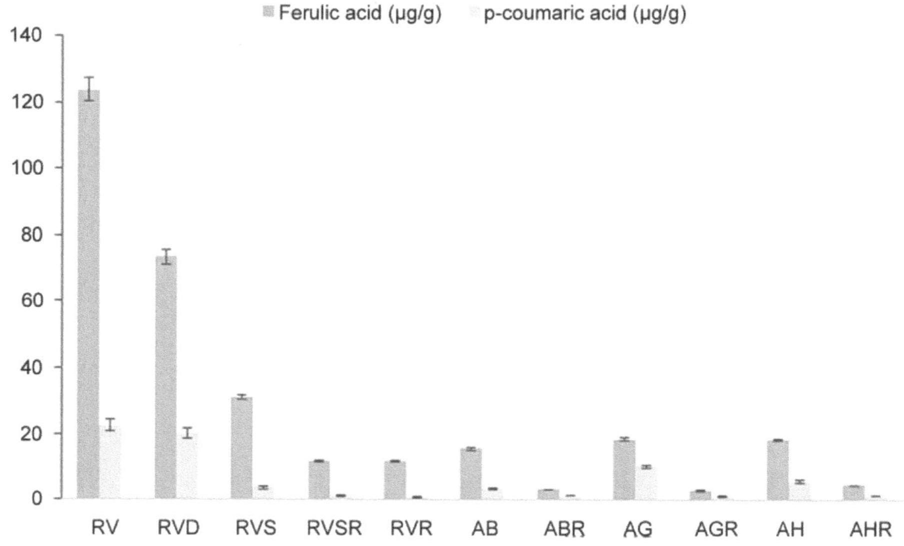

FIGURE 10.3 Concentration of ferulic and *p*-coumaric acids in *Coffea arabica* and *Coffea robusta* (μg g[-1]). See Table 10.1 for abbreviation legend.

increased three times after the decaffeination process (Table 10.2).

It was previously found that the level of chlorogenic acids in decaffeinated coffees depends on the solvent used. The decaffeination with supercritical carbon dioxide better preserves the original chemical composition of regular coffee and its flavor (Farah et al., 2006; Farah et al., 2010). A 16% decrease of chlorogenic acid was observed for *C. arabica* and 11% for *C. canephora* after treatment with dichloromethane (Toci et al., 2006). Likewise, Farah et al. (2010) showed that the water decaffeination method used for *C. arabica* led to a 20% decrease of the total chlorogenic acids in green coffee beans, and the supercritical carbon dioxide method removed even 1.2% more.

The steaming process is used for the Robusta coffees before the roasting to make coffee less irritating to the stomach as well as for better flavor and aroma of coffee. The steaming process decreased the level of chlorogenic acids and increased quinic acid content 1.8 times (Figure 10.2). Steamed coffee (RVS), as well as steamed and roasted coffee (RVSR), possessed similar concentrations of chlorogenic acids and quinic acid.

During the roasting of beans, chemical reactions and transformations such as hydrolysis, transesterification (acyl migration), lactonization, and isomerization of phenolic acids were observed. Some of the hydroxycinnamic acids were incorporated into the melanoidins during the Maillard reaction. The most of phenolic acids, such as ferulic and *p*-coumaric acids, in the Vietnamese Robusta coffee were degraded after the roasting process (Figure 10.3).

The same conclusion can be drawn for the quercetin content (Figure 10.4). Caffeic acid concentration was drastically

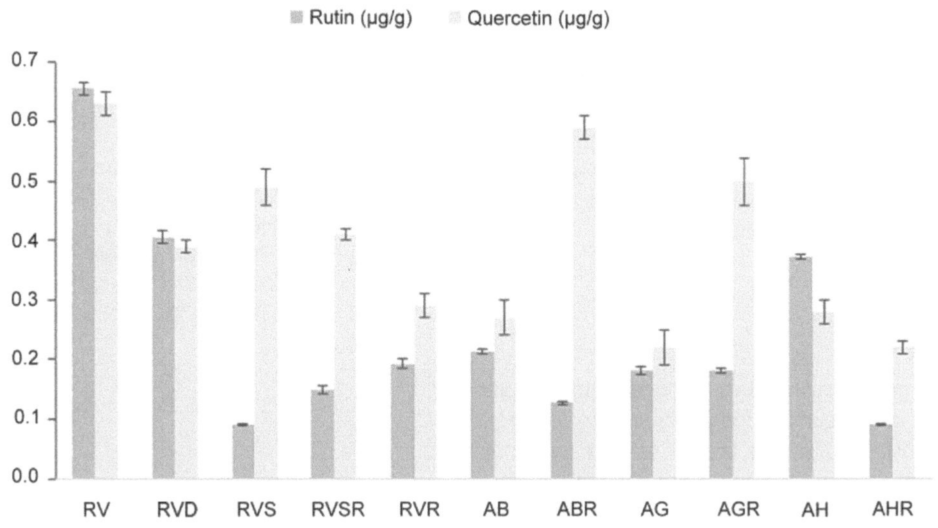

FIGURE 10.4 Concentration of rutin and quercetin in *Coffea arabica* and *Coffea robusta* (µg g^{-1}).

See Table 10.1 for abbreviation legend.

degraded after the roasting process in Arabica and Robusta coffees in particular for Vietnamese Robusta and Brazilian Arabica coffees (Table 10.1).

Khamitova et al. (2020) found that the roasted Robusta coffee brews contained a higher level of caffeic acid than roasted Arabica coffee brews. Also, caffeic, ferulic, and dihydrocaffeic acids determined in roasted Robusta coffee (*C. canephora* cv. Conillon) were in greater concentrations than in *C. arabica* coffees cv. Mundo Novo, Red Catuai, and Yellow Bourbon (Perrone et al., 2012). The total content of phenolic acids determined by the HPLC-DAD method (5-, 4-, and 3-CQA, *p*-coumaric, sinapic, and ferulic acids, among others) in Indian green Robusta coffee brews was the greatest (208.9 ± 2.7 mg 100 mL^{-1}) in comparison to the Arabica Indian green (183.9 ± 4.5 mg 100 mL^{-1}) and Arabica Brazilian green (171.8 ± 3.0 mg 100 mL^{-1}) coffee brews (Górnaś et al., 2016).

It was previously found that the greatest content of CQAs and derivatives was determined in the light roasted coffees in comparison to the dark roasted coffees (Budryn et al., 2009; Dawidowicz and Typek, 2016; Hečimović et al., 2011; Król et al., 2020). Also, the degradation of chlorogenic acids, rutin, quercetin, and kaempferol during 12 months of storage was observed (Król et al., 2020).

10.3.2 Influence of coffee bean processing on the phenolic acids and flavonoids in Arabica coffees

In the *Coffea arabica* varieties, the level of chlorogenic acids was from 4755 µg g^{-1} to 6768 µg g^{-1}, and after roasting the level of these compounds was in the range of 4816 µg g^{-1} to 5795 µg g^{-1} (Figure 10.2). The concentration of these compounds in Brazilian green and roasted coffees was similar. Jeon et al. (2019) found that coffee from Brazil (light degree of roasting) contains the greatest level of chlorogenic acids. Medium roasted coffee originating from Ethiopia showed the greatest level of total 5-, 4-, and 3-CQAs (1872 mg L^{-1}) in comparison to Brazilian (1477 mg L^{-1}) and Guatemalan (1430 mg L^{-1}) coffee brews (Jeon et al., 2017).

The variances between 5-CQA, 4-CQA, and 3-CQA concentrations in coffee beverages prepared by eight different methods were found (Angeloni et al., 2019). The total chlorogenic acid level in specialty espresso (ECS) was 9.16 mg mL^{-1}, followed by espresso coffee (EC) at 8.85 mg mL^{-1}, and the lowest concentration (1.05 mg mL^{-1}) was determined for the French press method. The content of chlorogenic acids was in the order: ECS > EC > ECF (Caffè Firenze) > Cold brew > Mokka > V60 filtration > Aeropress > French press.

Moreover, the concentration of 5-, 4-, and 3-CQAs in hot brew coffee (the French press) in comparison to cold brew (7 hours of brewing with room-temperature water in the French press) was the same: 1536 ± 46 mg L^{-1} and 1535 ± 14 mg L^{-1}, respectively (Rao et al., 2020).

Between analyzed Arabica coffees, the level of quinic acid in Brazilian roasted coffee was the highest, reaching 7507 µg g^{-1} (Figure 10.2). A similar concentration of quinic acid in light roasted coffees (roasted at 170°C in 10, 20, and 30 minutes) was previously observed, and the higher temperature and longer time of roasting influenced the decrease of the quinic acid content (Dawidowicz and Typek, 2016).

Gallic acid content in Arabica roasted coffees was three- to fourfold higher in comparison to the green ones (Table 10.1). Rutin is a glucoside of quercetin, and therefore this compound could be degraded resulting in an increased concentration of quercetin. This hypothesis, however, was only confirmed for Brazilian Arabica coffee (Figure 10.4). Kaempferol was only determined in Vietnamese unprocessed and processed coffees. The level of this compound in Arabica coffees was below the limit of detection (Table 10.2).

Muñoz et al. (2020) found a significant decrease in chlorogenic acids and tyrosol concentrations after the roasting process in Colombian Arabica coffees; however, caffeic, coumaric, and vanillic acids contents were on the same level as determined in the green coffees.

Mojica et al. (2018) showed that the cinnamon coffee extract (light roasted coffee) prepared from Colombian beans had the highest levels of gallic, caffeic, and chlorogenic acids in comparison to the city, full-city, and full-city plus dark roasted coffees. This coffee extract also possessed the highest antiproliferative effects in colon and oral cancer cells.

10.4 Conclusions

The light roasting process as well as the decaffeination and steaming treatment influenced the composition of coffee beans and then the composition of coffee brews. After roasting, the phenolic acids' composition decreased: chlorogenic acids (total of 5-, 4-, and 3-caffeoylquinic acids) from 14%–55%, caffeic (59–94%), ferulic (75–91%), and *p*-coumaric (64–95%) acids, and increased the level of quinic (113%–177%), gallic (1929%–4830%), succinic (100%–326%), and protocatechuic (174%–402%) acids in Arabica and Robusta coffee brews. The roasting process mainly decreased the level of flavonoids (rutin, 40%–75%) but increased the concentrations of myricetin and kaempferol. Decaffeination and steaming processes also decreased the level of chlorogenic (12% and 38%), caffeic (26% and 65%), and ferulic (41% and 75%) acids as well as rutin (38% and 86%), and the decaffeination process did not affect the concentration of quinic and *p*-coumaric acids. The dominating compounds were chlorogenic and quinic acids in Arabica and Robusta coffees. Kaempferol was determined in Vietnamese unprocessed and processed coffees in comparison to the level below the limit of detection in Arabica coffees. The flavan-3-ols: (+)-catechin, (–)-epicatechin, (–)-catechin-3-O-gallate, and (–)-epicatechin-3-O-gallate were also determined in Arabica and Robusta coffees in trace amounts.

10.5 Acknowledgments

This research was funded by the National Science Centre – NCN (Poland) under the program MINIATURA 4, grant number: 2020/04/X/NZ9/00736.

REFERENCES

Abe S. K., Saito E., Sawada N., Tsugane S., Ito, H., Lin Y., Tamakoshi A., Sado J., Kitamura Y., Sugawara Y., Tsuji I., Nagata C., Sadakane A., Shimazu T., Mizoue T., Matsuo K., Naito M., Tanaka K., Inoue M., & Research Group for the Development and Evaluation of Cancer Prevention Strategies in Japan. *Prev Med.* 2019; 123: 270–277.

Akash M.S.H., Rehman K., Chen S. *Nutrition* 2014; 30: 755–763.

Alonso-Salces R.M., Serra F., Reniero F., Heberger K. *J Agric Food Chem.* 2009; 57: 4224–4235.

Angelino D., Tassotti M., Brighenti F., Del Rio D., Mena P. *Sci Rep.* 2018; 8: 17874.

Angeloni S., Navarini L., Sagratini G., Torregiani E., Vittori S., Caprioli G. *J Mass Spectrom.* 2018; 53: 842–848.

Angeloni G., Guerrini L., Masella P., Bellumori M., Daluiso S., Parenti A., Innocenti M. *Food Res Int.* 2019; 116: 1327–1335.

Angeloni S., Navarini L., Khamitova G., Maggi F., Sagratini G., Vittori S. *Food Chem.* 2020a; 325: 126924.

Angeloni S., Navarini L., Khamitova G., Sagratini G., Vittori S., Caprioli G. *Int J Food Sci Nutr.* 2020b; 71: 193–200.

Asamenew G., Kim, H.W., Lee M.K., Lee S.H., Lee S., Cha Y.S., Lee S.H., Yoo S.M., Kim J.B. *Food Chem X.* 2019; 2: 100033.

Budryn G., Nebesny E., Podsędek A., Żyżelewicz D., Materska M., Jankowski S., Janda B. *Eur Food Res Technol.* 2009; 228: 913–922.

Dawidowicz A.L. and Typek R. *Eur Food Res Technol.* 2016; 243: 379–390.

Duarte G.S., Pereira A.A., Farah A. *Food Chem.* 2010; 118: 851–855.

Farah A. and de Paula Lima J. Publisher: Royal Society of Chemistry: London, UK, 2019; pp. 584–610.

Farah A., de Paulis T., Trugo L.C., Martin P.R. *J Agric Food Chem.* 2005; 53: 1505–1513.

Farah A., de Paulis T, Moreira D.P., Trugo L.C., Martin P.R. *J Agric Food Chem.* 2006; 54: 374–381.

Farah A., Perrone D., Fernandes J., Silanes J. In: *23rd International Conference on Coffee Science-ASIC*, Bali, Indonesia, 2010; 367–372.

Gökcen B.B. and Şanlier N. *Crit Rev Food Sci Nutr.* 2019; 59: 336–348.

Górnaś P., Dwiecki K., Siger A., Tomaszewska-Gras J., Michalak M., Polewski K. *Eur Food Res Technol.* 2016; 242: 641–653.

Hečimović I., Belščak-Cvitanović A., Horžić D., Komes D. *Food Chem.* 2011; 129: 991–1000.

ICO. 2021. www.ico.org (accessed January 01, 2021).

Jaiswal R., Patras M. A., Eravuchira P. J., Kuhnert N. *J Agric Food Chem.* 2010; 58: 8722–8737.

Jeon J.S., Kim H.T., Jeong I.H., Hong S.R., Oh M.S., Park K.H., Shim J.H., Abd El-Aty A.M. *J Chromatogr B.* 2017; 1064: 115–123.

Jeon J.S., Kim H.T., Jeong I.H., Hong S.R, Oh M.S., Yoon M.H., Shim J.H., Jeong I.H., Abd El-Aty A.M. *J Adv Res.* 2019; 17: 85–94.

Jeszka-Skowron M. and Zgoła-Grześkowiak A. *Food Anal Methods.* 2017; 10: 1245–1251.

Jeszka-Skowron M., Zgoła-Grześkowiak A., Grześkowiak T. *Eur Food Res Technol.* 2015; 240: 19–31.

Jeszka-Skowron M., Sentkowska A, Pyrzyńska K, De Peña M.P. *Eur Food Res Technol.* 2016a; 242: 1403–1409.

Jeszka-Skowron M., Stanisz E., De Peña M.P. *LWT.* 2016b; 73: 243–250.

Jeszka-Skowron M., Zgoła-Grześkowiak A., Frankowski R. *J Food Compos Anal.* 2018; 74C: 71–81.

Jeszka-Skowron M., Frankowski R., Zgoła-Grześkowiak A. *LWT.* 2020; 125: 109344.

Jeszka-Skowron M., Zgoła-Grześkowiak A., Grześkowiak T. Publisher: Springer, 2021; pp. 31–81.

Khamitova G., Angeloni S., Fioretti L., Ricciutelli M., Sagratini G., Torregiani E., Vittori S., Caprioli G. *Food Res Inter.* 2020; 133: 109220.

Król K., Gantner M., Tatarak A., Hallmann E. *Eur Food Res Technol.* 2020; 246: 33–39.

Ludwig I.A., Clifford M.N., Lean M.E., Ashihara H., Crozier A. *Food Funct.* 2014a; 5: 1695–1717

Ludwig I.A., Mena P., Calani L., Cid C., del Rio D., Lean M.E., Crozier A. *Food Funct.* 2014b; 5: 1718–1726.

Mehari B., Redi-Abshiro M., Chandravanshi B. S., Combrinck S., Atlabachew M., McCrindle R. *J Food Compos Anal.* 2016; 45: 16–25.

Mojica B.E., Fong L.E., Biju D., Muharram A., Davis I.M., Vela K.O., Rios D., Osorio-Camacena E, Kaur B., Rojas S., Forester S.C. *J Food Sci.* 2018; 83: 1125–1130.

Muñoz A.E., Hernández S.S., Tolosa A.R., Burillo S.P., Olalla-Herrera M. *LWT.* 2020; 128: 109457.

Perrone D., Farah A., Donangelo C.M., de Paulis T., Martin P. R. *Food Chem.* 2008; 106: 859–867.

Perrone D., Farah A., Donangelo C.M. *J Agric Food Chem.* 2012; 60: 4265–4275.

Rao N.Z. Fuller M., Grim M.D. *Foods* 2020; 7: 902.

Romualdo G.R., Rocha A.B., Vinken M., Cogliati B., Moreno F.S., Chaves M.A.G., Barbisan L.F. *Food Res Int.* 2019; 123: 567–589.

Socała K., Szopa A., Serefko A., Poleszak E., Wlaź P. *Int J Mol Sci.* 2021; 22: 107.

Toci A., Farah A., Trugo L.C. *Quim Nova.* 2006; 29: 965–971.

Ukers W. The Tea and Coffee Trade Journal, New York, 1922; pp. 5–9.

11

The metabolic fate and mechanisms of action of coffee polyphenols

Tunjung Mahatmanto and Wenny Bekti Sunarharum

CONTENTS

11.1 Introduction

As one of the most popular drinks worldwide, coffee is well known for its beneficial health effects and is considered a medicinal beverage (Valduga et al., 2019). The associated health benefits from regular coffee consumption include lower risk of cardiovascular disease, liver conditions, and specific cancers (Poole et al., 2017). The health effects of coffee stem from the bioactive components contained in and extracted from coffee beans, the majority being phenolic compounds, alkaloids, and diterpenes (de Melo Pereira et al., 2020; Rodrigues and Bragagnolo, 2013). Upon ingestion, these bioactive compounds are subject to bioconversions by the gut microbiota before being absorbed and distributed throughout the body. Drawing an example from polyphenols (Rechner et al., 2002; van Duynhoven et al., 2011), the plasma concentrations of intact bioactive compounds would probably be too low to exert significant health effects. Thus, the associated health benefits from coffee consumption presumably arise from the derivatives of the bioactive compounds, their interactions with the gut microbiota, and other mechanisms. The aim of this chapter is to provide a general overview of and a brief update on the metabolic fate and mechanisms of action of coffee polyphenols that may exert beneficial health effects.

11.2 The fate of coffee polyphenols upon ingestion

Polyphenols represent the largest group of natural compounds found in plants, with more than 8000 known structures (Tsao et al., 2010). Although structurally diverse, polyphenols can simply be classified into two main classes, flavonoids and phenolic acids (Abbas et al., 2017) or further be divided into subclasses based on their different structural features (Singla et al., 2019). In coffee, the most abundant polyphenols are chlorogenic acids (CGAs), which are esters of certain hydroxycinnamic acids and quinic acid (Farah and Donangelo, 2006; Rostagno et al., 2015). Currently, there are more than 300 CGAs reported, with the majority in coffee being caffeoylquinic acids (CQAs; Farah and de Paula Lima, 2019; Narita and Inouye, 2015). In coffee brew, around one-third of the CGAs consumed (100–600 mg CGAs/day) are absorbed in the gastrointestinal tract, while the remaining two-thirds are unabsorbed (Farah and de Paula Lima, 2019). The metabolism of coffee CGAs upon ingestion has been extensively reviewed (Clifford, 2000; Clifford et al., 2020; Farah and Duarte, 2015) and thus will only be discussed briefly here.

In general, following ingestion, CGAs contained in coffee will undergo digestion, absorption, and metabolism by the liver before distribution, whereas the remaining unabsorbed fraction will undergo metabolism by the gut microbiota in the intestine before being absorbed. The fate of coffee CGAs can be categorized into being absorbed in the stomach or small intestine (either intact or hydrolyzed and further metabolized in the liver) and being absorbed (with or without further conversions) following metabolism by the microbiota in the colon (Clifford, 2000). In short, CGAs can be digested and absorbed in the stomach, small intestine, and colon with or without conversion and distributed throughout the body before being excreted or recycled in the gastrointestinal tract (Farah and Duarte, 2015).

DOI: 10.1201/9781003043133-14

11.2.1 Digestion

Studies on CGAs in natural and artificial digestive (salivary, gastric, and intestinal) fluids show that the stability of CGAs is affected by the type of the digestive fluids and the CGA as well as the duration of the exposure to the digestive fluids (Monente et al., 2015; Olthof et al., 2001; Stalmach et al., 2010; Vilas-Boas et al., 2020). In general, the majority of CGAs are relatively stable in salivary and gastric fluids but tend to decline in intestinal fluids. Different CGAs are affected differently by different digestive fluids. The longer the exposure to the digestive fluids, the lower the concentration of CGAs that remain. Although large variations exist because of differences in the ionic forms and enzyme concentrations in the digestive fluids, it is estimated that around 90% of coffee CGAs would reach the mucosa of the gastrointestinal tract, where they will be absorbed (Farah and Duarte, 2015).

11.2.2 Absorption and liver metabolism

Studies on CGA absorption show that CGAs are readily absorbed in the stomach following coffee consumption (Farah et al., 2008; Monteiro et al., 2007). The remaining CGAs in the chyme that reach the small intestine will first be exposed to the pancreatic enzymes and liver bile in the duodenum before reaching the jejunum and ileum, where they will be absorbed and metabolized in the liver (together with the CGAs that were absorbed in the stomach) before being distributed and excreted. Notable changes that occur to the CGAs during metabolism in the small intestine and in the liver include hydrolysis and conjugation to sulfate or glucuronic acid (Farah and Duarte, 2015). These changes are mostly apparent in plasma and urine, where the metabolic derivatives of CGAs are found in the sulfate or glucoronate form (Nardini et al., 2002; Stalmach et al., 2009). It should be noted that there are many other forms of CGA metabolites (Farah & Duarte, 2015) and that intact CGAs have also been found in circulation albeit at low concentrations (Erk et al., 2012; Monteiro et al., 2007; Stalmach et al., 2009).

11.2.3 Gut microbiota metabolism

The metabolism of polyphenols, in general, by the gut microbiota has been reviewed in detail (Clifford et al., 2020; Williamson and Clifford, 2010, 2017), but the principle changes to coffee CGAs will be discussed briefly here. As mentioned earlier, the remaining majority of CGAs that are not absorbed in the small intestine will be subject to metabolism by the microbiota in the intestine before being absorbed and distributed. The changes that occur to the CGAs will depend on the composition of the gut microbiota they encountered. Although there have been many studies on the metabolism of CGAs by the gut microbiota (e.g., Couteau et al., 2001; Gonthier et al., 2006; Gonthier et al., 2003; Plumb et al., 1999; Rechner et al. 2004), details regarding the role of each microbial species in the metabolism of CGAs remains limited and will be outlined later.

The main challenge to understanding the role of microorganisms in metabolizing CGAs (and other nutrients in general) is the enormous population size of the microorganisms that live in the human intestine. The gut microbiota amount to 100 trillion (Bäckhed et al., 2005) and comprise more than 2000 species (Almeida et al., 2019). The composition of this diversity is primarily shaped by environmental factors such as diet and lifestyle (Rothschild et al., 2018). The most abundant and most diverse genus is the *Bacteroides* (Arumugam et al., 2011). With regard to the metabolism of CGAs and other polyphenols, some genera have been identified to be involved (e.g., *Eubacterium, Enterococcus, Micrococcus, Fusobacterium, Streptococcus, Peptostreptococcus*, and *Clostridium*; Farah and Duarte, 2015).

Different microbial species are known to perform distinct metabolic reactions. The metabolic capacity of the gut microbiota in processing CGAs and other polyphenols includes *O*- and *C-deglycosylation*, hydrolysis of esters and amides, and deglucuronidation; dehydroxylation, demethoxylation, and demethylation of the aglycons; and hydrogenation and α- and β-oxidation of the aliphatic intermediates (Williamson and Clifford, 2010). The metabolites produced from the gut microbiota metabolism are predominantly aromatic and phenolic acids with zero to three phenolic hydroxyls (Williamson and Clifford, 2010). In addition, aliphatic metabolites can also be produced, and together with all the other CGA metabolites, they will be absorbed, distributed, and excreted through the digestive fluids or recycled (Farah and Duarte, 2015). The metabolism of CGAs will determine their metabolite forms, whether converted or intact, and in turn will determine their bioavailability and bioactivity within the human circulatory system.

11.3 The actions of coffee polyphenols and their physiological outcomes

The consumption of coffee has been associated with lower incidence of various degenerative and non-degenerative diseases (Farah and de Paula Lima, 2019). These associated health effects are generally thought to arise from the polyphenols contained in coffee, but because the health effects cannot be directly attributed to the CGAs, such a correlation should still be viewed as a hypothesis (Clifford et al., 2020). Nevertheless, we view it important to discuss the possible mechanisms of action of coffee polyphenols that may be associated with the beneficial health effects supported by results from in vitro, in vivo, ex vivo, interventional, and epidemiological studies. For simplicity, we will describe the mechanisms of polyphenol action as being either direct (acting on human cells or the cells of the gut microbiota) or indirect (modulating the extracellular environment). For more detailed discussion on coffee polyphenols and their health effects, readers are referred to recent reviews by Farah and de Paula Lima (2019) and Tajik et al. (2017).

11.3.1 The actions of polyphenols on human cells

The health benefits of regular coffee consumption have been attributed to the antioxidant and anti-inflammatory activities of the polyphenols contained in coffee and their roles in glucose and lipid metabolism (Tajik et al., 2017). With regard to antioxidants, there has been intense debate as to whether dietary antioxidants are effective at modulating redox balance in humans and provide protection against excessive oxidative cell damage (Halliwell, 1999, 2011). The root of the controversy appears to lie around the dual (antioxidant and pro-oxidant) behavior of antioxidants that depends on their concentration and nature of the neighboring molecules (Villanueva and Kross, 2012). Despite this controversy, accumulating data suggest that both antioxidant and pro-oxidant activities are vital for homeostasis (Carocho and Ferreira, 2013) and that the health effects of antioxidants could arise from other mechanisms including the modulation of signaling events, gene expression, epigenetic mechanisms, and gut microbiota (Hrelia and Angeloni, 2020).

The antioxidant activity of polyphenols is closely related to their anti-inflammatory activity. It is known that oxidative stress can activate differential expression of genes involved in the inflammatory pathway; polyphenols can counter this effect through their radical scavenging activities, attenuation of signal transduction, and modulation of pro-inflammatory molecules in cells (Hussain et al., 2016; Zhang and Tsao, 2016). Polyphenols can also exert anti-inflammatory effects through interaction with the immune cells (Shimizu, 2017). In addition to their antioxidant and anti-inflammatory activities, polyphenols can inhibit or activate enzymes (and their expression), regulate receptors, and modulate molecules involved in glucose and lipid metabolism. Thus, it has been postulated that polyphenols may be beneficial for diseases related to glucose and lipid metabolism disorders such as cardiovascular diseases, diabetes, obesity, and cancer (Tajik et al., 2017).

11.3.2 The modulation of the gut microbiota by polyphenols

As mentioned earlier, another mechanism of action of polyphenols that may exert beneficial health effects is the modulation of gut microbiota. Interest in exploring the potential use of gut microbiota for rational intervention to alter host phenotype has grown, and the past two decades have seen significant progress in our understanding of the contribution of gut microbiota to metabolism, health, and the pathogenesis of diseases (Clemente et al., 2012; Fan and Pedersen, 2020; Flint et al., 2012; Heintz-Buschart and Wilmes, 2018; Kinross et al., 2011; Marchesi et al., 2016; Round and Mazmanian, 2009; Sekirov et al., 2010; Shreiner et al., 2015). Despite this progress, the characteristics of a healthy gut microbiome remain to be defined, but an ecologic stability supported by high taxa diversity and gene richness has been considered (Bäckhed et al., 2012; Huttenhower et al., 2012). This fundamental definition is important to consider when we seek to establish a correlation between intervention of dietary compounds and their health effects through the modulation of gut microbiota. In this regard, the relationship between polyphenols and gut microbiota is still poorly understood (Cardona et al., 2013; Ozdal et al., 2016).

Studies on the modulation of gut microbiota by coffee (and their polyphenols) have mainly been focused on the effect of their intake on the growth and activities of specific members of the gut microbial community. For example, Nakayama and Oishi (2013) showed that coffee consumption decreased *Escherichia coli* and *Clostridium* spp. but increased *Bifidobacterium* spp. Others have also reported similar results and observed that coffee consumption induced the growth and metabolic activity of *Bifidobacterium* spp. (Jaquet et al., 2009; Mills et al., 2015). Recently, Nishitsuji et al. (2018) observed that the intake of coffee and its major constituents, including CGAs, modulates gut microbiota and repairs the short chain fatty acid profile in plasma. Interestingly, previous study suggests that the proliferation of certain gut microbiota and the production of short chain fatty acids may be stimulated by the metabolic derivates of polyphenols and not by the polyphenols themselves (Parkar et al., 2013). In the long-term, heavy coffee consumption (45–500 mL day^{-1}) has been associated with the increase of the population of *Bacteroides, Prevotella*, and *Porphyromonas* (González et al., 2020). In general, these studies have provided insights into one of the possible mechanisms of action of dietary polyphenols where they display a prebiotic-like activity, selectively stimulating the growth of beneficial bacteria while inhibiting pathogenic ones (Cardona et al., 2013; Hervert-Hernández and Goñi, 2011; Ozdal et al., 2016).

11.3.3 The modulation of the extracellular environment by polyphenols

In the digestive system, polyphenols can interact with other dietary substances such as lipids, proteins, and carbohydrates before they reach the cells. This interaction not only affects their bioaccessibility and bioavailability but also how other substances are absorbed and thus affect the body (Jakobek, 2015). For example, polyphenols (including CGAs) can interact with heme and non-heme iron as well as lipase and lipoxygenase and inhibit lipid oxidation (Cao et al., 2020; Lorrain et al., 2010, 2012). The inhibition of lipase by CGAs may lead to the reduction of fat absorption, which could benefit people with obesity (Jakobek, 2015). Polyphenols may also interact with the products of lipid peroxidation and thus could potentially protect cells from oxidative damage (Kanner et al., 2012). In addition to lipase and lipoxygenase, CGAs have also been shown to inhibit α-amylase, α-chymotrypsin, and trypsin (Rohn et al., 2002, 2003) and may affect the digestion of carbohydrates and proteins. The interaction between polyphenols and carbohydrates may interrupt the interaction between polyphenols and proteins, which in turn may prevent the negative effects of the polyphenol-protein interaction. However, it may also affect carbohydrate fermentation (Jakobek, 2015). Thus, the consumption of polyphenols should consider their interactions with other dietary substances to avoid the potential negative effects.

11.4 Conclusion

The associated health effects of coffee polyphenols may be attributed, to some extent, to their antioxidant and anti-inflammatory activities. The mechanisms of action of coffee polyphenols may be performed by their intact and converted forms following metabolism. The actions of polyphenols and their metabolites may be direct (acting on human cells or gut microbiota) or indirect (acting through the modulation of the extracellular environment). Because the health effects (and potential negative effects) of coffee polyphenol consumption can arise from their complex interactions with other dietary substances and their metabolisms by human cells and gut microbiota, future studies should seek to reconstruct such a complex system to provide a holistic and mechanistic understanding of how coffee exerts its beneficial effects. To achieve this task, we may need to borrow a synthetic ecology approach to model our natural digestive system (Vrancken et al., 2019) and put it into use.

REFERENCES

Abbas M, Saeed F, Anjum FM, Afzaal M, Tufail T, Bashir MS, Ishtiaq A, Hussain S, Suleria HAR. Int J Food Prop. 2017; 20:1689–1699.

Almeida A, Mitchell AL, Boland M, Forster SC, Gloor GB, Tarkowska A, Lawley TD, Finn RD. Nature. 2019; 568:499–504.

Arumugam M, Raes J, Pelletier E, Le Paslier D, Yamada T, Mende DR, Fernandes GR, Tap J, Bruls T, Batto J-M, Bertalan M, Borruel N, Casellas F, Fernandez L, Gautier L, Hansen T, Hattori M, Hayashi T, Kleerebezem M, Kurokawa K, Leclerc M, Levenez F, Manichanh C, Nielsen HB, Nielsen T, Pons N, Poulain J, Qin J, Sicheritz-Ponten T, Tims S, Torrents D, Ugarte E, Zoetendal EG, Wang J, Guarner F, Pedersen O, de Vos WM, Brunak S, Doré J; MetaHIT Consortium, Antolín M, Artiguenave F, Blottiere HM, Almeida M, Brechot C, Cara C, Chervaux C, Cultrone A, Delorme C, Denariaz G, Dervyn R, Foerstner KU, Friss C, van de Guchte M, Guedon E, Haimet F, Huber W, van Hylckama-Vlieg J, Jamet A, Juste C, Kaci G, Knol J, Lakhdari O, Layec S, Le Roux K, Maguin E, Mérieux A, Melo Minardi R, M'rini C, Muller J, Oozeer R, Parkhill J, Renault P, Rescigno M, Sanchez N, Sunagawa S, Torrejon A, Turner K, Vandemeulebrouck G, Varela E, Winogradsky Y, Zeller G, Weissenbach J, Ehrlich SD, Bork P. Nature. 2011;473:174–180.

Bäckhed F, Ley RE, Sonnenburg JL, Peterson DA, Gordon JI. Science. 2005;307(5717):1915 LP – 1920.

Bäckhed F, Fraser CM, Ringel Y, Sanders ME, Sartor RB, Sherman PM, Versalovic J, Young V, Finlay BB. Cell Host & Microbe. 2012; 12:611–622.

Cao Q, Huang Y, Zhu Q-F, Song M, Xiong S, Manyande A, Du H. Food Chem. 2020; 333:127528.

Cardona F, Andrés-Lacueva C, Tulipani S, Tinahones FJ, Queipo-Ortuño MI. J Nutr Biochem. 2013; 24:1415–1422.

Carocho M, Ferreira, ICFR. Food Chem Toxicol. 2013; 51:15–25.

Clemente JC, Ursell LK, Parfrey LW, Knight R. Cell. 2012;148;1258–1270.

Clifford MN. J Sci Food Agric. 2000; 80:1033–1043.

Clifford MN, Kerimi A, Williamson G. Compr Rev Food Sci F. 2020; 19:1299–1352.

Couteau D, McCartney AL, Gibson GR, Williamson G, Faulds CB. J Appl Microbiol. 2001; 90:873–881.

de Melo Pereira GV, de Carvalho Neto DP, Magalhães Júnior AI, do Prado FG, Pagnoncelli MGB, Karp SG, Soccol CR. Advances in Food and Nutrition Research. Academic Press. 2020; 91:65–96.

Erk T, Williamson G, Renouf M, Marmet C, Steiling H, Dionisi F, Barron D, Melcher R, Richling E. Mol Nutr Food Res. 2012; 56:1488–1500.

Fan Y, Pedersen O. Nat Rev Microbiol. 2020;1–17.

Farah A, de Paula Lima J. Beverages. 2019; 5:11.

Farah A, Donangelo CM. Phenolic compounds in coffee. Braz J Plant Physiol. 2006; 18:23–36.

Farah A, Duarte G. Coffee and Health Disease Prevention. Elsevier: Cambridge. 2015;789–802.

Farah A, Monteiro M, Donangelo CM, Lafay S. J Nutri. 2008; 138:2309–2315.

Flint HJ, Scott KP, Louis P, Duncan SH. Nat Rev Gastroenterol Hepatol. 2012; 9:577.

Gonthier M-P, Verny M-A, Besson C, Rémésy C, Scalbert A. J Nutri. 2003; 133:1853–1859.

Gonthier M-P, Remesy C, Scalbert A, Cheynier V, Souquet J-M, Poutanen K, Aura A-M. Biomed Pharmacother. 2006; 60:536–540.

González S, Salazar N, Ruiz-Saavedra S, Gómez-Martín M, de los Reyes-Gavilán CG, Gueimonde M. Nutrients. 2020;12(5).

Halliwell B. Free Radic Res. 1999; 31:261–272.

Halliwell B. Trends Pharmacol Sci. 2011; 32:125–130.

Heintz- Buschart A, Wilmes P. Trends Microbiol. 2018;26(7): 563–574.

Hervert- Hernández D, Goñi I. Food Rev Int. 2011;27(2):154–169.

Hrelia S, Angeloni C. New mechanisms of action of natural antioxidants in health and disease. Multidisciplinary Digital Publishing Institute. 2020.

Hussain T, Tan B, Yin Y, Blachier F, Tossou MCB, Rahu N. Oxid Med Cell Longev. 2016;7432797

Huttenhower C, Gevers D, Knight R, Abubucker S, Badger JH, Chinwalla AT, Creasy HH, Earl AM, FitzGerald MG, Fulton RS, Giglio MG, Hallsworth-Pepin K, Lobos EA, Madupu R, Magrini V, Martin JC, Mitreva M, Muzny DM, Sodergren EJ. Nature. 2012; 486:207–214.

Jakobek L. Interactions of polyphenols with carbohydrates, lipids and proteins. Food Chem. 2015;175:556–567.

Jaquet M, Rochat I, Moulin J, Cavin C, Bibiloni R. Int J Food Microbiol. 2009;130:117–121.

Kanner J, Gorelik S, Roman S, Kohen R. J Agric Food Chem. 2012;60:8790–8796.

Kinross JM, Darzi AW, Nicholson JK. Genome Med. 2011;3:14.

Lorrain B, Dangles O, Genot C, Dufour C. J Agric Food Chem. 2010;58:676–683.

Lorrain B, Dangles O, Loonis M, Armand M, Dufour C. J Agric Food Chem. 2012;60:9074–9081.

Marchesi JR, Adams DH, Fava F, Hermes GDA, Hirschfield GM, Hold G, Quraishi MN, Kinross J, Smidt H, Tuohy KM. Gut. 2016;65:330–339.

Mills CE, Tzounis X, Oruna-Concha M-J, Mottram DS, Gibson GR, Spencer JPE. Br J Nutr. 2015;113:1220–1227.

Monente C, Ludwig IA, Stalmach A, de Peña MP, Cid C, Crozier A. Int J Food Sci Nutr. 2015;66:657–664.

Monteiro M, Farah A, Perrone D, Trugo LC, Donangelo C. J Nutri. 2007;137:2196–2201.

Nakayama T, Oishi K. FEMS Microbiol Lett. 2013;343(2):161–168.

Nardini M, Cirillo E, Natella F, Scaccini C. J Agric Food Chem. 2002;50:5735–5741.

Narita Y, Inouye K. Chlorogenic acids from coffee. In Preedy VR (Ed.). Coffee in Health and Disease Prevention. Academic Press. 2015;189–199.
Nishitsuji K, Watanabe S, Xiao J, Nagatomo R, Ogawa H, Tsunematsu T, Umemoto H, Morimoto Y, Akatsu H, Inoue K, Tsuneyama K. Sci Rep. 2018;8:16173.
Olthof MR, Hollman PCH, Katan MB. J Nutr. 2001;131:66–71.
Ozdal T, Sela DA, Xiao J, Boyacioglu D, Chen F, Capanoglu E. Nutrients. 2016;8:78.
Parkar SG, Trower TM, Stevenson DE. Anaerobe. 2013;23:12–19.
Plumb GW, Garcia-Conesa MT, Kroon PA, Rhodes M, Ridley S, Williamson G. J Sci Food Agric. 1999;79:390–392
Poole R, Kennedy OJ, Roderick P, Fallowfield JA, Hayes PC, Parkes J. BMJ. 2017;359:j5024.
Rechner AR, Kuhnle G, Bremner P, Hubbard GP, Moore KP, Rice-Evans CA. Free Radic BiolMedic. 2002;33:220–235.
Rechner AR, Smith MA, Kuhnle G, Gibson GR, Debnam ES, Srai SKS, Moore KP, Rice-Evans CA. Free Radic BiolMedic. 2004;36:212–225.
Rodrigues NP, Bragagnolo N. J Food Compos Anal. 2013;32: 105–115.
Rohn S, Rawel HM, Kroll J. J Agric Food Chem. 2002;50:3566–3571.
Rohn S, Rawel HM, Wollenberger U, Kroll J. Food/Nahrung. 2003;47:325–329.
Rostagno MA, Celeghini RMS, Debien ICN, Nogueira GC, Meireles MAA. In Preedy VR (Ed.), Coffee in Health and Disease Prevention. Academic Press. 2015;137–142.
Rothschild D, Weissbrod O, Barkan E, Kurilshikov A, Korem T, Zeevi D, Costea PI, Godneva A, Kalka IN, Bar N, Shilo S, Lador D, Vila AV, Zmora N, Pevsner-Fischer M, Israeli D, Kosower N, Malka G, Wolf BC. Nature. 2018;555:210–215.
Round, Mazmanian. Nat Rev Immunol. 2009;9:313–323.
Sekirov I, Russell SL, Antunes LCM, Finlay BB. Physiol Rev. 2010;90:859–904.
Shimizu M. J Food Drug Anal 2017;25:93–99.
Shreiner AB, Kao JY, Young VB. Curr Opin Gastroenterol. 2015;31:69–75.
Singla RK, Dubey AK, Garg A, Sharma RK, Fiorino M, Ameen SM, Haddad MA, Al-Hiary M. Journal of AOAC International 2019;102:1397–1400.
Stalmach A, Mullen W, Barron D, Uchida K, Yokota T, Cavin C, Steiling H, Williamson G, Crozier A. Drug Metab Dispos. 2009; 37:1749 LP – 1758.
Stalmach A, Steiling H, Williamson G, Crozier A. Arch Biochem Biophys. 2010; 501:98–105.
Tajik N, Tajik M, Mack I, Enck P. Eur J Nutr. 2017; 56: 2215–2244.
Tsao R. Nutrients. 2010; 2:1231–1246.
Valduga AT, Gonçalves IL, Magri E, Delalibera Finzer JR. Food Res Int. 2019;120:478–503.
van Duynhoven J, Vaughan EE, Jacobs DM, Kemperman RA, van Velzen EJJ, Gross G, Roger LC, Possemiers S, Smilde AK, Doré J, Westerhuis JA, Van de Wiele T. Proc. Natl. Acad. Sci. 2011;108(Supplement 1):4531 LP – 4538.
Vilas-Boas AA, Oliveira A, Jesus D, Rodrigues C, Figueira C, Gomes A, Pintado M. Food Res Int. 2020; 134:109223.
Villanueva C, Kross RD. Int J Mol Sci. 2012;13:2091–2109.
Vrancken G, Gregory AC, Huys GRB, Faust K, Raes J. Nature Reviews Microbiology. 2019;17:754–763.
Williamson G, Clifford MN. Brit J Nutr. 2010;104:S48 – S66.
Williamson G, Clifford MN. Biochem Pharmacol. 2017; 139:24–39.
Zhang H, Tsao R. Curr Opini Food Sci. 2016;8:33–42.

12

The effects of contact time, brewing method, and bean roast on the chemistry of cold brew coffee

Niny Z. Rao, Frank H. Wilkinson, and Brian G. Yust

CONTENTS

12.1 Introduction

In the beginning, there was coffee. And the coffee was good. But exactly how good? Also, we drink a lot of coffee.

Worldwide coffee consumption has increased in recent years, driven by the ever-growing popularity and availability of this iconic beverage. From 2016–2020, global consumption grew by 7.5% to 166 million 60-kg bags annually, breaking records every year from 2016–2019, while production climbed by 8.5% to 175 million bags per year (Foreign Agricultural Service 2020). With coffee consumption in the United States and European Union making up 44% of global consumption per year, and with consumer spending on coffee, tea, and other beverages ballooning to $16.4 billion annually ("Real Personal Consumption Expenditures by Type of Product, Chained Dollars" n.d.), changes in coffee preparation and consumption habits can have a large impact on the economy.

While hot brewing remains the most popular method for making coffee in the United States, cold brew techniques have recently captured the attention of both the marketplace and the general public. Specifically, cold brew methods have grown in popularity among specialty coffee drinks, with 20% of American coffee drinkers now regularly or occasionally choosing to drink cold brew ("National Coffee Association USA > Industry Resources > Market Research > NCDT > NCDT Infographic" n.d.). According to Mintel, a reputable and long-standing market analysis company, US sales of cold brew coffee grew 115% from 2014 to 2015, reaching $7.9 million in sales ("US Cold Brew Coffee Sales Grow 115% from 2014–2015 | Mintel.com" n.d.). A beverage analyst at Mintel, Elizabeth Sisel, reported, "While cold brew represents a small portion of the overall category, our research indicates curiosity about trying a new style of coffee is driving current consumer demand" ("US Cold Brew Coffee Sales Grow 115% from 2014–2015 | Mintel.com" n.d.). American consumers are also purchasing cold brew coffee more frequently than iced coffee, another popular cold coffee beverage. Orders for cold brew coffee in the US grew from 19% less than those for iced coffee in 2016 to 42% more in 2018, according to the Specialty Coffee Association and millions of transaction records from Square, Inc., a mobile payment company used by many coffee vendors ("Square + ☕☕☕ = ♡" n.d.).

Large commercial vendors, such as Starbucks and Dunkin' Donuts, have capitalized on this growing consumer demand with an array of ready-to-drink cold brew coffee beverages. Advertisers argue that the colder, slower brewing process improves flavor, aroma, and the presence of bioactive compounds ("Website" n.d.). Starbucks marketing touts cold brew coffee as sweeter, smoother, and more full-bodied than conventionally brewed coffee ("Starbucks® Cold Brew Coffee" n.d.). Dunkin' Donuts claims that "cold brew is less acidic and naturally sweeter than regular coffee, so it can easily be consumed black" ("Cold-Brew | Dunkin' Donuts" n.d.).

Despite this marked increase in consumer and market interest, and despite the growing perception of cold brew's superior qualities, few scientific studies have been published on cold brew coffee. Beyond the difference in preparation method, we know little about how cold brew differs from its hot brewed counterpart. New research aims to better characterize the composition of cold brew coffee and investigate whether it differs significantly from that of hot brewed coffee.

DOI: 10.1201/9781003043133-15

12.2 Brewing temperature and brewing time

Coffee, hot and cold brewed, is prepared by an aqueous extraction process. In both types of brewing, various compounds diffuse out of the coffee grounds and into the water, imbuing it with the characteristic taste, aroma, and color of coffee. The composition of the finished product will depend on water variables such as the temperature, volume, and duration of contact (Cordoba et al. 2020). The primary differences between cold and hot brewing methods are water temperature and the duration of water-coffee contact. These may affect the kinetics of compound extraction from the coffee bean matrix.

Temperature often significantly influences the solubility of compounds in water, especially flavor compounds, so brewing temperature may result in significantly different chemical compositions between hot and cold brewed coffees. As cold brewing has increased in popularity, most reported methods are carried out at or near room temperature (20–25°C or colder) with the ground coffee being steeped for 8- to 24-hour periods. In contrast, the optimal hot brewing temperature is often reported as 93°C, with a much shorter contact time of a few minutes (Batali, Ristenpart, and Guinard 2020). Polar, soluble compounds, such as furans, ketones, acids, and sugar, are readily solubilized at lower temperature. Less polar compounds, often associated with bitterness, require higher temperatures to be extracted (Mestdagh, Glabasnia, and Giuliano 2017). The lower temperatures of cold brewing may limit the solubility of these less polar, bitter compounds. Additionally, the loss of volatile aromatic compounds due to evaporation is less prominent for cold brew than for high-temperature extraction. Limited studies on the sensory profile of cold brew coffee noted sweetness as its distinct characteristic (Cordoba et al. 2019; Angeloni et al. 2019).

The longer brewing times needed for cold brew coffee may affect its final composition if the diffusion of compounds across the grind matrix is a kinetically limiting phenomenon. This process of diffusion involves complex interactions between the solid bean–based matrix and the larger volume of water, and it can vary over time. It is known, for example, that faster chemical extraction occurs at the surface or near surface of the granules constituting the coffee grounds (Moroney et al. 2015; Moroney et al. 2017). The longer brewing times used in cold brew methods may therefore lead to extraction of chemicals from deeper inside the coffee bean matrix.

12.3 Roast level, grind size, and bean varietal

In addition to water temperature and time of contact, other important factors in the chemical, sensory, and flavor profiles of a finished coffee product include roasting temperature and duration, grind size, and water chemistry. Because bean varietal, growing region, and climatological growing conditions also affect the composition of the final beverage, single-origin coffee beans are preferred in comprehensive studies, including many of those cited here. The effects of these variables on the chemistry of cold brew coffee are not well studied, though numerous papers in this vein have been published on hot brewed coffee. Recent studies on cold brew coffee have investigated the relationship between coffee preparation parameters (namely, roast level, grind size, brew temperature, and brew time) and the chemical profile of the final beverage, specifically levels of pH, total titratable acidity (TTA), total dissolved solids (TDS), browned compounds, caffeine concentration, total caffeoylquinic acid (CQA) concentration, and total antioxidant capacity (TAC).

In considering how grind size may influence the extraction profile of cold brew coffee, much can be hypothesized from previous analyses of hot brewed coffee. For instance, mathematical modeling of a continuous flow system predicts that both grind size (coarse vs. fine) and bed shape (cylindrical vs. conical) will have significant effects on extraction uniformity and taste (Kevin M. Moroney et al. 2019). Coffee bed density, which arises from differential particle size, was found to play a critical role in bed permeability, with direct effects on extraction (Corrochano et al. 2015). Moreover, pore size and distribution in whole beans directly affect heat transfer and chemical changes during the roasting process, leading to differences in extractable compounds (Oliveros et al. 2017). Finally, bean varietal, climatic conditions of growing regions, and coffee processing each have their own significant contributions to flavor profiles (Seninde and Chambers 2020). It is anticipated that many of these parameters will also translate to differences in extraction and sensory profiles among cold brew coffee products.

12.4 Antioxidant levels and implications for health

Aside from sensory effects, health consequences may also arise from varying extraction profiles in coffee. While the connection between coffee consumption and human health has been explored for many years, studies have only recently reported a generally positive relationship. In a review of 22 early studies by Thelle et al., the authors concluded that associations between serum cholesterol and coffee consumption were inconclusive (Thelle, Heyden, and Fodor 1987). In one study, consumption of boiled coffee was found to increase plasma cholesterol, and this was attributed to diterpenes, cafestol, and kahweol (Ranheim and Halvorsen 2005), though it should be noted that diterpenes are usually removed in filtered coffee (Gross, Jaccaud, and Huggett 1997; Urgert et al. 1995). A large study that accounted for genetic factors found a U-shaped relationship between coffee consumption, cardiovascular disease, and mortality (Nordestgaard and Nordestgaard 2016). A similar study conducted on a different population found a J-shaped association between coffee consumption and risk for a first cardiovascular disease event (Kouli et al. 2018). Beyond cardiovascular effects, other studies have indicated that mortality (Malerba et al. 2013; Preedy 2014), occurrences of some cancers (Sang et al. 2013; Li et al. 2013; Je and Giovannucci 2012), diabetes mellitus (F. Shang, Li, and Jiang 2016; Chrysant 2017), and depression (L. Wang et al. 2016) show an inverse relationship with coffee consumption. Because these studies focused on consumption of hot brewed coffee, it is not yet evident whether the different extraction profile of cold brewed coffee would amplify or attenuate these health effects.

Other studies have characterized the extractable antioxidant activity in coffee. The difference in total antioxidant capacity between hot and cold brew coffees, though small, stands out as an important finding. Antioxidants can act to negate the harmful health effects of reactive oxygen species (ROS). ROS are reduced products of molecular oxygen, which appear to arise from incomplete reaction in the respiratory chain and NADPH oxidase enzymes (Brieger et al. 2012). High concentrations of ROS can cause chemical alterations to DNA, proteins, and lipids (Dröge 2002). Chemical alteration to DNA can lead to dysregulated gene expression and cancer (Wiseman and Halliwell 1996). To manage ROS, cells deploy a variety of defense mechanisms, including enzymes (superoxide dismutase and catalase) and antioxidant molecules. People may also acquire antioxidants from typical dietary sources, such as vitamin C, vitamin E, and myriad plant-derived molecules (Brieger et al. 2012). The paleolithic diet may have delivered significantly more antioxidants than modern diets (Benzie 2003).

Because photosynthetic organisms are exposed to high levels of oxygen, the defenses used by plant species to mitigate oxygen toxicity may be acquired by animal consumers to support their own physiological needs. Currently available evidence shows that processing variables can affect the extraction of antioxidant compounds from plants. Extraction of antioxidants from mengkudu (*Morinda citrifolia*) was found to be dependent on time and temperature (Thoo et al. 2010). Phenolics, but not flavonoids, showed increased extraction and antioxidant activity over time. However, extraction of both phenolics and flavonoids increased with temperature (Thoo et al. 2010). Shang et al. found that pressurized liquid extraction recovered more antioxidant activity from black bamboo than from reflux. The extracted antioxidants included phenolics and flavonoids and showed a 300% increase over low-temperature extraction (Y. F. Shang, Kim, and Um 2014).

Multiple studies have been conducted on the extraction of antioxidants specifically from coffee. Phenolics extracted from spent coffee grounds were found to increase with both time and temperature (Xu et al. 2015). Antioxidant activity, measured as ABTS ((2,2′-Azino-bis(3-ethylbenzo-thiazonile-6-sulfonic acid) diammonium salt) scavenging activity showed an increase with temperature but a biphasic curve with time (maximum at 45–50 minutes). DPPH (2,2-diphenyl-1-picrylhydrazyl) scavenging activity was largely constant with temperature but declined significantly with time (Xu et al. 2015). Limited studies comparing cold and hot brew coffee indicated that both antioxidant activities and total phenolic content were higher in hot brew coffee than in cold brew counterparts (Cordoba et al. 2019; Rao and Fuller 2018; Bilge 2020).

It should be noted that roast level can affect the availability of phenolics and extractable antioxidant activity. Total extractable antioxidant activity was highest in medium roast coffee, represented by a mixture of extracted phenolics and non-phenolics (Sacchetti et al. 2009). Coffee prepared from roasted beans showed an increased availability of non-phenolic antioxidant activity over coffee prepared from green beans. However, dark roast coffee showed a loss of antioxidant activity from degradation of phenolic compounds (Sacchetti et al. 2009). In short, the extraction of functional antioxidants does not necessarily correlate with total extractables at any temperature or time point but varies instead with roast level.

12.5 Role of brewing method

Cold brew coffee is typically prepared in one of two ways: the steeping method or the drip method (Figure 12.1). The steeping method is the simplest and most common way of making cold brew. The coffee grind is soaked in cold water (25°C and below) for a prolonged period of time, typically 16–24 hours, but sometimes as long as 48 hours (Cordoba et al. 2019; Megan Fuller and Rao 2017). The brewing process can take place either on the counter at room temperature or in a chilled environment such as a refrigerator. Once the brewing process is complete, the coffee extract is filtered and stored for future consumption. No specialized equipment is required for the steeping method, only a sufficiently large container and a filter to remove coffee grounds after brewing (Cordoba et al. 2020). As the popularity of cold brew coffee grows, manufacturers are producing specialized, inexpensive cold brew carafes that consist of a chamber for coffee grounds, a built-in filter, and a vessel for the brewed coffee, further simplifying the process for consumers (Neace and Maas 2010).

The drip coffee method is another brewing technique that has gained popularity for cold brew in recent years. In a drip coffee setup, cold water (often chilled with ice) is slowly added, drop by drop, to a bed of ground coffee beans. The water is pulled by gravity through first the grounds and then a filter. The filtered coffee extract is collected and stored for future consumption. Brewing time varies with flow rate of the water drip, which is adjustable (Angeloni et al. 2019). A typical drip coffee setup consists of three vertically-stacked chambers (Figure 12.1) and, depending on the complexity and size of the setup, a drip coffee apparatus can cost as much as $600 USD.

(a) (b)

FIGURE 12.1 (a) Schematic of drip (*left*) and (b) steeping (*right*) setups for cold brew coffee.

Previous studies on hot brewed coffee have demonstrated that brewing methods significantly impact the chemical attributes and flavor profile of the final product (Caporaso et al. 2014; Gloess et al. 2013; Ludwig et al. 2012; Pérez-Martínez et al. 2010; Sanchez and Chambers 2015). For cold brew coffee, the relationships among brewing methods, chemical composition, and sensory characteristics are less understood. Some recent results are summarized later.

Studies on cold brew coffee have found differences in caffeine concentration arising from brewing methods. As shown in Table 12.1, higher caffeine concentrations were measured in cold brew coffee prepared by drip rather than by steeping (Kim and Kim 2014; Angeloni et al. 2019) This observed difference was attributed to the dynamic nature of the drip method. In a dynamic extraction process, fresh solvent is continuously added to the system, creating a concentration gradient in the solvent. This concentration gradient largely drives the mass transfer process of soluble compounds from the solid matrix to the solvent (Perry, Green, and Maloney 1997). This concentration gradient is absent in a steeping system, since the total volume of solvent is in contact with the solid matrix for the entire extraction time. The drip method is therefore a more efficient extraction technique than the steeping method, especially for molecules with low solubility in cold water.

Unlike caffeine, other chemical species in cold brew coffee exhibited little variation across brewing methods. Chemical attributes such as 5-chlorogenic acid (5-CQA; Angeloni et al. 2019), acrylamide (Kang et al. 2020; Han, Boo, and Chung 2020), furan (Kang et al. 2020; Han, Boo, and Chung 2020), total phenolic content (TPC; Kang et al. 2020; Han, Boo, and Chung 2020), and antioxidant activity (AA; Kang et al. 2020; Han, Boo, and Chung 2020) were less sensitive to choice of brewing method (Table 12.2). Furan levels and antioxidant activities reported by Han et al. were slightly higher for the steeping method than the drip method (Han, Boo, and Chung 2020). A similar trend in acrylamide was observed by Kang et al. (Kang et al. 2020). More research is required to establish what effects, if any, these small chemical differences may have on the sensory profile of cold brew coffee.

Research on the sensory profile of cold brew coffees prepared by drip versus steeping has yielded conflicting results. Kim and Kim reported that the drip method produced a beverage that was preferred by a panel of evaluators, who rated the drip brew higher in all seven sensory characteristics: chocolatey, winey, sweetness, acidity, bitterness, body, and aftertaste (Kim and Kim 2014). In contrast, cold brew studies by Angeloni et al. and Han et al. reported that coffee prepared using the drip method scored lower in sweetness and higher in bitterness than coffee prepared by steeping (Angeloni et al. 2019; Han, Boo, and Chung 2020). Han et al. also reported that the steeping method was preferred over the drip method. The high intensity of bitterness in coffee produced by the drip method is correlated with its higher concentration of caffeine. It has been shown that the caffeine concentration of a brewed coffee can influence its perceived strength, bitterness, and body (Gloess et al. 2013; Macrae 1985). It is evident from analysis of these reported studies that additional work is needed to clarify the effects of drip and steeping methods on the chemical composition and sensory profile of cold brew coffee.

Finally, Angeloni et al. reported that the brewing methods do not affect the physical properties of cold brew coffee. These include refractive index, pH, density, viscosity, and total dissolved solids (Angeloni et al. 2019).

TABLE 12.1

Average Caffeine Concentration of Cold Brew Coffee as a Function of Extraction Methods

Drip	Steeping	Reference
1.03 ± 0.19 mg mL^{-1}	0.85 ± 0.15 mg mL^{-1}	Angeloni et al. 2019
3.46 ± 0.30 mg mL^{-1}	0.64 ± 0.08 mg mL^{-1}	Kim and Kim 2014

12.6 Role of extraction time

The chemical profile of brewed coffee depends heavily on the contact time between the coffee grounds and water. The extraction process for espresso, French press, and drip coffee are completed in a matter of seconds to minutes. Multiple studies have reported the extraction efficiencies of various compounds

TABLE 12.2

Select Chemical Attributes of Cold Brew Coffee Comparing Extraction Methods

Compound	Drip	Steeping	Reference
5-CQA	0.36 ± 0.07 mg mL^{-1}	0.29 ± 0.06 mg mL^{-1}	Angeloni et al. 2019
Furan	14.2 ± 0.4 ng mL^{-1}	13.9 ± 0.3 ng mL^{-1}	Kang et al. 2020
Furan	16.5 ± 0.4 ng mL^{-1}	17.5 ± 0.4 ng mL^{-1}	Han, Boo, and Chung 2020
Acrylamide	4.6 ± 0.4 ng mL^{-1}	5.2 ± 0.2 ng mL^{-1}	Kang et al. 2020
Acrylamide	4.6 ± 0.3 ng mL^{-1}	4.8 ± 0.3 ng mL^{-1}	Han, Boo, and Chung 2020
TPC[a,b]	10.76 ± 0.49 µmol mL^{-1} GAE	10.91 ± 0.48 µmol mL^{-1} GAE	Kang et al. 2020
TPC[a,c]	~10.2 µmol mL^{-1} GAE	~10.8 µmol mL^{-1} GAE	Han, Boo, and Chung 2020
TAC[d]	3747.36 ± 289.1 mg VcE/L	3817.61 ± 297.6 mg/VcE/L	Kang et al. 2020
TAC[d,c]	~1000 mg VcE/L	~1100 mg VcE/L	Han, Boo, and Chung 2020

Notes: [a] TPC values were reported as µmol gallic acid equivalent (GAE) per mL of coffee brew; [b] reported value were measured at 20°C by Kang *et al.* (Kang et al. 2020); [c] values estimated for extraction at 20°C for 12 hours from figures by Han *et al.* (Han, Boo, and Chung 2020); [d] TAC values were reported as mg vitamin C equivalent (VcE) per liter of coffee brew.

at high temperatures. Highly soluble compounds, such as caffeine, sugar, and organic acids, are extracted rapidly within seconds of brewing (Severini et al. 2015), while less soluble compounds, such as chlorogenic acid, lactones and other bitter compounds, are extracted after longer contact time or with a greater volume of water (Frank, Zehentbauer, and Hofmann 2006; Mestdagh, Glabasnia, and Giuliano 2017). At lower temperatures, the solubility of many compounds decreases significantly. Low-polarity compounds do not extract well at low temperatures and require longer contact time for complete extraction (Mestdagh, Glabasnia, and Giuliano 2017). Thus, the prolonged extraction time necessary for preparing cold brew coffee will likely have a large impact on the sensory profile of the final beverage.

Kim and Kim first reported that both caffeine and CQA concentrations in cold brew coffee varied with extraction time. For the drip method, caffeine and CQA concentrations were inversely correlated to extraction time, while for the steeping method they were directly correlated to extraction time. They also noted that the sensory profile of drip method cold brew coffee with an extraction time of 18 hours was preferable to coffees brewed with shorter extraction times (Kim and Kim 2014). Angeloni et al., on the other hand, did not find any significant correlation between a cold brew coffee's extraction time and its chemical or sensory profile (Angeloni et al. 2019).

Sensory studies conducted by Cordoba et al. evaluated cold brew coffees extracted for 14 and 22 hours. They reported that cold brew coffees brewed using a 14-hour extraction time received higher overall scores than other types (Cordoba et al. 2019). Han et al. found that samples brewed with 3 hours of extraction time were preferred. They also noted that the concentration of furan, one class of aromatic compounds present in coffee, decreased with increasing extraction time (Han, Boo, and Chung 2020). Such an observation is not surprising due to the volatile nature of the compound. Significant loss of volatile compounds is expected after 12–24 hours.

Both studies by Cordoba et al. and Han et al. remarked on the increased bitterness of coffees that were prepared with longer extraction times. This observation correlates with higher concentrations of caffeine (Gloess et al. 2013). The extraction kinetics of caffeine in cold brew coffee has been observed to be first order by Fuller and Rao (2017b; Figure 12.2). Caffeine concentration increased rapidly over the first 180 minutes of extraction and then gradually slowed until reaching equilibrium after 400 minutes. Previous models attributed the initial fast extraction of soluble, polar compounds to the extraction of compounds from the surface and near-surface volume of the solid coffee grind matrix (Moroney et al. 2015). After 180 minutes, the slower extraction of additional caffeine over longer time scales is likely due to mass transfer of the compound through intragrain pores into intergrain pores, and ultimately into the bulk liquid phase.

Mathematical models for the kinetics of caffeine extraction at room temperature found the diffusion of caffeine from the intragranular pore space, defined as the pore space within each grain of coffee, to the intergranular network, or the space between grains of coffee, to be the rate-limiting step in the extraction process (Spiro and Selwood 1984; Moroney et al. 2015; Moroney et al. 2017). As a result, the time required for the extraction process to reach equilibrium increases with grind size, even at high extraction temperatures. In the cold brew process, the extraction time frame is on the order of hours instead of seconds. Because such long extraction time scales allow for the slow diffusion from intragranular to intergranular pores, this process is not rate limiting in cold brew methods. Thus, caffeine concentrations tend to be higher in cold brew coffee prepared using longer extraction times (Fuller and Rao 2017b; Han, Boo, and Chung 2020) and smaller grind sizes (Cordoba et al. 2019), resulting in beverages with higher intensities of bitterness.

Antioxidant levels and total antioxidant capacities (TAC) of cold brew coffee have also been found to vary as a function of extraction time. The extraction of 5-CQA has been reported to

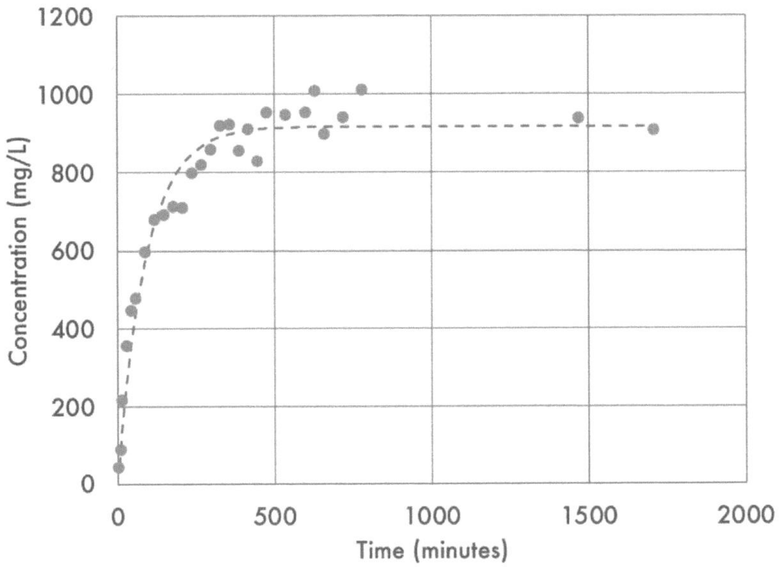

FIGURE 12.2 Caffeine concentration of cold brew coffee as a function of time.

Source: Fuller and Rao (2017a, unpublished data).

follow a first-order kinetic model, rapidly increasing over the initial 180 minutes and reaching near steady-state concentration after approximately 400 minutes (Fuller and Rao 2017b). Han et al. observed that both total phenolic compounds (TPC) and TAC increased with increased extraction time (Han, Boo, and Chung 2020). Overall, longer extraction times lead to cold brew coffees with higher levels of antioxidants (Han, Boo, and Chung 2020; Fuller and Rao 2017b). The sensory profile of cold brew coffee favors shorter extraction time to reduce bitterness, astringency, and aftertaste (Han, Boo, and Chung 2020; Cordoba et al. 2019).

12.7 Role of roasting temperature

The roast level of coffee beans is one of the most important factors influencing the chemistry and sensory properties of the final coffee beverage (Vignoli et al. 2014). The relationship between coffee chemistry and the roast level of coffee beans is well studied (Dias et al. 2018; Gloess et al. 2014; Masi et al. 2013; Petisca et al. 2013; Schenker and Rothgeb 2017; Schenker et al. 2000; Clarke and Macrae 1985). The roasting process produces a profile of gustatory and fragrant compounds, contributing to the overall flavor and aroma of the coffee.

During roasting, coffee beans are heated via controlled air flow to a minimum internal temperature of 190°C for at least 3 minutes (Schenker and Rothgeb 2017). As heating continues, CO_2 gas and volatile organic compounds (VOCs) are released as a result of degradation to the cellular tissue of the beans (Andueza, De Peña, and Cid 2003; Schenker and Rothgeb 2017). As much as 18% of coffee bean mass may be lost during the roasting process (Clarke and Macrae 1985). The high temperatures also initiate the synthesis of new compounds from precursor compounds found in green coffee beans, as well as pyrolysis of some organic material. When the internal bean temperature rises to above 180°C, exothermic reactions occur in the polysaccharides, proteins, and phenolic acids to produce compounds that contribute to the distinct flavor and aroma of coffee. These compounds include furans, pyrroles, pyridines, pyrazines, and ketones (Petisca et al. 2013).

At the same time, chlorogenic acids (CQAs) are lost during roasting due to isomerization and degradation (Clifford 1985; Hucke and Maier 1985; Schrader et al. 1996; Bennat et al. 1994; Trugo and Macrae 1984; Farah et al. 2005). Some of the CQAs are partially incorporated into the production of melanoidins (Bekedam et al. 2008; Nunes and Coimbra 2010), which give coffee its brown color and contribute to its overall antioxidant activity (Wang, Qian, and Yao 2011; Liu and Kitts 2011). Beyond phenolic acid changes, the quantities of other important acidic compounds also vary as a function of roast level. Small organic acids present in green coffee beans, such as malic and citric acids, are decomposed during the roasting process (Degenhardt et al. 2007), while concentrations of other acids, including formic, lactic, glycolic, and acetic acids, increase as a result of carbohydrate degradation (Ginz et al. 2000). The roasting process also degrades the coffee cellular matrix, which may cause chemicals such as carbohydrates, denatured proteins, and celluloses to be solubilized (Bicho et al. 2011; Mendonça and Pereira 2005).

In addition to chemical changes, coffee beans also undergo physical alterations during roasting. These physical changes contribute to chemical transformations in the beans by governing heat transfer and water loss. Differential rates of heat transfer and water loss lead to kinetic differences in the chemical transformations that alter the extractable compounds in the final product (Schenker et al. 2000; Uman et al. 2016). The application of heat modifies the solid matrix of the coffee bean while reducing its density. As the cellular-level organization of the bean is altered by changes in the matrix, oils are liberated to form droplets or a sheen on the bean's surface (Schenker et al. 2000). In general, as roast level increases, the porosity and brittleness of the roasted beans increase while their moisture content decreases. These physical changes affect the grinding of roasted beans, which in turn impacts the final quality of the brewed coffee (Schenker and Rothgeb 2017).

The grinding process breaks down coffee beans even further into small particles. The particle size of the coffee grounds controls chemical extraction rates (Spiro and Selwood 1984) and, ultimately, the final quality of the brewed coffee. Both the moisture content (Baggenstoss, Perren, and Escher 2008) and the degree of roasting (Andueza, De Peña, and Cid 2003; Lingle 2011) affect the particle size of the coffee grounds. While the large particle size of a coarse grind increases extraction water velocity and reduces extraction efficiency (Cappuccio and Liverani 1999), the smaller particle size of a fine grind increases exposure time to extraction water, risking a final beverage that suffers from over-extraction (Baggenstoss, Perren, and Escher 2008). Roasting can also produce undesirable effects, such as the oxidation of polyunsaturated fatty acids (Budryn et al. 2012). It is evident from these studies that roast level and grind size are critical variables in determining the availability of sensory compounds in the final product.

Although the effects of roast level on hot coffee beverages have been studied extensively, little research exists on how roast level affects the chemical and sensory characteristics of cold brew coffee products. However, recent studies are beginning to address these questions. In an analysis of coffee acidity, Fuller and Rao first noted that medium and dark roast Kona coffee had comparable pH values (Table 12.3; Fuller and Rao 2017b). Later studies by Bilge (2020) and by Rao et al. (Rao et al. 2020) indicated that the pH of cold brew coffee generally increases with roast level, regardless of bean origin, consistent with previous hot brew data reported by Moon et al. (Moon et al. 2009). This reduction in acidity is likely due to the soluble protonated acidic compounds decomposing during pyrolysis or being consumed in synthesis reactions during the roasting process (Ginz et al. 2000). Rao et al. (2020) further noted that variations in pH were more sensitive to roast level than to brewing temperature and should be considered as a criteria when crafting low-acidity beverages. Additional work will be needed to resolve the empirical inconsistencies reported thus far.

Coffee, like many food products, has visual aspects that must be considered in tandem with its aromatic and gustatory profile. Certainly, visual appeal has long been central to advertising food and beverages, including coffee (Mohebbi 2014). For example, color has been found to influence consumers' flavor perception (Spence et al. 2010). Research at the neurological

TABLE 12.3

Average pH of Cold Brew Coffee of Various Roast Levels and Bean Origin

Roast	Bean origin	Average reported pH	Reference
Medium	Kona	5.47 ± 0.02	Fuller and Rao 2017b
Dark	Kona	5.47 ± 0.02	Fuller and Rao 2017b
Light	Brazil	5.7 ± 0.04	Bilge 2020
Medium	Brazil	5.49 ± 0.06	Bilge 2020
Dark	Brazil	5.78 ± 0.04	Bilge 2020
Light	Colombia	5.12 ± 0.04	Bilge 2020
Medium	Colombia	5.2 ± 0.07	Bilge 2020
Dark	Colombia	5.64 ± 0.07	Bilge 2020
Light	Peru	5.24 ± 0.05	Bilge 2020
Medium	Peru	5.28 ± 0.04	Bilge 2020
Dark	Peru	5.75 ± 0.03	Bilge 2020
Light	Colombia	5.00 ± 0.08	Rao et al. 2020
Medium	Colombia	5.30 ± 0.01	Rao et al. 2020
Dark	Colombia	5.75 ± 0.02	Rao et al. 2020

level is beginning to determine specific brain regions activated by food imagery and potential connections to hunger and sensory perception areas (van der Laan et al. 2011). Anecdotally, one author reported consuming more coffee than usual while writing this chapter (FW, personal communication). One recent study has evaluated the taste response of participants sampling coffee with and without artistic decorations in latté foam. Drinkers reported significant changes in the perceived acidity and sweetness, but not bitterness, between decorated and undecorated lattés (Hsu and Chen 2021).

The characteristic brown color of coffee is, in part, the result of melanoidin production during the roasting process (Morales 2005; Nunes and Coimbra 2007). A class of high molecular weight compounds produced by the Maillard reaction, many melanoidins are not water soluble at low temperature. In addition to color, melanoidins are also known to contribute to a coffee's flavor and antioxidant capacity (Delgado-Andrade, Rufián-Henares, and Morales 2005; Bartel, Mesias, and Morales 2015; Liu and Kitts 2011; Smrke et al. 2013; Budryn et al. 2009; Cämmerer and Kroh 2006; Borrelli et al. 2002). Rao et al. found that higher concentrations of browned compounds in cold brew coffee, characterized by absorbance at 420 nm, were associated with higher degrees of roast (Table 12.4; Rao et al. 2020), consistent with previous studies (Liang et al. 2016; Ludwig et al. 2013; Budryn et al. 2009; Castillo, Ames, and Gordon 2002). Further research is needed to investigate the visual dimensions of cold brew coffee and their effects on consumer preference and sensory perception.

Chlorogenic acids (CQA) are abundant in coffee and contribute significantly to the flavor profile (Sunarharum, Williams, and Smyth 2014). These acids can contribute bitter and astringent flavors (Buffo and Cardelli-Freire 2004; Sunarharum, Williams, and Smyth 2014), but they also function as antioxidants (Clifford 1999, 2000; Oestreich-Janzen 2010) as described elsewhere in this chapter. Accordingly, factors that affect the composition and extractability of these compounds will have direct effects on a coffee's flavor and health implications. Roasting decreases the amount of CQA in the grind through chemical transformation (Trugo and Macrae 1984; Moon, Yoo, and Shibamoto 2009). Using mass spectroscopy, CQA were found to follow among four reaction schemes, including acyl migration, dehydration, epimerization, and lactonization (Jaiswal et al. 2012). Therefore, roast level is expected to affect CQA concentration in cold brew coffee. Fuller and Rao first reported that the 5-CQA concentration in Kona coffee was higher at medium roast than dark roast (Table 12.3). The average 5-CQA concentration was reported to be 485 ± 47 mg L^{-1} and 355 ± 51 mg L^{-1} for medium roast and dark roast, respectively, after brewing for 400 minutes (Fuller and Rao 2017b), consistent with previous studies (Clifford 1985; Hucke and Maier 1985; Schrader et al. 1996; Bennat et al. 1994; Trugo and Macrae 1984; Farah et al. 2005). This finding was further supported by a later study by the same authors. They found that 5-CQA concentration in Colombian coffee generally decreased as degree of roast increased: 757 ± 27 mg L^{-1}, 353 ± 15 mg L^{-1}, and 147 ± 14 mg L^{-1} for light, medium, and dark roasts, respectively, after 420 minutes of brewing (Rao et al. 2020). Similar trends were also observed for total CQA concentration (Rao et al. 2020). In addition to the flavor effects, the extracted CQA will affect the antioxidant activity.

CQA and their molecular products, both of which are produced during roasting, contribute to the antioxidant activity in coffee. The total antioxidant capacity (TAC) of cold brew coffee as a function of roast level has been measured using various methods. Rao et al., using ABTS radical cation decolorization assay, reported that TAC decreased as the degree

TABLE 12.4

Select Chemical Attributes of Cold Brew Coffee of Various Roast Levels and Bean Origins

Roasted samples	Bean origin	Browned compounds (Abs_{420})	Caffeine (mg L^{-1})	Total CQA (mg L^{-1})	5-CQA (mg L^{-1})	TAC (mmol Trolox/L coffee)	Reference
Light	Colombia	0.358 ± 0.004	1114 ± 56	1535 ± 28	757 ± 27	13.09 ± 0.22	Rao et al. 2020
Medium	Colombia	0.386 ± 0.006	1036 ± 19	733 ± 11	353 ± 15	11.11 ± 0.33	Rao et al. 2020
Dark	Colombia	0.422 ± 0.015	962 ± 41	308 ± 10	147 ± 14	10.13 ± 0.59	Rao et al. 2020
Medium[a]	Kona		1095 ± 55		485 ± 47		Fuller and Rao 2017b
Dark[a]	Kona		950 ± 51		355 ± 51		Fuller and Rao 2017b

Note: [a] values corresponding to 400 minutes extraction time are reported here.

of roast increased. Similar trends in cold brew coffee were also found by Bilge using the 2,2-diphenyl-1-picrylhydrazyl (DPPH) decolorization assay (Bilge 2020). However, Bilge noted that total phenolic content (TPC), another group of antioxidants, did not correlate well with degree of roast. This result was unsurprising, as the relationship between TAC and roast level is complex, and previous studies on hot brew coffee have yielded inconclusive results (Sacchetti et al. 2009; Vignoli et al. 2014; J. A. Vignoli, Bassoli, and Benassi 2011; Catelani et al. 2017). Discrepancies among various TAC quantification methods used for hot brew coffee have also been documented (Herawati et al. 2019; Liang et al. 2016; Opitz et al. 2014; Vignoli et al. 2014; Cämmerer and Kroh 2006; Delgado-Andrade, Rufián-Henares, and Morales 2005; Borrelli et al. 2002). The antioxidant activities of both hot and cold brew coffee can be influenced by factors such as bean origin (Kamiyama et al. 2015), coffee varietal (Hečimović et al. 2011), roasting procedure (Opitz et al. 2014), and extraction methods (Ludwig et al. 2012; Pérez-Martínez et al. 2010), making it difficult to directly compare TAC results from various studies (Catelani et al. 2017).

The effect of roast level on caffeine concentrations in cold brew coffee is very understudied. Available data show that caffeine concentration decreases slightly as the degree of roast increases (Fuller and Rao 2017b; Rao et al. 2020). However, Fuller and Rao also noted that changes in caffeine concentration are less sensitive to roast level than to extraction time. They reported that after 12 hours, both medium and dark roast Kona coffee had comparable levels of caffeine concentration (Fuller and Rao 2017b).

12.8 Conclusion

In light of its popularity, the growth of the cold brew coffee market is expected to continue. However, research on chemical and sensory profiles of cold brew coffee *is* still lacking. When comparing steeping and drip methods of brewing, the drip method is known to be more efficient at extracting chemical compounds from the bean matrix, with marked differences in concentration of caffeine as well as those compounds with low solubility in water. Other chemical characteristics, including chlorogenic acid concentration, total phenolic content, and antioxidant activity showed little to no variation between the two brewing methods. Higher concentrations of caffeine correlated with perceptions of increased bitterness, but other sensory characteristics and preferences of cold brew drinkers need to be further explored.

Rates of chemical extraction from ground coffee beans are primarily determined by the particle size of the grounds. Because many chemicals become less soluble as water, cold water extraction of these compounds tends to take longer than in hot brewing techniques, necessitating longer contact times for cold brewing methods. Caffeine extraction can persist during cold brewing for up to 24 hours, occurring most rapidly for the first 180 minutes before becoming more gradual. Concentrations of volatile organic compounds, such as furan, decreased with longer brewing times – an unsurprising result, since the volatile nature of such chemicals causes them to decompose naturally over time. Many VOCs are known to contribute to the sensory characteristics and aroma of coffee, while caffeine tends to amplify its bitterness. However, the relationship between sensory experience and length of brewing time remains inconclusive.

As in hot brewing, how coffee beans are roasted and prepared can have a large effect on the chemical and sensory characteristics of cold brew coffee. Once coffee beans reach an internal temperature above 180°C during the roasting process, exothermic reactions convert polysaccharides, proteins, and phenolic acids into VOCs and chlorogenic acids into melanoidins. Carbohydrates also degrade into other acids. Like VOCs, melanoidins contribute to the distinctive flavor and aroma of coffee, and they additionally impart to coffee its brown color.

It has been shown that coffee beans generally become more porous, dry, and brittle at higher levels of roast. Higher roast levels also tend to lead to lower the pH of the final coffee product, regardless of bean origin. Variations in cold brew coffee pH were observed to be more sensitive to roast level than to brewing temperature, and ought to be considered when lower acidity is desired. As chlorogenic acids have been shown to act as antioxidants, the reduction in CQA concentration due to degree of roast should be considered when weighing the flavor profile and potential health benefits bestowed to the coffee drinker. Additionally, some studies have shown that the total antioxidant capacity of cold brew coffee tends to decrease as roast level increases. However, a complete understanding regarding antioxidants in cold brew coffee does not appear to be settled, as total phenolic content, another group of antioxidants, has not been shown to correlate to degree of roast, and discrepancies in previous attempts to quantify TAC have been reported in the literature.

Recent studies of the chemical and sensory characteristics of cold brew coffee have advanced our understanding of cold brew extraction kinetics, the physical significance of grind size, and the chemical transformations that occur during coffee bean roasting. However, there are still many aspects of the final coffee product for which the contributing factors and variables are not completely understood. A small number of studies have compared cold brewing methods or investigated the effects of roast level in cold brew coffee, but many questions regarding the role of bean varietal, growing region, and climatological conditions still remain. Additionally, differences in the sensory, chemical, and flavor profiles of cold brew coffees remain understudied. Meanwhile, interest in the potential health benefits of coffee have continued to grow in both the scientific literature and mainstream media. Yet, nearly all studies completed to date have either focused on hot brewed coffee or not differentiated among coffee preparation techniques. Given that the studies cited herein demonstrate measurable differences in the characteristics of the final coffee product based on the brewing temperature, particularly total antioxidant activity, many fruitful avenues remain for future research on dietary and medical effects.

REFERENCES

Andueza, S et al. *J Agri Food Chem* 2003; 51: 7034–7039.
Angeloni, G et al. *J Agri Food Chem* 2019; 99: 391–399.

Baggenstoss, J et al. *European Food Research and Technology = Zeitschrift Fur Lebensmittel-Untersuchung Und -Forschung. A* 2008, 227 (5): 1357–1365.

Bartel, C et al. *Food Research International* 2015, 67: 356–365.

Batali, M et al. *Scientific Reports* 2020; 10: 16450.

Bekedam, EK et al. *J Agri Food Chem*, 2008; 6: 7138–7145.

Bennat, C et al. *Zeitschrift Für Lebensmittel-Untersuchung Und Forschung* 1994; 199: 17–21.

Benzie, IFF. *Comparative Biochemistry and Physiology. Part A, Molecular & Integrative Physiology* 2003; 136 (1): 113–126.

Bicho, NC et al. *European Food Research and Technology = Zeitschrift Fur Lebensmittel-Untersuchung Und -Forschung. A* 2011, 233 (2): 303–311.

Bilge, G. Journal of Food Science and Technology, 2020; 1–10.

Borrelli, RC et al. *Journal of Agricultural and Food Chemistry* 2002; 50 (22): 6527–6533.

Brieger, K et al. *Swiss Medical Weekly* 2012; 142: w13659.

Budryn, G et al. *European Food Research and Technology = Zeitschrift Fur Lebensmittel-Untersuchung Und -Forschung. A* 2009; 228: 913–922.

Budryn, G et al. *European Journal of Lipid Science and Technology: EJLST* 2012; 114: 1052–1061.

Buffo, RA and Cardelli-Freire, C. *Flavour and Fragrance Journal*. 2004, John Wiley & Sons, London: Library.

Cämmerer, B and Kroh, L. *European Food Research and Technology = Zeitschrift Fur Lebensmittel-Untersuchung Und -Forschung*. A 2006; 223 (4): 469–474.

Caporaso, N et al. *Food Research International* 2014; 61: 152–160.

Cappuccio, R and Liverani, F. In *Proceedings of the 18th International Scientific Colloquium on Coffee*. Association Scientifique Internationale du Café. 1999. www.sidalc.net/cgi-bin/wxis.exe/?IsisScript=CAFE.xis&method=post&formato=2&cantidad=1&expresion=mfn=025555.

Castillo, MD et al. *Journal of Agricultural and Food Chemistry* 2002; 50 (13): 3698–3703.

Catelani, TA et al. *Food and Bioprocess Technology* 2017; 10: 630–638.

Chrysant, SG. *Expert Review of Cardiovascular Therapy* 2017; 15: 151–156.

Clarke RJ and Macrae R, eds. *Coffee: Volume 1: Chemistry*. 1985, Elsevier Science Publishers, LTD, Essex, England.

Clifford MN. "Chlorogenic Acids." In *Coffee: Volume 1: Chemistry*, edited by RJ Clarke and R Macrae, 153–202. 1985, Essex, England: Elsevier Science Publishers, LTD.

Clifford MN. *Journal of the Science of Food and Agriculture* 1999; 79: 362–372.

Clifford MN. *Journal of the Science of Food and Agriculture* 2000; 80: 1033–1043.

"Cold-Brew | Dunkin' Donuts." n.d. Accessed July 26, 2017. www.dunkindonuts.com/en/food-drinks/iced-drinks/cold-brew.

Cordoba, N et al. *Scientific Reports* 2019; 9: 8440.

Cordoba, N et al. *Trends in Food Science & Technology* 2020; 96: 45–60.

Corrochano, BR et al. *Journal of Food Engineering* 2015; 150: 106–116.

Degenhardt, A et al. In *21st International Conference on Coffee Science, Montpellier, France, 11-15 September, 2006*, pp. 384–387. Association Scientifique Internationale du Café (ASIC), 2007.

Delgado-Andrade, C et al. *Journal of Agricultural and Food Chemistry* 2005; 53: 7832–7836.

Dias, RCE et al. *Food Chemistry* 2018; 255: 132–138.

Dröge W. 2002. *Physiological Reviews* 2002; 82: 47–95.

Farah, A et al. *Journal of Agricultural and Food Chemistry* 2005; 53: 1505–1513.

Foreign Agricultural Service. 2020. "Coffee: World Markets and Trade." United States Department of Agriculture. https://usda.library.cornell.edu/concern/publications/m900nt40f?locale=en.

Frank, O et al. *European Food Research and Technology*. 2006. https://doi.org/10.1007/s00217-005-0143-6.

Fuller, M and Rao, NZ. 2017a. "Unpublished Data."

Fuller, M and Rao, NZ. *Scientific Reports* 2017b; 7: 17979.

Ginz, M et al. European Food Research and Technology 2000; 211, no. 6: 404–410.

Gloess, AN et al. *European Food Research and Technology = Zeitschrift Fur Lebensmittel-Untersuchung Und -Forschung. A* 2013; 236 (4): 607–627.

Gloess, AN et al. *International Journal of Mass Spectrometry* 2014; 365–366: 324–337.

Gross, G, Jaccaud E, and Huggett AC. Food and Chemical Toxicology: *An International Journal Published for the British Industrial Biological Research Association* 1997; 35 (6): 547–554.

Han JW, Boo H, and Chung MS. *Food Science and Biotechnology* 2020; 29 (8): 1071–1080.

Hečimović, I et al. *Food Chemistry* 2011; 129: 991–1000.

Herawati, D et al. *Food Science and Biotechnology* 2019; 28: 7–14.

Hsu L, and Chen YJ. *British Food Journal* 2021; 123 (5): 1931–1946.

Hucke J and Maier HG. 1985. *Zeitschrift Für Lebensmittel-Untersuchung Und -Forschung*. https://doi.org/10.1007/bf02044298.

Jaiswal, R et al. *Food & Function* 2012; 3: 976–984.

Je Y and Giovannucci E. *International Journal of Cancer. Journal International Du Cancer* 2012; 131: 1700–1710.

Kamiyama, M et al. *Journal of Agricultural and Food Chemistry* 2015; 63: 1996–2005.

Kang, DE et al. *Food Science and Biotechnology* 2020; 29: 141–148.

Kim AR and Kim JS. *Journal of The Korea Society for Coffee Industry* 2014; 3: 1–9.

Kouli, GM et al. *European Journal of Nutrition* 2018; 57: 1677–1685.

Laan, LN et al. *NeuroImage* 2011; 55: 296–303.

Li, G et al. *Public Health Nutrition* 2013; 16: 346–357.

Liang, N et al. *Food Chemistry* 2016; 213: 251–259.

Lingle TR. *The coffee brewing handbook: a systematic guide to coffee preparation*. Santa Ana: Specialty Coffee Association of America, 2011.

Liu Y, Kitts DD. *Food Research International* 2011; 44: 2418–2424.

Ludwig, I et al. *Food Research International* 2012; 48: 57–64.

Ludwig, I et al. *LWT – Food Science and Technology* 2013; 51: 553–559.

Macrae R. In *Coffee – Volume 1 Chemistry*, edited by RJ Clarke and R Macrae, 115–152. 1985, Elsevier Science Publishers, LTD, Essex, England.

Malerba, S et al. *European Journal of Epidemiology* 2013; 28: 527–539.

Masi, C et al. *Journal of Food Science* 2013; 78: S1290–1300.

Mendonça, LMVL and Pereira, RGFA. 2005. *Food Science and Technology* 2005; 25: 239–243.

Mestdagh, F et al. In *The Craft and Science of Coffee*, 2017; 355–380. Elsevier.

Mohebbi, B. *International Journal of Organizational Leadership* 2014; 3: 92–102.

Moon, JK et al. *Journal of Agricultural and Food Chemistry* 2009; 57: 5365–5369.

Morales F. *Analytica Chimica Acta*, 2005; 534 (1): 171–176.

Moroney, KM et al. *Chemical Engineering Science* 2015; 137: 216–234.

Moroney, KM et al. *Journal of Mathematics in Industry* 2017; 7: 3.

Moroney, KM et al. *PloS One* 2019; 14: e0219906.

"National Coffee Association USA > Industry Resources > Market Research > NCDT > NCDT Infographic." n.d. Accessed January 25, 2021. www.ncausa.org/Industry-Resources/Market-Research/NCDT/NCDT-Infographic.

Neace, RJ and Maas, TH. USPTO 7858133. *US Patent*, filed April 18, 2006, and issued December 28, 2010. https://patentimages.storage.googleapis.com/9c/d3/5d/c91f0eb38e2cf3/US7858133.pdf.

Nordestgaard, AT and Nordestgaard, BG. *International Journal of Epidemiology* 2016; 45: 1938–1952.

Nunes, FM and Coimbra, MA. *Journal of Agricultural and Food Chemistry* 2007; 55: 3967–3977.

Nunes, FM and Coimbra, MA. In *Comprehensive Natural Products II*, edited by H-W (ben) Liu and L Mander, 2010; 1085–1117. Oxford: Elsevier.

Oestreich-Janzen, S. In *Comprehensive Natural Products II*, edited by H-W (ben) Liu and L Mander, 2010; 1085–1117. Oxford: Elsevier.

Oliveros, N et al. *Journal of Food Engineering* 2017; 199: 100–112.

Opitz, S et al. *Foods (Basel, Switzerland)* 2014; 3: 586–604.

Pérez-Martínez, M et al. *Journal of Agricultural and Food Chemistry* 2010; 58: 2958–2965.

Perry, RH et al. *Perry's Handbook of Chemical Engineering*. Seventh, International edition, New York, McGraw-Hill 1997.

Petisca, C et al. *Food and Bioproducts Processing* 2013; 91: 233–241.

Preedy VR. *Coffee in Health and Disease Prevention* 2014. Academic Press.

Ranheim, T and Halvorsen B. *Molecular Nutrition & Food Research* 2005; 49: 274–284.

Rao, NZ and Fuller, M. *Scientific Reports* 2018; 8: 16030.

Rao, NZ et al. *Foods* 2020; 9: 902.

"Real Personal Consumption Expenditures by Type of Product, Chained Dollars." n.d. Bureau of Economic Analysis. Accessed January 25, 2021. https://apps.bea.gov/iTable/iTable.cfm?reqid=19&step=3&isuri=1&1910=x&0=-99&1921=underlying&1903=2018&1904=2002&1905=2017&1906=a&1911=0.

Sacchetti, G et al. *Journal of Food Engineering* 2009; 90: 74–80.

Sanchez, K and Chambers, E. *Journal of Sensory Studies* 2015; 30: 499–511.

Sang, LX et al. *BMC Gastroenterology* 2013, 13: 34.

Schenker S and Rothgeb T. In *The Craft and Science of Coffee*. 2017; 245–271. Elsevier.

Schrader, K et al. *Journal of the Science of Food and Agriculture* 1996; 71: 392–398.

Schenker, S et al. *Journal of Food Science* 2000; 65: 452–457.

Seninde, DR and Chambers, E. *Beverages* 2020; 6 (3): 44.

Severini, C et al. *Journal of Agricultural and Food Chemistry* 2015; 63: 2321–2327.

Shang, F et al. *Food Chemistry* 2014; 154: 164–170.

Shang, YF et al. *Diabetes & Metabolism* 2016; 42: 80–87.

Smrke, S et al. *Food & Function* 2013; 4: 1082–1092.

Spence, C et al. *Chemosensory Perception* 2010; 3: 68–84.

Spiro, M and Selwood, R. *Journal of the Science of Food and Agriculture* 1984; 35: 915–924.

"Square + ☕☕☕ = ♡." n.d. Accessed January 25, 2021. https://squareup.com/us/en/townsquare/coffee-day-2018.

"Starbucks® Cold Brew Coffee." n.d. Accessed July 26, 2017. http://lp.starbucks.com/coldbrew.

Sunarharum, WB et al. *Food Research International* 2014; 62: 315–325.

Thelle, DS et al. *Atherosclerosis* 1987; 67: 97–103.

Thoo, YY et al. *Food Chemistry* 2010; 120: 290–295.

Trugo, LC and Macrae, R. *Food Chemistry* 1984; 15: 219–227.

Uman, E et al." *Scientific Reports;* 2016; 6: 24483.

Urgert, R et al. *Journal of Agricultural and Food Chemistry* 1995; 43: 2167–2172.

"US Cold Brew Coffee Sales Grow 115% from 2014–2015 | Mintel.com." n.d. Accessed May 8, 2017. www.mintel.com/press-centre/food-and-drink/us-cold-brew-coffee-sales-grow-115-from-2014-2015.

Vignoli, JA et al. *Food Chemistry* 2011; 124: 863–868.

Vignoli, JA et al. *Food Research International* 2014; 61: 279–285.

Wang, HY et al. *Food Chemistry*. 2011. https://doi.org/10.1016/j.foodchem.2011.03.075.

Wang, L et al. *The Australian and New Zealand Journal of Psychiatry* 2016; 50: 228–242.

"Website." n.d. Starbucks Cold Brew Coffee. Accessed May 8, 2017. http://lp.starbucks.com/coldbrew.

Wiseman H and Halliwell B. *Biochemical Journal* 1996; 313 (Pt 1): 17–29.

Xu, H et al. *Industrial Crops and Products* 2015; 76: 946–954.

Section IV

The intervention of biotechnology for improvement of *Coffea*

13

Latest developments on coffee discrimination by species, quality, and geographic origin

Adriana S. Franca and Leandro S. Oliveira

CONTENTS

13.1 Introduction

Coffee is one of the most consumed hot beverages worldwide. The beverage is prepared from roasted and ground coffee beans, which, in their raw state, are harvested from coffee plants all around the world. Raw (commonly termed "green") coffee beans are one of the most traded commodities in international markets, with *Coffea arabica* and *Coffea canephora* (or simply Arabica and Robusta, respectively) as the major commercial species. There are significant trait differences between these two species, with Arabica being considered of higher quality than Robusta and, thus, being commercialized at higher prices than the latter. Apart from the inherent differences between these species, the quality of coffee as a final product of consumption is determined by a number of factors, including edaphoclimatic conditions in which the plants are cultivated, the types of harvesting and of post-harvesting processing the beans are subjected to, and the roasting conditions of the beans. Ultimately, the preparation method affects the quality of the beverage.

Given that the quality of the final product of consumption (i.e., roasted and ground coffee) is significantly affected by these many factors, rigorous quality control in the coffee business should be provided to consumers in ways to ensure coffee traceability and authenticity (e.g., species, provenance, defects, adulteration). In order to provide this type of assurance, rigorous analytical methodologies should be developed and duly employed for discrimination of coffees by species and geographical origin, and for the verification of the presence of defects and of adulterants. The importance of developing analytical methodologies for such tasks was first emphasized in the early 1990s in the works by Kemsley et al. (1995) and Briandet et al. (1996). However, publications on these topics really gained momentum midway through the first decade of the twenty-first century. Thus, the majority of works in this area were published between 2005 and 2020, as the need for rigorous quality control in the coffee business has intensified due to both consumers' and businesses' increased awareness of and interest in product quality and authenticity. These recently published works constitute the focus of this chapter. However, this review does not intend to be exhaustive of the literature, but rather comprehensive and representative of the diversity of methodologies developed for the task at hand.

13.2 Discrimination by species and variety

Coffee belongs to the genus *Coffea* of the Rubiaceae family, with hundreds of catalogued species (Charrier and Berthaud, 1985). However, the most widely cultivated and commercially relevant species are *Coffea arabica* and *Coffea canephora*, representing approximately two-thirds and one-third of the world coffee production, respectively (Franca and Oliveira, 2019).

Coffee beans, the main product of commercial interest from coffee plants, represent 50%–55% of the dry matter of the fruit and vary in size, shape, and density according to the species (genotype) and agricultural practices. *C. arabica* beans are deep green to bluish-green and shaped like a half triaxial ellipsoid with a longitudinal sigmoidal slit on their flat sides, whereas *C. canephora* beans are brownish-yellow, rounder, with straight central slits and are usually smaller (Figure 13.1). Hence, Arabicas and Robustas, in their whole raw state, can be easily differentiated solely by their physical traits. However, after roasting and grinding, differentiation between the two species by physical traits is an impossible task.

DOI: 10.1201/9781003043133-17

(a) (b)

FIGURE 13.1 (a) Arabica and (b) Robusta coffee beans.

Regarding the organoleptic characteristics of final product of consumption, Arabica is considered to produce a milder, fruitier, and more acidic beverage than Robusta, which has been characterized as weak flavored and with an occasional pronounced bitterness. Therefore, Arabicas are usually commercialized at prices 20%–25% higher than those of Robustas (Esteban-Díez et al., 2007).

The commercially available roasted and ground coffees are produced from Arabica and Robusta beans or blends of these two species. Arabica is usually present in higher amounts in blends due to its higher quality and more enhanced aroma (a relevant trait) than Robusta. Robustas are usually added to the blends to increase the body and foam of coffee beverages (e.g., espresso), and it is heavily present in soluble and instant coffees because it produces more soluble solids than Arabicas when roasted. Given that Arabica coffees are deemed of higher quality than Robustas and thus are commercialized at higher values, the development of suitable means to ensure coffee authenticity to consumers is necessary to avoid fraudulent admixtures being sold as pure products to unaware consumers. To achieve such a task, analytical methodologies that consider the chemical differences between the two species, prior to and after roasting of the beans, have been developed.

Kemsley et al. (1995) have developed a Fourier transform infrared (FTIR) spectroscopy methodology for species discrimination of Arabica and Robusta ground roast coffees, as a rapid alternative to existing authentication methods, then considered time-consuming or difficult to implement successfully. A Diffuse Reflectance Infrared Fourier Transform (DRIFT) type of sampling was employed, and the spectral data were processed by principal component analysis (PCA), revealing an explicit grouping of the spectra according to species. A classical discriminant analysis (DA) was carried out based on the PC scores and yielded 100% successful discrimination. However, the major drawback of the work was the rather limited number of samples available for the analyses, 20 Arabicas and eight Robustas. In a follow-up to this work, Briandet et al. (1996) applied PCA to spectra of freeze-dried instant coffees, acquired by DRIFT and by attenuated total reflection (ATR) sampling techniques, resulting in clear clustering of the coffee species. Linear discriminant analysis (LDA) of the PC scores yielded 100% correct classifications for both training and test samples. Partial least squares (PLS) regression was applied to spectra of Arabica and Robusta blends to determine the relative content of each species, and internal cross-validation gave a correlation coefficient of 0.99 and a standard error of prediction of 1.20% (w/w), demonstrating ATR-FTIR to be suitable for offline quality control of instant coffees.

Arabica and Robusta roasted coffees from several geographical origins were analyzed by Casal et al. (2000) employing a high-performance liquid chromatography (HPLC)/diode-array detector technique targeting their contents of caffeine, trigonelline, and nicotinic acid. Twenty-nine samples were submitted to the same roasting procedure to eliminate possible processing variations. Multivariate and nonparametric processing of the chromatographic results demonstrated the two coffee species could be clearly discriminated by their trigonelline and caffeine contents and that nicotinic acid could not be used as a species discriminator. Costa Freitas et al. (2001) compared the aroma of nine coffee varieties of both Arabica and Robusta species by means of electronic aroma detection using an AromaScan apparatus equipped with a multisampler headspace device. Aroma patterns were evaluated by the Sammon mapping technique, allowing for both the separation of Arabica and Robusta coffees into two distinct groups and their discrimination by the respective geographic origins. The results for the sensor array technique were compared with those for solid-phase microextraction and gas chromatography (SPME-GC). Both methods provided clear discrimination of Arabica and Robusta species; however, the SPME-GC method results did not provide discrimination by geographic origin. The major advantage attributed to the sensor array technique was that it was faster than the SPME-GC method, taking 7 minutes and 1 hour for the analysis, respectively.

Spaniolas et al. (2006) employed a molecular genetics approach to discriminate green Arabica and Robusta coffees. A polymerase chain reaction–restriction fragment length polymorphism method was used to monitor a single-nucleotide polymorphism (SNP) within the chloroplastic genome in a lab-on-a-chip capillary electrophoresis system. The chloroplastic target trnL(UAA)-trnF(GAA) intergenic spacer regions was found to be discriminatory of all the Arabica and Robusta varieties studied. However, further studies were deemed necessary to confirm that plastid copy number remains relatively constant across a wider range of varieties and is not influenced significantly by edaphoclimatic conditions in the coffee plant cultivation.

A near-infrared spectroscopy (NIRS) technique was used by Esteban-Díez et al. (2007) to discriminate Arabica and Robusta pure coffee species and blends of varied compositions of the species in roasted and ground coffees. Eighty-three roasted coffee samples from different provenances and species (36 Arabica and 47 Robusta coffees) processed under different roasting conditions were employed, together with 108 blends of Arabica and Robusta coffee species prepared in the laboratory by combining the three coffee samples most representative of each species that were previously selected. The direct orthogonal signal correction pre-processing method was applied on a set of 191 roasted coffee NIR spectra from both pure species and blends. The Robusta content in the blends varied from 0 to 60% (w/w), so the response variable for the orthogonal signal correction was the percentage of Robusta coffee in each sample, which were already known for calibration samples. The potential functions method as used as a class-modeling technique. The applied classification strategy, which relied solely on NIR measurements, was demonstrated suitable for authenticity assessment of coffee in regard to the content of Robusta coffee in samples of roasted and ground coffees.

Electrospray ionization–mass spectrometry (ESI-MS) was employed by Mendonça et al. (2009) for the characterization of aqueous extracts of green and roasted Arabica and Robusta coffees and to verify the feasibility of this technique for the discrimination of the two species. ESI-(+)-MS also provided discrimination of roasted beans by species, whereas profiles obtained in the negative mode (ESI(−)-MS) provided separation between Arabica and Robusta green coffees due to the higher content of chlorogenic acids (CGAs) in Robustas in comparison to Arabica coffees. Alves et al. (2009) evaluated the feasibility of using the tocopherol profiles of Arabica and Robusta coffee beans, both green and roasted, to discriminate the two species. A solid-liquid micro-extraction was used, and the extract was analyzed by normal-phase HPLC/diode-array/fluorescence detection. Green Arabicas presented mean contents of α-tocopherol of 2.7 ± 0.4 mg/100 g and 8.0 ± 0.9 mg/100 g for β-tocopherol, whereas green Robustas presented mean values of 1.7 ± 0.3 and 2.1 ± 0.3 mg/100 g for α- and β-tocopherol, respectively. It was determined that more than 90% of tocopherols remained after roasting of the beans, except for β-tocopherol in Robustas, which presented a degradation of about 25% (m/m d.b.). The results demonstrated tocopherol profiles to be suitable for employment in the discrimination of Arabica and Robusta coffees, either green or roasted. Fourier-transform Raman spectroscopy was used by Keidel et al. (2010) to discriminate Arabica and Robusta green coffees based on their Kahweol contents. A spectral kahweol index (σka) was quantitatively determined as the relative contribution of the kahweol in individual beans by component analysis of the spectra. Unambiguous distinction between Arabica and Robusta samples was verified from single-bean measurements, as the σka values are greater than 10 for Arabicas and less than 10 for Robusta coffees.

Gas chromatography–mass spectrometry (GC-MS) was employed by Hovell et al. (2010) to analyze coffee samples and determine the most relevant peaks for the discrimination of Arabica and Robusta species. The peak tables from GC-MS analysis were aligned and pretreated for the subsequent multivariate analysis. Three types of data pretreatment – namely normalization, logarithmic and square root transformations, and respective combinations – were tested to verify their influence on the variables selected as most important by the regression coefficients of partial least squares–discriminant analysis (PLS-DA). Square root transformation followed by normalization (RN) was determined as the most appropriate data pretreatment. The results of the multivariate analysis indicated four furans, three phenols, a pyrrole, pyridine, an alcohol, an aromatic compound, and a non-identified substance as the most relevant for the discrimination of Arabica and Robusta coffees. Bicho et al. (2011) evaluated pH, soluble solids, caffeine, trigonelline, total monocaffeoylquinic acids (3-O-, 4-O-, and 5-O-caffeoylquinic), total dicaffeoylquinic acids (3,4-O-, 3,5-O-, and 4,5-O-dicaffeoylquinic), and total feruloylquinic acids (3-O- and 5-O feruloylquinic) as discriminators of Arabica and Robusta coffees submitted to three roasting levels. By means of multivariate statistics, it was demonstrated that these chemicals form four clusters composed of caffeine, trigonelline, total dicaffeoylquinic acids and total feruloylquinic acids, respectively, with the latter deemed a relevant group for T3 roasting level discrimination in both coffee species. It was also determined that the specific discriminators for roasting intensity for Arabica coffee are caffeine, trigonelline, 3-O-caffeoylquinic acid, and 4-O-caffeoylquinic acid, whereas for Robustas the discriminators are trigonelline, 3-O-caffeoylquinic, 4-O-caffeoylquinic, 5-O-caffeoylquinic, 3,4-O-dicaffeoylquinic, 3,5-O-dicaffeoylquinic, 4,5-O-dicaffeoylquinic, 3-O-feruloylquinic, and 5-O-feruloylquinic acids.

Wermelinger et al. (2011) analyzed the extracted oils of pure Robusta, pure Arabica, and blends thereof consisting of different Robusta and Arabica contents by means of Raman spectroscopy. The determination of the amount of Robusta in a given blend was demonstrated feasible by the intensity ratio of the Raman peak characteristic for kahweol and the one characteristic for fatty acids (FAs), with the ratio being linearly dependent on the kahweol content of the Robusta fraction. Kahweol was not detectable in the blends with Robusta contents above 75% (w/w). The Raman data were calibrated with results obtained by the HPLC-based DIN 10777 procedures. The authors pointed out that the major drawback of the developed methodology was the photo instability of the main analyte, kahweol, suggesting avoidance of long laser irradiation of the samples. El-Abassy et al. (2011) applied visible micro Raman spectroscopy and PCA for the discrimination of green Arabica and Robusta coffees, based on their CGA and lipid contents. The PCA analysis of the complete Raman spectrum provided a clear separation of Arabica and Robusta, with the most significant spectral bands located in the range of 1000–1750 cm^{-1}, which are CGA-dominated bands in the Raman spectrum. Discrimination of the two coffee species could also be reliably achieved by restricting the PCA analysis to the lipid-dominated spectral range of 2700–3050 cm^{-1}. Garrett et al. (2012) employed direct-infusion electrospray-ionization–quadrupole–time of flight–mass spectrometry

(ESI(±)–QTOF) and ESI(±) Fourier transform ion cyclotron resonance (FT-ICR) MS to analyze hot aqueous extracts of certified Arabica and Robusta coffee samples. The ESI-MS data were processed with a PLS multivariate calibration technique to detect and quantify Robusta in admixtures with Arabica coffee, and 16 PLS models were built. The model that produced the most accurate coffee blend percentage prediction was the one using the 30 more abundant ions detected by ESI(+) FT-ICR MS, which was further employed to successfully predict blend compositions of Robusta and Arabica coffees.

A metabolomics technique based on ^{13}C nuclear magnetic resonance (NMR) was developed by Wei et al. (2012) for discriminating green coffee bean samples of Arabica and Robusta by species and provenance. Samples from six different geographic regions were used in the study. Multivariate statistical models identified significantly different levels of 14 metabolites (e.g., sucrose, caffeine, CGAs, choline, amino acids, organic acids, and trigonelline) between the two species, allowing for both species and geographical origin be successfully discriminated by ^{13}C NMR-based metabolite profiling. Garrett et al. (2013a) successfully discriminated Arabica coffee Sarchimor and Catuaí cultivars grown under the same edaphoclimatic conditions, in two different regions, by applying a metabolomic approach to ESI FT-ICR MS data. Twenty coffee metabolites (e.g., phenolics, FAs, sucrose, and diterpene glycosides) were identified by negative ESI FT-ICR MS with a mass error <2 ppm. The multivariate statistics techniques PCA and PLS-DA also revealed the correlations of the identified coffee metabolites with the respective cultivars and provenances, thus being deemed a valuable tool for traceability and certification processes.

Romano et al. (2014) evaluated the feasibility of discriminating Arabica and Robusta coffees in blends of the two species by the respective FA compositions. The PCA-selected parameters total monounsaturated FAs (ΣMUFA), linolenic acid (cis18:3n − 3) concentration, 18:0/cis18:1n − 9c ratio, and ΣMUFA/ΣSFA ratio showed a linear relationship with the percentage of Arabica contained in the coffee blends, allowing for determination of the relative amounts of Arabica and Robusta in coffee blends in the range of 0 to 60% (w/w) Arabica. It was also verified that linoleic and α-linolenic acid were more abundant in Arabica than in Robusta coffee, whereas a greater amount of oleic acid was observed in Robusta. Taveira et al. (2014) investigated potential metabolite markers to describe differences in *C. arabica* L. genotypes (Mundo Novo and Bourbons) produced in the Brazilian municipalities of Lavras, Santo Antônio do Amparo, and São Sebastião da Grama. Green beans of distinct samples harvested in the same agricultural crop season were frozen, ground, and lyophilized, and the resultant powders were extracted with methanol and the extracts derivatized. The derivatized coffee extracts were analyzed by GC-MS, and the data was used to build a PLS-DA model. Forty-four metabolites were identified, and the ones most influential in differentiating genotypes were 5-CQA, oxalic acid, galactinol, nicotinic acid, caffeine, and caffeic acid for the Bourbon variety; and myo-inositol, quinic acid, malic acid, fructose, and D-glucose for the Mundo Novo variety. The model built for discriminating the Bourbon and Mundo Novo genotypes showed 100% accuracy, indicating these genotypes to have distinct metabolite profiles.

Self-organizing maps (SOMs) artificial neural network (ANN) was employed by Link et al. (2014a) for the geographical and genotypic discrimination of coffees grown in the coffee region of Paraná State, Brazil. Fourteen Arabica coffee genotypes were collected and analyzed for density, caffeine, CGAs, tannins, total and reducing sugars, proteins, and lipids. Discrimination of the 14 genotypes was achieved by SOMs, with segmentation of the geographical origin also observed. Reducing sugars, caffeine, and CGA were the most important variables for separation of the region of cultivation of Arabica coffee genotypes. Blends with different proportions of Arabica and Robusta coffees, roasted to three distinct degrees, were studied by Dias and Benassi (2015), who evaluated the feasibility to discriminate the species by color parameters and the levels of chlorogenic (5-CQA) acids, nicotinic acid, caffeine, and trigonelline. Caffeine was deemed a suitable discriminator, regardless of roasting degree, whereas the efficiency of discrimination of the remaining hydrosoluble compounds was influenced by the roasting degree. A model for the estimation of roasting degree was developed using color parameters, 5-CQA/caffeine contents ratio, and sum of nicotinic acid and trigonelline contents. The discrimination of coffee species was proposed to be carried out in two steps: first, the roasting degree is characterized; second, the appropriate parameters are defined for each roasting degree. The combined use of color parameters and hydrosoluble compounds content ratio was considered suitable for the differentiation of coffee species in blends of roasted samples.

Buratti et al. (2015) investigated the feasibility of using NIRS and electronic nose (e-nose) and electronic tongue (e-tongue) as instrumental tools for the classification of coffee species subjected to different types of processing (i.e., washed Arabica, and natural Arabica and Robusta coffees). Analyses were performed on green and ground and roasted coffee, and on coffee beverages. Spectral and sensory data were analyzed by PCA and LDA to uncover correlations between samples and variables and to classify the samples according to the type of processing they were subjected to. LDA led to 100% of correctly classified samples with the external test set validation performed with NIRS data. A reasonable percentage of correct classification in cross-validation was obtained for the electronic devices with an average value of correctly classified samples of 81.83% for e-nose and 78.76% for e-tongue.

Authentication of Arabica and Robusta coffee blends was achieved by Schievano et al. (2014) by quantitative determination of esterified 16-O-methylcafestol in roasted coffees extracts by high-resolution proton NMR. Detection of the presence of Robusta in blends with Arabica was demonstrated possible at Robusta percentages lower than 0.9%, and the proposed methodology was deemed faster, more sensitive, and more reproducible than the corresponding DIN standard method. Proton NMR spectroscopy was used by Monakhova et al. (2015) to determine the content of Robusta in blends with Arabica coffee. Lipophilic extracts of authentic roasted and green coffees showed the presence of established markers. The markers 16-O-methylcafestol (16-OMC) and kahweol were identified for Robusta and Arabica, respectively, and quantification of 16-OMC was used to estimate the amount of Robusta in coffee blends. The method was successfully applied for the analysis of samples of coffee pods, capsules, and green beans. Discrimination was made possible by multivariate analysis identification of differences based on shift variation of resonances in the coffee matrices.

Cossignani et al. (2016) reported the characterization of Arabica and Robusta species by structural analysis of the respective triacylglycerol (TAG) fractions. Lipids were Soxhlet extracted from ground and roasted coffee with petroleum ether, and the FAs were determined as their corresponding methyl esters. The results of the employed chemical-enzymatic-chromatographic method were processed by a chemometric procedure using LDA, providing a 100% classification when the TAG stereospecific results of pure samples and samples of 90:10 Arabica/Robusta coffee mixture were analyzed. HPLC–ESI-MS/MS was employed by Babova et al. (2016) for the identification and quantification of CGAs and caffeine in extracts of green Arabica and Robusta coffees from different provenances (Brazil, Colombia, Ethiopia, Honduras, Kenya, Mexico, Peru, Uganda, and Vietnam). PCA and cluster analyses were used to identify chemical patterns discriminating the different species and respective geographical origins. All the Robustas presented higher caffeine contents than the Arabicas, whereas the Arabica from Kenya presented higher CGAs and lower caffeine contents. The antioxidant capacity of the green coffee extracts, assayed by the reducing power and DPPH assays, correlated with the CGA content.

Correia et al. (2016) used univariate- and multivariate-processed data from ESI(–)-FT-ICR-MS and ATR-FTIR analyses to quantify Robusta in blends with Arabica coffee. The univariate calibration model for the ESI(–)-FT-ICR-MS data employed 12 samples of Arabica coffee adulterated with different proportions of Robusta coffee and doped with an internal standard. For the ATR-FTIR results, a PLS regression model was built with 23 adulterated samples. The proposed methodologies successfully enabled the prediction of Robusta contents in adulterated commercial Arabica coffee samples. Moreover, the ESI(–)-FT-ICR-MS technique provided chemical profiles of Robusta and Arabica coffees, allowing for the identification of major compounds in coffee samples.

Colzi et al. (2017) successfully evaluated the application of proton transfer reaction–time of flight–mass spectrometry (PTR-ToF-MS) to discriminate Arabica and Robusta coffees along the entire processing chain (green beans, roasted beans, ground coffee, brews). Mass spectra for the volatile organic compounds of seven Arabica and six Robusta commercial coffee samples were processed by multivariate statistics; the results demonstrated that, regardless of the processing step, the volatile composition of coffee was highly influenced by the species. However, except for green beans, PTR-ToF-MS technique was able to correctly discriminate Arabica and Robusta samples. The best discrimination of species was obtained for roasted and ground coffees, suggesting that such a matrix is the most suitable target for used in coffee species authentication.

Dankowska et al. (2017) investigated the potential of fluorescence and UV-Vis spectroscopies and of data fusion (DF) of both spectroscopies for the quantification of roasted Arabica and Robusta in coffee blends. PCA was used to reduce data multidimensionality, and LDA was applied to fluorescence intensities and UV spectra of pure coffee samples and blends thereof to verify classification ability. The best performance of PCA-LDA analysis was observed for DF of UV and fluorescence intensity measurements at a wavelength interval of 60 nm. PCA-LDA demonstrated that DF could achieve 100% of correct classifications in the training set and over 96% of correct classifications in the test set. Fluorescence and UV-Vis spectroscopies were deemed complementary to each other in respect to quantification of roasted Arabica and Robusta coffees in blends of the two species. Zhang et al. (2018) evaluated the possibility of discriminating four coffee cultivars from China, two Arabica varieties (e.g., Typica and Catimor Arabica from Yunnan Province) and two Robustas (e.g., Fushan and Xinglong Robusta coffee from Hainan Province), applying laser-induced breakdown spectroscopy (LIBS) combined with chemometric methods. Wavelet transform was used to reduce LIBS spectra noise. Classification models were built using PLS-DA, radial basis function (RBF) neural network, and support vector machine (SVM). Spectral variables contributing most to the discrimination of coffee varieties and species were selected based on PCA loadings, and 21 wavelength variables were selected, corresponding to C I, Mg I, Mg II, Al II, CN, H, Ca II, Fe I, K I, Na I, N I, and O I. Models based on SVM and RBF neural network presented the best performances with a classification accuracy of over 80% in the prediction set for both the full spectra and the selected variables. Although the overall results confirmed the feasibility of using LIBS and chemometric methods to discriminate coffee varieties, further studies were deemed necessary with larger samples to produce more robust classification models. ATR-FTIR, PLS calibration, and different variable selection methods were evaluated by Assis et al. (2018) for the development of models to determine the amounts of Robusta in blends with Arabica coffee blends. The analytical range employed was from 0 to 33% (w/w) of Robusta in the blends. Samples of different origins were roasted at light, medium, and dark degrees. Specific models were built for the individual roasting degrees, and a robust model was built for all the 120 samples included. The variable selection methods genetic algorithm (GA), ordered predictors selection (OPS), successive projections algorithm (SPA), and interval PLS (iPLS) were investigated. The best results were obtained with the discrete methods OPS and GA, which greatly improved the predictions in regard to that of full-spectra models. In the optimized models, the number of selected variables varied from 6.5% to 17% of the number of original variables, demonstrating the relevance of selecting a limited number of wave numbers.

Figueiredo et al. (2019) evaluated the ability of Raman spectroscopy to discriminate four genotypes of wet-processed Arabica coffee: one Mundo Novo line (G1) and three Bourbon lines (G2, G3, and G4). Unsupervised and supervised chemometric classification tools were used for processing the Raman data. For the unsupervised analysis (PCA), a tendency of agglomeration of samples G1 and G3 was verified and was attributed to the spectral bands characteristic of FAs and kahweol. Based on this information, a PLS-DA classification model was built to successfully discriminate Mundo Novo and Bourbon coffee genotypes. Raman spectroscopy allowed the building of an adequate model to differentiate between coffee genotypes. Strict sensorial analysis was not able to discriminate these coffee genotypes as well as the Raman-multivariate statistics method. Interspecific SNPs in the rbcL DNA barcode have been validated by Nakagawa et al. (2019) as a genotyping marker to discriminate *C. arabica* and *C. canephora*, for both green and roasted coffee beans, and also for instant coffee powder. The genomic information of *C. arabica* and *C. canephora* was compared and the nucleotide sequences of the matK, rpoB, and rpoC1 genes, and atpF-atpH, psbK-psbI, and trnH-psbA ISRs, and trnL (UAA) 5′ exontrnF (GAA) ISR were determined in addition to rbcL gene. All these loci were deemed useful to discriminate the two species as well as rbcL DNA barcode. Standard candidate reference materials (RMc) comprising blends of medium-roasted Arabica and Robusta coffees were developed by Couto et al. (2019) and analyzed by real-time PCR, using a specific primer, ARA, developed in a previous work. All the samples containing Arabica coffee presented amplification, whereas the RMc material with 100% Robusta did not, indicating the specificity of ARA primers for Arabica coffees. The detection system assay was deemed a promising molecular tool to identify and quantify percentages of Arabica coffees in blends with Robustas.

Badmos et al. (2019) employed NMR to confirm coffee species authenticity and a LC-ESI-MS technique to profile and quantify CGAs in 54 different samples of the two coffee species from distinct origins. Mono-caffeoyl quinic acids (CQA) contents presented no significant variations among coffee species, whereas feruloyl quinic acids and dicaffeoyl quinic acids (diCQA) contents presented statistically significant variations between Robusta and Arabica species, but only for green beans. Isomer ratios were also investigated as potential markers for coffee species authenticity with Robusta 5-CQA and 3,4-diCQA showing high positive correlation. A high correlation coefficient was also observed for 5-p-coumaroyl quinic acids (pCoQA) in Robusta and 3-pCoQA in Arabica coffee.

The number of analytical techniques employed for the chemical discrimination of coffee species and varieties, both in the raw and roasted states, is significant. A few attempts were made to find specific markers for the discrimination, but the majority of the developed methodologies was based on multivariate statistical analysis of chemical profiles (e.g., spectral data, chromatograms). The levels of success varied among the diverse techniques, mostly due to a number of limited samples employed in the analysis, but most of them were quite satisfactory in performing the task.

13.3 Discrimination by geographic origin

Coffee is a perennial evergreen dicotyledon plant, cultivated exclusively within the region delimited by the Tropic of Cancer (23.43695°N) and the Tropic of Capricorn (23.43695°S), commonly termed the "Coffee Belt" (Figure 13.2). Coffee plants require warm climates with average temperatures of 18–24°C and abundant rainfall in a 9-month period for their development and a relatively dry 3-month period for their maturation and harvest. Arabica species requires cultivation at altitudes higher than 1000 m above sea level, where the climate is cooler, whereas Robusta species are successfully cultivated at altitudes lower than 800 m above sea level, tolerating higher temperatures and heavier rainfalls than Arabicas. The edaphoclimatic conditions in which coffees are cultivated have a significant impact on their specific quality attributes; hence, the geographic origin of a coffee plays a significant role in defining the organoleptic characteristics of the final product of consumption, the coffee beverage. In addition to edaphoclimatic conditions, geographically specific cultivation, harvesting, and processing practices can turn coffee into a trademark product with distinctive organoleptic features that will attract a niche of consumers willing to pay a higher premium for the product, which translates into higher revenues for the producers.

Consumers awareness of and interest in origin-specific organoleptic features generated a demand for implementation of coffee certification, inducing producers to market certified provenance-labeled coffees. To that end, producers worldwide have adopted the certification labeled protected designation of origin (PDO) to protect and ensure the quality and provenance of their coffee, and consequently charge higher premiums for it. As with any internationally traded commodity, coffee

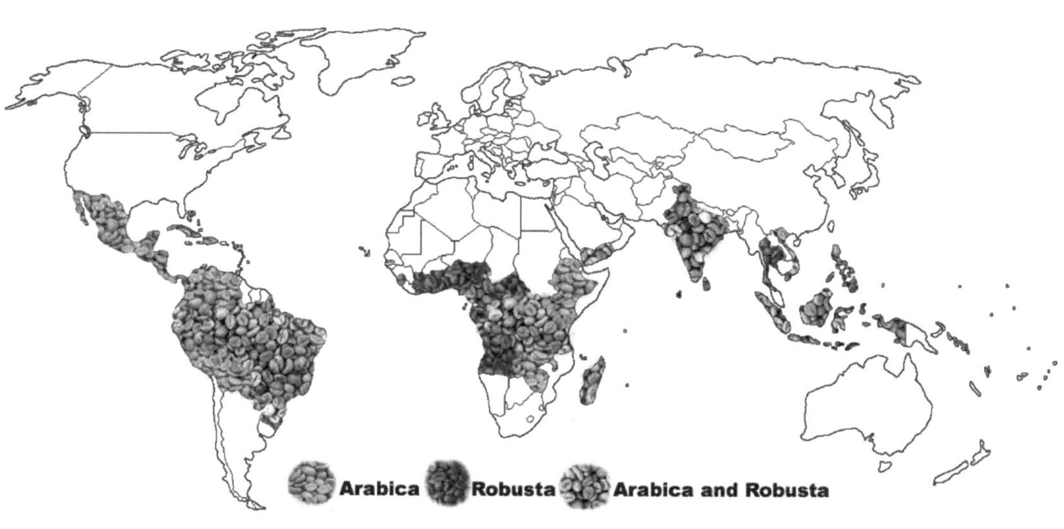

FIGURE 13.2 The "Coffee Belt," delimited by the Tropics of Cancer and Capricorn.

is prone to fraudulent practices and, because its organoleptic attributes are complex and subjectively defined, there is a paramount need for the development of analytical methodologies that allow an objective and reliable discrimination of coffee provenance to ensure authenticity of its associated quality attributes to the consumers. A significant number of publications have dealt with such subject in the last decades, and a few are reviewed next.

Anderson and Smith (2002) employed inductively coupled plasma–atomic emission spectrometry (ICP-AES) to analyze the elemental profile (K, Mg, Ca, Na, Al, V, Cr, Mn, Fe, Co, Ni, Cu, Zn, Mo, S, Cd, Pb, and P) of 160 samples of coffee from Costa Rica, Colombia, Panama, Guatemala, Sulawesi, Sumatra, Kenya, and Ethiopia. Multivariate statistics (PCA, discriminant function analysis, and neural network modeling) was applied to the elemental data sets, allowing for the discrimination of the samples into three major geographic regions of coffee production: Indonesia, East Africa, and Central/South America. Individual inspection of elemental concentrations did not provide differentiation of geographic origin. Neural network models together with discriminate function analysis successfully discriminated coffees by their sub-regional growing origins (70%–86% successful classification). Risticevic et al. (2008) developed a headspace solid-phase microextraction (SPME)–gas chromatography–time of flight mass-spectrometry method for traceability of geographical origin of coffees. A two-level full factorial first-order design was applied for the optimization of extraction-influencing parameters, identifying the variables extraction time and sample temperature as the most influential, with the optimum conditions for the two variables being 12 minutes and 55°C, respectively. For the solid-phase microextraction, a divinylbenzene/carboxen/polydimethylsiloxane metal fiber was employed, and the GC-MS data of the corresponding extracted volatiles were submitted to PCA to verify the geographical origins of the samples. Geographical origin discriminations were successfully established for coffees originating from South and Central America, Africa, and Asia. The study also succeeded in detecting compositional changes underwent by coffees over extensive storage conditions.

Borsato et al. (2011) analyzed 172 coffee samples from Parana State, Brazil for moisture, proteins, CGAs, tannins, total acidity, total lipids, caffeine, total and reducing sugars, and minerals, and the resulting data set was tabulated and presented to ANN multilayer perceptron to correlate to their respective provenances. The ANN correctly classified and tested 100% of the samples by their respective geographic origins. The multilayer perceptron-type ANN showed 99% accuracy in training and 100% accuracy in the stage of testing and validation. Wei et al. (2012) employed ^{13}C NMR-based metabolomics as a tool for successfully discriminating species and provenance of green coffee beans from six different geographic regions. Multivariate statistical models were applied to the NMR data set and were able to identify significantly different levels of 14 metabolites of green coffee beans, including sucrose, caffeine, CGAs, choline, amino acids, organic acids, and trigonelline, allowing for the discrimination of both species and geographic origins.

Consonni et al. (2012) used ^1H-NMR spectroscopy to investigate the chemical composition of roasted *C. arabica* samples from America, Africa and Asia. PLS-DA was applied to the NMR data, and the resultant models were able to discriminate the samples according to their provenances. Samples from America were characterized by their FA contents, African samples by CGAs and lactates, and samples from Asia by acetate and trigonelline. The multivariate statistical tools PCA and hierarchical cluster analysis were used by Garret et al. (2013b) to process paper spray mass spectrometry (PS-MS) data for coffee and successfully discriminate green Arabica coffee beans from three different origins in Brazil. In addition, direct analysis of a single coffee bean using a bean slice as the substrate for electrospray ionization was reported by the authors.

Link et al. (2014b) developed a RBF network and applied it to FTIR spectral data of Arabica coffees to classify their geographic and genotypic origins. Sequential simplex optimization was applied to the development of networks allowing for the determination of the optimal set of parameters, maximizing performance and minimizing the size of networks in a heuristic manner. The optimized RBFs used the range 1900–800 cm^{-1} of the spectrum providing 100% correct classification of geographic origin and 94.44% correct classification of genotypic type. The aroma profiles of roasted *C. arabica* coffees from Brazil, Ethiopia and Guatemala were analyzed by PTR-ToF-MS, and the resultant data set was processed by unsupervised and supervised multivariate statistics, which provided successful classification of the coffees according to their respective origins. PCA provided separation of the three coffees by their provenance and subsequent PLS-DA classification showed completely correct predictions.

NMR fingerprinting was used by Arana et al. (2015) to discriminate coffee samples from one country against all others, including its closest neighbors. A total of 192 samples of roasted coffees were used in the study: 22 from Asia, 12 from Africa, and 158 from America. Of the 158 samples from America, 140 samples were from Colombia, collected from seven different regions located at the slopes of the Andean Cordillera. High classification rates were reported with discrimination of coffees from Colombia largely relying on FAs, acetate, and caffeine. Misclassification occurred for samples of distinct countries prepared via wet process, the process most used in Colombia. An attempt was made by Antoine et al. (2016) to distinguish the growing origins of Jamaican coffees and coffees from other provenances (e.g., Colombia, Brazil, Costa Rica, Mexico, Kenya, Ethiopia, Rwanda, Sumatra) and identify intra-island growing regions by elemental analysis. Neutron activation analysis and inductively coupled plasma optical emission spectroscopy was employed to analyze 24 samples of roasted and ground coffee and soluble coffee, with 16 elements (Al, Br, Ca, Cr, Cu, Fe, K, Mg, Mn, Na, Ni, P, Rb, Sc, Sr, and Zn) selected for statistical evaluation using agglomerative hierarchical clustering and PCA. Soluble clustered discretely from roasted and ground samples and Jamaican samples clustered discretely from all the non-Jamaican coffees. However, separation of the growing regions within the Jamaican island could not be attained.

NIRS was used by Marquetti et al. (2016) to analyze different coffee genotypes cultivated in distinct municipalities (e.g., Cornélio Procópio, Paranavaí, Mandaguari, and Londrina) in

Paraná State, Brazil, and the spectral data were processed by PLS-DA. The best-built model correctly identified 94.4% of validation samples for both geographic and genotypic origin, with sensitivity and specificity for calibration and prediction being greater for geographic than for genotypic discrimination.

Trigonelline, theophylline, theobromine, and caffeine compositions of 99 green Arabica coffees comprising eight varieties (Harar, Jimma, Kaffa, Wollega, Sidama, Yirgachefe, Benishangul, and Finoteselam) from the four major production regions of Ethiopia (26 samples from the east, six from the northwest, 49 from the south, and 18 from the west) were investigated by Mehari et al. (2016) using HPLC. The concentrations of the alkaloids varied significantly with their geographical origins, and theophylline was not detected in any of the samples. Theobromine was not detected in Harar coffees, grown in the east, and coffees from the northwest region were presented higher concentrations of caffeine. LDA provided 75% correct classification of samples with respect to production regions, with a 74% prediction success rate.

Toledo et al. (2017) analyzed 25 coffee samples by HS-SPME-GC/MS and/or HSPTR-ToF-MS to identify of key compounds that would allow their discrimination in respect to their continental or country provenance. A preliminary study on the identification/selection of a suitable compound to be used as standard for results normalization showed that pyridine is the best choice. Three DA functions were built based on 18 compounds and used to classify coffees according to African, Asian, South American, or Central American origin. The model was based on molecules like 2,3-butanedione, 2,3-pentanedione, 2-methylbutanal and 2-ethyl-6-methylpyrazine, furfural, and 1-hydroxy-2-butanone. A single function model was also built to specifically differentiate Brazilian coffees from others. The compounds were relatively the same as those included in the three-function model, however, with a rearrangement of their relative relevance. Furthermore, a model based solely on eight of the previous compounds with the highest odor activity values was built to differentiate Brazilian coffees from American ones. Accurate classification capability was demonstrated for each model, and robustness of the proposed discriminant functions was demonstrated in cases of insufficient characterization (lack of data). Flambeau et al. (2017) employed both e-nose and e-tongue for the discrimination of washed specialty Bourbon coffee (*C. arabica*) from four geographical subregions (northern, southern, eastern, and western) of Rwanda and a coffee from Ethiopia and another from Brazil. PCA was applied to the e-nose and e-tongue data set for geographical origin discrimination, and a discriminant factorial analysis (DFA) model was developed to predict the geographical origin of coffee samples based on their intrinsic flavor. Discrimination was achieved with both e-nose and e-tongue, with clear grouping of coffee samples from areas within the same geographical sub-regions. The e-nose DFA model successfully predicted the geographical origin of the studied coffees, whereas the e-tongue model did not.

Fluorescence spectroscopy and multivariate statistics were employed by Botelho et al. (2017) to develop a geographical discrimination model for coffees produced in the four major producing areas (Cerrado Mineiro, Matas de Minas, Norte de Minas, and Sul de Minas) of Minas Gerais State, Brazil. Parallel factor analysis (PARAFAC), N partial least squares (NPLS-DA), and unfolded partial least squares (UPLS-DA) were tested for the discrimination, with UPLS-DA presenting the best results for classification. The developed models performed well in the classification of coffees produced in different regions of Minas Gerais State. Bona et al. (2017) developed a methodology for geographical classification of distinct Arabica coffee genotypes based on infrared spectroscopy and SVM. Spectral data were collected in the range of near and mid-infrared. The SVM model was built using radial basis as kernel function and the one-versus-all multiclass approach. The SVM model for the NIRS data correctly classified all samples according to their provenances with a sensitivity and specificity of 100%, whereas SVM model for FTIR data presented a slightly poorer performance.

An untargeted metabolomics approach based on ultra-performance liquid chromatography coupled to hybrid quadrupole–ToF–high-resolution mass spectrometry (HRMS) was developed by Ossa et al. (2018) and applied to 41 green coffee samples of different genotypes, collected from five distinct growing regions of Colombian. PLS-DA and data-driven soft independent modeling of class analogy (DD-SIMCA) were used to build models for the discrimination of coffees using a variable importance in projection variable selection method. Thirteen compounds were responsible for the correct separation of green coffees according to their origins. The developed models provided 94% of correctly classified samples using both cross-validation and an additional set with coffee samples from different harvest year. Proton-transfer reaction mass spectrometry (PTR-MS) and NIRS were applied by Monteiro et al. (2018) to compare their performances for the authentication of geographical and farming system origins of Brazilian coffees. Nineteen organic (ORG) and 26 conventional (CONV) coffees from distinct producing regions (Minas Gerais, São Paulo, Espírito Santo, Paraná, and Bahia States) were analyzed. Differences in the 44 and 71 ion intensities were verified between the producing regions and farming systems, respectively. PCA did not perform well in separating the studied coffees according to the farming system or geographical origin regardless of the analytical technique used. A PLS-DA model was built and presented the best discriminating performance compared to several other chemometric techniques, with the farming system being thoroughly discriminated. Classification performance (>80%) was independent of the data acquisition method, promoting the use of NIRS over PTR-MS for discrimination farming systems due to its higher speed and lower cost. Coffee samples from Minas Gerais and São Paulo were clearly differentiated from those of the other regions.

Toci et al. (2018) developed an ^{1}H-NMR methodology coupled with multivariate statistics for the discrimination of Brazilian coffees according to their major producing regions (Minas Gerais, São Paulo, Paraná, and Bahia States). DA provided a clear discrimination of the samples, with catechol, trigonelline, caffeine and n-methylpyridine being the most relevant for the differentiation. Schipilliti et al. (2019) used the carbon isotope ratio (δ13C) of caffeine in green and in the respective roasted coffee together with the value of δ13C of the whole volatile fraction of the roasted samples to differentiate coffee species and geographic origins. Gas chromatography coupled to carbon isotope ratio mass spectrometry (GC-C-IRMS) was employed for the analyses, and the results were analyzed by a comprehensive isotopic data evaluation (CIDE) model. Clear separation of the samples was attained based on the typical climate of the region where the coffee was cultivated and on the respective coffee species. Robusta samples produced in tropical regions in India are grouped separately from Robusta produced in subtropical and equatorial areas. Arabica samples from the equatorial area (Ethiopia) were discriminated from those of the southern tropical hemisphere (Santos) and from the San Salvador samples produced in the tropical regions. All the Arabica samples were distinctively separated from all the Robusta samples.

Peng et al. (2019) analyzed 67 Brazilian coffees for their stable isotope ratios (δ13C, δ18O, δ2H, and δ15N) in an attempt to discriminate them in respect to their producing regions (São Paulo, Minas Gerais, Paraná, Espírito Santo, and Bahia States). The samples comprised 25 organic and 42 conventionally produced coffees. LDA, k-nearest neighbors (k-NN), and SVM were used to classify the coffees based on their provenances. Higher δ15N values for organic coffee were determined in comparison to those for conventional coffees, whereas no differences were observed for δ13C, δ18O, δ2H. None of the employed supervised chemometric methods was able to differentiate the organic coffees, whereas all chemometric methods were able to classify conventional coffees, with accuracy >85%. The distinct Brazilian producing regions could not be fully discriminated using the studied stable isotope ratios. Inductively coupled plasma (ICP)-based and x-ray fluorescence spectrometry (XRF)-based techniques and elemental analyzer–isotope ratio mass spectrometry were used by Worku et al. (2019) to analyze 103 green Arabica coffee samples from distinct geographical origins in Ethiopia (eastern [Harar], southeastern, southwestern, and northwestern). The linear discriminant model for the XRF-based multi-elements was the most effective in discriminating the Ethiopian geographical origins of coffee, with classification accuracy of 89%. The coffee origin diagnostic power of XRF-based multi-elements was further refined by including δ13C values of coffees, increasing classification accuracy by 3% and 8% for four and three regions, respectively.

A plethora of analytical methodologies (e.g., NIRS, FTIR, NMR, ICP, XRF) was developed for the discrimination of coffees by their geographical origins. The majority of them employed multivariate statistics to process the analytical data and produce discriminating models with varying degrees of success. Some of the methodologies provided discrimination of coffees by continents, others by countries, and a few by subregions within a specific country. A tendency of employing analytical methods for isotopic elemental analysis can be observed for this purpose, since it is quite reasonable to assume that isotopes are non-uniformly distributed in different regions within the same country or even throughout a region of a country.

13.4 Discrimination by quality

The overall quality of green coffee beans is assessed by a wide variety of criteria, including bean species and variety, size, color, shape, processing method, crop year, cup quality (flavor), and defect count (type classification), with the last two being the most important in terms of price variations for coffee trading (Franca and Oliveira, 2008). These criteria will also reflect on the quality of the final product of consumption, roasted coffee. Furthermore, ground roasted coffee is vulnerable to adulteration, because its physical characteristics (particle size, texture, and color) can be easily reproduced by roasting and grinding a variety of biological materials (cereals, seeds, roots, parchments, etc.). Therefore, the quality of ground roasted coffee must be also monitored with respect to fraudulent admixtures with cheaper materials, such as twigs, coffee berry skin and parchment, spent coffee, roasted barley, maize, cocoa, soybean, and others (Oliveira et al., 2009; Reis et al., 2013a, 2013b, 2013c; Sezer et al., 2018).

Sensory analysis is one of the most important criteria employed for determining green coffee prices and is still the official method worldwide for establishing coffee quality. Different producing countries employ specific methodologies and classification for coffee sensory attributes. In Brazil, for example, coffees are officially categorized using the following flavor scale (Franca and Oliveira, 2008):

1. Strictly soft: pleasant feel of the mouth, low acidity, mellow sweetness
2. Soft: same characteristics as strictly soft, but less accentuated
3. Softish: same characteristics as soft, but less accentuated
4. Hard: slight astringency, lacks sweetness and softness
5. Rioysh: iodine, inky flavor from microbe-tainted beans
6. Rio: same characteristics as Rioysh, but more accentuated.

However, this methodology is time-consuming in terms of panel training and alignment and can be subjective. It is challenging to define a relationship between chemical profile and sensory impact, but it might provide a more objective evaluation. Thus, many studies have focused on trying to establish a correlation between specific chemical markers and sensory profile. Some of the evaluated chemical parameters include amines (Oliveira et al., 2005), caffeine and acidity (Franca et al., 2005), CGAs (Farah et al., 2006), and FAs (Oliveira et al., 2006), whereas recent studies have focused on chemical profiles (fingerprint) employing spectroscopic and chromatographic methods.

A gas chromatography method using pulsed split injector was proposed by Novaes et al. (2015) for quantification of the diterpenes cafestol, kahweol, and isokahweol in methanolyzed Arabica coffee oils. The methanolyzed oils from 13 samples of green Brazilian Arabica coffees of different qualities (soft, hard, and rio) were analyzed, and the authors observed that the highest values (>1.2) of the concentration ratio of cafestol/kahweol corresponded to the higher-quality coffees, classified as soft. Thus, it was concluded that the ratio between cafestol and kahweol could be used to predict cup quality. Craig et al. (2018) evaluated the potential of mid-infrared spectroscopy

(FTIR) for prediction of cup quality of Arabica coffees submitted to different roasting degrees. Arabica coffee samples were classified by five trained cuppers as soft, hard, rioysh, rio, and rio zona. A two-level PLS-DA hierarchical strategy was employed. The coffee samples were classified as high or low quality in the first level and then separated according to cup quality in the second level. The second-level models exhibited 100% sensitivity and specificity in the training sets. For the test sets, sensitivity values ranged from 67% (rio zona) to 100% (soft), whereas specificity ranged from 71% (rio) to 100% (rioysh, hard). The results demonstrated that the developed methodology could be a useful tool for quality control and inspection purposes.

Among the available methods for coffee sensory analysis, those of the Specialty Coffee Association of America (SCAA) classification are deemed the most appropriate for high-quality (specialty) coffees (Belchior et al., 2019). These protocols are based on objective assessment of sensory parameters including aroma, flavor, aftertaste, acidity, body, and sweetness (SCAA, 2009).

Piccino and coworkers (2014) employed solid-phase extraction coupled to GC-MS for identification of volatile components related to coffee sensory analysis. Samples of the specialty coffee Bourbon Pointu (*C. arabica* var. *laurina*) were evaluated according to three trade classifications: grand cru, sublime, and authentique. Samples classified as grand cru received the highest scores for the following criteria: preference, fruity, aromatic quality, acidity, and body. On the other hand, coffee brews that belong to the authentique category received highest scores for negative descriptors such as roughness, astringency, and bitterness. Coffees classified as sublime received higher positive scores in comparison to authentique in terms of positive descriptors such as aromatic intensity, aromatic quality, fruity, and preference. Evaluation of the volatiles composition indicated a predominance of aldehydes (e.g., (E,E)-nona-2,4-dienal) in the high-quality (grand cru) coffees, of 2-phenylacetaldehyde (270) in sublime coffees, and of pyrazine compounds (e.g., 2-methyl-5-propylpyrazine) in the lower-quality (authentique) coffees.

Tolessa et al. (2016) evaluated the use of NIR spectra of green coffee beans for prediction of coffee cup quality. PLS regression was employed for developing prediction models for total specialty cup quality and its individual quality attributes (overall cup preference, acidity, body and aftertaste). The models presented high correlations between measured (sensory) and predicted (NIR) cupping scores, with r values ranging from 0.9 (total quality) to 0.72 (body and aftertaste). Prediction models were not as effective for aroma and flavor ($r < 0.6$), probably due to the complex nature of the attribute and the fact that aroma is a result of a number of chemical reactions that occur during roasting. In a later study, the same research group (Barbosa et al., 2019) employed NIR in association with the contents of proteins, lipids, caffeine, total CGAs, total titratable acidity (TTA), total sugars, sucrose, kahweol, and cafestol. The goal was to verify if specific chemical attributes could be associated to distinct sensory profiles based on the SCAA protocol. Coffee brews with lower quality scores presented high levels of caffeine, protein, CGAs, and TTA. High cafestol, sucrose/acid, and cafestol/kahweol ratios in the green coffee beans were usually associated with higher quality scores for the coffee brews.

A recent study (Bressanello et al., 2018) aimed to define the chemical signature of coffee sensory notes in association to its volatile profile determined by HS-SPME-GC-MS, with prediction models being developed by a chemometric-driven approach. The most reliable models were the ones obtained for woody, bitter, and acidic properties, where selected volatiles reliably represented the sensory note fingerprints. For instance, guaiacol, 1-methylpyrrole-2-carboxaldehyde, (5H)-5-methyl-6,7-dihydrocyclopentanpyrazine, 2-furanmethanethiol, and difurfuryl disulfide were directly involved in defining the woody note. In general, the strategy was discriminative and informative, identifying aroma compounds characteristic of the selected sensory notes.

Belchior and collaborators (2019) employed FTIR for discrimination of espresso coffee samples in terms of sensory parameters such as aroma, flavor, aftertaste, acidity, and body. PCA provided grouping of samples in accordance with their descriptors of aroma and flavor. PLS-DA classification models were obtained for each attribute and presented good values of sensitivity and specificity for the calibration and validation sets. Evaluation of the spectra indicated that acidity was associated with the presence of carboxylic acids, alcohols, and CGAs. Flavor, body, and aftertaste were affected by the composition of lipids, proteins, trigonelline, caffeine, carbohydrates, carboxylic acids, and CGAs. Aroma was related to the presence of aldehydes, esters, ketones, and acids in general.

Sittipod et al. (2019) applied untargeted LC-MS flavoromic for identification of compounds that positively impact coffee quality. The chemical profiles of 18 coffee samples and the corresponding SCAA cupping scores were modeled by orthogonal PLS. Four chemical compounds that presented high positive correlation to cup score were subsequently isolated and purified by multi-dimensional preparative LC-MS fractionation. Sensory recombination analysis by certified SCAA Q-graders confirmed that three out of these four compounds significantly increased cup score when added to a control coffee. These were identified as 3-*O*-caffeoyl-4-*O*-3-methylbuta noylquinic acid, the corresponding lactone 3-*O*-caffeoyl-4-*O*-3-methylbutanoyl-1,5-quinide, and unknown phenolic derivative containing a 3-methylbutanoyl moiety ($[M-H]^{-1}$, m/z 671). It was concluded that these compounds act as flavor modifiers, given that no direct flavor activity was observed for any individual compound.

The term "defect" is employed in reference to the presence of either defective (black, sour or brown, immature, insect damaged or bored) beans or of extraneous matter (twigs, husks, skins, and stones) in a given coffee sample. The presence of defective coffee beans is relevant to coffee quality, because most of these defects are associated to problems that occur during harvesting and post-harvesting operations. Black beans result from beans that are dead within the coffee cherries or from beans that fall on the ground by action of rain or overripening. Sour or brown beans are usually associated with "overfermentation" that occurs during processing with inappropriate processing conditions or overripe cherries. Immature beans are the ones that come from immature fruits. These defects are known to negatively affect beverage quality (Franca and Oliveira, 2009), and therefore some studies have focused on developing analytical methodologies for discrimination of defective and non-defective coffee beans. Earlier studies have focused on caffeine, trigonelline, and CGAs (Franca et al., 2005), FAs (Oliveira et al., 2006), amines (Vasconcelos et al., 2007), and electrospray soft ionization–mass spectrometry (Mendonça et al., 2009), among others. Overall, it was concluded that, prior to roasting, it is possible to discriminate defective from non-defective beans by color, size, acidity levels, sucrose levels, and the presence of histamine (Franca and Oliveira, 2008). In the case of roasted coffee, it was expected that the volatile profile (obtained by gas chromatography analysis) could provide the means for differentiation, although it was observed that the roasting parameters (time, temperature) affected the volatile components regardless of coffee quality (Mancha Agresti et al., 2008; Franca and Oliveira, 2009).

Craig et al. (2012) evaluated the feasibility of employing FTIR for discrimination between defective and non-defective coffees. Green coffee samples were separated into black, sour (dark and light), immature, and non-defective. PCA and agglomerative hierarchical clustering (AHC) results indicated that each type of defective coffee samples could be separated into distinct groups, based on transmittance or diffuse reflectance (DR) spectra. PCA and AHC results based on normalized ATR-FTIR spectra indicated the separation of the samples into two major groups: non-defective/light sour and black/dark sour/immature. In a later study, Craig et al. (2015) compared the performances of FTIR and NIR spectroscopies for the quantification of defective beans in roasted coffees. Mixtures of defective and non-defective roasted and ground coffees were produced and analyzed, with the content of defects ranging from 0 to 30%. PLS models were able to quantify the amount of defective roasted coffees mixed with non-defective ones, with high values of correlation coefficients for both FTIR (0.891) and NIR (0.953). It was concluded that NIR spectroscopy provided more robust models.

Dias et al. (2018) employed infrared-photoacoustic spectroscopy (FTIR-PAS) for discrimination of defects in roasted and ground coffee. Mixtures of specific defects (black, broken, and sour beans, skin, and coffee woods) were blended with healthy (defect-free) Arabica and Robusta coffees in distinct ratios, forming different classes of blends. While PCA allowed for discrimination of defects in blends, PLS-DA provided the development of a predictive model using six classes of blends. The model was able to classify 100% of the samples into four classes, with specificity values above 0.9. Application of this methodology for discrimination and classification of blends of roasted coffee was shown to be accurate, easy, and quick, providing a green alternative to other analytical methods.

Ground roasted coffee presents physical characteristics (particle size, texture, and color) that can be easily reproduced by roasting and grinding a wide variety of biological materials (e.g., cereals, seeds, parchments). Therefore, it has been the target of fraudulent admixtures with several materials, such as the previously mentioned lower-quality coffees or defective beans, and a variety of spurious materials, including twigs, coffee berry skin and parchment, spent coffee grounds, roasted barley, corn, and other cheaper grains (Oliveira et al., 2009; Reis et al., 2013b, 2013c). Some of the analytical techniques employed for determination of coffee adulteration include multispectral imaging (Sano et al., 2003), SPME/GC-MS (Oliveira et al., 2009), FTIR (Reis et al., 2013a, 2013b, 2013c), HPLC (Domingues et al., 2014), and UV-Vis spectroscopy (Souto et al., 2015). An overview of earlier studies is presented by Toci et al. (2016), and a summarized description of recent work is presented as follows.

Ribeiro and collaborators (2017) evaluated the potential of ^{1}H-NMR to analyze adulteration in roasted coffee. Four adulterants were tested: corn, coffee husks, barley, and soybean. After superposition of the adulterant spectra with the fingerprint of roasted coffee, the following peaks were selected as markers: a high intensity singlet chemical shift at 5.03 (s), for detection of corn; signals at 5.30 (s) and 3.15 (s) associated to the presence of barley; chemical shifts at 5.30 (J = 4.15 Hz) (dd) and 4.87 (t) for soybean; and signals at 5.08 (J = 5.92 Hz) (d) and 4.98 (s) for the presence of coffee husks. Preliminary quantitative analysis indicated that peak intensity changed with the amount of adulterant, but the lowest level evaluated was 30% – quite high in comparison to the 1% maximum limit established by Brazilian legislation.

A recent study by Reis et al. (2017) compared two different sampling approaches in FTIR analyses (namely, DR and ATR) for detection and identification of multiple adulterants (barley, corn, spent coffee grounds, and coffee husks) in roasted and ground coffee. Although previous studies have already shown DR as an effective method for discrimination and quantification of roasted coffee adulterants (Reis et al., 2013a, 2013b, 2013c), ATR does not require any sample pretreatment, whereas for DR the sample must be mixed with KBr prior to analysis. The performance of both techniques was compared using hierarchical models, and DF was employed to combine the data from DR and ATR. The use of DF to combine the data obtained by both DR and ATR resulted in an improvement of adulterant discrimination models when compared with each sampling technique alone. DF models could detect adulterated samples even when a mixture of up to four different adulterants was used. The method worked well for different roasting levels (light, medium, and dark) and for adulteration levels as low as 1%.

The use of capillary electrophoresis–tandem mass spectrometry was proposed by Daniel et al. (2018) for detection of coffee adulteration with soybean and corn. The limit of quantification (LOQ) for the monosaccharides (fructose, galactose, arabinose, glucose, rhamnose, xylose, mannose, and ribose) was lower than 0.01 mmol L^{-1}, in accordance with the tolerable limits for coffee. Results indicated that fucose, xylose, and glucose could be used as potential markers to identify this type of adulteration in coffee samples. A recent study by Correia and collaborators (2018) evaluated the potential of microNIR in quality control of Arabica coffee, by identification and quantification of adulterants such as Robusta coffee (in different roasting levels), as corn, peels, and sticks. A total of 125 blends were prepared, employing 13 different concentrations of the adulterants (corn and peels/sticks, and the Robusta coffee) in the range of 1 to 100 wt%. The developed PCA and PLS models were also applied to monitor the quality of 16 commercial coffee samples. The results obtained indicated that the proposed method was more efficient in terms of predicting adulterations by corn (limit of detection [LOD] = 1.6 wt%, LOQ = 5.2 wt%) in comparison to peels and sticks (LOD = 2.6 wt%, LOQ = 8.8%). The efficiency of the method varied considerably for discrimination of Robusta coffees and decreased significantly as the roasting degree increased (LOQ = 18.4 wt% for dark roasts).

The study by Morais et al. (2019) proposed a simple and low-cost analytical approach to detect adulterations in ground roasted coffee by employing a single voltammetric sensor. The voltammogram of a coffee extract (prepared like a house-made coffee cup) was submitted to pattern recognition analysis, preceded by variable selection, in order to detect the addition of coffee husks and sticks, with adulteration levels ranging from 1.6% to 6.5%. The following recognition methods were tested: LDA with variable selection by SPA or GA; and PLS-DA. Both LDA models presented satisfactory results in terms of discrimination of adulterated and unadulterated samples. The voltammograms were also evaluated for the quantitative determination of the content of adulterants in ground roasted coffees. The detection and quantification limits were estimated at 0.9% and 2.7%, respectively. Arrieta and collaborators (2019) also employed a voltammetric sensor to evaluate coffee adulteration. A polypyrrole sensor array was employed, and the methodology was tested for roasted coffee samples mixed with corn and soybean, with adulteration levels ranging from 2% to 20%. PCA and hierarchical cluster analysis (HCA) provided a clear discrimination of adulterated samples. PLS models provided quantitative determinations, with satisfactory correlation factors ($r^2 = 0.973$ for corn and $r^2 = 0.941$ for soybean). Given that the limit for establishing adulteration in roasted coffees is usually 1%, the proposed methods seemed to be more appropriate for classification rather than quantification of adulterants.

13.5 Discrimination by other parameters

Most of the coffee discrimination studies have focused on separation by species, geographic origin, and quality, as previously discussed. However, some other parameters have been evaluated such as brand authenticity, and processing parameters including fermentation and roasting conditions, among others. A compilation of recent studies is presented in this section.

Sarraguça et al. (2013) developed a methodology for rapid assessment of coffee brand authenticity based on NIRS, using the Portuguese market as case study. The goal was to verify the potential of NIRS for discrimination of different coffee brands, demonstrating the utility of the approach to be employed by coffee roasters for brand control. For one of the tested brands, a zero misclassification error was achieved, whereas for another the error was around 28%, this being attributed to the specific manufacturer also selling the same coffee to other generic brands that were included in the analysis. The discrimination occurred mostly near the 7000 cm^{-1} spectral region and was attributed to differences in concentrations of CGAs and sucrose, and also to differences in the lipid fraction.

GC-MS with headspace SPME was employed by Yang and collaborators (2016) for identification of volatile aroma compounds associated with five common roast coffee defects (light, scorched, dark, baked, and underdeveloped). Coffee beans from the same origin were roasted using six different time-temperature profiles, and 37 aroma compounds were selected. Among these, one marker compound was chosen for each defect type, based on its relative abundance: indole for light defect; 4-ethyl-2-methoxyphenol for scorched defect; phenol for dark defect; maltol for baked defect; and 2,5-dimethylfuran for underdeveloped defect. However, variations in coffee quality with respect to origin or other parameters was not taken into account.

An amperometric sensory approach based on boron-doped diamond (BDD) electrodes decorated with transition metals nano-catalysts was developed by Zribi et al. (2019) for discrimination of coffee capsules. The detection of electro-active species in coffee was investigated, using a BDD electrode as well as BDD electrodes modified with platinum/iridium (PtIr), platinum/gold (PtAu), gold/ruthenium (AuRu), and ruthenium (Ru). The transition metal nanoparticles were deposited onto the electrodes by sputtering thin metal layers onto polycrystalline BDD by a physical vapor deposition method, followed by a heat treatment at 700°C under oxygen-free atmosphere. The tested coffee capsules were coded as follows: black (intensity 12), orange (intensity 5), blue (decaffeinated, intensity 6), and caramel (caramel flavored). PCA and HCA analyses indicated a clear separation of the low-intensity coffee (orange) from the remaining samples. Separation of the decaffeinated coffee from the remaining samples occurred at 80% similarity. The authors suggested that selectivity could be enhanced by addition of transducers that are sensitive to non-electroactive species.

Kim and Kang (2018) proposed a method based on colorimetric sensor array (CSA) for discrimination of roasted coffee samples with respect to species (Arabica vs. Robusta) and roasting degree (light vs. medium/dark) as well as coffee beverage preparation method (espresso vs. aero-press). The array was fabricated by printing 12 sensing materials selective to key coffee aroma odorants on a hydrophobic membrane. A unique color profile was obtained for each coffee sample, allowing for differentiation by the naked eye without any statistical analysis. The CSA color responses were evaluated by PCA and HCA, and the samples were successfully distinguished according to their categories. Results indicated that CSA is a potentially promising methodology for rapid discrimination of different coffees.

Pérez-Míguez et al. (2019) developed a methodology based on capillary electrophoresis-mass spectrometry (CE-ESI-QToF-MS) for metabolic fingerprinting of *C. arabica* samples. The goal was to evaluate changes occurring in the metabolic profiles of coffee during roasting. The analyses were carried out in positive ESI mode, aiming to detect highly polar cationic metabolites present in coffee beans. The different parameters that affected electrophoretic separation (background electrolyte, temperature, voltage, and injection time), MS detection (temperature and flow of drying gas, sheath gas of jet stream, temperature, and capillary, fragmentator, nozzle, skimmer, and octapole voltages), and the sheath liquid (composition and flow rate) were evaluated. Results from multivariate statistical analysis provided a clear discrimination of the following coffee groups: green, light roast, medium roast, and dark roast. Different families of compounds, including pyridines, pyrroles, betaines, and indoles, were pointed out as markers of the coffee roasting process.

Makimori and Bona (2019) employed an e-nose, with seven metal-oxide-semiconductor (MOS) sensors, for the analysis of samples of six different commercial instant coffees produced by the same industry. Chemometric methods such as common dimension (CD) analysis and LDA were applied to classify the samples. Four CDs were used to represent the e-nose data, and the scores from the first four CDs were used as input to construct LDA classifiers. All models presented sensitivity and specificity of 100% in leave-one-out cross-validation. Therefore, the proposed methodology was deemed feasible for application in the coffee industry as an efficient aromatic quality control method.

Kim et al. (2019) employed headspace SPME coupled with GC-MS to identify volatiles capable of discriminating green and roasted coffee beans that underwent distinct fermentation processes: digestive (luwak coffee) and monsooning (monsoon coffee). The most representative flavor compounds of the luwak green coffee were pentamethyl heptane and 3-methyl-1-butanol, whereas a wider variety of compounds was observed for the monsooned green beans, namely alkyl butanals, pinenes, benzaldehyde and butanediols. After roasting, the fermented form of both beans contained similar types of volatiles, except for α-pinene in luwak coffee and 2-methyl furan in monsooned beans. Results from PCA and PLS-DA analyses provided clear separation of all groups. Based on these analyses, the most significant compounds contributing to the identification of the each group were as follows: 3-methyl-1-butanol (luwak, green); β-pinene, ethylbenzene, and benzaldehyde (monsoon, green); furfural, α-pinene, and 3-carene (luwak, roasted); and pyrazine, alkyl pyrazines, pyridine, and alkyl pyridines (monsoon, roasted).

Tsukui and co-workers (2019) evaluated the potential of ESI(−) FT-ICR MS for discrimination of coffees with respect to post-harvest processing (dry vs. semi-dry), employing coffees from two subsequent crops. Coffee fruits submitted to dry processing were dried under the sun on a cement terrace to 12% moisture and later peeled. In the semi-dry process, the coffee cherries were washed mechanically, peeled, and then dried also under the sun to 11% moisture. After drying, the coffee samples were pulped and kept under 11% moisture. Analysis was carried out using methanol:water extracts. Despite the expected variation in climate and other parameters from one crop to the other, PLS-DA analysis indicated a clear separation according to processing method (dry vs. semi-dry), with atractyligenin derivatives being the substances responsible for discrimination.

An untargeted metabolomic method based on UPLC-QTOF was employed by Xu et al. (2019) to investigate differences among the following coffee brewing methods: boiling, pour-over, and cold brew. PCA and HCA showed clear separation of the three groups. Analysis of variance, fold change, and orthogonal projection to latent structures discriminant mode were conducted to find characteristic potential markers. Nine potential markers were identified using general chemical databases, and five of them were further confirmed by reference standards. Results showed that the contents of two selected markers, norharman and harman, were higher in the thermally processed samples (pour-over and boiling) in comparison to the cold-brew method.

Velásquez et al. (2019) evaluated the feasibility of discriminating coffees at different maturity stages with respect to the volatile and sensory profiles after roasting. Coffee cherries from two coffee cultivars (Caturra and Catimor) were sorted into seven maturity stages, from fully immature (Stage 1, green) up to fully overripe (Stage 7, purple). Although the volatile profile of Stage 1 roast coffees presented lower concentrations of carbohydrate degradation products and higher concentrations of N-heterocycles and phenols, the remaining stages did not present significant differences with respect to volatile profiles. A similar behavior was observed in terms of sensory analysis. Given that the cupping panel reported few differences among Stages 2–7 in comparison to significantly lower sensory scores for Stage 1, it was concluded that partially mature and overripe cherries could be appropriate for specialty coffee.

13.6 Conclusion

Distinct analytical methodologies have been recently developed to discriminate coffees by species, geographical origins, and quality with different levels of success. Relatively simple spectroscopic techniques (e.g., FTIR, NIRS) and sophisticated hyphenated techniques (e.g., LC-MS, GC-MS) were employed for such task. A few methodologies were based on targeted approaches, in which the detection and quantification of target chemical markers was sought after (e.g., caffeine, CGAs, tocopherol, polyunsaturated FAs, and others). However, due to the complexity of the coffee matrix, both in the raw and in the roasted state, the majority of them was based on untargeted approach, in which fingerprint chemical profiles were used for the discrimination (e.g., FTIR spectra, MS spectra, and others). Regardless of the approach, multivariate statistics was necessarily employed to allow for the development of discriminating models. Analytical methodologies that are fast and require little or no sample preparation at all (e.g., NIRS, FTIR) are suitable for employment in routine analysis for coffee discrimination, for they are usually of lower cost than the sophisticated hyphenated ones. However, sophisticated techniques, such as NMR and mass spectrometry, although more labor intensive and expensive than NIRS or FTIR, are necessary to proper calibrate the other analytical methods and ensure their functionality for discriminating coffees.

13.7 Acknowledgment

The authors acknowledge financial support from CNPq (Conselho Nacional de Desenvolvimento Científico e Tecnológico), Brazil.

REFERENCES

Alves RC, Casal S, Alves MR, Oliveira MB. Food Chem. 2009;114: 295–299.

Anderson KA, Smith BW. J. Agric. Food Chem. 2002;50: 2068–2075.

Antoine JMR, Fung LAH, Grant CN. J. Radioanal. Nucl. Chem. 2016;309:525–534.
Arana VA, Medina J, Alarcon R, Moreno E, Heintz L, Schäfer H, Wist J. Food Chem. 2015;175:500–506.
Arrieta AA, Arrieta PL, Mendoza JM. Acta Sci. Pol. Technol. Aliment. 2019;18:35–41.
Assis C, Oliveira LS, Sena MM. Food Anal. Methods 2018;11: 578–588.
Babova O, Occhipinti A, Maffei ME. Phytochemistry 2016;123: 33–39.
Badmos S, Lee SH, Kuhnert N. Food Res. Int. 2019;126:108544.
Barbosa MSG, Scholz MBS, Kitzberger CSG, Benassi MT. Food Chem. 2019;292:275–280.
Belchior V, Botelho BG, Oliveira LS, Franca AS. Food Chem. 2019;273:178–185.
Bicho NC, Leitão AE, Ramalho JC, Lidon FC. Eur Food Res. Technol. 2011;233:303–311.
Bona E, Marquetti I, Link JV, Makimori GYF, Arca VC, Lemes ALG, Ferreira JMG, Scholz MBS, Valderrama P, Poppi RJ. LWT – Food Sci. Technol. 2017;76:330–336.
Borsato D, Pina MVR, Spacino, KR, Scholz MBS, Androcioli Filho A. Eur. Food Res. Technol. 2011;233:533–543.
Botelho BG, Oliveira LS, Franca AS. Food Control 2017;77:25–31.
Bressanello D, Liberto E, Cordero C, Sgorbini B, Rubiolo P, Pellegrino G, Ruosi MR, Bicchi C. J. Agric. Food Chem. 2018;66:7096–7109.
Briandet R, Kemsley EK, Wilson RH. J. Agric. Food Chem. 1996;44:170–174.
Buratti S, Sinelli N, Bertone E, Venturello A, Casiraghi E, Geobaldo F. J. Sci. Food Agric. 2015;95:2192–2200.
Casal S, Oliveira MBPP, Alves MR, Ferreira MA. J. Agric. Food Chem. 2000;48:3420–3424.
Charrier A and Berthaud J. In MN Clifford and KC Willson (Eds.), *Coffee: Botany, Biochemistry and Production of Beans and Beverage*. 1985. (pp. 13–47). Westport, CT: The AVI Publishing Company, Inc.
Colzi I, Taiti C, Marone E, Magnelli S, Gonnelli C, Mancuso S. Food Chem. 2017;237:257–263.
Consonni R, Cagliani LR, Cogliati C. Talanta 2012;88:420–426.
Correia RM, Loureiro LB, Rodrigues RRT, Costa HB, Oliveira BG, Filgueiras PR, Thompson PJ, Lacerda Jr. V, Romão W. Anal. Methods 2016;8:7678.
Correia RM, Tosato F, Domingos E, Rodrigues RRT, Aquino LFM, Filgueiras PR, Lacerda V, Romão W. Talanta. 2018;176:59–68.
Cossignani L, Montesano D, Simonetti MS, Blasi F. J. Anal. Methods Chem. 2016:Article ID 7482620.
Costa Freitas AM, Parreira C, Vilas-Boas L. J. Food Compos. Anal. 2001;14:513–522.
Couto CC, Santos TF, Mamede AMGN, Oliveira TC, Souza AM, Freitas-Silva O, Oliveira EMM. Food Res. Int. 2019;115: 227–233.
Craig AP, Franca AS, Oliveira LS. Food Chem. 2012;132: 1368–1374.
Craig AP, Franca AS, Oliveira LS, Irudayaraj J, Ileleji K. Talanta 2015;134:379–386.
Craig AP, Botelho BG, Oliveira LS, Franca AS. Food Chem. 2018;245:1052–1061.
Daniel D, Lopes FS, Santos VB, Lago CL. Food Chem. 2018;243: 305–310.
Dankowska A, Domagała A, Kowalewskic W. Talanta 2017;172: 215–220.
Dias R, Benassi M. Beverages 2015;1:127–139.
Dias RCE, Valderrama P, Março PH, Scholz MBS, Edelmann M, Yeretzian C. Food Chem. 2018;225:132–138.
Domingues DS, Pauli ED, Abreu JEM, Massura FW, Cristiano V, Santos MJ, Nixdorf SL. Food Chem. 2014;146:353–362.
El-Abassy RM, Donfack P., Materny A. Food Chem. 2011;126: 1443–1448.
Esteban-Díez I, González-Sáiz JM, Sáenz-González C, Pizarro C. Talanta 2007;71:221–229.
Farah A, Monteiro MC, Calado V, Franca AS, Trugo LC. Food Chem. 2006;98:373–380.
Figueiredo LP, Borém FM, Almeida MR, Oliveira LFC, Alves APC, Santos CM, Rios PA. Food Chem. 2019;288:262–267.
Flambeau KJ, Lee WJ, Yoon J. Food Sci. Biotechnol. 2017;26: 1245–1254.
Franca AS, Oliveira LS. In Z Pan, R Zhang, S Zicari (Eds.), *Integrated Processing Technologies for Food and Agricultural By-Products*. 2019. (pp. 413–438). London: Academic Press.
Franca AS, Oliveira LS. In EN Koeffer (Ed.), *Progress in Food Chemistry*. 2008. (pp. 105–138). New York: Nova Publishers.
Franca AS, Oliveira LS, Mendonça JCF, Silva XA. Food Chem. 2005;90:89–94.
Franca AS, Mendonça JCF, Oliveira SD. LWT-Food Sci. Technol. 2005;38:709–715.
Franca AS, Oliveira LS, Oliveira RCS, Mancha Agresti PC, Augusti R. J. Food Eng. 2009;92:345–352.
Garrett R, Vaz BG, Hovell AMC, Eberlin MN, Rezende CM. J. Agric. Food Chem. 2012;60:4253–4258.
Garrett R, Schmidt EM, Pereira LFP, Kitzberger CSG, Scholz MBS, Eberlin MN, Rezende CM. LWT – Food Sci. Technol. 2013a;50:496–502.
Garrett R, Rezende CR, Ifa DR. Anal. Methods 2013b;5:5944.
Hovell AMC, Pereira EJ, Arruda NP, Rezende CM. Anal. Chim. Acta 2010;678:160–168.
Keidel A, Von Stetten D, Rodrigues C, Máguas C, Hildebrandt P. J. Agric. Food Chem. 2010;58:11187–11192.
Kemsley EK, Ruault S, Wilson RH. Food Chem. 1995;54:321–326.
Kim S-Y, Kang B-S. Sens. Actuators B Chem. 2018;275:277–283.
Kim, SJ, Lee, S, Bang, E, Lee, S, Rhee, JK, Na, YC. Flavour Fragr. J. 2019;34:365–376.
Link JV, Lemes ALG, Marquetti I, Scholz MBS, Bona E. Food Res. Int. 2014a;59:1–7.
Link JV, Lemes ALG, Marquetti I, Scholz MBS, Bona E. Chemom. Intell. Lab. Syst. 2014b;135:150–156.
Makimori GYF, Bona E. Food Anal. Methods. 2019;12:1067–1076.
Mancha Agresti PC, Franca AS, Oliveira LS, Augusti R. Food Chem. 2008;106:787–796.
Marquetti I, Link JV, Lemes ALG, Scholz MBS, Valderrama P, Bona E. Comput. Electron. Agric. 2016;121:313–319.
Mehari B, Redi-Abshiro M, Chandravanshi BS, Atlabachew M, Combrinck S, McCrindle R. Food Anal. Methods 2016;9: 1627–1637.
Mendonça JCF, Franca AS, Oliveira LS, Afonso RJCF. In JK Lang (Ed.), *Handbook on Mass Spectrometry: Instrumentation, Data and Analysis, and Applications*. 2009 (pp. 269–290). New York: Nova Publishers.
Monakhova YB, Ruge W, Kuballa T, Ilse M, Winkelmann O, Diehl B, Thomas F, Lachenmeier DW. Food Chem. 2015;182:178–184.
Monteiro PI, Santos JS, Brizola VRA, Deolindo CTP, Koot A, Boerrigter-Eenling R, van Ruth S, Georgouli K, Koidis A, Granato D. Food Control 2018;91:276–283.

Morais TCL, Rodrigues DR, Polari Souto UTC, Lemos SG. Food Chem. 2019;273:31–38.

Nakagawa T, Doi M, Nishi K, Sugahara T, Nishimukai H, Asano M. Biosci. Biotechnol. Biochem. 2019;83:1829–1836.

Novaes FJM, Oigman SS, Alves de Souza ROM, Rezende CM, Aquino Neto, FR. Talanta. 2015:139:159–166.

Oliveira LS, Franca AS, Mendonça JCF, Barros-Júnior MC. Food Chem. 2005;90:287–291.

Oliveira RCS, Oliveira LS. Franca AS, Augusti R. LWT-Food Sci. Technol. 2006;39:235–239.

Oliveira SD, Franca AS, Glória MBA, Borges MLA. J. Food Composit. Anal. 2009;22;257–261.

Ossa DEH, Gil-Solsona R, Peñuela GA, Sancho JV, Hernández, F. Food Chem. 2018;250:89–97.

Peng CY, Zhang YL, Song W, Cai HM, Wang Y, Granato D. Food Chem. 2019;297:124963.

Pérez-Míguez R, Sánchez-López E, Plaza M, Marina ML, Castro-Puyana M. J. Chromatogr. A. 2019;1605:360353.

Piccino S, Boulanger R, Descroix F, Sing, ASC. Food Res. Int. 2014;61:264–271.

Reis N, Franca AS, Oliveira LS. LWT-Food Sci. Technol. 2013a;50: 715–722.

Reis N, Franca AS, Oliveira LS. LWT-Food Sci. Technol. 2013b; 53:395–401.

Reis N, Franca AS, Oliveira LS. Talanta. 2013c;115:563–568.

Reis N, Botelho BG, Franca AS, Oliveira LS. Food Anal. Methods 2017;10:2700–2709.

Ribeiro MVM, Boralle N, Pezza HR, Pezza L, Toci AL. J. Food Compos. Anal. 2017;57:24–30.

Risticevic S, Carasek E, Pawliszyn J. Anal. Chim. Acta. 2008; 617:72–84.

Romano R, Santini A, Grottaglie L, Manzo N, Visconti A, Ritieni A. J. Food Compos. Anal. 2014;35:1–9.

Sano EE, Assad ED, Cunha SAR, Correa, TBS, Rodrigues, HR. J Food Qual. 2003;26:123–134.

Sarraguça MC, Santos JR, Rangel AOSS, Lopes, JA. Food Anal. Methods 2013;6:892–899.

SCAA: Cupping Protocols of Specialty Coffee Association of America, 2009. www.scaa.org/?page=resources&d=cupping-protocols. Accessed 04.16.20.

Schievano E, Finotello C, Angelis E, Mammi S, Navarini L. J. Agric. Food Chem. 2014;62:12309–12314.

Schipilliti L, Bonaccorsi I, Buglia AG, Mondello L. Food Anal. Methods 2019;12:121–127.

Sezer B, Apaydin H, Bilge G, Hakki Boyaci IH. Food Chem. 2018;264:142–148.

Sittipod S, Schwartz E, Paravisini L, Peterson DG. Food Chem. 2019;301:125250.

Souto UTCP, Barbosa MF, Dantas HV, Pontes AS, Lyra WS, Diniz PHGD, Araújo MCU, Silva EC. LWT-Food Sci. Technol. 2015;63:1037–1041.

Spaniolas S, May ST, Bennett MJ, Tucker GA. J. Agric. Food Chem. 2006;54:7466–7470.

Taveira JHS, Borém FM, Figueiredo LP, Reis N, Franca AS, Harding SA, Tsai CJ. Food Res. Int. 2014;61:75–82.

Toci AT, Farah A, Pezza HR, Pezza L. Crit. Rev. Anal. Chem. 2016;46:83–92

Toci AT, Ribeiro MVM, Toledo PRAB, Boralle N, Pezza HR, Pezza L. Food Sci. Biotechnol. 2018;27:19–26.

Toledo PRAB, Melo MMR, Pezza HR, Toci AT, Pezza L, Silva CM. Food Control 2017;73:164–174.

Tolessa K, Rademaker M, Baets B, Boeckx P Talanta. 2016;150: 367–374.

Tsukui A, Vendramini PH, Garrett R, Scholz MBS, Eberlin MN, Bizzo HR, Claudia M. Rezende CM. LWT-Food Sci. Technol. 2019;103:205–211.

Vasconcelos ALS, Franca AS, Glória MBA, Mendonça JCF. Food Chem. 2007;101:26–32.

Velásquez S, Peña N, Bohórquez JC, Gutierrez N, Sacks GL. Food Chem. 2019;274:137–145.

Wei F, Furihata K, Koda M, Hu F, Kato R, Miyakawa T, Tanokura M. J. Agric. Food Chem. 2012;60:10118 – 10125.

Wermelinger T, D'Ambrosio L, Klopprogge B, Yeretzian C. J. Agric. Food Chem. 2011;59:9074–9079.

Worku M, Upadhayay HR, Latruwe K, Taylor A, Blake W, Vanhaecke F, Duchateau L, Boeckx P. Food Chem. 2019;290: 295–307

Xu L, Lao F, Xu Z, Wang X, Chen F, Liao X, Chen A, Yang S. Food Chem. 2019;286:106–112.

Yang N, Liu C, Liu X, Degn TK, Munchow M, Fisk I. Food Chem. 2016;2012:206–214.

Zhang C, Shen T, Liu F, He Y. Sensors 2018;18:95.

Zribi B, Dragoe D, Scorsone E. Sens. Actuators B Chem. 2019;290:147–154.

14

New frontiers in coffee processing for health and enhanced flavor molecules

Varun Eranna and Pushpa Srinivas Murthy

CONTENTS

14.1 Introduction

Botanically, coffee belongs to the family Rubiaceae, genus *Coffea*. More than 80 species have been identified worldwide, but only two have gained economic importance. Currently, worldwide coffee consumption is 70% and is consistently growing at an annual rate of 2.5%. It also contributes to an increase of 17% per annum in its agricultural production (Sette 2017). Currently, coffee is referred to as a functional food because of its bioactive molecules exhibiting antioxidant, antimicrobial, anticarcinogenic, and host of other benefits.

The important groups of biologically essential components are caffeine, caffeic acid, chlorogenic acids, trigonelline, melanoidins, and diterpenes (cafestol and kahweol). The composition of bioactive compounds depends mainly on pre-harvest (40%) and post-harvest (60%) factors, thereby affecting the physicochemical and sensory characteristics of coffee in a complex ways (Wintgens 2009). Due to the dynamic nature of coffee during all stages of processes, innovative methods are required to hold the nutritional and health bio-components until the peak. The selection of a post-harvest processing system is culture driven and depends upon the coffee type and origin. Along with, there is also a trend to introduce modulation

DOI: 10.1201/9781003043133-18

in routine processing methodologies to bring novel changes and advancements in coffee flavor. Thus advance research on bioactive roles of coffee components is essential.

14.2 Coffee bioactives and their health benefits

14.2.1 Bioactive components of coffee

Bioactive compounds have gained functionally more importance and vary according to plant genetics, processing, roasting, and brewing. These non-volatiles play a major role and help as aroma precursors and substrates for the Maillard reaction during roasting. The important biologically essential components are shown in Figure 14.1 and detailed below.

Caffeine (1,3,7-trimethylxanthine) is a colorless and odorless alkaloid that influences the bitterness and viscosity of brewed coffee (Vitzthum and Werkhoff 1976). Caffeine is a nitrogenous secondary metabolite, water soluble, and gives a physiological stimulating effect (Higdon and Frei 2006) and exhibits neuroprotection, anticarcinogenic, and antioxidant properties.

Chlorogenic acid (CGA) is the esterified product of caffeic acid and L-quinic acid, which imparts astringent taste to the coffee brews (Moon et al. 2009). CGAs are one of the major polyphenols observed in coffee and exhibit antioxidant, anti-obesity, antimicrobial, and neuroprotective activities. During roasting, chlorogenic acids and quinic acid may form chlorogenic lactones by reacting with volatile phenols, such as guaiacol and 4-vinylguaiacol, which influences increased bitterness of the coffee brew (Ginz and Engelhardt 2001). Additionally derivatives of CGA – such as 3-CQA, 3-FQA, 3,4-diCQA, 3,5-diCQA, 4-CQA, 4,5-diCQA, 5-CQA, 5-FQA, diCQA, and FQA – have exhibited hypolipidemic effects and similar pharmacological properties to the parent CGA compound. Caffeic acids (CA) are metabolites of chlorogenic acid that have strong

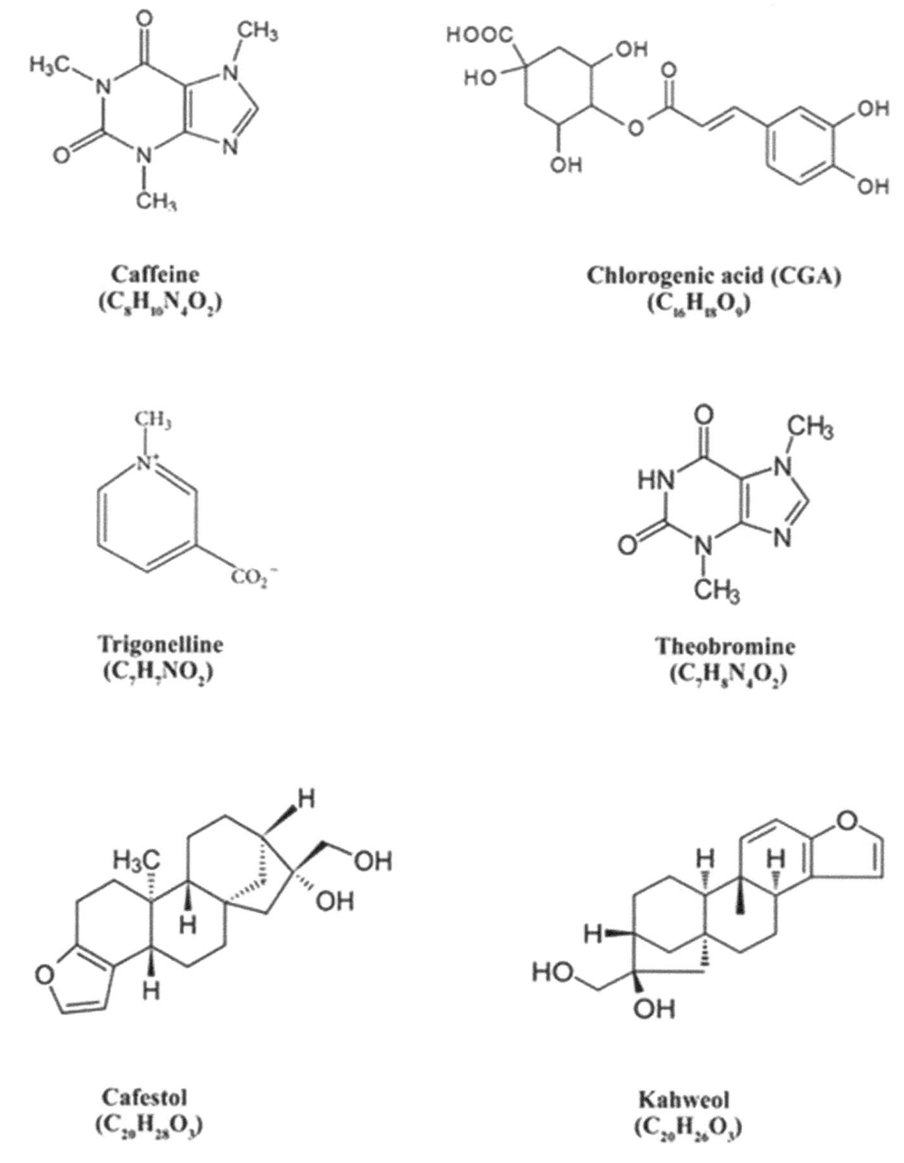

FIGURE 14.1 Major bioactive compounds of coffee.

antioxidant properties. CA has been shown to possess anti-inflammatory, anticarcinogenic, and enzyme-inhibiting properties (e.g., lipoxygenases, cyclooxygenases). It also has been found to reduce the risk of type 2 diabetes.

Trigonelline (N-methylpyridinium-3-carboxylate) is an alkaloid formed from the nitrogen methylation of nicotinic acid (niacin or B_3; Zhou et al. 2012). The roasting level in both Arabica and Robusta coffees can be discriminated using trigonelline. These are associated with perception of overall aroma of both roasted coffee beans and a brewed coffee beverage. Apart from caffeine, trigonelline is found at higher levels in the Arabica cultivar compared to others.

Diterpenes are a group of terpenoids obtained from the coffee oil, which is the lipid fraction that migrates to the bean's surface during roasting (Savonitti 2005). These lipid fractions are extracted into the coffee brew and contribute to viscosity, perceived texture, and mouthfeel of the brew. The kahweol and cafestol are the major diterpenes observed to have a positive effect on health in relation to cholesterol (Clarke and Vitzthum 2008). In addition, cafestol is found in both Arabica and Robusta, whereas kahweol is mainly present in Arabica and 16-O-methylcafestol is found only in Robusta (Narita and Inouye 2013). Kahweol is a fat-soluble factor present in Arabica coffee bean oil, and its concentration ranges from 0.1 to 7 mg mL^{-1}. It acts both as an adverse and chemoprotective factor. It also protects against carcinogens such as aflatoxin B_1, 2-amino-l-methyl-6-phenylimidazo[4,5-b]-pyridine, and 2-amino-l-methyl-6-phenylimidazo[4,5-b]-pyridine. Cafestol ranges from 0.4% to 0.7% in coffee and can regulate a variety of inflammatory mediators to reduce inflammation. In addition, this diterpene can prevent cancer from occurring by blocking the activation of carcinogens and improving liver detoxification functions. Cafestol also exhibit anti-osteoclastogenesis activities.

Theobromine (xantheose) is a water-soluble, bitter alkaloid used as a myocardial stimulant, vasodilator, and smooth muscle relaxant (Windholz et al. 1983). Theobromine is an isomer of theophylline. Theophylline (1,3-dimethylxanthine) is a methylxanthine drug used in therapy under various brand names for respiratory diseases such as chronic obstructive pulmonary disease and asthma.

The minor bioactive components includes melanoidins formed during the roasting process by the Maillard and caramelization reactions and have been found to possess antimicrobial, anticarcinogenic, anti-inflammatory, antihypertensive, and antiglycative properties (Borrelli et al. 2002). Nicotinic acid is a form of vitamin B_3, an essential human nutrient produced during the roasting of coffee by the decomposition of trigonelline present in the raw beans. Cinnamic acid is a white crystalline substance which functions as precursors to flavor components, imparting sweetness to the coffee. Cinnamic acids play key roles in the formation of other more complex phenolic compounds (Figure 14.2). Ferulic acid, p-coumaric acid, gallic acid, and many other micro-components exhibits good antioxidant properties.

Also, many volatile compounds contribute to bioactivity of coffee with beneficial health effects. Thus in the context of consumer health and physiological effects, data acquisition on bioactive compounds of the brews are gained most importance. From the above discussion, it is evident that the prime elements of the beneficial health effects of coffee are chemically complex and multi-factorial.

14.2.2 Modulation of coffee bioactives and their health implications

Consumption of coffee has exhibited many assured beneficial effects in various human and animal examinations, such as diseases related to the cardiovascular, nervous, and reproductive systems and also reduces cholesterol levels, diabetes, DNA injury, and chronic liver disease, among others, because

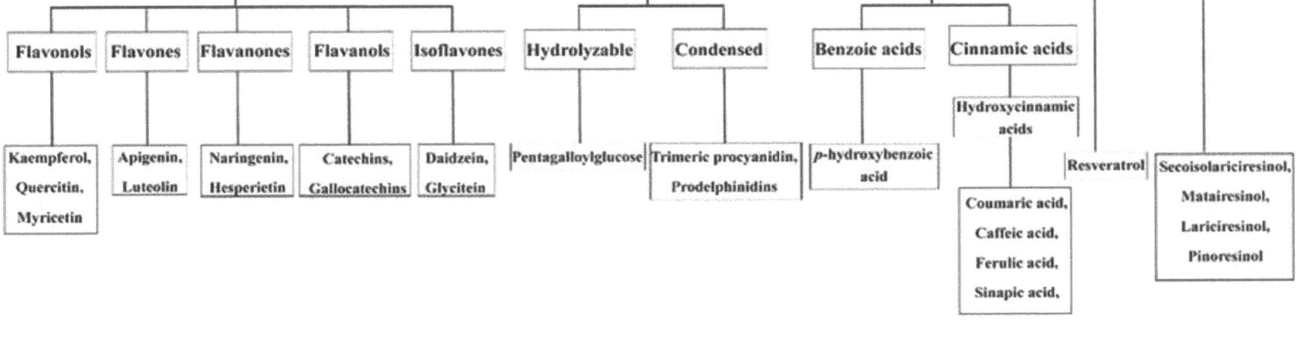

FIGURE 14.2 Phenolic compounds in coffee.

of its constituents. The aforementioned bioactive compounds are interwoven in various biological effects in the human body.

14.2.2.1 Anticardiovascular activity

The prime components of coffee responsible for anticardiovascular activities are CGAs, followed by caffeine, trigonelline, and melanoidins. CGAs have been observed to be antihypertensive and antithrombotic agents (Suzuki et al. 2006). The possible mechanism may be the activity of ferulic acid by improving the bioavailability of nitric oxide in the inner walls of blood vessels and regulating the tension of blood vessels. Additionally, compounds such as 1,3- and 1,4-dimethylpyridine formed during the roasting of coffee have influenced antithrombotic activity in a rat model (Kalaska et al. 2014). Many studies showed that coffee consumption at a moderate level could lower hypertension (Grosso et al. 2016), prevent cerebrovascular accident (CVA), and cardiovascular diseases (Liebeskind et al. 2016), and reduce the risk of type 2 diabetes (Santos and Lima 2016). The most of the effects of coffee consumption are thought to be associated with antioxidant property, depleting low-density lipoprotein (LDL) oxidation, and enhancing the ability of 5-methyl tetrahydrofolate to scavenge superoxide radicals, and increasing nitric oxide production (Ranheim and Halvorsen 2005).

14.2.2.2 Neuroprotection

Caffeine is the major bioactive compound involved in neuroprotection and, to a certain extent, kahweol. Caffeine acts as antagonist of adenosine receptors, which in turn releases excitatory transmitters including dopamine, and is demonstrated to increase calcium signaling and neurogenesis (Nehlig 2016). Another bioactive of coffee called trigonelline has also proved to be an effective drug for treatment of peripheral neuropathy, a common result of diabetes, nerve injuries, and infections (Arlt et al. 2013). Many studies on neurodegenerative diseases have reported that regular coffee/caffeine intake is associated with reduced risk of the development of Parkinson's and Alzheimer's diseases (Cappelletti et al. 2015) and prevention of memory decline during aging (Hashimoto et al. 2003). In individuals with Parkinson's, caffeine can be used to manage both motor and non-motor symptoms by attenuating dopaminergic neurodegeneration (Chen, 2019). Also, the effect of crude caffeine was found to lower the accumulation of β-amyloid peptides in mouse model studies (Chu et al. 2012). A meta-analysis of coffee drinkers and non-coffee drinkers in the United States, Europe, and Asia found a 25% lower risk of Parkinson's disease (Costa et al., 2010). Alzheimer's disease is a neurological disorder most common in people over age 60; the disease disorganize a person's ability to function independently and causes progressive cognitive decline. Presently, there is no cure for Alzheimer's disease (Waite 2015), but there are preventive methods, including consumption of foods rich in polyphenols and caffeine (Blum et al., 2019).

14.2.2.3 Anticancer activity

Trigonelline and diterpenes are the prime components exhibiting anti-carcinogenic properties. Trigonelline has been found to be an inhibitor of Nrf2 gene transcription, which causes pancreatic tumor cells to be more susceptible to apoptosis. In diterpenes, the oridonin and abietane types have been observed to be cytotoxic/antiproliferative against leukemia-derived Jurkat cells and the human pancreatic cancer cell line MIA PaCa-2, respectively (Pascual et al. 2002; Yeretzian et al. 2012). Coffee consumption has reported to lower the risk of primary live cancer and hepatocellular carcinoma (Heliövaara et al. 2000), antiproliferative against some human cancer cell lines (Clark and Landolt 2017) and a reduction in oropharyngeal (Al-Dakkak 2011), esophageal (Yu and Bao, 2011), gastroesophageal (Li et al. 2015), colorectal (Michels et al. 2005), hepatocellular (Bravi et al. 2017), pancreas (Dong et al. 2011), urothelial (Sugiyama et al. 2017), prostate (Wilson et al. 2011), endometrial (Je and Giovannucci 2012), and brain cancers (Holick et al. 2010).

14.2.2.4 Anti-diabetic activity

It is observed that there is a reduction in development of human diabetes by coffee intake, in which ingredients such as caffeine, CGAs, and trigonelline play major roles in prevention (Kobayashi et al. 2017). A study on mice subjected to streptozotocin- and niacinamide-induced type 2 diabetes found that caffeine metabolism protected the pancreas from development of type 2 diabetes (Fernandez-Gomez et al. 2016). Apart from this study, Cheng and co-workers (2011) reported that caffeic acid and caffeine intake could inhibit the aberrant misfolding of human amylin (or islet amyloid polypeptide [IAPP]), which reduces the incidence of type 2 diabetes (Cheng et al. 2011). Chlorogenic acids have found to be pancreas protective, similar to caffeine. They can also enhance the body's insulin response and normalize the acute glucose response. Henry-Vitrac and team (2010) observed that the coffee extract Svetol has inhibited hydrolysis of glucose-6-phosphatase in human liver microsomes, which can achieve antidiabetic effects by reducing the production of hepatic glucose (Henry-Vitrac et al. 2010). Stefanello and coworkers (2014) have found that CGA and caffeic acid (CA) have prevented lipid peroxidation, improved memory and reduced anxiety in diabetic rats by inhibiting acetylcholinesterase (AChE) activity (Stefanello et al. 2014). Trigonelline can regulate key enzymes of lipid and glucose metabolism (like fatty acid synthase, glucokinase, glucose-6-phosphatase, etc.), by which it can improve auditory threshold changes caused by diabetic neuropathy and slow down the latency of auditory evoked potentials in mice (Yoshinari et al. 2009). However, trigonelline (50 mg kg^{-1} daily for 4 weeks) was also found to increase bone mineral density and spongy bone strength in nicotinamide/streptozotocin-treated rats (Folwarczna et al. 2016).

14.2.2.5 Antimicrobial activity

Coffee exhibits antimicrobial activities on different microbial entities. CGAs, dicarbonyl compounds, melanoidins, polysaccharides, and some other trace ingredients also play great roles in the antibacterial activity of coffee. CGA has reported hepatitis B virus inhibition in a duckling model (Wang et al. 2009), as well as potential antiviral activity against H1N1 influenza (Urushisaki et al. 2011). The analog of CGAs has also exhibited appreciable anti-HIV activities (Ma et al. 2010) along with additional antibacterial and antifungal properties. CGA has observed inhibition of all tested bacterial pathogens at minimum inhibitory concentrations (MIC) of 20–80 μg mL^{-1}. Additionally, TRG of *C. canephora* extracts noted an observed decrease in biofilm formation via bacteriostatic action on *Streptococcus mutans* at an MIC of 0.8 mg mL^{-1} (Antonio et al. 2010).

14.2.2.6 Antioxidant and anti-inflammatory activities

Crude caffeine was examined to possess both hydrophilic (145 μmol TE (Trolox equivalent)/g) and lipophilic antioxidant activity (66 μmol TE/g), and its application has led to the inhibition of cyclo-oxygenase-2 enzyme better than aspirin. Caffeine anti-inflammatory potential has also been observed in many studies, which is linked to modifications in production of cell-signaling molecules and the release of anti-inflammatory cytokines, including interleukins (IL-10 and IL-2; Tauler et al. 2016). Caffeine also mediates immune-suppression of release of cytokines, including TNF-α (tumor necrosis factor alpha) and IFN-γ (interferon-gamma), which has a central role in initiation and propagation of autoimmune disease (Ritter et al. 2005). In some studies, antioxidant function of CGAs has indicated anti-HIV potential (Cavalcanti et al. 2011).

14.2.2.7 Anti-obesity activity

There are several reports that coffee intake reduces body fat accumulation and triglyceride levels (Carson et al. 1994; Thom 2007). Coffee also influences the gut microbiota in obese animals and humans. CGA has been found to reduce cholesterol and triglyceride concentrations in serum and liver (de Sotillo and Hadley 2002). Green coffee prepared using unroasted beans has also been noted in weight loss (Onakpoya et al. 2010).

Bioactive contents are modulated by pre-harvest and post-harvest methodologies. Various primary and secondary processing methods have been emphasized for optimum conditions; their influences on bioactive composition are explained in the next section.

14.3 Coffee processing for enhanced coffee quality

The quality of coffee is a multidimensional trait, and evidently 60% of the coffee quality attributes are regulated by post-harvest processing (Hameed et al. 2018). Every process has a unique and important role for producing a splendid quality brew. The wet and the dry methods are two main ways commonly used to process coffee fruits (Figure 14.3). The fermentation of coffee is the major factor, which can influence

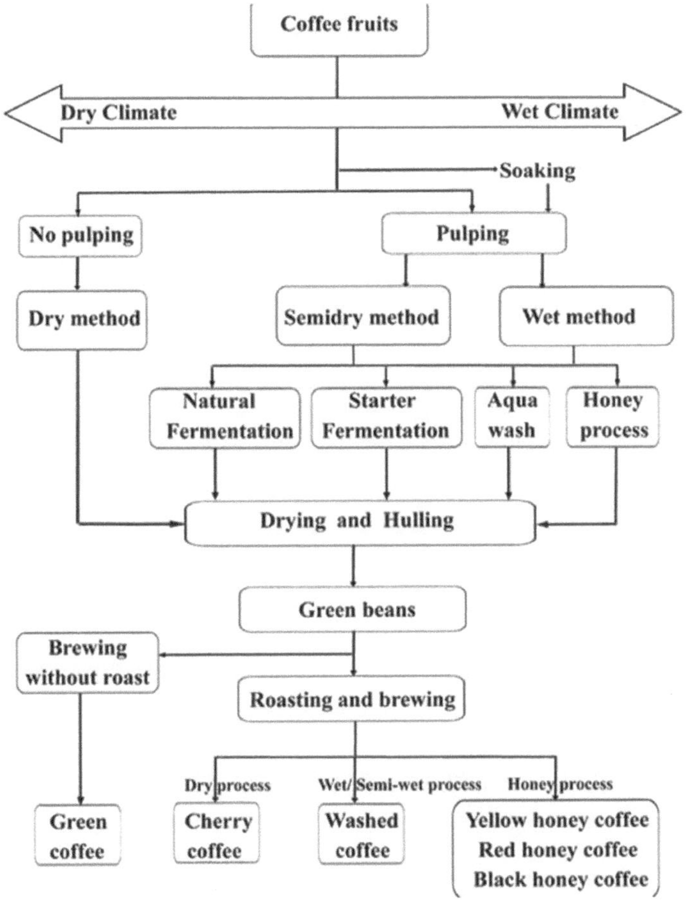

FIGURE 14.3 Coffee processing methods used according to regional climatic characteristics.

coffee aroma and organoleptic properties. During fermentation, microflora mainly secrete pectinases and cellulase enzymes, resulting in release of microbial metabolites that significantly influence the formation of volatile flavor notes. Volatile compounds are the most important quality determinant, which include hydrocarbons, aldehydes, alcohols, ketones, pyrazines, pyridines, pyrroles, esters, furans, sulfur compounds, phenols, and furanones (Sunarharum et al. 2014). The nonvolatile compounds include alkaloids (caffeine and trigonelline), CGAs, carboxylic acids, proteins, lipids, carbohydrates, polymeric polysaccharides, and melanoidins (Buffo and Cardelli-Freire 2004).

14.3.1 Primary processing

14.3.1.1 Dry processing

The drying of whole fruit involves sun-drying/fermenting, which is a long process and imparts heavy in body, sweet, smooth, and complex properties to the coffee. Of the Arabica coffee produced in Brazil, Indonesia, Ethiopia, Paraguay, Ecuador, and India, 95% is are sun-dried (Silva et al. 2008). A higher number of dry-processed beans are found defective with musty, earthy, greenish coffee defects and are no longer considered better than wet-processed beans (Tadesse et al. 2016). The drying surfaces (flat surfaces or raised beds) on which cherries are dried depend on temperature distribution, which is also considered a critical conditions to maintain high-quality drying. Studies also reveal that better cup quality is conserved for beans which slow dried (60/40°C) on patios with hot air rather than on flat-ground surfaces (Da Rosa et al. 2015). Additionally, coffee membranes are damaged at temperatures of 60°C or higher at 20% to 30% (w.b.) moisture level (Coradi 2007). During drying, a thicker cherry layer of >20 kg m^{-2} restricts the movement of air and uniformity of drying, consequently requiring a longer duration of fermentation. This overfermentation develops an acidic environment and a higher microbial load, which results in degradation of CGAs and increases a mold mycotoxin called ochratoxin A (OTA), respectively (Gonzalez-Rios et al. 2007). The production of good-quality dried coffee results in larger accumulation of proteins, which engage in Maillard reactions with reducing sugars to produce enhanced flavor and aroma profiles (do Livramento et al. (2017). A study by do Livramento et al. (2017) examined the influence of different drying systems on proteomics of green coffee beans and concluded that natural sun-dried coffee has the highest content of protein after wet-processed pulped coffee (do Livramento et al. 2017).

14.3.1.2 Wet processing

During wet processing, coffee fruits are mechanically de-pulped by leaving mucilage to adhere to the beans. These de-pulped coffee beans are subjected to fermentation for 6–72 hours in large tanks, during which the mucilage undergoes microbial degraded. The beans are then finally sun-dried (Evangelista et al. 2014a). The processing manner adopted in wet processing determines the chemical composition and final cup quality of brews. Additionally, changes in the amount of volatile compounds also rise due to modifications in wet processing methods (Gonzalez-Rios et al. 2007). The sugar-rich pulp and mucilage are a good source of microbes (yeasts and bacteria); the metabolism of these microbial communities results in production of organic acids, predominantly lactic acid (up to 2.33 g kg^{-1}), followed by acetic acid (up to 0.02 g kg^{-1}), and traces of citric, malic, and succinic acids (Evangelista et al. 2014b). The usage of mature fruits and persistence with accuracy are the two major factors that exclusive decide the quality attributes of coffee during processing (Selmar et al., 2006). Many findings recorded that physiological environment, germination, extent of stress, and microbial community variations play vital roles in determining the quality of final cup (Selmar et al. 2014).

14.3.1.3 Starter culture fermentation

Fermentation phase is crucial in coffee processing; if well controlled, it can influence the creation of essential sensory quality characteristics (Lee et al. 2015). Starter culture technology can also be used for better control of fermentation by using specific microorganisms to degrade coffee mucilage and impart quality attributes. Furthermore, the microbial community and its diversity are also influenced by the environmental conditions of fermentation. Poltronieri and coworkers (2016) have considered six factors of fermentation crucial for desirable attributes: water activity (Aw), sugar concentration (measured in degrees Brix), availability of oxygen, temperature, acidity, and duration (Poltronieri and Rossi 2016). In the studies of Vilela et al. (2010), out of 144 microbial isolates, only nine microbial strains belonging to *Pichia fermentans, Saccharomyces* sp., *P. kluyveri, P. guilliermondii, Hanseniaspora opuntiae*, and *Candida glabrata* were identified as having a good impact on producing distinct aromatic compounds (Vilela et al. 2010). Further, these strains were verified and found responsible for producing 14 volatile compounds that impart a characteristic fruity flavor to washed coffee (Czerny and Grosch 2000). *P. kluyveri* YC7.16 and *P. fermentans* YC5.2 produce a substantial quantity of ethyl acetate and isoamylacetate, which are responsible for fruity flavors such as banana and pineapple.

An expansive study in this respect was carried out by Massawe and Lifa (2010), who selected and used three strains of *Pichia anomala* and *P. kluyveri*; six strains of LAB as a starter culture in the wet-process fermentation. They found a contribution towards the suppression/inhibition of other microorganisms, including the fungi producing OTA (Massawe and Lifa 2010). Inclusion of LAB in the starter results in a considerable amount of lactic acid production and endows pleasant acidic body on final brew, whereas butyric, acetic, and propionic acids impart oniony and foul flavor development (Lara et al. 2014). Based on the above criteria, LAB strain *L. plantarum* LPBR01 was selected to use as inoculum for pulped coffee fermentation and compared with a natural fermented coffee. The fermentation time was reduced to 12 hours, with efficient conversion of hexoses to lactic and fumaric acids by enhancing the aroma compounds, giving a fresh fruity flavor., et al., LWT. 2019; 114: 108382 has studied further the starter culture application in Arabica coffee by optimizing the mixture of cultures such as *S. cerevisiae, L. plantarum*, and *B. sphaericus* (1:1:1) in comparison with natural fermentation.

The inoculum concentration of 10% was found significant in de-mucilization of coffee beans. In silico studies were also performed on the fermentation mechanism and noted minimal interaction between pectin and pectinase at lower energy level. The best interaction was observed in 1CZF (binding energy: –3.92 kJ mol^{-1}) with potent enzymatic activity and desired fermentation. Additionally, the sensory profile of starter consortia fermentation scored 7.0 compared to natural fermented (5.6) on 10-point scale (Siridevi et al. 2019).

14.3.1.4 Honey processing

During the past years, coffee processors in several countries have been experimenting with natural sun-dried methods and hybrid methods like the honey process, in which most of the mucilage is left on the bean after washing. In honey processing, coffee fruits were de-pulped to remove skin by retaining mucilage of 20%–80%. Then the de-pulped beans were dried quickly in the sun or slowly in the shade, yielding yellow, red, and black honey coffee. This processing style often results in an array of tastes that had never before been associated with coffee (Boot 2013). The quality of honey-processed coffee is also expressed in terms of percentage of retained mucilage level. A coffee with approximately 25% of mucilage retained, typically dried under little cloud/shade cover in order to speed up the drying time to about 8 days, will gain a yellow color and be called yellow honey coffee. Similarly, coffee with approximately 50% of mucilage retained, typically dried with cloud cover or shading for about 12 days is known as red honey coffee. Coffee with 100% mucilage retained with a prolonged drying period of about 30 days is called black honey coffee. In each area, the initial moisture level of coffee cherries and the weather conditions contribute to the duration of the period required for the drying process (Poltronieri and Rossi 2016).

14.3.1.5 Enzyme application

Pectinases play an vital role in coffee fermentation and are sometimes added to remove the mucilage layer of the bean. Cellulases and hemicellulases also present in the enzyme preparation aid the degradation of the mucilage. A diluted commercial enzyme was applied at a dose of 2 ± 10 g ton^{-1} at 15–20°C and was observed to accelerate fermentation by reducing fermentation time by 25–50% (Amorim and Amorim 1977). Murthy and Naidu (2011) produced pectinase using *Aspergillus niger* CFR utilizing coffee pulp by solid-state fermentation. The exogenous pectinase has observed 7539 ±70 U g^{-1} of enzyme activity (Murthy and Naidu 2011). Since, the large-scale application of commercial pectinases is costly and uneconomical, the utilization of coffee waste for the source of microbial pectic enzymes is much more effective. Fungal pectinases are also used, and the dosage of the enzyme must be adjusted carefully to avoid damage to the quality of the beans (Oberman et al. 1985).

14.3.1.6 Genetics and breeding applications

Biotechnological techniques such as somatic embryogenesis, protoplast culture, micro-propagation, and in vitro selection are presently at an improved stage of development and are being used to enhance coffee quality. The molecular phylogeny of *Coffea* species has been generated using DNA sequencing. The chloroplast DNA variation (Lashermes et al. 1996) and the internal transcribed spacer (ITS 2) region of the nuclear ribosomal DNA (Lashermes et al. 1997) were successfully used to infer phylogenetic relationships of *Coffea* species. The genetic diversity in *C. arabica* appeared extremely reduced in comparison to the diversity observed in *C. canephora* (Lashermes et al. 2000). In terms of genetic linkage maps, a complete linkage map of *C. canephora* (Lashermes et al. 2001) and a partial interspecific linkage map *C. pseudozanguebarie* (*C. liberica*) have so far been reported (Ky et al. 2000). These data can be helpful particularly in mapping important genes (e.g., the S-locus controlling self in compatibility) as well as underlying loci (Lashermes et al. 1996). The geographical distributions of the major groups are also indicated, in addition to quantitative traits (i.e., QTL), such as biochemical compounds implied in development of coffee cup quality (Barre et al. 1998). The genes encoding caffeine biosynthetic pathway have been amply studied; 7-methylxanthine methyltransferase and xanthosine-N7-methyltransferase have been isolated (Maoisyadi et al. 1997). Seed storage proteins have also gained particular attention, and the 11S-globulin gene coding has been cloned and characterized (Rogers et al. 1999).

14.3.2 Secondary processing

14.3.2.1 Roasting and blending technology

Roasting is a crucial step in the generation of characteristic volatile flavor and aroma compounds from a combination of pyrolysis, Maillard, and caramelization products. The roasting of coffee affects the level of antioxidant activity; it has been noticed that medium roasted coffee provides maximum antioxidant activity because of the most favorable conditions for decomposition of phenolic compound. Antioxidant activity can be optimized by blending light roast and dark roast beans. It has been reported that thermally unstable chlorogenic acids degrade progressively during roasting phase; up to 93% degradation for dark roasted coffee has been shown (Farah et al. 2006). Additionally, the production of nicotinic acid is dependent on the extent of roasting; dark roast coffee contains more nicotinic acid than light roast coffee. Normal samples of roasted coffee contained 95–263 μg g^{-1} of nicotinic acid. Thus, it is clear that roasting technique affects many bioactive compounds, and this is different between Robusta and Arabica coffees. The compounds generated by the result of roasting includes methyl pyridinium and dimethyl pyridiniums by thermal degradation of trigonelline, and acrylamide formation by Maillard reaction between asparagine and reducing sugars (Clifford 2012).

14.3.2.2 Decaffeination

Decaffeinated coffee is beneficial for consumers who have health problems such as insomnia, anxiety, nervousness, and stress (Shlonsky 2003). The market for decaffeinated coffee is increasing worldwide due to increasing health consciousness among the population (Mazzafera et al. 2009). Caffeine is removed from coffee by different extraction method, and organic solvents like ethyl acetate or water have been used for cleaning the beans before and after extraction. Coffee beans should be dried after decaffeination to optimum moisture content as it was before extraction of caffeine. It has been utilized to prepare different consumable products such as chocolates cold drinks, in addition to medicine preparation (Farah et al. 2006).

14.3.2.3 Monsooned coffee

Monsooned coffee is one of the world's specialty coffees processed only in India. Monsooned Malabar (MM) and Monsooned Robusta (MR) are processed from native Arabica and Robusta coffees (Murthy and Manonmani 2009). Monsooning is a process of exposing coffee beans to the moist winds of the monsoon season for several weeks, which yellows the bean and reduces acidity, imparting a heavy, syrupy flatness reminiscent of aged coffee (Ahmad 2000). During this process the moisture content of coffee beans increases from an initial 10%–12% to 18%–22%, and they double in size. The increased bitterness in Monsooned coffee could be associated with the enhanced hydrolysis of chlorogenic acid to caffeic acid, whereas the characteristic spicy aroma of the processed product can be attributed to the release of isoeugenol and 4-vinylguaiacol from its glycosidic precursors (Variyar et al. 2003).

14.3.3 Novel processing technologies

14.3.3.1 Green coffee and its extraction

In recent years, green coffee has become an emerging and novel technology in coffee industry. Green coffee beans are obtained after fermentation followed by hulling either by dry or wet processing, more precisely before roasting. Green coffee beans are rich in CGAs, and there has been an observed inhibitory effect on fat accumulation and body weight gain in mice (Shimoda et al. 2006). Green coffee bean extract has also been reported to have antihypertensive (Ochiai et al. 2004), lipolytic (Flanagan et al. 2014), and antibacterial activity (Bharath et al. 2015). Further, to increase the bioavailability of green coffee biomolecules, CGA extracted from green coffee was subjected to maltodextrin-nanoencapsulation by nano-spray drying. The encapsulation of the green coffee extract into nanoparticles significantly increased stability, thus preventing the loss of polyphenols, especially CGA, over storage period of 150 days. The application of this nanotechnology could offer interesting perspectives for the potential use of green coffee extract as an innovative nutraceutical (Desai et al. 2020).

14.3.3.2 Civet and Ivory coffee

Besides the aforementioned varied processing techniques, the recently noticed novel way of modulating coffee volatile profile is digestive bio-processing. The ripened coffee cherries are fed to civets (Indonesia) or elephants; beans obtained from the dung of these animals are brewed to produce the world's most expensive coffees, called Kopi Luwak and Black Ivory, respectively (Thammarat et al. 2018). The characteristic flavors developed are mainly due to various conditions of the animals' gastrointestinal tracts, causing unique changes in the amino acid profile of the beans, which in turn brings changes in aroma precursors on roasting (Marcone 2004).

14.3.3.3 Enrichment of coffee

Desai and coworkers (2019) decaffeinated Arabica and Robusta varieties by ethyl lactate and extracted using a polar solvent to obtain CGA enriched green coffee extract (GCE). The stability has been enhanced by further microencapsulation of GCE with maltodextrin (1:1) and obtained 86% encapsulation efficiency (Desai et al. 2019). Additionally, coffee enriched with mono-oligosaccharides content exerts prebiotic effects in animals (Asano et al. 2004) and humans (Asano et al. 2007). This coffee seems to lower fat absorption and decreases abdominal fat in humans, as clinically studied in a Japanese population (Huxley et al. 2009).

14.4 Conclusion and future insights

Coffee production and consumption is rapidly increasing throughout the world, and demand is increasing day to day. The economic importance and special constituent bioactive substances are promoting an upsurge in consumer interest. The applications of many biotechnological tools, such as molecular assisted selection (MAS), in vitro propagation, and genetic transformation, promise to increase the efficiency of breeding programs of coffee. Additionally, many ongoing technological developments such as automation, DNA chips, and allele-specific diagnostics will make large-scale MAS approaches much more effective. Essentially, coffee as a whole has the potential for novel functional/nutraceutical food products. It is important to study reactions of bioactive components in appropriate matrices, which might behave differently in a complex biological system.

REFERENCES

Ahmad R, *Indian Coffee*. 2000; 64(4): 3–7.

Al-Dakkak I, *Evid Based Dent*. 2011; 12(1): 23–24.

Amorim HV, Amorim VL, *Am Chem Soc*. 1977; 47: 27–56.

Antonio, A. G., Moraes, R. S., Perrone, D., Maia, L. C., Santos, K. R. N., Iório, N. L., & Farah, A. *Food Chem*. 2010; 118(3): 782–788.

Arlt, A., Sebens, S., Krebs, S., Geismann, C., Grossmann, M., Kruse, M., Schreiber, S. & Schäfer, H. *Oncogene*. 2013; 32(40): 4825–4835.

Asano, I., Umemura, M., Fujii, S., HOSHINO, H., & IINO, H. *Food Sci Technol Res*. 2004; 10(1): 93–97.

Asano, I., Ikeda, Y., Fujil, S., & Iino, H. *Food Sci Technol Res*. 2007; 10(3): 273–277.

Barre, P., Akaffou, S., Louarn, J., Charrier, A., Hamon, S., & Noirot, M. *Theor Appl Genet*. 1998; 96(2): 306–311.

Bharath, N., Sowmya, N. K., & Mehta, D. S. *Contemp Clin Dent*. 2015; 6(2): 166.

Blum D., Farah A., & Lopes L V. *Coffee*. RSC, London, UK, 2019; pp. 196–210.

Boot W, *Coffee Process Handb*. 2013; 173–192.

Borrelli, R. C., Visconti, A., Mennella, C., Anese, M., & Fogliano, V. *J Agric Food Chem*. 2002; 50(22): 6527–6533.

Bravi F., Tavani A., Bosetti C., Boffetta P., & La Vecchia C. *Eur J Cancer Prev*. 2017; 26(5): 368–377.

Buffo RA, Cardelli-Freire C, *Flavour Fragr J*. 2004; 19(2): 99–104.

Cappelletti S., Daria P., Sani G., & Aromatario M. *Curr Neuropharmacol*. 2015; 13(1): 71–88.

Carson CA, Caggiula AW, Meilahn EN, Matthews KA, & Kuller LH. *Int J Epidemiol*. 1994; 23(3): 523–527.

Cavalcanti DN, de Oliveira M-AR, De-Paula JC, Barbosa LS, Fogel T, Pinto MA, de Palmer Paixão ICN, Teixeira VL. *J Appl Phycol*. 2011; 23(5): 873–876.

Chen J-F, *Coffee*. RSC, London, UK, 2019; pp. 171–195.

Cheng B, *J Agric Food Chem*. 2011; 59(24): 13147–13155.

Chu Y-F, *Food Chem*. 2012; 135(3): 2095–2102.

Clark I, Landolt HP. *Sleep Med Rev*. 2017; 31: 70–78.

Clarke R, Vitzthum OG, *Coffee: Recent Developments*. John Wiley & Sons, New York, 2008.

Coradi, PC, *Coffee Science* 2007; 2(1): 38–47.

Costa J, *J Alzheimer's Dis*. 2010; 20(s1): S221–S238.

Czerny M, Grosch W, *J Agric Food Chem*. 2000; 48(3): 868–872.

Desai NM, Gilbert Stanley J, & Murthy PS. *Prep Biochem Biotechnol*. 2019; 49(4): 400–406.

Desai NM, Haware DJ, Basavaraj K, & Murthy PS. *J Microencapsul*. 2020; 37(1): 52–64.

Dong J, Zou J, Yu X-F. *World J Gastroenterol WJG*. 2011; 17(9): 1204.

Evangelista, S. R., Miguel, M. G. D. C. P., de Souza Cordeiro, C., Silva, C. F., Pinheiro, A. C. M., & Schwan, R. F., *Food Microbiol*. 2014a; 44: 87–95.

Evangelista, S. R., Silva, C. F., da Cruz Miguel, M. G. P., de Souza Cordeiro, C., Pinheiro, A. C. M., Duarte, W. F., & Schwan, R. F., *Food Res Int*. 2014b; 61: 183–195.

Farah A, Monteiro MC, Calado V, Franca AS, & Trugo LC. *Food Chem*. 2006; 98(2): 373–380.

Fernandez-Gomez B, Ramos S, Goya L, Mesa MD, del Castillo MD, & Martín MÁ. *Food Res Int*. 2016; 89: 1015–1022.

Flanagan J, Bily A, Rolland Y, & Roller M. *Phyther Res*. 2014; 28(6): 946–948.

Folwarczna J, Janas A, Pytlik M, Cegieła U, Śliwiński L, Krivošíková Z, Štefíková K, & Gajdoš M. *Nutrients*. 2016; 8(3): 133.

Ginz M, Engelhardt UH, I *Eur food Res Technol*. 2001; 213(1): 8–11.

Gonzalez-Rios O, Suarez-Quiroz ML, Boulanger R, Barel M, Guyot B, Guiraud J-P, & Schorr-Galindo S. *J Food Compos Anal*. 2007; 20(3–4): 289–296.

Grosso G, Stepaniak U, Polak M, Micek A, Topor-Madry R, Stefler D, Szafraniec K, & Pajak A. *Eur J Clin Nutr*. 2016; 70(1): 109–115.

Hameed, A., Hussain, S. A., Ijaz, M. U., Ullah, S., Pasha, I., & Suleria, H. A. R. *Compr Rev food Sci food Saf*. 2018; 17(5): 1184–1237.

Hashimoto M, Rockenstein E, Crews L, & Masliah E. *Neuromolecular Med*. 2003; 4(1–2): 21–35.

Heliövaara M, Aho K, Knekt P, Impivaara O, Reunanen A, & Aromaa A. *Ann Rheum Dis*. 2000; 59(8): 631–635.

Henry-Vitrac C, Ibarra A, Roller M, Merillon J-M, & Vitrac X. *J Agric Food Chem*. 2010; 58(7): 4141–4144.

Higdon JV, Frei B, *Crit Rev Food Sci Nutr*. 2006; 46(2): 101–123.

Holick CN, *Cancer Epidemiol Prev Biomarkers*. 2010; 19(1): 39–47.

Huxley R, Lee CMY, Barzi F, Timmermeister L, Czernichow S, Perkovic V, Grobbee DE, Batty D, & Woodward M. *Arch Intern Med*. 2009; 169(22): 2053–2063.

Je Y, Giovannucci E, *Int J cancer*. 2012; 131(7): 1700–1710.

Kalaska B, Piotrowski L, Leszczynska A, Michalowski B, Kramkowski K, Kaminski T, Adamus J, Marcinek A, Gebicki J, & Mogielnicki A. *J Agric Food Chem*. 2014; 62(13): 2853–2860.

Kobayashi M, Kurata T, Hamana Y, Hiramitsu M, Inoue T, Murai A, & Horio F. *J Nutr Sci Vitaminol (Tokyo)*. 2017; 63(3): 200–207.

Ky C-L, Barre P, Lorieux M, Trouslot P, Akaffou S, Louarn J, Charrier A, Hamon S, & Noirot M. *Theor Appl Genet*. 2000; 101(4): 669–676.

Lara J, Soccol VT, Pereira GVD, Gollo AL, Pandey A, Medeiros ABP, & Soccol CR. Inter J of Microbiol 2014; 188: 60–66

Lashermes P, Cros J, Combes M-C, Trouslot P, Anthony F, Hamon S, & Charrier A. *Theor Appl Genet*. 1996; 93(4): 626–632.

Lashermes P, Combes M-C, Trouslot P, & Charrier A. *Theor Appl Genet*. 1997; 94(6–7): 947–955.

Lashermes P, Combes M-C, Topart P, Graziosi G, Bertrand B, & Anthony F. *Theor Appl Genet*. 2000; 100(1), 139–146.

Lashermes P, Combes M-C, Prakash NS, Trouslot P, Lorieux M, & Charrier A. *Genome*. 2001; 44(4): 589–595.

Lee LW, Cheong MW, Curran P, Yu B, & Liu SQ. *Food Chem*. 2015; 185: 182–191.

Li L, Gan Y, Wu C, Qu X, Sun G, & Lu Z. *BMC Cancer*. 2015; 15(1): 733.

Liebeskind DS, Sanossian N, Fu KA, Wang H-J, & Arab L. *Nutr Neurosci*. 2016; 19(9): 406–413.

do Livramento KG, Borém FM, José AC, Santos AV, do Livramento DE, Alves JD, & Paiva LV. *Food Chem*. 2017; 221: 1874–1882.

Ma C-M, Kawahata T, Hattori M, Otake T, Wang L, & Daneshtalab M. *Bioorg Med Chem*. 2010; 18(2): 863–869.

Thammarat, P., Kulsing, C., Wongravee, K., Leepipatpiboon, N., & Nhujak, T. *Molecules*. 2018; 23(8), 1910

Marcone MF, *Food Res Int*. 2004; 37(9): 901–912.

Massawe GA, Lifa J, *Int J Postharvest Technol Innov*. 2010; 2(1): 41–82.

Mazzafera P, Baumann TW, Shimizu MM, & Silvarolla MB. *Trop Plant Biol*. 2009; 2(2): 63–76.

Michels KB, Willett WC, Fuchs CS, & Giovannucci E. *J Natl Cancer Inst*. 2005; 97(4): 282–292.

Moon J-K, Yoo HS, & Shibamoto T. *J Agric Food Chem*. 2009; 57(12): 5365–5369.

Murthy PS, Manonmani HK, *Eur Food Res Technol*. 2009; 229(4): 645–650.

Murthy PS, Naidu MM, *Eur J Appl Sci*. 2011; 3(4): 130–139.

Narita Y, Inouye K, *J Agric Food Chem*. 2013; 61(4): 966–972.

Nehlig A, *Pract Neurol*. 2016; 16(2): 89–95.

Oberman H, Libudzisz Z, & Wood BJB. *Microbiology of Fermented Foods*. Elsevier Appl Sci Publ. 1985.

Ochiai R, Jokura H, Suzuki A, Tokimitsu I, Ohishi M, Komai N, Rakugi H, & Ogihara T. *Hypertens Res*. 2004; 27(10): 731–737.

Onakpoya I, Terry R, & Ernst E. *Gastroenterol Res Pract*. 2010; 2011.
Pascual EC, Goodman BA, & Yeretzian C. *J Agric Food Chem*. 2002; 50(21): 6114–6122.
Poltronieri P, Rossi F, Challenges in specialty coffee processing and quality assurance. *Challenges*. 2016; 7(2): 19.
Ranheim T, Halvorsen B, *Mol Nutr Food Res*. 2005; 49(3): 274–284.
Reineccius GA, *16th ASIC Colloq*. Kyoto, 1995.
Ritter M, Hohenberger K, Alter P, & Herzum M, *Cytokine*. 2005; 30(4): 177–181.
Rogers WJ, Bézard G, Deshayes A, Meyer I, Pétiard V, & Marraccini P. *Plant Physiol Biochem*. 1999; 37(4): 261–272.
Da Rosa SDVF, Oliveira PD, Giomo GS, & Isquierdo EP. *African J Agric Res*. 2015; 10(12): 1457–1466.
Santos RMM, Lima DRA. *Eur J Nutr*. 2016; 55(4): 1345–1358.
Savonitti O, *Elsevier Academic Press*, London, UK, 2005; pp. 241–245.
Selmar D, Bytof G, Knopp S-E, & Breitenstein B. *Plant Biol*. 2006; 8(02): 260–264.
Selmar D, Kleinwächter M, & Bytof G. *Cocoa coffee Ferment*. 2014; 73–81.
Sette, J., Country Coffee profile ITALY, International Coffee Organization, On line at http://www. ico. org/documents/ cy2016-17/icc-120-6e-profile-italy. pdf. 2017
Shimoda H, Seki E, & Aitani M. *BMC Complement Altern Med*. 2006; 6(1): 9.
Shlonsky AK, Klatsky AL, & Armstrong MA. *Ann Epidemiol*. 2003; 13(4): 273–279.
Silva CF, Batista LR, Abreu LM, Dias ES, & Schwan RF. *Food Microbiol*. 2008; 25(8): 951–957.
Siridevi GB, Devendra H, Basavaraj K, & Murthy PS. *LWT*. 2019; 114: 108382.
de Sotillo DVR, Hadley M, *J Nutr Biochem*. 2002; 13(12): 717–726.
Stefanello N, Schmatz R, Pereira LB, Rubin MA, da Rocha JBT, Facco G, Pereira ME, de Andrade Mazzanti CM, Passamonti S, & Rodrigues MV. *Mol Cell Biochem*. 2014; 388(1–2): 277–286.
Sugiyama K, Sugawara Y, Tomata Y, Nishino Y, Fukao A, & Tsuji I. *Eur J Cancer Prev*. 2017; 26(2): 125–130.
Sunarharum WB, Williams DJ, & Smyth HE. *Food Res Int*. 2014; 62: 315–325.
Suzuki A, Fujii A, Yamamoto N, Yamamoto M, Ohminami H, Kameyama A, Shibuya Y, Nishizawa Y, & Tokimitsu I, *FEBS Lett*. 2006; 580(9): 2317–2322.
Tadesse FT, et al., *J Nutr Ecol Food Res*. 2016; 3(1): 44–50.
Tauler P, Martinez S, Martinez P, Lozano L, Moreno C, & Aguiló *Int J Sport Nutr Exerc Metab*. 2016; 26(1): 8–16.
Thom E, *J Int Med Res*. 2007; 35(6): 900–908.
Urushisaki T, Takemura T, Tazawa S, Fukuoka M, Hosokawa-Muto J, Araki Y, & Kuwata K. *Evidence-Based Complement Altern Med*. 2011; 2011.
Variyar, P. S., Ahmad, R., Bhat, R., Niyas, Z., & Sharma, A *J Agric Food Chem*. 2003; 51(27): 7945–7950.
Vilela DM, Pereira GV de M, Silva CF, Batista LR, & Schwan RF. *Food Microbiol*. 2010; 27(8): 1128–1135.
Vitzthum OG, Werkhoff P, *Zeitschrift für Leb und Forsch*. 1976; 160(3): 277–291.
Waite LM, *Aust Prescr*. 2015; 38(2): 60.
Wang G-F, Shi L-P, Ren Y-D, Liu Q-F, Liu H-F, Zhang R-J, Li Z, Zhu F-H, He P-L, & Tang W. *Antiviral Res*. 2009; 83(2): 186–190.
Wilson KM, Kasperzyk JL, Rider JR, Kenfield S, van Dam RM, Stampfer MJ, Giovannucci E, & Mucci LA. *J Natl Cancer Inst*. 2011; 103(11): 876–884.
Willson KC, Clifford MN, *Tea: Cultivation to Consumption*. Springer Science & Business, Media, New York. 2012.
Windholz M, Budavari S, Blumetti RF, & Otterbein ES. *The Merck Index*, Merck & Co. Inc, Rahway, NJ. 1983; p. 1051.
Wintgens JN, *Coffee: Growing, Processing, Sustainable Production: A Guidebook for Growers, Processors, Traders, and Researchers*. Wiley-VCH, Weinheim. 2009.
Yeretzian C, Pascual EC, & Goodman BA. *Food Chem*. 2012; 131(3): 811–816.
Yoshinari O, Sato H, & Igarashi K. *Biosci Biotechnol Biochem*. 2009; 73(5): 1033–1041.
Yu, X., Bao, Z., Zou, J., & Dong, J., BMC cancer; 2011; 11(1): 1–11.
Zhou J, et al., *Curr Med Chem*. 2012; 19(21): 3523–3531.

15

Emerging technologies for the management of coffee processing wastewater

Ijanu Emmanuel Madu, Mohamad Anuar Kamaruddin, Mohd Suffian Yusoff, Noorzalila Muhammad Niza, Abdubaki Mohamed Hussen Shadi, and Faris Aiman Norashiddin

CONTENTS

15.1 Introduction

The global trend in population growth has placed more demand on effective ways of achieving good environmental management practices using sustainable technologies (Lalevic et al., 2012). Growing concerns about environmental issues underlie the increasing need for policies and strategies to solve these problems (Figueroa, Homann, & Rawel, 2016). This paradigm is also witnessed by a growing amount of research aimed at proffering solutions to coffee wastewater from various agro-based industries.

The coffee processing industry is one of the agro-based industries that contributes immensely to the pollution of the ambient environment (Woldesenbet, Woldeyes, & Chandravanshi, 2014). Scientifically, coffee belongs to the genus *Coffea* from the Rubiaceae family, and its seeds are called coffee beans or berries (Farah & Donangelo, 2006). Global coffee production has recently risen to almost 170 million bags for the year 2020/2021; hence, it is the most traded agro-based commodity (Amna, Mohammad, Amirul, & Abd, 2016; USDA Foreign Agricultural Service, 2019).

Coffee processing uses a large volume of water; consequently, solid waste by-products and wastewater are generated (Dadi et al., 2018; Tomizawa, Kurosu, Kobayashi, & Kawase, 2016). Although there have been many reported studies on the utilization of coffee by-products such as coffee husk, pulp, and grounds as adsorbents in wastewater treatment (Anastopoulos, Karamesouti, Mitropoulos, & Kyzas, 2017; Pagalan et al., 2020), pollution as a result of coffee wastewater has been a common concern in many coffee producing and processing countries (Figueroa et al., 2016). A detailed flow chart for processing coffee beans (both dry and wet method) into the coffee beverage and all the supernatant waste (solid and liquid) has been highlighted (Rattan, Parande, Nagaraju, & Ghiwari, 2015).

The constituents of coffee processing wastewater (CPW) include many toxic chemicals such as polyphenols, tannins, and caffeine (Torres-Valenzuela, Ballesteros-Gomez, Serna, Arango, & Rubio, 2020). The existence of sugars, protein, and minerals makes it a good substrate for organic pollution. In some major coffee-producing countries, coffee-processing industries are located close to water bodies, into which effluent is conveniently dumped (Dadi et al., 2018; Kebede, Kebede, Assefa, & Amsalu, 2010; Sengupta et al., 2020). Coffee effluent deteriorates the quality of the ambient water body, which is not in agreement with environmental guidelines for the discharge of wastewater established by the World Health Organization

DOI: 10.1201/9781003043133-19

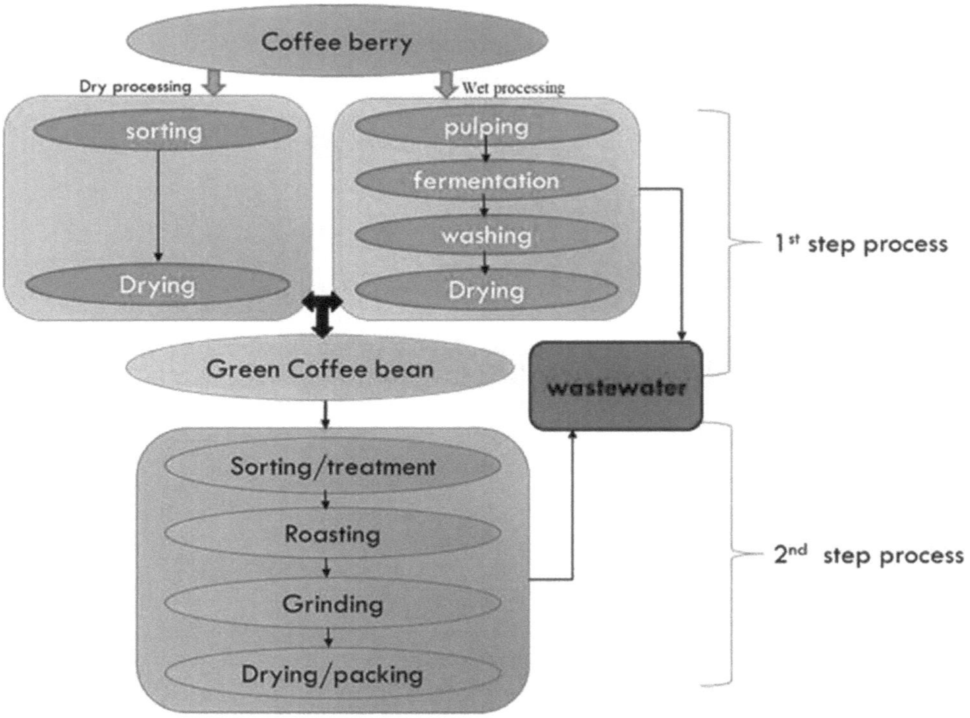

FIGURE 15.1 Schematic diagram of coffee processing wastewater.

(WHO) and the US Environmental Protection Agency (EPA; Torres-Valenzuela et al., 2020).

Effects of CPW are seen in the depletion of oxygen level in water essential for biological breakdown of organic waste, consequently creating anaerobic conditions to the detriment of aquatic habitats (Padmapriya, Tharian, & Thirunalasundari, 2015; Rattan et al., 2015; Woldesenbet et al., 2014). CPW is characterized by the high concentration of pollutants including high organic content (COD, BOD, TSS, TDS, and NH_3-N; Can, Gengec, & Kobya, 2019), organic load like sugars, protein, pectin, polyphenols, and alkaloids (caffeine), tannins and a considerable amount of phenols (Chagas, Torres, Silva, & Corrêa, 2015; Coelho et al., 2014). Toxicity studies on the effect of CPW show skin irritation and breathing problems in humans (Haddis & Devi, 2008), reduction in germination speed and retarded growth in plants (Louzada, Andrade-Vieira, Augusto, & David, 2016), reduction of aquatic diversity, and eutrophication that affect aquatic life unless properly treated before discharge (Beyene, Kassahun, & Addis, 2014).

15.2 Coffee processing

Wet coffee processing is often preferred to dry due to the quality produced after the process (Manoj Kumar, 2015). Coffee berries are collected for processing after harvesting and sorting out all adhering dirt such as leaves, twigs, and diseased coffee berries, and the thin, red outer skin and the pulp are removed. A large volume of organically polluted effluent is generated from these processes (Figure 15.1). This is followed by the removal of mucilage through fermentation by a microorganism (Villa Montoya et al., 2019). After decomposing the layer, the bean is further washed, thereby generating more wastewater (Rattan et al., 2015). Finally, the parchment is removed by soaking the bean in clean water for some time, followed by stirring and rinsing till all beans are devoid of the silver skin, and then drying takes place. When it's fully dried, it is packed for export (Enden & Calvert, 2014; Rattan et al., 2015). The second phase of processing (Figure 15.1) involves the sorting out and roasting of green beans. During roasting, several additives that give the aroma and enhance the quality of the product are added such as margarine and sugar (ceremization), thus giving it its dark brown color. It is followed by grinding (depending of the size of preference) and then drying and packing for the market (Alemayehu, Asfaw, & Tirfie, 2020). The roasting and grinding machines are washed every day after use, and a large volume of water is required to achieve this task successfully.

15.3 Characteristics of coffee wastewater

The characterization of coffee wastewater has been studied on two fronts over the years, based on the stages of the treatment process as shown in Figure 15.1. The first stage of the process (from coffee berries to dried green bean) generates higher organic pollution (Table 15.1); while the second stage procession generates relatively lower pollution, it does not meet the discharge standard limit permitted by the WHO or the Malaysian Department of Environment.

Wastewater from the fermentation of mucilage contains a certain amount of carbohydrate, which turns the water to a dark green color, and a high amount of organic matter measured in form of BOD and COD (Dadi et al., 2018; Haddis &

TABLE 15.1

Characteristics of Coffee Processing Wastewater after Roasting

Parameter	1st stage	2nd stage	Standard*	WHO limit
pH	3.92–4.99	3.5–4.2	5.5–9.0	6.8–8.5
Residual conductivity (μS cm^{-1})	382–390	221–354	–	1×10^{-3}
Temperature (°C)	20–25	30.0–34.0	40	20
Colour (pt.co)	–	2512–5342	200	300
BOD_5 (mg L^{-1})	10,000–20,000	1136–3200	50	100
COD (mg L^{-1})	20,000–50,000	2309–7328	200	300
Total dissolve solids (mg L^{-1})	4500	124.67–197.67	50	200
NH_3-N (mg L^{-1})	168–212	125–175	5	50

** Standard B of the environmental quality (industrial effluent) regulation 2013 requirement.*

Devi, 2008) The pH of CPW drops to become acidic when alcohol is converted to acetic acid or vinegar after the fermentation of sugars to alcohol and CO_2 (Asha & Kumar, 2016).

A characterization study carried out in a second step coffee processing industry in Malaysia over 1 year by Ijanu E.M. et al. (n.d.) confirms that CPW is highly polluted and does not meet the Malaysian Department of Environment effluent discharge limit, as shown in Table 15.1. The low pH value of (3.5–4.2) is a result of the high production of VFA from acidification caused by high organic matter compounds (Cruz-Salomón et al., 2017). High color and COD are the result of ceramization during roasting as well as the presence of tannin and pectin, protein, and sugar (Woldesenbet et al., 2015).

The variations in the concentrations of the parameters in the processing stages are due to reasons including the different volumes of water used during the process, the species of coffee processed, and so on (Shanthakumar et al., 2018). According to Ijanu, Kamaruddin, & Norashiddin (2020), the presence of toxic chemicals like tannins, phenolics, and alkaloids inhibits the biological degradation of organic materials in water bodies where the effluent has been discharged, thus leading to high BOD and COD.

Characteristics of CPW such as its phenol extract and other compounds formed from Maillard reaction during coffee roasting have been documented in the literature (Alemayehu et al., 2020; Enden & Calvert, 2014; Ijanu et al., 2020). This chapter discusses the major advancement in the methods that have been used in the treatment of coffee wastewater in recent years.

15.4 Traditional treatment method of CPW management

The most common and toxicologically threatening method of coffee processing effluent management is direct discharge into the ambient environment. Other notable methods include collecting the raw CPW in a sump tank and treating with either alum or talc, as reported by Rattan et al., (2015); this method does not take into consideration optimum conditions of the independent variable to achieve required discharge standard.

The adoption of biological, chemical, electrochemical, and physicochemical methods for CPW has been studied for many years. While many pieces of research advocate for biological treatment (Cruz-Salomón et al., 2017; Enden & Calvert, 2014; Gupta, 1998), According to a review by Ijanu et al., (2020), physicochemical treatment is preferred over biological not only because of the effectiveness in removal all pollutant but also because of the less time it takes in the process (Takashina, Leifeld, Zelinski, Mafra, & Igarashi-Mafra, 2018). Examples of physicochemical treatments that have been studied include, but are not restricted to, zero-valent (Tomizawa et al., 2016), photo-Fenton (Tokumura, Ohta, Znad, & Kawase, 2006), oxidation (Chagas et al., 2015), and activated carbon (Haddis & Devi, 2008; Table 15.2).

15.5 Combined treatment technologies

The single-treatment method of CPW is found to have many drawbacks. As a result, many researchers tried to combine treatment recently to achieve the desired national/international discharge standard limit (Yu, Han, & He, 2017). For example, studies revealed that coagulation-flocculation alone did not effectively remove the COD concentration in CPW (Wang, Chen, Ge, & Yu, 2007). Therefore, this chapter compiles studies on the improvement that has been made recently by researchers in achieving the desired goal of effective coffee wastewater treatment.

15.5.1 Combined coagulation/flocculation and electrochemical oxidation

Generally, coagulation is the process of destabilizing colloidal particles and formation of flocs, whereas during electrochemical oxidation stage pollutants are oxidized directly (at the anode) or indirectly (in the presence of Cl^-) employing oxidants species generated electrolytically (Primo, Rivero, Urtiaga, & Ortiz, 2009). The treatment performance of coagulation followed by electro-oxidation was examined in the removal of suspended particles and the dissolved recalcitrant organic matter of CPW (Can et al., 2019). In the electro-oxidation experiments, the performance of boron-doped diamond (BDD), Pt, and various mixed metal oxide (MMO) Ti-based electrode was examined, as well as the effect of the current density and

TABLE 15.2
Recent Treatment Methods of Coffee Processing Wastewater (CPW)

Method	Pollutant	Removal efficiency	Reference
Coagulation-flocculation followed by elect-oxidation	TOC AND COD removal	95% and 97%, respectively	(Can et al., 2019)
Coagulation/flocculation followed by Fenton reaction	TOC, COD and BOD	76.2%, 76.5%, 66.3% removal, respectively	(Gomes de Barros et al., 2020a)
Electrochemical AOP	Color TOC	73%–89% and 73%–84%	(Villanueva-Rodríguez, Bello-Mendoza, Wareham, Ruiz-Ruiz, & Maya-Treviño, 2014)
Coagulation-flocculation and photo-degradation processes	COD, color, and turbidity	87% in all cases	(Teresa, Gunther, & Fernando, 2007)
Integrated electrocoagulation-electro-oxidation process	Color, COD, and TOC	99%, 89%, and 72%, respectively	(Ibarra-Taquez, GilPavas, Blatchley, Gómez-García, & Dobrosz-Gómez, 2017)
Cation-anion sequence treatment	COD and NH_3-N	92.11%, and 93.87%, respectively	(Ijanu E. M., Kamaruddin, M. A. et al., 2020 unpublished)
O_3 and O_3/UV	Caffeine and color removal	O_3/UV achieved the removal of 98 and 99%, respectively	(Takashina et al., 2018)
Combined chemical flocculants and UV/H_2O_2, UV/O_3, and UV/ H_2O_2/O_3	COD	67%, 86%, 84%, and 87%, respectively	(Teresa et al., 2007)
Electro-Fenton using Fe_3O_4–Mn_3O_4	COD, color, and TOC	88%, 98%, and 93% of COD, color, and TOC, respectively	(Nguyen, Nguyen, Huynh, & Nguyen, 2019)
Coagulation-flocculation using moringa seed mesh	TSS, turbidity, and COD	69%, 88%, and 41% removal, efficiency	(Novita, Wahyuningsih, Pradana, Marsut, & Farisul, 2019)
Natural coagulant (moringa seed)	TSS, COD, total nitrogen	8%–54%, 26%–100%, and 20%–100%, respectively	(Garde, Buchberger, Wendell, & Kupferle, 2017)
Integrated aerobic and anaerobic treatment	BOD and COD	Removal of 85.4% and 78.0% of BOD and COD, respectively	(Selvamurugan, Doraisamy, & Maheswari, 2010a)
Two-stage sequence phytoremediation	TSS, color, and COD	94%, 79%, and 95% removal, respectively	(Said, Abdullah, Ismail, Hasan, & Othman, 2020)
Supramolecular solvents	BOD, DO, pH, TSS, and conductivity	Minimized	(Torres-Valenzuela et al., 2020)

the effect the flow rate for the most efficient electrode. From the results of the study, the highest removal of 95% TOC and 97% COD was attained by BDD (Can et al., 2019). Therefore, the removal efficiency of pollutants directly depends on the kind of anode electrode used, as well as the current density (Moreira, Boaventura, Brillas, & Vilar, 2017).

In a similar work, using aluminum and graphite electrodes, electrocoagulation-electro-oxidation (EC-EO) was proposed as an alternative for the treatment of CPW (Ibarra-Taquez et al., 2017). The result shows that after optimization of operational parameters, total decoloration, COD reduction of 89%, and TOC mineralization of 72% was attained, as depicted in Table 15.2. The work further affirms that expensive electrodes are necessary for higher performance of the system (Moreira et al., 2017). In another work, the performance of electrochemical advanced oxidation processes (ECOP) was compared against the conventional advanced oxidation processes (AOPs). The major drawback of AOPs is that in most cases, H_2O_2 is required to facilitate the reaction process (Tokumura et al., 2006). In the case of electrochemical advance oxidation, additional H_2O_2 was produced by the ECOP in situ during the treatment process (Villanueva-Rodríguez et al., 2014). The study tries to improve the effectiveness of CPW quality by relating four different ECOPs and compared the result to the conventional AOP used for CPW treatment in the past. The processes showed better color removal (73%–93%) and total organic removal (73%–84%; Villanueva-Rodríguez et al., 2014).

15.5.2 Photo-degradation process

The use of ozone (O_3) has been recommended by many researchers as a strong oxidant used to oxidize the numerous organic compound (Mainardis, Buttazzoni, De Bortoli, Mion, & Goi, 2020). Its advantages include the removal of fecal coliforms as well as some physicochemical parameters (Xu, Janex, Savoye, Cockx, & Lazarova, 2002). However, O_3 treatment alone is not efficient in the treatment of wastewater with high toxicity and complex recalcitrant organics such as CPW (Kharel et al., 2020; Malik, Ghosh, Vaidya, & Mudliar, 2020). There are two reaction mechanisms in the application of O_3; the first involves direct reaction by O_3 itself, and the second involves the indirect reaction of OH* radicals with the dissolved organic and inorganic compounds (Mainardis et al., 2020; Malik et al., 2020).

The Fenton process involves the reaction between H_2O_2 (as the oxidant) and Fe^{2+} (as a catalyst) to produce highly reactive specie such as OH radicals with high oxidation potential (Alemayehu et al., 2020; Bianco, De Michelis, & Vegliò, 2011). It is considered to be environmentally friendly because of its low financial implication, it operates at ambient temperature and pressure, and non-toxic reagent is employed for the process (Bianco et al., 2011; Gomes de Barros et al., 2020b). Although Fenton oxidation alone was reported to remove most of the pollutants in CPW, it performed poorly in the removal of nitrogen, hence additional treatment is required to achieve good results (Manoj Kumar, 2015). Removal of COD was determined using a combined effect of coagulation/flocculation and UV/O_3, UV/H_2O_2, and UV/ H_2O_2/O_3 under acidic conditions (Teresa et al., 2007); the order of effectiveness of COD removal shows that UV/ H_2O_2/O_3 > UV/ H_2O_2 > UV/O_3 > coagulation/flocculation 87%, 86%, 84%, and 67%, respectively. Performance of Fenton's oxidation, coagulation/flocculation (C/F), and the combination of C/F with Fenton's process for the treatment of bio-digested CPW was assessed by (Gomes de Barros et al., 2020b), at optimum operation parameter (T = 55°C, $[Fe^{3+}] = 1.8\ g\ L^{-1}$, pH = 5.0, and $[H_2O_2] = 9.0\ g\ L^{-1}$) combination of C/F with Fenton's process produce the best result (TOC – 76.2%, COD – 76.5 and BOD5 – 66.3%).

Nguyen et al., (2019), compared to the performance of electro-Fenton (EF) with particulate catalysts as against the conventional Fenton process; results show that EF produces less sludge, and fewer chemicals are needed for optimum process performance. Furthermore, 88%, 98%, and 93% of COD, color, and TOC were removed, respectively. In another study, to generate a reaction of hydroxyl radicals, oxidation of organic pollutants by the combining O_3/UV process was studied at neutral pH for effective organics removal, hence improving the low mineralization factor (Kondo et al.,2014) that comes with the use of O_3 alone for wastewater treatment. The process achieved 99% removal of color and 96% removal of caffeine (Takashina et al., 2018). We can therefore infer that pH is a major factor to reckon with when considering this type of AOP.

15.6 Biological treatment

15.6.1 Application of natural coagulant

Coagulation has been used over the years to remove impurities (chiefly suspended particles). Although inorganic coagulant has proven to be effective, its drawbacks have prompted the demand for natural coagulants to be explored by researchers. The major advantages of natural coagulants include their cost-effectiveness, environmental friendliness, low sludge generation, non-toxic nature, and high biodegradability (Ang & Mohammad, 2020). Moringa seed is one of the prominent natural coagulants that has been used in place of alum as a quick method to clean turbid and polluted water (Mahmood, Mugal, & Haq, 2010). Its efficiency as a good coagulant and anti-microbial species is due to its vast array of protein content (Kansal & Kumari, 2014). Furthermore, coagulant dosage is the major independent factor in treatment (Saritha, Karnena, & Dwarapureddi, 2019).

The performance of moringa seeds as natural coagulants for the treatment of CPW was studied by Novita et al. (2019) and Padmapriya et al. (2015). Treatment of coffee wastewater with moringa seed has a major drawback in the time it takes to treat the pollutant, which is very long as compared to other treatments (Padmapriya et al., 2015). Recently, this drawback was overcome when different particle sizes of moringa seed were studied in a jar test for the treatment of coffee wastewater; 60 mesh was the most efficient, resulting in the removal of above 60% of pollutants in few hours, as reflected in Table 15.2 (Novita et al., 2019). In another study, moringa seed doses varied from zero to 4 g L^{-1} to assess its performance in the removal of COD, TSS, and total nitrogen in CPW. Results show very good performance on the response, but the method has not been found efficient in removing the dissolved organic content of CPW (Garde et al., 2017).

15.6.2 Phytoremediation

Phytoremediation is the direct and natural use of plants to absorb pollutants (organic and inorganic) through roots and eventually transmitted to the upper part of the plant (Rezania, Taib, Md Din, Dahalan, & Kamyab, 2016). The potential of phytoremediation for wastewater treatment has been assessed by many researchers as an emerging technology and is summarized in a review by Galiulin, Bashkin, Galiulina, and Birch (2001). The efficiency of the technology is highly dependent on the growth rate of the plant in question, tolerance level to pollutants, and the particular species of plant. These have to be considered before the design of the system. According to Rezania et al. (2016), for successful pollutant removal, more than one phytoremediation technique should be adopted. Therefore, a two-stage plant system was studied in sequence to observe the treatment viability of CPW as earlier recommended by literature. In two phytoremediation plants

(*Phragmites karka* and *Eichhornia crassipes*), each with a different medium, the result shows that 74% of suspended solids, 79% of color, and 95% of COD concentration were successfully removed after 7 days (Said et al., 2020).

15.6.3 Adsorption

Adsorption by agricultural by-products has in recent times gained more attention than adsorption by activated carbon because the technology is economically and environmentally benign and it does not generate secondary waste (Zhang et al., 2017). The capability of agricultural waste as adsorbents has been proved by the elimination of different kinds of pollutants in industrial wastewater (Brinza, Geraki, Cojocaru, Holdt, & Neamtu, 2020; Corral-Bobadilla, González-Marcos, Alba-Elías, & Diez de Santo Domingo, 2020).

In the quest to remove the high concentration of the COD from CPW, Ramya, Jayasravanthi, Dulla, and Venkata (2015) utilized different agricultural wastes as low-cost adsorbents; namely, banana stem powder (BSP), sorghum stem powder (SSP), and casuarina fruit powder (CFP). With pH at a neutral level, the result shows that the adsorbents have a promising role in the removal of COD with BSP up to 80%, SSP 92.7%, and CFP 99.2%.

15.6.4 Bioreactors

The use of various types of bioreactors for CPW treatment has been established in literature (dos Santos, Cervantes, & van Lier, 2007; Selvamurugan, Doraisamy, & Maheswari, 2010a; Selvamurugan, Doraisamy, Maheswari, & Nandakumar, 2010b). However, there have been many conflicting outcomes ranging from the difference between lab-scale experiments and full-scale experiments, differences in geographical location, differences in concentration of CPW, and so on (Dinsdale, Hawkes, & Hawkes, 1997). For example, in a study carried out in Mexico using a lab-scale expanded granular scale sludge bed (EGSB) bioreactor, the optimum hydraulic retention time of the operation was determined to be 7–9 days for effective removal of 94%–98% of COD concentration at a neutral pH from CPW (Cruz-Salomón et al., 2017). In another study, the performance of anaerobic reactors filled with crushed stones, polyurethane foam, and blast furnace cinders was assessed for COD removal in CPW. The reactor filled with foam performed better than the other two after 30–50 days due to its porosity; as a result, better collection of biomass was noted within 26 hours (Fia, Matos, Borges, Fia, & Cecon, 2012).

Other work (Beyene, Yemane, Addis, Assayie, & Triest, 2014; Chen, Jiang, & Li, 2018; Chen et al., 2019) looked at the energy generation potential of CPW while removing the COD concentration. Results suggested that at optimum performance, 82.4% of COD was converted to methane at HRT of 10 days using an anaerobic membrane bioreactor in a thermophilic condition (Chen et al., 2019). On the other hand, with an HRT of 40–50 days, anaerobic digesters used to treat CPW removed up to 90% of organic load and total solids, generating large among of energy in the process (Beyene et al., 2014). Biomethanation with aeration studies was carried out by varying the hydraulic retention time of UAHR followed by wetland plant treatment, with the maximum pollutant removal 85.4% and 78.0% of BOD and COD respectively (Selvamurugan, Doraisamy, & Maheswari, 2010a). Important to note in this treatment is the production of methane, which can be converted to energy. Although coffee waste was found to have good potential for energy production, it is not effective in treating the pollutants. Fundamentally, the more HRT, the higher the treatment efficiency of the process (Asha & Kumar, 2016)

15.6.5 Valorization

As part of an effort to manage industrial wastes and by-products, researchers have recently looked into the possibility of recycling, reuse, and reparation of any waste and by-products from any process. This process promotes greater environmental and economic efficiency (Mekki, Anderson, Benzina, & Ammar, 2008). Valorization of bio-waste through viable value chains from various industries has been studied and compiled (Gregg et al., 2020). Other researchers have studied the treatment and valorization of different wastewaters and have recorded tremendous success in many quarters (Lachapelle-T., Labrecque, & Comeau, 2019; Mekki et al., 2008). The potential of valorizing coffee wastewater has been recommended. Taking advantage of its high organic load, electricity can be generated using a microbial fuel cell through anaerobic oxidation of the organics (Sengupta et al., 2020). Other CPW potentials include generation of bio-hydrogen, biogas, and methane.

Furthermore, CPW was valorized by supramolecular solvents (SUPRASs) to recover bioactive waste such as caffeine and antioxidant compounds like phenols, while removing high concentration of physicochemical pollutants BOD, COD, DO, pH, TSS, and conductivity (Torres-Valenzuela et al., 2020). Although the study was able to recover a considerable amount of bioactive waste, the pollutants were not successfully removed, and subsequent treatment was required for complete removal to meet the permissible discharge limit.

15.6.6 Cation-anion sequence treatment

Ion exchange is a reversible exchange of ions between a solid and a liquid phase (Rodrigues, 1986). Ion exchange resins (IER) are insoluble polymers that contain acidic or basic functional groups and have the ability to interchange counter-ions within aqueous solutions surrounding them (Nicolich, 2015). The effectiveness of ion exchange resins mainly depends upon their physical properties, such as particle size, porosity, acid-base strength, degree of cross-linking, stability, and purity (Inamuddin & Luqman, 2012; Mohammad & Luqman, 2008). Many researchers have reported the application of ion exchange for the management of different wastewater. Different types of ion exchange resins have been used to remove a wide range of pollutants including heavy metals, organic matter, and color (Bashir, Aziz, & Yusoff, 2010; Levchuk, Rueda Márquez, & Sillanpää, 2018; Nikoloski & Ang, 2014).

In the case of CPW, simultaneous removal of COD and NH_3-N from CPW by cation-anion sequence treatment was studied in a batch experiment (Ijanu E. M., Kamaruddin, M. A. et al., 2020) using design expert software. Operational conditions (shaking speed, resin dosage, and contact time) were optimized, and the experimentally achieved optimum removal levels of COD, and NH_3-N were 92.11% and 93.87%, respectively.

15.7 Conclusion

This chapter highlighted the improvements that have been made in recent years with regard to the treatment of CPW. It has been deduced that one technology is not sufficiently effective in CPW treatment; therefore, as part of the objective of this chapter, the following technologies were highlighted: biological and non-biological processes such as coagulation-flocculation followed by elect-oxidation, coagulation/flocculation followed by Fenton reaction O3 and O3/UV, ECOP, coagulation-flocculation and photo-degradation processes, integrated electrocoagulation-electro-oxidation process, cation-anion sequence treatment, coagulation-flocculation using moringa seed mesh, integrated aerobic and anaerobic treatment, two-stage sequence phytoremediation, and supramolecular solvents. At optimal conditions, the removal of pollutants ranging from 70%–99% was achieved. Therefore, for application of the CPW management options discussed above, one needs to consider the cost, environmental friendliness, nature, and characteristics of the CPW as major parameters before choosing the best combination or sequence treatment to adopt. This is not exhaustive, but it serves as a starting point for more study in this highly topical field of research.

REFERENCES

Alemayehu, Y. A., Asfaw, S. L., & Tirfie, T. A. (2020). *Journal of Material Cycles and Waste Management*, 22(2), 454–469. https://doi.org/10.1007/s10163-019-00953-y

Amna, M. A., & Amirul, M. (2016). *Exploring the Potentials of Coffee Industry in Malaysia*. Food and Fertilizer Technology Center for the Asian Pacific Region, Malaysia 1–5.

Anastopoulos, I., Karamesouti, M., Mitropoulos, A. C., & Kyzas, G. Z. (2017). *Journal of Molecular Liquids*, 229. https://doi.org/10.1016/j.molliq.2016.12.096

Ang, W. L., & Mohammad, A. W. (2020). *Journal of Cleaner Production*, 262, 121267. https://doi.org/10.1016/j.jclepro.2020.121267

Asha, G., & Kumar, B. (2016). *Journal of Scientific Research and Reports*, 9(5), 1–9. https://doi.org/10.9734/jsrr/2016/21830

Bashir, M. J. K., Aziz, H. A., & Yusoff, S. (2010). *Journal of Hazardous Materials*, 186, 92–102. https://doi.org/10.1016/j.jhazmat.2010.10.082

Beyene, A., Yemane, D., Addis, T., Assayie, A. A., & Triest, L. (2014). *International Journal of Environmental Science and Technology*, 11(7), 1881–1886. https://doi.org/10.1007/s13762-013-0339-4

Bianco, Barbara De Michelis, Ida Vegliò, Francesc. (2011) *Journal of Hazardous Materials*, 186, 1733–1738.

Brinza, L., Geraki, K., Cojocaru, C., Holdt, S. L., & Neamtu, M. (2020). *Chemosphere*, 238, 124652. https://doi.org/10.1016/j.chemosphere.2019.124652

Can, O. T., Gengec, E., & Kobya, M. (2019). *Journal of Water Process Engineering*, 28(September 2018), 28–35. https://doi.org/10.1016/j.jwpe.2019.01.002

Chagas, P. M. B., Torres, J. A., Silva, M. C., & Corrêa, A. D. (2015a). *International Journal of Biological Macromolecules*. https://doi.org/10.1016/j.ijbiomac.2015.08.061

Chen, R., Jiang, H., & Li, Y. Y. (2018). *Chemical Engineering Journal*, 334(August 2017), 444–452. https://doi.org/10.1016/j.cej.2017.10.052

Chen, R., Wen, W., Jiang, H., Lei, Z., Li, M., & Li, Y. Y. (2019). *Bioresource Technology*, 274, 127–133. https://doi.org/10.1016/j.biortech.2018.11.080

Coelho, C., Ribeiro, M., Cruz, A. C. S., Rosário, M., Domingues, R. M., Coimbra, M. A., . . . Nunes, F. M. (2014). *Journal of Agricultural and Food Chemistry*. https://doi.org/10.1021/jf501510d

Corral-Bobadilla, M., González-Marcos, A., Alba-Elías, F., & Diez de Santo Domingo, E. (2020). *Journal of Cleaner Production*, 264. https://doi.org/10.1016/j.jclepro.2020.121608

Cruz-Salomón, A., Ríos-Valdovinos, E., Pola-Albores, F., Lagunas-Rivera, S., Meza-Gordillo, R., & Ruíz-Valdiviezo, V. M. (2017). *Sustainability (Switzerland)*, 10(1). https://doi.org/10.3390/su10010083

Dadi, D., Mengistie, E., Terefe, G., Getahun, T., Haddis, A., Birke, W., Van der Bruggen, B. (2018). *Ecohydrology and Hydrobiology*, 18(2), 201–211. https://doi.org/10.1016/j.ecohyd.2017.10.007

Dinsdale, R. M., Hawkes, F. R., & Hawkes, D. L. (1997). *Water Research*, 31, 163–169.

dos Santos, A. B., Cervantes, F. J., & van Lier, J. B. (2007a). *Bioresource Technology*, 98(12), 2369–2385. https://doi.org/10.1016/j.biortech.2006.11.013

Enden, J. C. von, & Calvert, K. C. (2014). *Journal of Agricultural and Food Chemistry*, 42(5), 132–137. https://doi.org/10.11606/D.74.2016.tde-07042016-161058

Farah, A., & Donangelo, C. M. (2006). *Brazilian Journal of Plant Physiology*, 18(1), 23–36. https://doi.org/10.1590/S1677-04202006000100003

Fia, F. R. L., Matos, A. T., Borges, A. C., Fia, R., & Cecon, P. R. (2012). *Journal of Environmental Management*, 108, 14–21. https://doi.org/10.1016/j.jenvman.2012.04.033

Figueroa, G. A., T, H., & H.M., R. (2016). *Austin Food Sciences*, 1(1), 1014–1018.

Galiulin, Rauf & Bashkin, V. & Galiulina, Rumiya & Birch, Paul. (2001). A Critical Review: Protection from Pollution by Heavy Metals- Phytoremediation of Industrial Wastewater. *Land Contamination & Reclamation*. 9(4) https://doi.org/10.2462/09670513.568

Garde, W. K., Buchberger, S. G., Wendell, D., & Kupferle, M. J. (2017). *Journal of Hazardous Materials*, 329, 102–109. https://doi.org/10.1016/j.jhazmat.2017.01.006

Gomes de Barros, V., Rodrigues, C. S. D., Botello-Suárez, W. A., Duda, R. M., Alves de Oliveira, R., da Silva, E. S., . . . Madeira, L. M. (2020). *Environmental Pollution*, 259, 113796. https://doi.org/10.1016/j.envpol.2019.113796

Gregg, J. S., Jonas, J., & Happel, M. (2020). *Journal of Cleaner Production*, 267, 122093. https://doi.org/10.1016/j.jclepro.2020.122093

Gupta V. K. (1998). Equilibrium uptake Sorption, dynamics process, development, Column operations for the removal of copper and nickel from aqueous solution and waste- water using activated slag, a low-cost adsorbent. *Industrial & Engineering Chemistry Research*, 37(1), 192–202.

Haddis, A., & Devi, R. (2008). *Journal of Hazardous Materials*, 152, 259–262.

Ibarra-Taquez, H. N., GilPavas, E., Blatchley, E. R., Gómez-García, M. Á., & Dobrosz-Gómez, I. (2017). *Journal of Environmental Management*, 200, 530–538. https://doi.org/10.1016/j.jenvman.2017.05.095

Ijanu, E. M., Kamaruddin, M. A., & Norashiddin, F. A. (2020). *Applied Water Science*, 10. https://doi.org/10.1007/s13201-019-1091-9

Inamuddin, M., & Luqman, M. (2012) In *Igarss*. **Springer, Dordrecht, Heidelberg, New York, London.**

Kansal, S. K., & Kumari, A. (2014). *Chemical Reviews*, 114(9), 4993–5010. https://doi.org/10.1021/cr400093w

Kebede, Y. K., Kebede, T., Assefa, F., & Amsalu, A. (2010). *Ecohydrology and Hydrobiology*, 10(2–4), 259–270. https://doi.org/10.2478/v10104-011-0019-2

Kharel, S., Stapf, M., Miehe, U., Ekblad, M., Cimbritz, M., Falås, M., . . . Bester, K. (2020). *Science of the Total Environment*, 731.

Lachapelle-T, X., Labrecque, M., & Comeau, Y (2019). *Ecological Engineering*, 130, 32–44.

Lalevic, B., Raicevic, V., Kikovic, D., Jovanovic, L., Surlan-Momirovic, G., Jovic, J., & Morina, F. (2012). *International Journal of Environmental Research*, 6(1), 81–86. https://doi.org/10.22059/IJER.2011.474

Levchuk, I., Rueda Márquez, J. J., & Sillanpää, M. (2018). *Chemosphere*, 192, 90–104. https://doi.org/10.1016/J.CHEMOSPHERE.2017.10.101

Louzada, L., Andrade-vieira, L. F., Augusto, J., & David, D. O. (2016). *Ecotoxicology and Environmental Safety*, 133, 366–372. https://doi.org/10.1016/j.ecoenv.2016.07.019

Mahmood, K. T., Mugal, T., & Haq, I. U. (2010). *Journal of Pharmaceutical Sciences and Research*, 2(11), 775–781.

Mainardis, M., Buttazzoni, M., De Bortoli, N., Mion, M., & Goi, D. (2020) *Journal of Cleaner Production*, 258, 120781.

Malik, S. N., Ghosh, P. C., Vaidya, A. N., & Mudliar, S. N. (2018). *Journal of Water Process Engineering*, 35.

Manoj Kumar, B. (2015). *IOSR Journal of Environmental Science*, 9(9), 74–82. https://doi.org/10.9790/2402-09927482

Mekki H., Anderson, M., Benzina, M., & Ammar, E. (2009). *Journal of Hazardous Materials*, 158, 308–315.

Moreira, A. S. P., Nunes, F. M., Simões, C., Maciel, E., Domingues, P., Domingues, M. R. M., & Coimbra, M. A. (2017). *Data in Brief*, 13, 145–161. https://doi.org/10.1016/J.DIB.2017.05.027

Moreira, F. C., Boaventura, R. A. R., Brillas, E., & Vilar, V. J. P. (2017). *Applied Catalysis B: Environmental*, 202, 217–261. https://doi.org/10.1016/j.apcatb.2016.08.037

Nguyen, D. D., Nguyen, C. N. T., Huynh, K. A., & Nguyen, T. P. (2019). *Research on Chemical Intermediates*, 45(11), 5341–5356. https://doi.org/10.1007/s11164-019-03973-3

Nguyen, T. V., Zhang, R., Vigneswaran, S., Ngo, H., & Kandasamy, J. (2011). *Desalination* https://doi.org/10.1016/j.desal.2011.03.028

Nikoloski, A. N., & Ang, K. L. (2014). *Mineral Processing and Extractive Metallurgy Review*, 35(6), 369–389. https://doi.org/10.1080/08827508.2013.764875

Novita, E., Bagastyo, A. Y., & Sudarjanto, G. (2012). *CUTSE Conference*, 7, 131–134.

Novita, E., Wahyuningsih, S., Pradana, H. A., Marsut, W. D., & Farisul, A. F. (2019). *IOP Conference Series: Earth and Environmental Science*, 347(1). https://doi.org/10.1088/1755-1315/347/1/012019

Padmapriya, R., Tharian, J. A., & Thirunalasundari, T. (2013). *International Journal of Current Science*, 9, 83–91.

Pagalan, E., Sebron, M., Gomez, S., Salva, S. J., Ampusta, R., Macarayo, A. J., Arazo, R. (2020). *Industrial Crops and Products*, 145(November), 111953. https://doi.org/10.1016/j.indcrop.2019.111953

Primo, O., Rivero, M. J., Urtiaga, A. M., & Ortiz, I. (2009). *Journal of Hazardous Materials,* 164. https://doi.org/10.1016/j.jhazmat.2008.08.012

Ramya, P. M., Jayasravanthi, M., Dulla, B. J., & Venkata, N. R. (2015). *Global Nest Journal*, 17(2), 291–300. https://doi.org/10.30955/gnj.001492

Rattan, S., Parande, A. K., Nagaraju, V. D., & Ghiwari, G. K. (2015). *Environmental Science and Pollution Research*, 22(9), 6461–6472. https://doi.org/10.1007/s11356-015-4079-5

Rezania, S., Taib, S. M., Md Din, M. F., Dahalan, F. A., & Kamyab, H. (2016). *Journal of Hazardous Materials*, 318, 587–599. https://doi.org/10.1016/j.jhazmat.2016.07.053

Rodrigues de Araujo Felipe Rocha, N., Lima Ribeiro, R., Pereira Merlo, N., & Ribeiro Franco Junior, M. (2016). *Journal of Chemistry* and *Chemical Engineering*, 10, 161–166. https://doi.org/10.17265/1934-7375/2016.04.002

Said, N. S. M., Abdullah, S. R. S., Ismail, N., Izzati, Hasan, H. A., & Othman, A. R. (2020). *Environmental Technology and Innovation*, 17, 100502. https://doi.org/10.1016/j.eti.2019.100502

Saritha, V., Karnena, M. K., & Dwarapureddi, B. K. (2019). *Journal of Water Process Engineering*, 32(August), 100982. https://doi.org/10.1016/j.jwpe.2019.100982

Selvamurugan, M., Doraisamy, P., & Maheswari, M. (2010). *Ecological Engineering*, 36(12), 1686–1690. https://doi.org/10.1016/j.ecoleng.2010.07.013

Selvamurugan, M., Doraisamy, P., Maheswari, M., & Nandakumar, N. (2010). *Iranian Journal of Environmental Health Science & Engineering*, 7(2), 129–136.

Sengupta, B., Priyadarshinee, R., Roy, A., Banerjee, A., Malaviya, A., Singha, S., . . . Kumar, A. (2020). *Clean Technologies and Environmental Policy*. https://doi.org/10.1007/s10098-020-01841-y

Shanthakumar, S. G. and S. (2018). *Indian Journal of Environmental Protection*, 38(3), 213–220.

Takashina, T. A., Leifeld, V., Zelinski, D. W., Mafra, M. R., & Igarashi-Mafra, L. (2018). *Ozone: Science and Engineering*, 40(4), 293–304. https://doi.org/10.1080/01919512.2017.1417112

Teresa, Z. P., Gunther, G., & Fernando, H. (2007). *Journal of Environmental Sciences*, 19, 300–305.

Tokumura, M., Ohta, A., Znad, H. T., & Kawase, Y. (2006). *Water Research*, 40(20), 3775–3784. https://doi.org/10.1016/j.watres.2006.08.012

Tokumura, M., Znad, H. T., & Kawase, Y. (2008). *Water Research*, 42(18), 4665–4673. https://doi.org/10.1016/j.watres.2008.08.007

Tomizawa, M., Kurosu, S., Kobayashi, M., & Kawase, Y. (2016). *Journal of Environmental Management*, 183, 478–487. https://doi.org/10.1016/j.jenvman.2016.08.081

Torres-Valenzuela, L. S., Ballesteros-Gomez, A., Serna, J., Arango, A., & Rubio, S. (2020). *Environmental Science:*

Water Research and Technology, 6(3), 757–766. https://doi.org/10.1039/c9ew01095e

USDA Foreign Agricultural Service. Coffee: World Markets and Trade. *Coffee: World Markets and Trade*, (April 2019), 1–9. Retrieved from http://apps.fas.usda.gov/psdonline/circulars/coffee.pdf

Villa Montoya, A. C., Raissa Delforno, C. de. S., Delferno, T. P., Centurion, V. B., Sakamoto, I. K., de Oliveira, V. M., . . ., Varesche, M. B. (2019). *International Journal of Hydrogen Energy*, 44, 21434–21450. https://doi.org/10.1016/j.ijhydene.2019.06.115

Villanueva-Rodríguez, M., Bello-Mendoza, R., Wareham, D. G., Ruiz-Ruiz, E. J., & Maya-Treviño, M. L. (2014). *Water, Air, and Soil Pollution*, 225(12). https://doi.org/10.1007/s11270-014-2204-6

von Enden, J. C., Calvert, K. C., Sanh, K., Hoa, H., Tri, Q., Vietnam, S. R., & Consulting, C. E. O. R. (2002). *German Technical Cooperation Agency (GTZ)*, (January 2010), 1–10.

Wahab, N. A. A. M. N. and M. A. M. A. (2016). *Food and Fertilizer Technology Center for the Asian Pacific Region*, (Table 2), 1–5.

Wang, J. P., Chen, Y. Z., Ge, X. W., & Yu, H. Q. (2007). *Colloids and Surfaces A: Physicochemical and Engineering Aspects*. https://doi.org/10.1016/j.colsurfa.2007.02.023

Woldesenbet, A. G., Woldeyes, B., & Chandravanshi, B. S. (2014). *International Journal of Research in Engineering and Science (IJRES)*, 2(4), 1–5.

Xu, P., Xu, H., & Zheng, D. (2019). *Journal of Power Sources*, 421(February), 156–161. https://doi.org/10.1016/j.jpowsour.2019.03.033

Yu, L., Han, M., & He, F. (2017). *Arabian Journal of Chemistry*, 10, S1913–S1922. https://doi.org/10.1016/j.arabjc.2013.07.020

Zuhair Mohd Zain, M. (2017). *European Journal of Clinical and Biomedical Sciences*, 3(5), 97. https://doi.org/10.11648/j.ejcbs.20170305.13

16

Sequencing-based molecular markers for wild and cultivated coffee diversity exploration and crop improvement

Tram Vi, Pierre Marraccini, Alexandre de Kochko, Philippe Cubry, Ngan Giang Khong, and Valérie Poncet

CONTENTS

16.1 Introduction

Before the emergence of sequencing technologies, several PCR-based approaches and methods that rely on restriction sites were applied to establish molecular markers for coffee genetic studies, such as random amplified polymorphic DNA (RAPD), restriction fragment length polymorphism (RFLP), amplified fragment length polymorphism (AFLP), simple sequence repeat (SSR), and inter-simple sequence repeat (ISSR). Initial studies on the genetic diversity of *Coffea canephora* have defined two main groups (Congolese and Guinean) distributed in central and western Africa (Berthaud, 1986; Montagnon et al., 1992), and later five and more divergent groups of wild population were determined using RFLP, AFLP, and SSR data (Dussert et al., 1999; Cubry et al., 2008; Gomez et al., 2009; Ferrão et al., 2013; Kiwuka et al. 2021). Despite the fact that these markers, especially SSR, have high discriminant capacity in population genetics, they do not allow high-throughput production and only contribute to limited information in genetic and genomic studies. It is also difficult to link the different studies together because SSR markers are not easily transferable. The advancement of sequencing technologies has enabled not only high-throughput data analysis but also the availability of new molecular markers and thus has provided new insights at the whole-genome single-base level or structural level. Regarding sequencing methods, there are several, including whole-genome DNA sequencing, restriction-site associated DNA sequencing (RADSeq; Davey and Blaxter, 2010), genotyping-by-sequencing (GBS; He et al., 2014) and diversity arrays technology sequencing (DArTSeq). This chapter focuses on the molecular markers that can be detected by these sequencing approaches and their application and achievement on genetic and genomic studies of coffee species; also, their importance in coffee conservation and crop improvement will be further discussed.

16.2 Types of sequencing-based markers used in coffee

Earlier developed sequencing technologies, including Sanger, 454 sequencing, and especially next-generation sequencing (NGS), in combination with computational algorithms (bioinformatic tools), allow the detection of high-accuracy as well as high-density single nucleotide polymorphisms (SNPs) and insertions/deletions (Indels). In coffee genetics studies, SNP markers (mostly biallelic SNPs) might be considered as one of the most commonly used and powerful types of molecular markers to date. Whole-genome DNA sequencing is used for detection of abundant genome-wide variations, whereas RADSeq and GBS could be alternatives to reduce the variation redundancy. RADSeq limits the read sequences from the restriction sites to their 300–400 bp downstream positions (Davey and Blaxter, 2010). GBS covers only the highly divergent genomic regions that are predetermined. GBS markers were used, for example, to reconstruct well-resolved phylogeny for the *Coffea* genus (Hamon et al., 2017). Complex reduced SNP arrays have been discovered and scored in several crop species using DArTseq (Huttner et al., 2005). DArT also allows detecting variations in specifically targeted regions. The use of these methods has several privileges: (1) lower sequencing cost, (2) simultaneous screening of multiple genotypes, and (3) preselection of variations that could be used across arrays. Despite the complexity reduction, RADSeq is capable of offering highly informative data (Davey and Blaxter, 2010). Genome-wide SNPs, besides having discrimination capacity,

DOI: 10.1201/9781003043133-20

have been proved to be statistically associated with coffee quality traits such as caffeine content, aroma components (Sant'Ana et al., 2018; Tran et al., 2018a, 2018b), and resistance to diseases (Gimase et al., 2020). The wide application of SNP markers in genetic and genomic studies has led to various developments of robust bioinformatic methods for SNP detection, mostly based on Bayesian likelihood (Kumar et al., 2012; Xu et al., 2012). Long-read sequencing technologies, such as PacBio and Nanopore, because of the low coverage resulting in low-quality reads (Logsdon et al., 2020), are less effective in SNP calling.

Indels are less commonly used compared to SNPs, although they are relatively abundant along the genome. While SNPs are point mutations, Indels can concern a single base or a longer DNA sequence. They also play important roles in evolution and genetic differentiation. Indels were found to be the major type of variations in the chloroplast genome of Arabica (Min et al., 2019). Some Indels detected in the chloroplastic genome were supportive in classifying *C. arabica* and *C. canephora* (Nakagawa et al., 2019). Recently phylogenomic models have been developed to improve the use of Indels in phylogenetic analysis in species including plants (Donath and Stadler, 2018; Mahadani et al., 2018; Liu et al., 2019).

A large proportion of eukaryotic genomes is made of transposable elements (TEs) (Britten and Kohne, 1968), particularly in *C. canephora*, where the estimated proportion is over 50% (Denoeud et al., 2014). Studies of TEs in coffee species, however, have been mainly limited to their detection and classification, although some retrotransposons have shown abilities to resolve recent phylogenetic relationships (Hamon et al., 2011). There is currently numerous software for the identification and classification of TEs (Makałowski et al., 2019), but their accuracy is a constraint (Vendrell-Mir et al., 2019). Nonetheless, the improvement in read length and development in machine learning algorithms would leverage the efficiency in the discovery of TEs (Ewing, 2015; Arango-López et al., 2017).

Structural variation (SV), referring to Indels, duplications, interchromosomal/intrachromosomal translocations, and inversions that occur in approximately 50 bp or larger (Ho et al., 2020), is another type of molecular marker that can only be efficiently detected using whole-genome sequencing. Some of them are mediated by TEs. As is the case for TEs, long-read sequencing is more efficient in the identification of SVs compared to short-read sequencing (Mahmoud et al., 2019), but improvements are still needed in detection computational tools. SVs have been characterized and analyzed in evolutionary genetics for a number of plant species (Żmieńko et al., 2014; Wang et al., 2020); however, they are still poorly explored in coffee genomes.

Most of recent researches in coffee genomics, genetics, and phylogeny have been using SNPs because of their ease in detection and well-developed bioinformatic tools (Tranchant-Dubreuil et al., 2018), although other markers are evolutionally informative at certain levels.

16.3 Application in coffee phylogeny

Coffea is a genus belonging to the Rubiaceae family and strongly related with *Psilanthus* genus (Davis et al., 2006). These two genera have been assembled to include 124 described species (Davis et al., 2011), increasing the geographical range of *Coffea*. Before the inclusion of *Psilanthus, Coffea* was restricted to tropical Africa, Madagascar, the Comoros, and the Mascarenes. It also exists now in southern Asia, south tropical Asia, southeastern Asia, and Australasia.

The first nearly full-resolved phylogeny of the *Coffea* genus, including 81 species out of the 124 in the genus, was constructed based on 28,800 nuclear SNPs detected by GBS using Illumina sequencing (Hamon et al., 2017). The phylogeny analysis classified the studied species into two clades and four subclades which were geographical differentiated, and suggested that the biogeographical origin of the *Coffea* genus might be either Asia or Africa. This study also showed the evolution of caffeine in the *Coffea* genus and the transmission from free or low caffeine content in the ancestral species to moderate and high caffeine accumulators in west and central African species, especially *C. canephora*. Besides differentiated morphology from other coffee species, genetic distinction of *C. canephora* was confirmed by phylogenetic analysis of the 28,800 SNPs taken together with plastid phylogeny, which suggested that *C. canephora* might be still undergoing a sub-speciation process (Charr et al., 2020). Evolution of coffee has been also assessed by using TEs, for example, in *C. canephora* Pierre ex A. Froehner (Hamon et al., 2011; Dereeper et al., 2013), or in native *Coffea* species from Madagascar which provided results in accordance with analyses using SSR and/or SNPs (Charr et al., 2020; Roncal et al., 2016; Razafinarivo et al., 2013).

16.4 Application in genetic diversity analysis

Few *Coffea* species have been studied at the intra-specific level over their entire range, mainly the two cultivated species (*C. canephora*: Gomez et al., 2009; Cubry et al., 2013; *C. arabica*: Anthony et al., 2002; Scalabrin et al., 2020). Among studied *Coffea* species, *C. canephora* and *C. liberica* are the two species having widest genetic diversity, largely distributed in west and central Africa (Davis et al., 2006), and *C. arabica* has relatively low differentiation and a natural distribution centered on Ethiopia (Anthony et al., 2002).

SNP markers are widely used in genetic discrimination of populations within coffee species. Mérot-L'Anthoëne et al. (2019) have developed a high-density array of 8.5K SNPs that reveals a high level of polymorphism and was applicable for three species: *C. canephora, C. arabica*, and *C. eugenioides*. The SNP panel also allowed the discovery of two new *C. canephora* groups in the southern Democratic Republic of the Congo and Angola (Mérot-L'Anthoëne et al., 2019). Practically, by applying multiple SNPs selection methods such as filtering based on genome distribution, linkage disequilibrium, or minor allele frequency (ability to reveal polymorphism), only a reduced coreset of the 8.5K SNP array is sufficiently effective for the genetic diversity analysis of unidentified coffee collections (Akpertey et al., 2021; Zhang et al., 2021). To unravel the genetic origin and population structure of cultivated coffee species, one could also analyze complexity-reduced SNPs detected by DArTSeq (Garavito et al., 2016; Spinoso-Castillo

TABLE 16.1

Recent Studies on Local Genetic Diversity in Different Regions Using SNPs

Study	Region	Species	Source of plants	Markers
Garavito et al., 2016	Mexico and Vietnam	87 *C. canephora*	Wild and cultivated	4021 DArTSeq SNPs
Anagbogu et al., 2019	Nigeria	44 *C. canephora*, 1 *C. arabica*, 1 *"C. abeokutae"* (*C. liberica* var. *liberica*), 1 *C. liberica*, and 1 *C. stenophylla*	Cultivated	433,048 GBS SNPs
Garot et al., 2019	Reunion Island	25 *C. mauritiana*, 16 *C. canephora*, 1 *C. eugenioides*, 1 *C. liberica*, and 1 *C. bertrandii*	Wild	3953 DArTSeq SNPs
Mérot-L'Anthoëne et al., 2019	Africa	27 *C. canephora*, 17 *C. arabica*, and 6 *C. eugenioides*	Wild and cultivated	Coffee 8.5K SNP array (8580 SNPs)
Huang et al., 2020	China	48 *C. arabica*, 34 *C. canephora*, and 11 *"C. excelsa"* (*C. liberica* var. *dewevrei*)	Cultivated	15,367,960 whole-genome SNPs
Scalabrin et al., 2020	Ethiopia, Yemen, etc.	736 *C. arabica*	Wild and cultivated	698 GBS SNPs
Spinoso-Castillo et al., 2020	Mexico	3 *C. canephora*, 83 *C. arabica*, and 1 *C. liberica*	Cultivated	1739 DArTSeq SNPs
Akpertey et al., 2021	Ghana	400 *C. canephora*	Cultivated	192 KASPar SNPs
Gimase et al., 2021	Kenya	91 *C. canephora*	Cultivated	2280 DArTSeq SNPs
Zhang et al., 2021	Ethiopia, Yemen, etc.	130 *C. arabica*	Wild and cultivated	96 KASPar SNPs
de Aquino et al., 2022	Uganda	207 *C. canephora*	Wild	SNPs in 323 candidate genes

et al., 2020). These reduced SNP arrays are powerful in genetic classification; however, they do not fully cover the variations of the large coffee genome and therefore a portion of high-resolved information might be lost. More than 15 million whole-genome SNPs were identified in 93 accessions of three species (*C. arabica, C. canephora*, and *C. excelsa* [*C. liberica* var. *dewevrei*]) by whole-genome NGS and mapping on the *C. canephora* reference genome (Huang et al., 2020). In addition to genetic evidence strongly supported by an abundance of variations, SNPs via whole-genome sequencing could also enable detection of genome regions responsible for the divergence (Huang et al., 2020). Some limitations in whole-genome SNPs detection are high consumption of time, cost, and computational resources, and possibly some background noises in close-related genomes.

Genetic diversity of wild and cultivated coffee accessions in different regions of the world have been evaluated using various selections of SNPs (Table 16.1).

These studies have expanded the knowledge of the diversity of some species at the regional level, but they only provide a partial view. Coffee biodiversity has been partly preserved by in situ collections for wild species and ex situ collections for wild crop relatives (WCR; Bramel et al., 2017; Engelmann and Dulloo, 2007; Figure 16.1). Nevertheless, 60% of *Coffea* species are at high risk of extinction, and a large proportion of WCR is lacking ex situ collections (Davis et al., 2019). Therefore, understanding how much genetic diversity has been captured in present-day ex situ collections is indispensable to improve coffee conservation strategies. To determine genetic diversity and parentages within and between germplasms and compare them, a minimum set of key common markers and the inclusion of reference accessions would play an essential role to ensure the effective and efficient conservation and use of genetic diversity.

16.5 Application in genotype-phenotype association studies

The main traits of agronomic and organoleptic interest in coffee trees are resistance to pests and diseases (especially for the susceptible species *C. arabica*) and traits related to aroma and flavor in low cup quality species for *C. canephora*. Improvement in these traits is one of the keys in crop breeding and requires understanding of the relationship between phenotype and genotype.

Genome-wide association studies (GWAS), which are used to link molecular markers (mainly SNPs) to quantitative traits, have become routine in genomic analysis in plants (Purugganan and Jackson, 2021), including coffee. Several

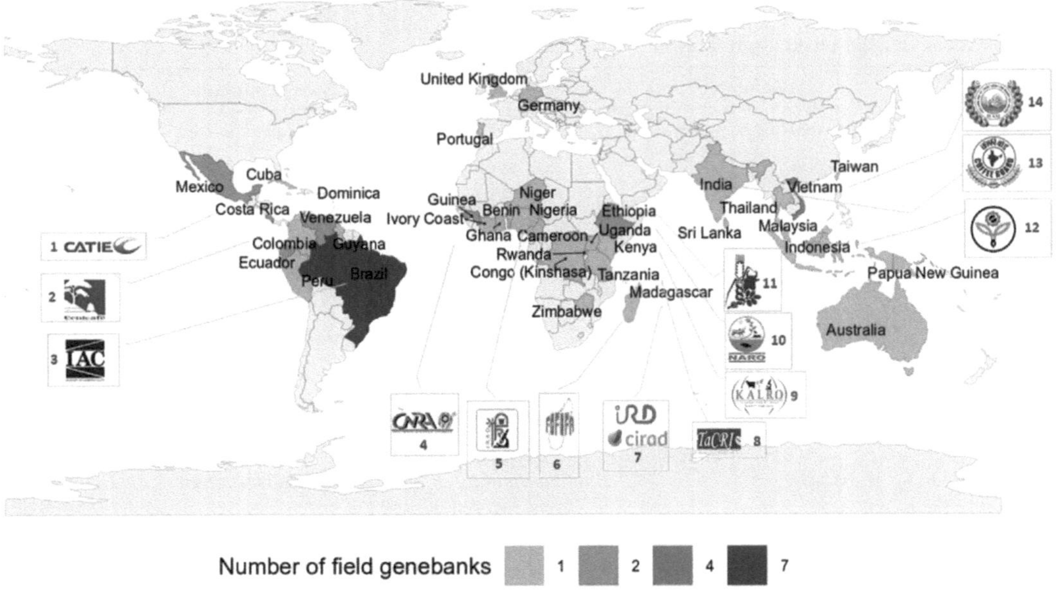

FIGURE 16.1 Ex situ field collections in countries from 1986 to 2016 (using data in Bramel et al., 2017) and major coffee gene banks. (1) Centro Agronómico Tropical de Investigación y Enseñanza (CATIE); (2) Centro Nacional de Investigaciones de Café (CENICAFE); (3) Instituto Agronômico de Campinas (IAC); (4) Centre National de Recherche Agronomique (CNRA); (5) Institut de Recherche Agronomique et de Développement (IRAD); (6) Centre National de Recherche Appliquée au Développement (FOFIFA); (7) Institut de recherche pour le développement (IRD)/Centre de coopération internationale en recherche agronomique pour le développement (CIRAD); (8) Tanzania Coffee Research Institute (TACRI); (9) Kenya Agricultural and Livestock Research Organization (KALRO); (10) National Agricultural Research Organisation (NARO); (11) Jimma Agricultural Research Center (JARC); (12) Indonesian Coffee and Cocoa Research Institute (ICCRI); (13) Central Coffee Research Institute (CCRI); (14) Western Highlands Agriculture and Forestry Science Institute (WASI).

attempts have been done to identify molecular markers responsible for specific traits in coffee, such as caffeine content and aroma compounds biosynthesis. In *C. arabica*, 1444 nonsynonymous SNPs were found to be related to caffeine content, of which 11 SNPs were linked to genes involved in the caffeine biosynthesis pathway (Tran et al., 2018b), and more than 20 SNPs exhibited association with aroma and flavor production (Sant'Ana et al., 2018). An equivalent number of caffeine-linked SNPs were identified in *C. canephora* and *C. eugenioides* (Tran et al., 2018a). A number of Indels also showed high association with caffeine biosynthesis in tea plants (Liu et al., 2019), but that has not yet been evaluated in coffee species.

SNPs are also robust markers for genetic linkage mapping and quantitative trait loci (QTL) mapping. Great efforts have been made, using a variety of molecular markers, to construct genetic maps of coffee species (Paillard et al., 1996; Ky et al., 2000; Pearl et al., 2004; Moncada et al., 2015), to locate QTL for productive traits (Leroy et al., 2011; Moncada et al., 2015), for resistance to diseases in *C. arabica* (Romero et al., 2014; Gimase et al., 2020), and for cup quality (Crouzillat et al., 2016; Figure 16.2). In addition, the availability of high-quality reference genomes could enable high-density genetic maps (Denoeud et al., 2014). The high-density SNPs can include markers that are in linkage disequilibrium with the QTL, and thus are useful in association studies of polygenic traits (van der Vossen et al., 2015; Nonato et al., 2021). However, major current QTL maps are based on SSRs and other PCR-based markers that are sparsely detected, while the advantage provided by high-density SNP markers has not yet been identified. Moreover, this approach is challenging in *C. arabica* because of its limited gene pools as a result of its probable recent single origin (Cui et al., 2020).

For genomics-assisted breeding (GAB), GWAS and molecular-marker-assisted selection (MAS) are useful in relating the phenotypes with the genetic markers. MAS underpinned by QTL mapping is only applicable in traits affected by strong-effect loci; however, with phenotypes such as resistance to diseases associated with a large number of minor-effect markers, GWAS gains more advantage as it takes all markers genome-wide into account (Nonato et al., 2021; Varshney et al., 2021).

16.6 Trending

Molecular markers potentially possess great information that has not been uncovered yet because of current limitations in mining and analysis tools.

Besides QTL mapping or linkage mapping, local ancestry inference and admixture mapping are further applications of genome-wide SNP identification, which are promising for association studies of polygenic traits and highly admixed populations such as in *C. canephora*. Such introgression analysis has been commonly used in tracking ancestral races in humans to study the complex admixture patterns in the Latino American population and their association with diseases

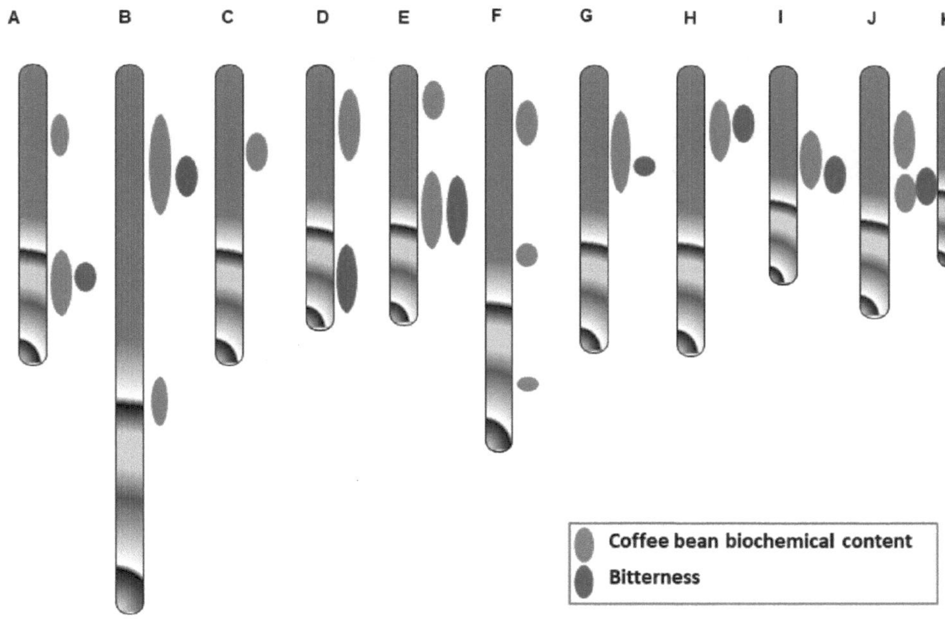

FIGURE 16.2 QTLs linked to cup quality in *C. canephora* (Crouzillat et al., 2016). The genetic map is realized from the cross of two elite clones from ICCRI BP409 X Q121 on an F1 progeny of 92 individuals. Coffee biochemical bean content was assessed by NIR and bitterness from sensory panel analysis. Colocalization of QTLs from sensory and biochemical data could indicate the putative origin of the sensory trait analyzed.

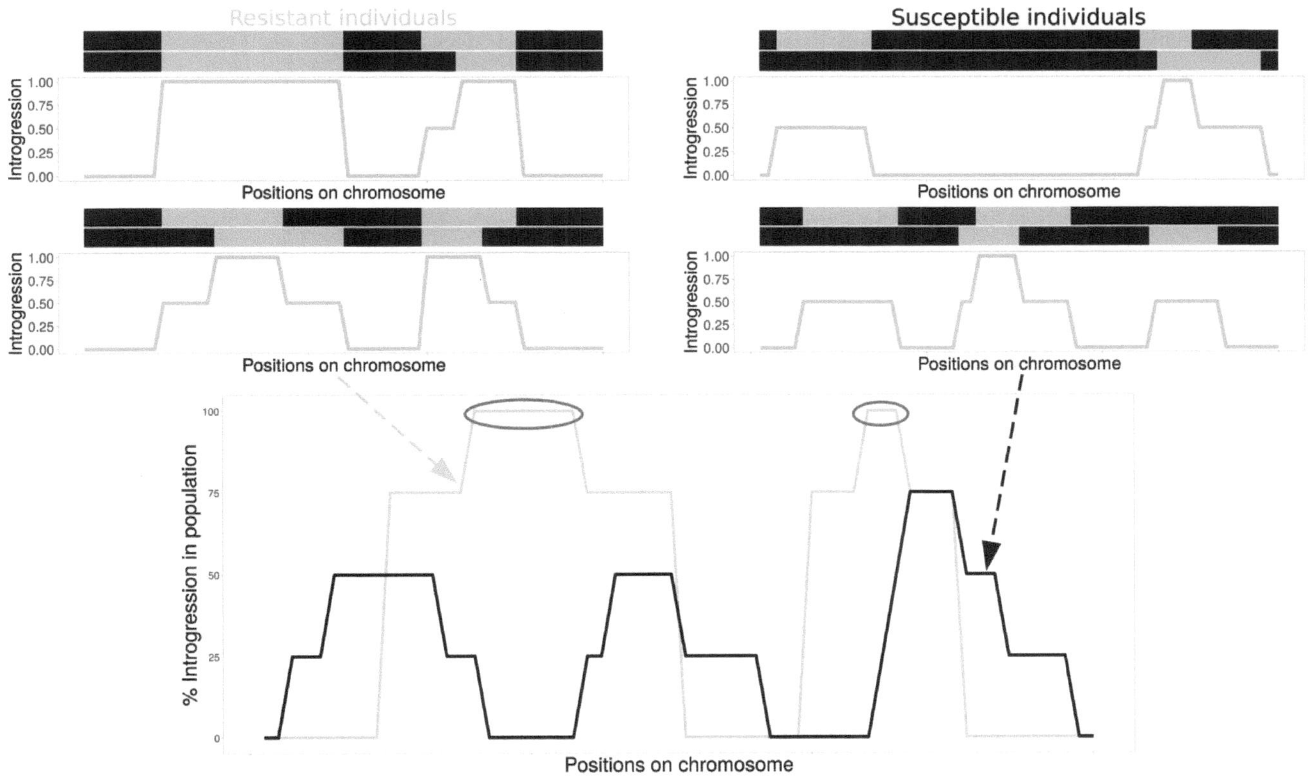

FIGURE 16.3 Application of introgression detection and admixture mapping in identification of ancestry associated with disease resistance. Introgression is detected at each locus (*green lines*) in all individuals. Admixture mapping is constructed by using average local ancestry inference in susceptible population (*black line*) and resistant population (*beige line*). Regions where the two populations are discriminated by ancestry are spotted (*red circles*). This approach could be used to assess the relationship between introgression and disease resistance traits.

(Shriner, 2013; Thornton and Bermejo, 2014; Mani, 2017), but this approach is still rarely applied in plants. A number of tools based on hidden Markov models were developed to assign the ancestry proportion at each locus by using haplotype-phased data (Leitwein et al., 2020), and some of them were evaluated to have high accuracy in plant species (Dias-Alves et al., 2018; Cottin et al., 2019). The investigation of mosaic genomes, in combination with linkage mapping, would enhance the knowledge of hybridization in self-incompatible coffee species and could be helpful in breeding programs. Most of GAB to improve adaptability in crops relies on introgression of QTL or resistance genes (Varshney et al., 2021). Introgression analysis might help to pinpoint QTL "hotspot" regions responsible for traits of interest (Figure 16.3).

The combination of genotypic data and phenotypic data, moreover, could be used for genomic prediction, which is a useful tool for crop breeding. The first prediction models, using Bayesian linear regression, for coffee essential traits (bean production, leaf rust incidence, and green bean yield) were built in *C. canephora* using up to 60K SNPs detected by GBS (Ferrão et al., 2019). More recently, leaf rust resistance in *C. arabica* was predicted based on decision trees; however, this only used SSR and AFLP markers without the assistance of linkage maps (de Sousa et al., 2020). Even though the accuracy of prediction was not high (< 70%), this is still a prospective approach in GAB and can be improved by using genome-wide dense SNP markers, increasing sample size and developing computational methods. Genomic prediction could be subsequent to genomic selection as a checkpoint to simulate and optimize the breeding efficacy (Cortés and López-Hernández, 2021). Moreover, using this approach, it would be possible to predict the adaptability of plants to different environments and climate change.

Furthermore, novel types of molecular markers could be possibly identified to fill the gap of the existing ones. Promisingly, development of a new category of molecular markers based on the sequencing of the methylome should allow a better understanding of the plant epigenetic response to biotic or abiotic stresses and the mechanisms involved in plant acclimatation and phenotypic plasticity (Bräutigam and Cronk, 2018; Corrêa et al., 2020; Li et al., 2020).

16.7 Conclusion

The ultimately main purpose of studying coffee genetics and genomics using molecular markers is to preserve and enhance the gene pool of high-quality coffee varieties and improve or produce varieties with desirable traits. Studying coffee evolution is important to understand their mechanism of adaptation to the changing environment and predict their divergent or convergent direction. Assessment of genetic diversity in wild and cultivated species plays a crucial role in preserving coffee species at high risk of extinction. Genomic selection is more robust with molecular marker assistance compared to conventional breeding, which is solely based on phenotyping; for example, localization of SNPs associated with disease resistance could help to optimize appropriate breeding strategies and produce prolonged resistant *C. arabica* varieties. However, there is still a big gap from genomic selection to breeding enhancement because phenotype is an output of complex genotype/environment interactions. Rapid development of biotechnology and bioinformatics is improving the identification of molecular markers and extending their application, allowing new aspects of the coffee genomics to be exposed and the revolution in breeding programs in the future.

REFERENCES

Akpertey A, Padi FK, Meinhardt L, Zhang D. *Frontiers in Plant Science*. 2021; 11:612593.

Anagbogu CF, Bhattacharjee R, Ilori C, Tongyoo P, Dada KE, Muyiwa AA, Gepts P, Beckles DM. *Genetics Resources and Crop Evolution*. 2019; 66:685–696.

Anthony F, Combes M-C, Astorga C, Bertrand B, Graziosi G, Lashermes P. *Theoretical and Applied Genetics*. 2002; 104:894–900.

Arango-López J, Orozco-Arias S, Salazar J A, Guyot R. *Colombian Conference on Computing*. Cham: Springer International Publishing, 2017; 156–170.

Berthaud J. *Conséquences pour l'application*. Paris: Orstom. ed, 1986.

Bramel P, Krishnan S, Horna D, Lainoff B, Montagnon C. Crop Trust and World Coffee Research. 2017.

Bräutigam K, Cronk Q. *Frontiers in Plant Science*. 2018; 9:1447.

Britten RJ, Kohne DE. *Science*. 1968; 161:529–540.

Charr J-C, Garavito A, Guyeux C, Crouzillat D, Descombes P, Fournier C, Ly SN, Raharimalala EN, Rakotomalala J-J, Stoffelen P, Janssens S, Hamon P, Guyot R. *Molecular Phylogenetics and Evolution*. 2020; 151:106906.

Corrêa RL, Sanz-Carbonell A, Kogej Z, Müller SY, Ambrós S, López-Gomollón S, Gómez G, Baulcombe DC, Elena SF. *Molecular Phylogenetics and Evolution*. 2020; 37:1866–1881.

Cortés AJ, López-Hernández F. *Genes*. 2021; 12(5):783.

Cottin A, Penaud B, Glaszmann J-C, Yahiaoui N, Gautier M. *G3: Genes, Genomes, Genetics*. 2019; 10:569–579.

Crouzillat D, Rigoreau M, Mérot-L'Anthoene V, Tranchant-Dubreuil C, Poncet V, de Kochko A. The 26th International Conference on Coffee Science (ASIC). Kunming. 2016.

Cubry P, Musoli P, Legnate H, Pot D, de Bellis F, Poncet V, Anthony F, Dufour M, Leroy T. *Genome*. 2008; 51(1):50–63.

Cubry P, de Bellis F, Pot D, Musoli P, Leroy T. *Genetic Resources and Crop Evolution*. 2013; 60:483–501.

Cui L, Hanika K, Visser RGF, Bai Y. *Agronomy*. 2020; 10:1928.

Davey JW, Blaxter ML. *Briefings in Functional Genomics*. 2010; 9:416–423.

Davis AP, Govaerts R, Bridson DM, Stoffelen P. *Botanical Journal Linnean Society*. 2006; 152:465–512.

Davis AP, Tosh J, Ruch N, Fay MF. *Botanical Journal Linnean Society*. 2011; 167:357–377.

Davis AP, Chadburn H, Moat J, O'Sullivan R, Hargreaves S, Lughadha EN. *Science Advances*. 2019; 5:eaav3473.

de Aquino SO, Kiwuka C, Tournebize R, Gain C, Marraccini P, Mariac C, Bethune K, Couderc M, Cubry P, Andrade AC, Lepelley M, Darracq O, Crouzillat D, Anten N, Musoli P, Vigouroux V, de Kochko A, Manel S, François O, Poncet V. *Molecular Ecology*. 2022; 31(6):1800–1819.

de Sousa IC, Nascimento M, Silva GN, Nascimento ACC, Cruz CD, Silva FF, Almeida DPD, Pestana KN, Azevedo CF, Zambolim L, Caixeta ET. *Scientia Agricola*. 2020; 7(4).

Denoeud F, Carretero-Paulet L, Dereeper A, Droc G, Guyot R, Pietrella M, Zheng C, Alberti A, Anthony F, Aprea G, Aury J-M, Bento P, Bernard M, Bocs S, Campa C, Cenci

A, Combes M-C, Crouzillat D, Da Silva C, Daddiego L, De Bellis F, Dussert S, Garsmeur O, Gayraud T, Guignon V, Jahn K, Jamilloux V, Joët T, Labadie K, Lan T, Leclercq J, Lepelley M, Leroy T, Li L-T, Librado P, Lopez L, Muñoz A, Noel B, Pallavicini A, Perrotta G, Poncet V, Pot D, Priyono, Rigoreau M, Rouard M, Rozas J, Tranchant-Dubreuil C, VanBuren R, Zhang Q, Andrade AC, Argout X, Bertrand B, de Kochko A, Graziosi G, Henry RJ, Jayarama, Ming R, Nagai C, Rounsley S, Sankoff D, Giuliano G, Albert VA, Wincker P, Lashermes P. *Science*. 2014; 345:1181–1184.

Dereeper A, Guyot R, Tranchant-Dubreuil C, Anthony F, Argout X, de Bellis F, Combes M-C, Gavory F, de Kochko A, Kudrna D, Leroy T, Poulain J, Rondeau M, Song X, Wing R, Lashermes P. *Plant Molecular Biology*. 2013; 83:177–189.

Dias-Alves T, Mairal J, Blum MGB. *Molecular Biology and Evolution*. 2018; 35:2318–2326.

Donath A, Stadler PF. *Algorithms for Molecular Biology*. 2018; 13:12.

Dussert S, Lashermes P, Anthony F, Montagon C, Trouslot P, Combes M-C, Berthaud J, Noirot M, Hamon S. *Diversity of cultivated tropical plants*. Montpellier: CIRAD. 1999; 175–194.

Engelmann F, Dulloo ME. *Complementary Strategies for ex situ Conservation of Coffee (Coffea arabica L.) Genetic Resources. A case study in CATIE, Costa Rica*. IRD Editions. 2007; 1–11.

Ewing AD. *Mobile DNA*. 2015; 6:4.

Ferrão LFV, Caixeta ET, Souza F de F, Zambolim EM, Cruz CD, Zambolim L, Sakiyama NS. *Plant Systematics and Evolution*. 2013; 299:225–238.

Ferrão LFV, Ferrão RG, Ferrão MAG, Fonseca A, Carbonetto P, Stephens M, Garcia AAF. *Heredity*. 2019; 122:261–275.

Garavito A, Montagnon C, Guyot R, Bertrand B. *BMC Plant Biology*. 2016; 16:242.

Garot E, Joët T, Combes M-C, Lashermes P. *Heredity*. 2019; 122:833–847.

Gimase JM, Thagana WM, Omondi CO, Cheserek JJ, Gichimu BM, Gichuru EK, Ziyomo C, Sneller CH. *Euphytica*. 2020; 216:86.

Gimase JM, Thagana WM, Omondi CO, Cheserek JJ, Gichuru EK. *African Journal of Plant Science*. 2021; 15(1):39–48.

Gomez C, Dussert S, Hamon P, Hamon S, de Kochko A, Poncet V. *BMC Evolution and Biology*. 2009; 9(1):167.

Hamon P, Duroy PO, Dubreuil-Tranchant C, Mafra D'Almeida Costa P, Duret C, Razafinarivo NJ, Couturon E, Hamon S, de Kochko A, Poncet V, Guyot R. *Molecular Genetics and Genomics*. 2011; 285:447–460.

Hamon P, Grover CE, Davis AP, Rakotomalala J-J, Raharimalala NE, Albert VA, Sreenath HL, Stoffelen P, Mitchell SE, Couturon E, Hamon S, de Kochko A, Crouzillat D, Rigoreau M, Sumirat U, Akaffou S, Guyot R. *Molecular Phylogenetics and Evolution*. 2017; 109:351–361.

He J, Zhao X, Laroche A, Lu Z-X, Liu H, Li Z. *Frontiers in Plant Science*. 2014; 5:484.

Ho SS, Urban AE, Mills RE. *Nature Review Genetics*. 2020; 21:171–189.

Huang L, Wang X, Dong Y, Long Y, Hao C, Yan L, Shi T. *Plant Molecular Biology*. 2020; 103:51–61.

Huttner E, Wenzl P, Akbari M, Caig V, Carling J, Cayla C, Evers M, Jaccoud D, Peng K, Patarapuwadol S, Uszynski G, Xia L, Yang S, Kilian A *Discovery to Delivery: BioVision 2004*. Alexandria: Bibliotheca Alexandrina. 2005; 145–155.

Kiwuka C, Goudsmit E, Tournebize R, de Aquino S O, Douma J C, Bellanger L, Crouzillat D, Stoffelen P, Sumirat U, Legnaté H, Marraccini P, de Kochko A, Andrade AC, Mulumba JW, Musoli P, Anten N, Poncet V. *PLoS ONE*. 2021; 16:e0245965

Kumar S, Banks TW, Cloutier S. *International Journal of Plant Genomics*. 2012; 2012:831460.

Ky CL, Barre P, Lorieux M, Trouslot P, Akaffou S, Louarn J, Charrier A, Hamon S, Noirot M. *Theoretical and Applied Genetics*. 2000; 101:669–676.

Leitwein M, Duranton M, Rougemont Q, Gagnaire P-A, Bernatchez L. *Trends in Ecology & Evolution*. 2020; 35:245–258.

Leroy T, De Bellis F, Legnate H, Kananura E, Gonzales G, Pereira LF, Andrade AC, Charmetant P, Montagnon C, Cubry P, Marraccini P, Pot D, de Kochko A. *Tree Genetics & Genomes*. 2011; 7:781–798.

Li R, Hu F, Li B, Zhang Y, Chen M, Fan T, Wang T. *Scientific Reports*. 2020; 10:8013.

Liu S, An Y, Tong W, Qin X, Samarina, L, Guo R, Xia X, Wei C. *BMC Genomics*. 2019; 20:935.

Logsdon GA, Vollger MR, Eichler EE. *Nature Review Genetics*. 2020; 21:597–614.

Mahadani AK, Sanyal G, Mahadani P, Bhattacharjee P. *International Journal of Data Mining Bioinformatics*. 2018; 21(1):43–66.

Mahmoud M, Gobet N, Cruz-Dávalos DI, Mounier N, Dessimoz C, Sedlazeck FJ. *Genome Biology*. 2019; 20:246.

Makałowski W, Gotea V, Pande A, Makałowska I. *Evolutionary Genomics*. New York: Springer, 2019; 177–207.

Mani A. *Circular: Cardiovascular Genetics*. 2017, 10(2):e001747.

Mérot-L'Anthoëne V, Tournebize R, Darracq O, Rattina V, Lepelley M, Bellanger L, Tranchant-Dubreuil C, Coulée M, Pégard M, Metairon S, Fournier C, Stoffelen P, Janssens SB, Kiwuka C, Musoli P, Sumirat U, Legnaté H, Kambale J-L, Ferreira da Costa Neto J, Revel C, de Kochko A, Descombes P, Crouzillat D, Poncet V. *Plant Biotechnology Journal*. 2019; 17(7):1418–1430.

Min J, Kim Y, Xi H, Heo KI, Park J. *Mitochondrial DNA Part B*. 2019; 4(2):2240–2241.

Moncada MDP, Tovar E, Montoya JC, González A, Spindel J, McCouch S. *Tree Genetics & Genomes*. 2015; 12:5.

Montagnon C, Leroy T, Yapo A. *Café Cacao Thé (France)*. 1992; 36:187–198.

Nakagawa T, Doi M, Nishi K, Sugahara T, Nishimukai H, Asano M. *Bioscience, Biotechnology, and Biochemistry*. 2019; 83:1829–1836.

Nonato JVA, Carvalho HF, Borges KLR, Padilha L, Maluf MP, Fritsche-Neto R, Fiho OG. *Euphytica*. 2021; 217:190.

Paillard M, Lashermes P, Pétiard V. *Theoretical Applied Genetics*. 1996; 93:41–47.

Pearl HM, Nagai C, Moore PH, Steiger DL, Osgood RV, Ming R. *Theorical Applied Genetics*. 2004; 108:829–835.

Purugganan MD, Jackson SA. *Nature Genetics*. 2021; 53:595–601.

Razafinarivo NJ, Guyot R, Davis AP, Couturon E, Hamon S, Crouzillat D, Rigoreau M, Dubreuil-Tranchant C, Poncet V, de Kochko A, Rakotomalala JJ, Hamon P. *Annals of Botany*. 2013; 111:229–248.

Romero G, Vásquez LM, Lashermes P, Herrera JC. *Plant Breeding*. 2014; 133:121–129.

Roncal J, Guyot R, Hamon P, Crouzillat D, Rigoreau M, Konan ON, Rakotomalala J-J, Nowak MD, Davis AP, de Kochko A. *Molecular Genetics and Genomics*. 2016; 291:155–168.

Sant'Ana GC, Pereira LFP, Pot D, Ivamoto ST, Domingues DS, Ferreira RV, Pagiatto NF, da Silva BSR, Nogueira LM, Kitzberger CSG, Scholz MBS, de Oliveira FF, Sera GH, Padilha L, Labouisse J-P, Guyot R, Charmetant P, Leroy T. *Scientific Reports*. 2018; 8:465.

Scalabrin S, Toniutti L, Di Gaspero G, Scaglione D, Magris G, Vidotto M, Pinosio S, Cattonaro F, Magni F, Jurman I, Cerutti M. *Scientific Reports*. 2020; 10(1):1–3.

Shriner D. *Current Protocols in Human Genetics*. 2013; 94: 1.23.1–1.23.8.

Spinoso-Castillo JL, Escamilla-Prado E, Aguilar-Rincón VH, Morales Ramos V, de los Santos GG, Pérez-Rodríguez P, Corona-Torres T. *Genetic Resources and Crop Evolution*. 2020; 67:1795–1806.

Thornton TA, Bermejo JL. *Genetic Epidemiology*. 2014; 38(S1): S5–S12.

Tran HTM, Furtado A, Vargas CAC, Smyth H, Slade LL, Henry R. *Tree Genetics & Genomes*. 2018a; 14:72.

Tran HTM, Ramaraj T, Furtado A, Lee LS, Henry RJ. *Plant Biotechnology Journal*. 2018b; 16:1756–1766.

Tranchant-Dubreuil C, Ravel S, Monat C, Sarah G, Diallo A, Helou L, Dereeper A, Tando N, Orjuela-Bouniol J, Sabot F. *bioRxiv*. 2018; 1:245480.

van der Vossen H, Bertrand B, Charrier A. *Euphytica*. 2015; 204:243–256.

Varshney RK, Bohra A, Yu J, Graner A, Zhang Q, Sorrells ME. *Trends in Plant Science*. 2021; 26:631–649.

Vendrell-Mir P, Barteri F, Merenciano M, González J, Casacuberta JM, Castanera R. *Mobile DNA*. 2019; 10:53.

Wang J, Lu N, Yi F, Xiao Y. *Evolutionary Bioinformatics*. 2020; 16:1176934320930263.

Xu F, Wang W, Wang P, Jun LM, Chung SP, Wang J. *Nature Communications*. 2012; 3:1258.

Zhang D, Vega FE, Solano W, Su F, Infante F, Meinhardt LW. *Conservation Genetics Resources*. 2021; 13(3):329–335.

Żmieńko A, Samelak A, Kozłowski P, Figlerowicz M. *Theoretical and Applied Genetics*. 2014; 127(1):1–18.

17

Insights into the coffee microRNAs: Present and future perspectives

Johns Saji and Shibin Mohanan

CONTENTS

17.1 Introduction

MicroRNAs (miRNAs) are small endogenous, non-coding regulatory RNA sequences consisting of 18–22 nucleotides derived from their precursor, sequences which negatively regulate the expression of target mRNAs by initiating mRNAS cleavage or inhibition of mRNA translation (Bartel, 2004; Zhang, Pan, Cannon, Cobb, & Anderson, 2006). The first miRNA was discovered in 1991, when Ambros and coworkers first isolated a *lin-4* mutant of *Caenorhabditis elegans* that was arrested at the first larval stage by downregulating the expression of *lin-14* protein (Lee, Feinbaum, & Ambros, 1993; Llave et al., 2002). Later on, the *let-7* mutation was isolated in the same system, which was responsible for development through the fourth larval stage. Both *lin-4* and *let-7* encode short 22-nucleotide mature RNAs and were called short temporal RNA because they control the temporal development program of *C. elegans*. The mature lin-4 RNA defines (negatively regulates) the mRNA expression of the lin-14 and lin-28 heterochronic genes with the antisense mediated repression mechanism of translation initiation and thus specifies the fate of cells during the first three larval stages. The first plant miRNAs was discovered in *Arabidopsis thaliana* (Llave et al., 2002; Reinhart, Weinstein, Rhoades, Bartel, & Bartel, 2002). Many investigations have experimentally proven that these miRNAs play important roles in different biological and metabolic processes in plants, including hormonal control, immune response, and adaptation to a variety of biotic and abiotic stresses (Bonnet et al., 2006; Ruiz-Ferrer & Voinnet, 2009; Sunkar et al., 2007; Voinnet, 2009). Some examples of these processes include the regulation of plant development (miR172), floral differentiation and (miR166) leaf polarity, root initiation and development (ath-miR164), signal transduction (i.e., miR159, miR160, miR164, and miR167), and plant environmental response (miR391 and miR395; Bonnet et al., 2006; Kidner & Martienssen, 2005).

Coffee is one of the most important crops in the world and is the second most traded commodity and an important source of income for several countries. The genus *Coffea* belongs to the family Rubiaceae and contains more than hundred species (Davis et al., 2006). Among all the species, only *Coffea arabica* and *C. canephora* are of economic importance, being responsible for 64% and 36% of world production, respectively. Many researchers suggest coffee consumption reduces the risk of Alzheimer's disease, Parkinson's disease, heart disease, type 2 diabetes mellitus, gout, and cirrhosis of the liver (Klatsky et al., 2006). It is also a valuable source of some secondary metabolites like polyphenols, caffeic acid, chlorogenic acid, and more importantly caffeine, which is an antioxidant and known to have medicinal properties against type 2 diabetes (Bae, 2021; Boonphang et al., 2021).

17.2 MicroRNAs in plants

MicroRNAs are 18 to 22 nucleotides long and play a key role in negative control of gene expression (Bonnet et al., 2006). MicroRNAs play an important role indifferent biological and

DOI: 10.1201/9781003043133-21

metabolic process in plants (Kidner & Martienssen, 2005; Reinhart et al., 2002; Samad et al., 2017). In plants, microRNAs function to control tissue differentiation and development, phase change from vegetative to reproductive growth, signal transduction, and respond to environmental conditions such as biotic (pathogens) and abiotic stress (e.g., drought, salinity; Samad et al., 2017; Sun et al., 2019). In 2002 the first miRNAs were discovered in plants (Park et al., 2002), but at present, several hundred miRNAs have been identified in plants by computational and experimental approaches.

MicroRNAs have been discovered by three basic approaches: direct cloning, forward genetic, direct cloning, forward genetic and bioinformatic prediction followed by experimental validation. The most direct method of microRNA discovery is to isolate and clone small RNAs from biological samples; several groups have used this approach to identify plant miRNAs (Dezulian, Remmert et al., 2006; Llave et al., 2002; Mette et al., 2002; Park et al., 2002; Reinhart et al., 2002; Sunkar et al., 2005; Xie et al., 2003). The cloning methods were adapted similar to those used to identify large numbers of animal miRNAs (Lagos-Quintana et al., 2001; Lau et al., 2001), which involve isolating small RNAs, followed by oligonucleotide adapter ligation, reverse transcription, amplification, and sequencing. Some protocols incorporate methods to enrich for Dicer cleavage products (i.e., molecules with 5′-phosphates and 3′-hydroxyls) and to concatemerize the short cDNAs so that several may be identified in a single sequencing read (Lau et al., 2001; Llave et al., 2002). The initial cloning experiments in *Arabidopsis* identified 19 miRNAs belonging to 15 families (Llave et al., 2002; Mette et al., 2002; Park et al., 2002; Reinhart et al., 2002), although hundreds of other small RNAs, such as siRNAs and RNA degradation fragments, were also cloned through the direct cloning methods. To select miRNAs from the pool of small RNAs, secondary structures of the RNA molecules were examined in compliance with the acknowledged structural features of miRNAs (Ambros et al., 2003). The secondary structures were validated using several ad hoc software tools, such as the Mfold or RNAfold programs (Reeder et al., 2006). Finally, the existence of mature miRNAs was determined by RNA gel blot hybridization. These direct cloning methods were successful in isolating many miRNAs in *Arabidopsis* (Reinhart et al., 2002; Sunkar & Zhu, 2004), rice (Sunkar et al., 2005), poplar (Lu et al., 2005), moss (Arazi, 2005), and tobacco (Billoud, 2005).

Although miRNAs were first discovered through forward genetic screens in worms (Lee et al., 1993; Reinhart et al., 2000), few *A. thaliana* miRNA gene families have been discovered by this method (Chen, 2008) in plants. miRNA involvement in plant mutant phenotypes were not properly inferred until the cloning experiments established that plant genomes contain numerous miRNAs (Park et al., 2002; Reinhart et al., 2002; Rhoades et al., 2002). A MiRNA loss-of-function allele has been identified by forward genetic screens: *early extra petala1* is caused by a transposon insertion about 160 bp upstream of the predicted miR164c stem loop and results in flowers with extra petals (Baker et al., 2005). The fact that loss-of-function miRNA mutants have been recovered so rarely using forward genetics may reflect a small target size for mutagenesis coupled with redundancy; nearly all evolutionarily conserved plant miRNAs are encoded by gene families (Table 17.1). Family members are likely to have overlapping functions, buffering against loss at any single miRNA locus. Overexpression screens can circumvent redundancy limitations. At least four plant miRNAs, miR319 (also known as miR-*JAW*), miR172 (also known as *EAT*), and miR166, were isolated in over expression screens for dominant mutants with developmental abnormalities (Aukerman & Sakai, 2003; J. Kim et al., 2005; Palatnik et al., 2003; Williams et al., 2005). Mutations in miRNA target sites, which can prevent the entire family of miRNAs from repressing a target gene, can also circumvent redundant functions of miRNA family members.

The dominant mutations in the *HD-ZIP* genes *PHB, PHV*, and *REVOLUTA (REV)* in *Arabidopsis* and *ROLLEDLEAF1* (*RLD1*) in maize, resulting in adaxialization of leaves and/or vasculature (McConnell & Barton, 1998; McConnell et al., 2001; Nelson et al., 2002; Zhong & Ye, 2004), and are all caused by mutations in miR 166 complementary sites (Emery et al., 2003; Jones-Rhoades & Bartel, 2004a; Mallory et al., 2004; Rhoades et al., 2002; Zhong & Ye, 2004).

In both plants and animals, cloning was the initial means of large-scale miRNA discovery (Lagos-Quintana et al., 2001; Lau et al., 2001; Lee & Ambros, 2001; Reinhart et al., 2002). However, cloning is biased toward RNAs that are expressed highly and broadly. MicroRNAs expressed at low levels or only in specific cell types or in response to certain environmental stimuli were more difficult to clone. Sequence-based biases in cloning procedures might also cause certain miRNAs to be missed. Because of these limitations, bioinformatic approaches to identify miRNAs have provided a useful complement to cloning.

A straightforward use of bioinformatics has been carried out to find homologs of known miRNAs, both within the same genome and in the genomes of other species (Lagos-Quintana et al., 2001; Lau et al., 2001; Lee & Ambros, 2001; Pasquinelli et al., 2000). A more difficult challenge is to identify miRNAs unrelated to previously known miRNAs. This was first accomplished for vertebrate, nematode, and fly miRNAs using algorithms that search for conservation of sequence and secondary structure (i.e., miRNA stem-loop precursors) between species, searching for patterns that are characteristic of miRNAs (Lai et al., 2003; Lim et al., 2005; Lim et al., 2003), although these methods identified numerous necessarily more relaxed in terms of allowed structures. Some approaches take advantage of the high complementarity of plant miRNAs to target messages, implementing the requirement that the candidate has conserved complementarity to mRNAs (Adai et al., 2005; Jones-Rhoades & Bartel, 2004). This additional filter has been useful for distinguishing authentic plant miRNAs from false positives, which later extended to mammalian miRNA gene prediction (Xie et al., 2005).

Another criteria used by most algorithms in the filters is evolutionary conservation (Dezulian et al., 2006; Rhoades et al., 2002). In plants, mature miRNA sequences are conserved to a higher degree than their precursor sequences. For example, *Arabidopsis* and rice diverged more than 130 million years ago (Friis, 2004), but some miRNAs are highly conserved in these plant species (Reinhart et al., 2002). Furthermore, various parameters for the secondary structures, such as free

TABLE 17.1

MicroRNA Families Conserved in Plants

miRNA	*Arabidopsis thaliana*	*Oryza sativa*	*Vitis vinifera*	*Solanum lycopersicum*	*Solanum tuberosum*	*Brassica napus*	*Gossypium hirsutum*	*Zea mays*
Total no. of miRNA families	**148**	**302**	**135**	**175**	**78**	**67**	**189**	**194**
Total no. of miRNA loci	**221**	**699**	**289**	**324**	**169**	**388**	**631**	**465**
Major miRNA families								
miR156/157	27	13	8	13	14	28	26	13
miR158	2	0	0	0	0	6	0	0
miR159/319	6	7	11	10	8	11	16	14
miR160	3	6	5	2	2	15	12	7
miR161/162	3	2	1	2	2	10	2	1
miR163/164	4	6	4	4	1	15	14	7
miR165	2	0	0	0	0	11	0	0
miR166	7	13	8	10	11	14	30	14
miR167	3	10	4	4	4	7	12	10
miR168	2	1	1	2	2	20	2	2
miR169	14	16	21	15	8	30	39	18
miR170/171	5	8	11	12	5	17	19	16
miR172	5	4	5	8	4	21	19	5
miR173	1	0	0	0	0	0	0	0
miR390	2	1	2	3	3	5	9	2
miR391/393	3	2	3	2	2	12	7	3
miR394	2	1	3	2	0	13	12	2
miR395	6	19	16	16	0	9	2	16
miR396	2	7	6	3	3	3	16	8
miR397	2	2	1	1	1	4	5	2
miR398	3	1	3	2	4	3	7	2
miR399	6	12	9	13	15	22	25	11
miR400	1	0	0	0	0	3	0	0
miR403	1	0	6	1	1	2	4	0
miR408	1	1	1	2	1	3	2	1
miR447	2	0	2	3	0	0	9	0

Note: All known miRNA families that are conserved between more than one plant species are listed together with the number of genes identified in the sequenced genomes. The following families contain miRNA genes annotated with more than one number: miR156 (miR156 and miR157), miR159/319 (miR159 and miR319), miR166 (miR165 and miR166), miR171 (miR170 and miR171), and miR390 (miR390 and miR391).

energy, the number of paired residues within a miRNA, or the number and size of bulges, are also used to validate the predictions (Adai et al., 2005; Jones-Rhoades & Bartel, 2004; Wang et al., 2004). Most miRNAs are conserved throughout flowering plants (Table 17.1), but many more were found only in a single sequenced genome thus considered to be of a more recent evolutionary origin. The extended homology between few non-conserved miRNAs precursors and their target genes provides strong evidence that these relatively young miRNAs arose from the duplication of the target gene segments (Allen et al., 2004). Several non-conserved miRNAs, including miR161, miR163, miR173, miR447, miR475, and miR476, are known to direct cleavage of target transcripts (Adai et al., 2005; Allen et al., 2004; A. Li & Mao, 2007; Sunkar et al., 2005), thus it is difficult to confidently predict targets for many because it is not possible to use complementary site as a filter against false-positive target predictions. It is difficult to confidently annotate non-conserved miRNAs as miRNAs rather than siRNAs. The established minimal standard for miRNA annotation is a small RNA with detectable expression and the potential to form a stem-loop when joined to a flanking genomic sequence (Kasschau et al., 2003). In practice, these requirements are too loose to definitively categorize many small RNAs cloned from plants. Many plant siRNAs are detectable on blots (Vazquez et al., 2004), and hundreds of thousands of non-miRNA genomic sequences can be predicted to fold into secondary structures that resemble structures of plant miRNA precursors (Jones-Rhoades & Bartel, 2004a). Therefore, without conservation of both sequence and secondary structure, it is difficult to be confident that a given cloned RNA originated from a stem-loop (i.e., is a miRNA) rather than from a double-stranded RNA (i.e., is a siRNA). In fact, many of the thousands of small RNAs cloned from *Arabidopsis* (Gustafson et al., 2005) would probably meet the literal requirements for annotation as miRNAs. A few of these sequences might be miRNAs, but others that meet the literal criteria probably are not so. The challenges of annotating non conserved small RNAs were evidenced by three related small silencing RNAs that were originally annotated as miRNAs that turned out to be *trans-acting* siRNAs (tasiRNAs).

17.3 MicroRNAs in coffee

There have been very few studies on the MicroRNAs in coffee. Nine studies have predicted miRNAs from *C. canephora* using different approaches (Akter et al., 2014; Bibi, Barozai, & Din, 2017; Chaves et al., 2015; Dos Santos et al., 2019; Ivamoto et al., 2017; Loss-Morais et al., 2014; Mohanan & Chandrashekar, 2012; Mohanan et al., 2013; Noronha Fernandes-Brum et al., 2017). Most of the studies focused only on prediction of conserved microRNAs the first attempt to study microRNAs were done by expressed sequence tag (EST) analysis for conserved microRNAs (Mohanan & Chandrashekar, 2012). Two studies have used small RNA sequencing to confirm transcriptional activity of miRNAs (Guedes et al., 2018; Loss-Morais et al., 2014). An intronic microRNA with a probable role in regulating a secondary metabolite production was also reported from *C. canephora* (Mohanan et al., 2013). A comprehensive non-coding RNA catalog was prepared for *C. canephora*, which described and characterised miRNAs along with snRNAs, snoRNAs, miRNAs, tRNAs, rRNAs, and incRNAs. The authors had used the latest criteria for curation of precursors (Axtell & Meyers, 2018; Taylor et al., 2017), and those predicted were of high quality (Lemos et al., 2020a).

17.4 Techniques used for identification of *Coffea* microRNA

17.4.1 Homology search and potential miRNA identification

Most of the coffee microRNAs were identified using a homology search method based on the point criteria (Ambros et al., 2003; Unver et al., 2009), summarized in Figure 17.1, with modification and different BLAST software like NCBI BLAST (Altschul et al., 2009; Altschul et al., 1997) or mpi-BLAST which is an open source parallel implementation of NCBI BLAST running a similar algorithm to BLAST-2.2.22 (Darling, Carey, & Feng, 2003). The EST and GSS (genome survey sequence) and nucleotide sequences deposited in the NCBI database were used as references.

The Zuker folding algorithm mFOLD is the most commonly used to analyze the secondary structure of RNAs (Zuker, 2003). Another approach is to retrieve sequences that form hairpin-like structures using from coffee nucleotide databases using Einverted (from EMBOSS) and BLASTN tools. Before identifying candidate miRNAs, call protein coding sequences were removed by using BLASTX analysis against NCBI non-redundant protein database. The e-value threshold was kept at 0.001, minimal match length 25 nucleotides, and 80% for identity, score value for 100, alignment of 1000, and except value of 1000. Low complexity was chosen as the sequence filter; the number of descriptions and alignments was raised to 1000. The candidate sequences were not allowed to have more

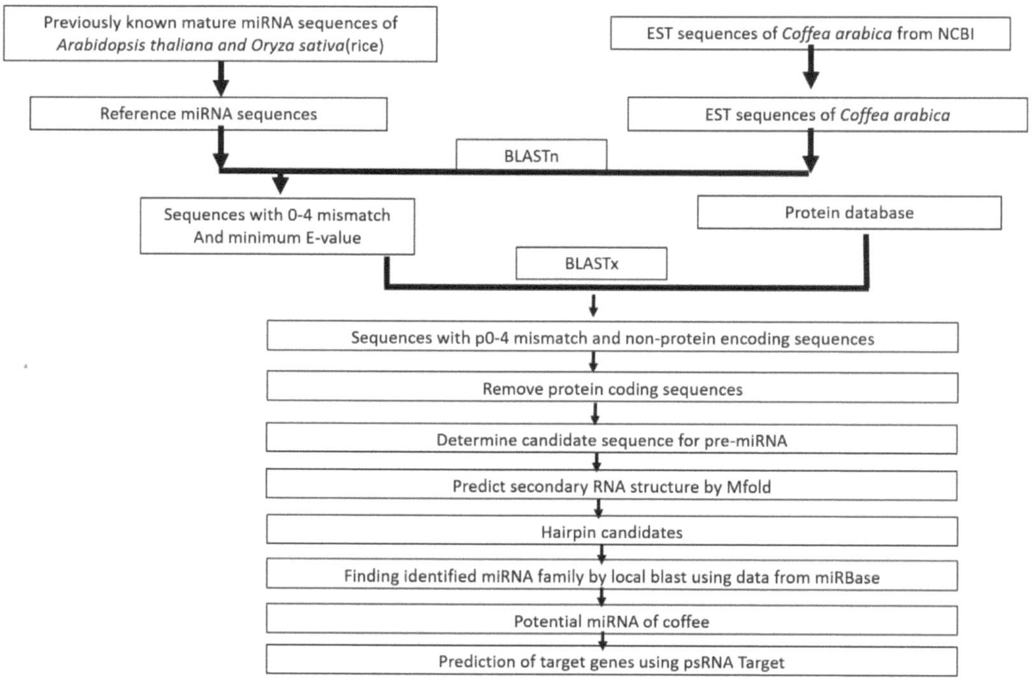

FIGURE 17.1 Schematic representation of the coffee miRNA search procedure used to identify homologs of known plant miRNA.

than four mismatches. The candidate sequences had length of 60–400 nt and were filtered. The following criteria is usually used for filtering of the microRNA sequences.

The secondary structure of candidate pre-miRNA sequences of these potential miRNA homologs was predicted using the Zuker folding algorithm with MFOLD-3 (Zuker, 2003). All parameters were set to default values. All outputs obtained from mfold were recorded into a spreadsheet, which included the EST ID numbers, respective miRNA homologs, total length of the sequences, the number of each nucleotide (A, G, C, and U), the number of arms per structure, location of the matching regions, percentage (%) of (A + U/T) and (G + C) content, and minimal folding free energy (MFE; DG in kcal/mol). Then, the adjusted minimal folding free energy (AMFE) and the minimal folding free energy index (MFEI) were calculated according to the following equation from a previous report (Zhang et al., 2006):

$$AFME = \left(MFE / \left(lengtlength\, of\ the\ potential\ pre \right.\right.$$

$$MFEI = \left(100 \times MFE\right) ?\, Length\, ofRNA / \left(\left(\right.\right.$$

miRNA precursor sequences have significantly higher negative MFEs and MFEIs than other types of RNA (e.g., tRNA, rRNA). To avoid designating other RNAs as miRNA candidates, these two characteristics and G + C content were considered when predicting secondary structures. Briefly, an EST, GSS, or nucleotide was considered as a potential miRNA candidate if it fulfills the following criteria described by Wang (Wang et al., 2004): (1) the predicted mature miRNAs had no more than three nucleotide substitutions compared with a known reference mature miRNA; (2) the EST sequence must fold into an appropriate and proper stem-loop hairpin secondary structure; (3) the mature miRNA must be localized in one arm of the stem-loop structure; (4) there were no more than six mismatches between the predicted mature miRNA sequence and its opposite miRNA* sequence in the secondary structure; and (5) the predicted secondary structure must contain high negative MFE and high MFEI values.

17.4.2 Phylogenetic analysis

Plant microRNAs are highly conserved, and discovery of plant microRNAs is done through bioinformatics analysis. Homology search of candidate coffee miRNA was done against all plant miRNAs using computational approaches, and sequence alignment tools like the Bioedit sequence alignment tool (Akter et al., 2014). The precursor sequences of homolog small RNAs were identified and collected.

17.5 MicroRNAs identified from coffee

The first attempt to identify coffee microRNA using in silico methods was done using EST sequences for *C. canephora* and identified 18 microRNAs belonging to 12 families. The nine main families are miR156a/b, miR172, miR393, miR395, miR414, miR834, miR838, miR854 and miR829. All the identified miRNAs are considered valid, as they fulfill almost all the criteria of empirical formula for microRNA formation and biogenesis (Ambros et al., 2003). miR414 and mir854 are rare microRNAs reported from few species. These are speculated to be involved in speciation (Gao et al., 2015). Many highly conserved miRNAs found in different plant species, such as miR162 and miR164, were not identified in this study (Mohanan & Chandrashekar, 2012). A similar homology search yielded 88 potential miRNAs candidates from which 18 miRNAs from 174275 EST sequences belonging to 11 families, while seven does not belong to any Rfam families (5658a, 5658b, 1171a, 1171b, 1171b2, 5368 and 5532; Rebijith et al., 2013). The 41 targets identified for the all the potential miRNA coded for transcription factors involved in plant development and physiological process. Other targets included proteins involved in metabolism, signal transduction, and stress responses. miR-2673a had multiple gene targets involved in nucleic acid binding DNA repair, nucleic acid binding, DNA repair, helicase activity, proline rich family proteins involved in actin polymerization, and zinc knuckle protein activity. Zinc knuckle protein is involved in transcriptional activation metabolism, cell signaling, and apoptosis (Rebijith et al., 2013).

Using an identical number of EST sequences, 67 unique ESTs showing homology with known reference plant miRNAs were observed, and one potential microRNA with sequence homology to miRNA-393 along with six potential target genes was identified. Genes targeted by the identified miRNA contain transport inhibitor response (TIR)-like protein such as auxin transporter protein I, which plays an important role in regulatory pathway like auxin polar transport, ethylene signaling pathway involved in auxin influx transmembrane transporter activity (Akter et al., 2014).

Transcriptional factors are important components in the transcriptional process and play an important role in a variety of biological functions including plant development, hormone signaling, and metabolism. First, transcriptional factor WRKY is an important protein involved in biological processes such as camalexin biosynthetic process, cellular heat acclimation, defense response to bacterium and fungus, positive regulation of autophagy, response to chitin, cold salt stress, water deprivation, and DNA-dependent transcription. Second, transcriptional factor bHLH7 is involved in DNA binding. Another important target of coffee miRNA is GRR 1-like protein, which has also some important biological and molecular functions such as auxin mediated signaling pathway, negative regulation of transcription by glucose, stamen development, and inositol hexakisphosphate binding. Two auxin signaling proteins are auxin signaling, F-box 2 and auxin signaling, F-box 3. These two proteins are the component of SCF (ASK-cullin-F-box) E3 ubiquitin ligase complexes, which may mediate the ubiquitination and subsequent proteasomal degradation of target proteins. These are also involved in embryogenesis regulation by auxin and some plant defense mechanisms like antibacterial resistance (Cui et al., 2020; Lai et al., 2011; Libault et al., 2007; Robert-Seilaniantz et al., 2011).

Another study identified 26 microRNAs and total of 616 and 362 potential miRNA targets for *C. arabica* and *C. canephora*. The 36 miRNA families identified (16 possible

TABLE 17.2

Updated Criteria for Plant miRNA Annotations

	2008 criteria	2018 criteria
1	One or more miRNA/miRNA* duplexes with two-nucleotide 39 overhangs	Add requirements that exclude secondary stems or large loops in the miRNA/miRNA* duplex and limit precursor length to 300 nucleotides
2	Confirmation of both the mature miRNA and its miRNA miRNA/ miRNA* duplex contains #4 mismatched bases	Disallow confirmation by blot; sRNA-seq only; disallow confirmation by blot; sRNA-seq only, which are nucleotides in asymmetric bulges
3	The duplex has at most one asymmetric bulge containing at most two bulged nucleotides	Up to five mismatched positions, only three of which are nucleotides in asymmetric bulges
4	75% of reads from exact miRNA or miRNA*	Include one-nucleotide positional variants of miRNA and miRNA* when calculating precision
5	Replication suggested but not required	Required; novel annotations should meet all criteria in at least two sRNA-seq libraries (biological replicates)
6	Homologs, orthologs, and paralogs can be annotated without expression data, provided all criteria met for at least one locus in at least one species	Homology-based annotations should be noted as provisional, pending actual fulfillment of all criteria by sRNA-seq
7	miRNA length not an explicit consideration	No RNAs <20 or >24 nucleotides should be annotated as miRNAs; annotations of 23- or 24-nucleotide miRNAs require extremely strong evidence

Source: Adapted from Axtell and Meyers (2018).

miRNAs characterized for *C. arabica* and 20 for *C. canephora*; Table 17.2) were much less than the 33 families previously reported, which may be due to the much stringent and robustness that was used when compared to previous studies (Chaves et al., 2015). Out of these, seven were unique and not reported for *C. arabica* and *C. canephora* (car-MIR7122, car-MIR835, cca-MIR368, cca-MIR3169, cca-MIR5293, cca-MIR6459, and cca-MIR845).

Similar study involving deep sequencing approach to identify and classify miRNAs in *C. canephora* and *C. arabica*, 58 unique mature miRNAs that were divided into 33 families. The recent change in criteria as described in Table 17.2 (Axtell & Meyers, 2018) is more stringent and has helped in removing false positives from the miRNA from different plants. Of all the MicroRNAs identified up to 2018 using in silico methods, only 60 precursors and 120 mature sequences passed scrutiny (Lemos et al., 2020b). In summary, the *C. canephora* set of high-standard miRNAs was composed of 72 precursors and 145 mature miRNAs belonging to 27 families. Most of the identified miRNA families are variants of miR-156, miR-159, miR-162, miR-164, miR-166, miR-167, miR-168, miR-169, miR-171, miR-172, miR-319, miR-390, miR-393, miR-394, miR-395, miR-396, miR-397, miR-398, miR-399, miR-403, miR-408, miR-477, miR-482, miR-530, and miR-2111, all of which are conserved in plants.

Most of the microRNAs identities were found and are known to target transcription factors (Table 17.3). MYB transcription factors represent families of proteins with a conserved MYB DNA-binding domain and are involved in regulation of secondary metabolism, control of cellular morphogenesis, and control of meristem formation. This domain was considered to be involved in regulation of secondary metabolism, control of cellular morphogenesis, and regulation of meristem formation and the cell cycle (Jin & Martin, 1999). This MYB protein is targeted by miR-3434 in *Coffea*. Another transcription factor, homeobox-leucine zipper protein ATHB-14, is involved in the determination of adaxial-abaxial polarity in the ovule primordium and specifies adaxial leaf fates (Kim et al., 2008). In *Coffea*, miR-165 directs the regulation of ATHB- 14. Aside from MYB and ATHB-14, there are several transcription factor groups that have been detected to be targets of miRNAs, like zinc ion binding (miR-2938, miR-3516), homeobox-leucine zipper protein REVOLUTA (miR-165), and zinc finger protein (miR-5725, miR-8166, miR-9408).

17.6 Present challenges and future perspective

Significant progress has been achieved in the analysis and characterization of plant miRNAs in recent years, with a growing number of research papers on the critical role of miRNAs in agricultural plants. As mentioned previously, these small nucleic acids play an important role in plant growth and development, the transition from vegetative to reproductive phase, hormone signaling, signal transduction pathways, and homeostatic responses to various abiotic and biotic stresses (Sun et al., 2012). Plant miRNAs have also been implicated as molecular regulators of plant immune and defense responses in a number of recent studies (Balmer & Mauch-Mani, 2013; Djami-Tchatchou & Dubery, 2015; Y. Li et al., 2010). miRNAs are emerging as the next generation of targets for genetic engineering to improve agricultural agronomic characteristics, owing to mounting evidence of their relevance in crop plants.

Manipulation of miRNA expression levels might be a useful technique in this respect for enhancing crop plant responses to environmental stress, pathogen and parasite assault, and plant growth and development (Zhang & Wang, 2015). Different transgenic methods, focusing on important miRNAs and the associated known target genes, can be employed in practice. These include constitutive miRNA overexpression, stress-induced or tissue-specific miRNA or target expression, production of miRNA-resistant target genes, artificial target

TABLE 17.3

Targets of Genus *Coffea* miRNAs Identified and Description and Function of Target Genes

Coffea **miRNA**	**Target acc.**	**Target description**	**Function**
cca-miR160, car-iR161, 162, 164, cca-miR165, car-iR417, cca-miR476, ccamiR479, car-miR837, 847, 1919, 2938, 3434, 516, cca-miR4364, 4386, car-miR4392, 5018, 5141, 5380, 5653, 5717, 5725, 6021, 8049, 8051, 8166, cca-mi8172, car-miR9408, 9558	NP305750, TC362356, TC362295, TC361205, SGN-U619883, TC400535, TC358231, TC401289, NM_001328854, XM_017374849, XM_017238780, SGN-U620162, TC370407, NP1646759, XM_007556415, SGN-U616271, C374351, BX838928, NP2705764, TC358153, TC386720, NP456897, TC362407, TC371231, NP1659848, SGN-U619489, TC394910, NP232525, TC363392, SGN-U618032, TC374246, SGN-U615369, TC368992, SGN-U625715, SGN-U615687, SGN-U619936, TC373661, SGN-U616929, SGN-U618455, NP13352118, XM_009411299, SGN-U618580, SGN-U615789, XM_017119852, XM_010099976, NM_001036965, XM_014626289, XM_007914167	Putative auxin response factor, auxin response factor 17, auxin response factor 10, auxin response factor 16, pentatricopeptide (PPR) repeat-containing protein, endoribonuclease Dicer homolog, guanine nucleotide exchange factor, NAC5 protein, homeobox-leucine zipper protein REVOLUTA, homeobox-leucine zipper protein ATHB-14, mediator of RNA polymerase II transcription, indoleacetic acid-induced protein 9; transcription factor, similarity to ankyrin-like protein, ATPase/nucleoside-triphosphatase, transcriptional co-activator YAP1, small nuclear ribonucleoprotein D2, ATP binding, vacuolar protein sorting-associated protein 45, domain-containing protein, ubiquitin- protein ligase, zinc ion binding, non-LTR retroelement reverse transcriptase, retrotransposon-like protein, MYB transcription factor, oxido reductase NAD- binding domain-containing protein, zinc ion binding, contains similarity to heat shock transcription factor, similar to RNA helicases, NAC domain-containing protein 78, zinc phosphodiesterase ELAC protein 2, PHD finger protein-related, RNA recognition motif (RRM)-containing protein, ATPase subunit 1, calcium-binding EF hand family protein, homeobox-leucine zipper protein, RNA- binding protein, GDSL-motif lipase/hydrolase-like protein, zinc finger protein, DNA binding, ATP binding, pentatricopeptide repeat-containing protein, ZFN1(ZINCFINGERPROTEIN1), F-box family protein, zinc finger protein, isoleucine--tRNA ligase, tRNA synthetase class I (I, L, M, and V) family protein, finger protein 12, transcriptional activator protein acu-15	Transcription factor
car-miR162, 417, 837, 847, 861, 862, 1919, 2108, 2938, 3434, 4245, cca-miR4364, car-miR4392, 5380, 5634, 5640, 5653, 5670, 5717, 5725, 6021, 6151, 6483, 8051, 8140, 8166, cca-miR8172	EG513934, XM_010326324, XM_015654998, XM_011523105, XM_014029716, SGN-U620027, SGN-U615000, SGN-U615355, SGN-U615009, SGN-U613213, NP237657, SGN-U614620, SGN-U618704, BX837933, TC383691, NP305087, SGN-U615294, TC379708, TC398934, SGN-U615360, SGN-U613503, SGN-U615324, SGN-U616867, SGN-U615295, SGN-U626186, SGN-U620263, NP229721, TC366012, SGN-U615711, SGN-U626543, SGN-U625139, SGN-U620278, XM_016557612, TC397703, SGN-U615916, SGN-U620515, SGN-U618360	Inositol oxygenase 4, serine racemase, putative protein arginine N-methyltransferase, domain containing oxidoreductase, cysteinyl-tRNA synthetase, caffeic acid3-O-methyltransferase, pectinmethylesterase 1, glutamyl-tRNA reductase 1, lipase class 3 family protein, vacuolar ATP synthase subunit E, similarity to mutator-like transposase, trehalose-6-phosphate synthase, nucleotidase family protein, N-acetylglucosamine deacetylase precursor, alpha-glucosidase, putative glucosidase I, ubiquitin-conjugating enzyme variant, glutaredoxin-C5, citrate synthase, ubiquinol-cytochrome C reductase, calmodulin-related protein, aldo/ketoreductase AKR, xyloglucan galactosyltransferase, aspartyl protease family protein, alpha-amylase, (+)-delta-cadinene synthase, terpene synthase/cyclase family protein, cysteine proteinase, putative, cysteine synthase, anthocyanidine rhamnosyl-transferase, glucose-6-phosphate dehydrogenase, vitamin K–dependent gamma-carboxylase, monothiol glutaredoxin-S9, stroma ascorbate peroxidase precursor, glucose-6- phosphate 1-dehydrogenase, methionylaminopeptidase	Metabolism
car-miR161, 171, 397, cca-miR476, car-miR861, 862, 1888, 2936, 2938, 3434, 3521, cca-miR4364, 4386, car-miR5380, 5717,6021, 6151, 8049, 8140, 8166, 8170, cca-miR8172	SGN-U626881, SGN-U625098, SGN-U626324, SGN-U627786, SGN-U625609, SGN-U614001, SGN-U620177, SGN-U617583, SGN-U619299, BP836641, TC374054, SGN-U618312, TC390997, SGN-U615813, SGN-U615557, NP306739, SGN-U618454, SGN-U614989, TC359052, NP183968, TC375992, SGN-U627832, SGN-U617079, SGN-U618361	Predicted protein, hypothetical protein, predicted protein, hypothetical, hypothetical protein, hypothetical protein isoform 1, hypothetical protein, hypothetical protein, hypothetical protein, uncharacterized protein, uncharacterized protein, hypothetical protein, uncharacterized protein, predicted protein, hypothetical protein, unknown protein, unnamed protein, hypothetical protein, uncharacterized protein, hypothetical protein, protein COBRA precursor, hypothetical protein, unnamed protein, hypothetical protein	Hypothetical

(*Continued*)

TABLE 17.3 (Continued)

Coffea miRNA	Target acc.	Target description	Function
car-miR164, 397, cra-miR408, car-miR2111, 2936, 3434, cca-miR4386, car-miR5640, 8049, 8051, 9558	TC359120, SGN-U613356, SGN-U626870, SGN-U614604, SGN-U613822, TC373116, DR366527, SGN-U619538, SGN-U613972, TC391373, NP224198, XM_012991812, XM_018249558	Phytochrome c, transducin family protein, ubiquitin family protein, protein kinase family protein, 2,3-diketo-5-methylthio-1-phosphopentanephosphatase, MAP kinase, serine/threonine protein phosphatase, microsomal signal peptidase 25 kDa, kinase family protein, calcineurin B-like protein, receptor protein kinase-like, phosphomevalonate kinase, serine /threonine-protein kinase 40	Signal transduction
car-miR837, 3434, cca-miR4386, car- miR5653, 5725, 9558	XM_014750876, TC362916, SGN-U615371, SGN-U627672, SGN-U614909, XM_010184592, XM_017973497	centromere-associated protein E, kinesin-like protein, cyclic nucleotide-gated ion channel, coatomer protein complex, potassium efflux antiporter, carbonic anhydrase 5B, glycosyl transferase-like domain containing 1	Transporter
car-miR162, 164, cca-miR476, car-miR837, 6151, 8051	TC370085, TC379483, TC370018, SGN-U613155, XM_016328266, SGN-U614441, XM_014917224	Expansin-A1 precursor, embryo defective 1381, NAM (no apical meristem)-like protein, Aux/IAA protein, auxin response factor 9, auxin-induced protein, centrosomal protein	Growth and development
car-miR1888, 2086, 2111, 5141, 5380, 5725	SGN-U620022, SGN-U619259, SGN-U628254, TC380075, SGN-U613294, SGN-U614101	Heat shock cognate 70 kDa protein 1, heat shock protein, similar to heat shock protein, dehydration stress-induced protein, chaperone protein, similar to chaperonin-like	Stress-related
car-miR841, 3521, 5020, 5380	SGN-U616689, SGN-U619632, SGN-U619690, SGN-U620019, SGN-U620861	Putative histone H2A, ribosomal protein, 60s acidic ribosomal protein P1, TUA4 (tubulin alpha-4 chain), small nuclear ribonucleoprotein	Structural constituent
car-miR1919, 2111, 2938	NP2698013, SGN-U613205, TC385823	Similar to disease resistance protein, wound-responsive family protein, enhanced disease susceptibility 5	Disease- related

Source: Adapted from Bibi et al. (2017).

mimics, and artificial miRNAs (Gupta, 2015; Zhou & Luo, 2013).

17.6.1 More functional research on coffee plant miRNAs is required

Shibin and Chandrashekar (2012) used computational methods to identify 18 miRNAs belonging to 12 families from *C. canephora* ESTs, which was one of the first attempts to investigate coffee miRNAs. These researchers discovered expression for seven families in *C. canephora* leaves using RT-PCR (mirl56, mirl69, mirl72, mir319, mir393, mir395, and mir396). This was followed by others who discovered miRNAs in *C. arabica* and *C. canephora* using the same computational approach, revealing that the majority of their potential targets were mRNA coding proteins involved in transcriptional regulation and signal transduction pathways (Akter et al., 2014; Bibi et al., 2017; Chaves et al., 2015; Dos Santos et al., 2019; Ivamoto et al., 2017; Loss-Morais et al., 2014; Mohanan & Chandrashekar, 2012; Mohanan et al., 2013; Noronha Fernandes-Brum et al., 2017). With the aid of computer tools like the miRDeepFinder (Xie et al., 2012), next-generation deep sequencing (NGS) technology allows for the rapid identification of conserved and novel miRNAs from huge deep sequencing data sets. miRNA studies based on deep sequencing have not only found a large number of miRNAs, but they can also estimate miRNA expression levels based on sequencing read counts. A putative miRNA (called mir393) found in *C. arabica* was discovered through ESTs, which showed that this sequence contained many genes coding transcription factors (e.g., bHLH7 and WRKY TFs) or proteins involved in the auxin signaling cascade and plant defense responses as potential targets (e.g., auxin signaling F-box 2 and auxin transporter protein 1) (Akter et al., 2014). The involvement of miRNAs in diverse cellular processes and gene regulatory networks in both coffee plant types were disclosed by miRNA target genes predicted from the discovered miRNAs. These newly discovered *Coffea* miRNA families influenced plant development, hormone responses, mature mRNA synthesis, stress responses, and cellular signaling pathways, among other biological activities (Chaves et al., 2015). These computer simulations give information that is mostly descriptive and lacks mechanistic insights. To confirm the in silico findings, just a few hypothesis-driven experiments have been carried out. As a result, the present descriptive research in the field of *Coffea* miRNA has to be expanded to include more mechanism-based studies.

The most potent and effective approach to research miRNA functions, similar to protein-coding genes, is to create transgenic plants in which a single miRNA gene is overexpressed or knocked out/knocked down. Because miRNAs target particular transcripts, partial function of a miRNA may also be investigated by producing mutants of their target sites. In the early phases of miRNA research, all of these approaches were used to investigate plant miRNA functions. Because of the ease of the floral-dip transformation approach, most research have focused on model plant species, notably *Arabidopsis.* These well-studied miRNAs are expected to have comparable roles in other plant species because they are all evolutionarily conserved. miRNAs have been found in crop species cultivated for food (Zhao et al., 2015), textiles (Zhang et al., 2007), and biofuel (Gentile et al., 2015)], in addition to model plants. There have been several attempts to characterize miRNAs in coffee. Deep-sequencing libraries have also been created utilizing *C. canephora* genomic (Denoeud et al., 2014) and transcriptome data (Loss-Morais et al., 2014; Sharma et al., 2015). For researching miRNA functions, it is important to establish accurate transformation procedures for recalcitrant plants. The Plant Genome Program of the National Science Foundation (NSF) in the United States identified this need in 2017, designating plant transformation as a high priority. Hopefully, in the not too distant future, more simple ways for transforming a wide range of plant species, particularly agriculturally essential crops, will be available. Implementing such a tool will make miRNA research easier and lead to more functional studies.

17.6.2 For miRNA functional research, better genetic tools are required

The research of miRNA function is based on the availability of several genetic techniques. Deep sequencing, quantitative real-time PCR (qRT-PCR), and conventional gene expression analyses are being used to determine the spatial and temporal patterns of miRNA gene expression in plants throughout growth and development. Although miRNA gene expression data can be linked to a specific function, it does not give direct evidence of gene function. Traditional transgenic techniques should be used to explore miRNA function. However, manipulating the short 20-nucleotide sequence is generally challenging because it necessitates precise sequence length and nucleotide composition. Alternatively, the lengthy precursor sequences can be introduced into plants, which will trigger their own miRNA biogenesis mechanisms to produce the desired miRNA sequence. This method will be really useful. First, we can avoid the problem of generating precise miRNA genes by transforming sequences that include pre-miRNA sequences into mature miRNAs. Second, as miRNA sequences are conserved among plant species, a model plant species' miRNA can be employed. More specifically, the structural frame of a known miRNA may be used to create an artificial miRNA (amiRNA) that targets a single protein-coding gene. The gene sequence can then be utilized to create a mature amiRNA sequence that is complementary to the mRNA of this particular protein-coding gene. Plant cells may then generate this amiRNA on a continual basis, targeting the gene of interest. Artificial miRNAs that suppress target gene expression of a protein-coding mRNA of interest are thought to be one of the most useful techniques for crop improvement. Many significant miRNAs' actions in live cells, however, may not be observable due to insufficient concentration or expression. Before developing artificial miRNAs aimed at producing plants with desirable agronomical traits, a quantitative study of the natural plant miRNA(s) present in cells, as well as that required for observable impacts, would be required. Despite the fact that this artificial miRNA approach has been shown to be efficient (Khraiwesh et al., 2008), the evaluation and differentiation of mRNA cleavage and translational inhibition still needs to be clarified for better use. Knowing that such synthetic nucleic acids produce a single miRNA capable of silencing a single target gene, all potential undesirable target genes may be predicted and avoided during the experimental design phases (Schwab et al., 2010). Engineered decoys' capacity to alter miRNA regulatory networks by modifying miRNA activity is yet another method for attaining a desired result (Ivashuta et al., 2011). Endogenous RNAs that may negatively control miRNA activity, known as miRNA decoys, are a flexible and robust tool for understanding the function of miRNA families and targeted gene expression engineering in plants (Banks et al., 2012). This method allows for precise control of native miRNA targets and the creation of a desirable range of plant phenotypes by changing the sequence of miRNA decoy sites, which reduces miRNA inactivation. However, achieving a desired amount of miRNA inactivation to minimize off-type effects, which might result in a false positive outcome, is one of the challenges in successfully using this method for crop development. In addition, the interaction between miRNA and the miRNA decoy, as well as the molecular complex that may form as a result of this interaction, is unknown, which might contribute to miRNA instability in plants. As a result, further information on the mechanism behind decoy-associated miRNA turnover and the function of decoys in global gene expression control in plants is needed for this strategy to be useful in crops. With the recognition of miRNA manipulation's potential as a viable tool for plant crop development, it is also vital to remember that such genetic alteration techniques might have unexpected consequences (Kamthan et al., 2015). If the expression of a miRNA or the target gene is changed, it might cause pleiotropic changes in plant growth and morphology that are undesired. As a result, understanding the mechanisms of miRNA control of plant growth and development, as well as plant responses to diverse abiotic and biotic stressors, is critical. Inhibiting miRNA expression and studying miRNA function may both be accomplished using RNA interference (RNAi) technology. Target mimicry technology can potentially be used to investigate miRNA function and to inhibit mature miRNA activity. This will make it easier to develop appropriate methods that provide the required characteristics while minimizing trade-offs in the modified crops.

Despite the fact that all existing technologies have been quite beneficial in elucidating miRNA functions, there is still a lot of space for development. Many members make up the majority of conserved miRNA genes. There are some mature sequences that are nearly similar. It is unclear why a single miRNA family has so many members in the plant genome. During plant growth and development, and in response to various environmental stresses, each miRNA member may perform a distinct role. Some members may be expressed during early development in response to biotic and abiotic stresses. All existing methods, however, are unable to differentiate between members of the same miRNA family. New genetic techniques are desperately needed to explore miRNA functions, especially the activities of a single miRNA gene. The CRISPR/Cas system (clustered regularly interspaced short palindromic repeats/CRISPR-associated genes) genome editing technique (Liu et al., 2017; Maxwell, 2007; Yin et al., 2017) is a potent tool for understanding the function of specific miRNA genes within the same miRNA family. CRISPR/Cas 9 technology has been widely used to delete numerous protein-coding genes in recent years (Liu et al., 2017; Yin et al., 2017; Y. Zhao et al., 2016). Around the protospacer adjacent motif sequence, CRISPR/Cas 9 can delete a couple of nucleotides (Liu et al., 2017; Yin et al., 2017). A nucleotide frame switch occurs when a few nucleotides are removed from protein-coding genes, resulting in gene silencing. In most situations, however, deleting a few nucleotides will not effectively silence miRNA genes. As a result, there have been few reports of miRNA genes being effectively knocked out using CRISPR/Cas9. The main explanation is that miRNA biogenesis differs from that of protein-coding genes. The newly discovered cpf1 enzyme is capable of knocking out a single gene with fewer off-target effects (Begemann et al., 2017; Zaidi, Mahfouz, & Mansoor, 2017; X. Zhang et al., 2017), but it still suffers from the same issue in which deletion of a few nucleotides has no effect on miRNA biogenesis. Designing two guide RNAs (gRNA) and creating two cleavage sites on the two sides of the mature miRNA sequence is one possible method for knocking off a miRNA gene. The mature miRNA sequences can be entirely deleted this manner. Although this technique has recently been effective in silencing a single miRNA gene (Y. Zhao et al., 2016), it still has numerous flaws. To begin, CRISPR/Cas9 technology is mostly based on plant tissue culture and regeneration. Many plant species have slow transformation and plant regeneration rates. As a result, scientists working in the disciplines of genome editing and miRNA should collaborate to develop a better method for knocking out a single miRNA gene with multiple distinct characteristics.

17.6.3 Improving agricultural production and quality through regulating miRNAs and the gene networks

Although numerous studies have demonstrated that miRNAs have great potential for enhancing crop output and plant tolerance to abiotic and biotic environmental stresses, several challenges must be overcome before they can be employed in agricultural practices. In switchgrass, moderate overexpression of miR156 increased plant biomass substantially, but excessive overexpression resulted in transgenic lines with decreased plant growth and biomass (Fu et al., 2012). Another factor to consider is that because a single miRNA might have numerous targets, it may have an impact on the gene network. Downstream expression of genes that make up a gene network may be affected by miRNA targeting protein-coding genes. As a result, it is essential to look at not only the miRNA-targeted genes but the entire impacted gene network in order to understand the possible benefits and drawbacks of miRNA manipulation as well as the effects on plant growth and development. *Coffea* miRNAs have also been discovered to target proteins important in plant growth and development. miR-162, miR-476, miR-837, and miR-6151, are miRNA families that control auxin/IAA protein and auxin-induced protein. miR-837 was used to target auxin-induced protein, which is important for growth and development (Barozai, 2012a, 2012b). The more complete miRNA-mRNA network will help to design new techniques for employing miRNAs as a target for crop improvement and answer some basic concerns about miRNA activities during plant growth and development. To create the miRNA-mRNA gene network, a variety of currently available tools and techniques may be used, including but not limited to gene pathway analysis using various modeling approaches (Fei et al., 2018; Galatenko et al., 2018; X. Lai, Wolkenhauer, & Vera, 2016; P. Zhang et al., 2017).

17.7 Conclusion

Coffee is an important crop for many countries and an source of income for a large population. The plant is susceptible to pests. Further studies is needed for functional validation of microRNAs from both the economically import coffee species. The studies on coffee miRNAs needs to be expanded to provide a better understanding of their essential roles in the miRNA based regulation process in *Coffea*, their processing pathways, and their gene expression so as to pave the way for future investigations.

REFERENCES

Adai, A., Johnson, C., Mlotshwa, S., Archer-Evans, S., Manocha, V., Vance, V., & Sundaresan, V. (2005). Computational prediction of miRNAs in Arabidopsis thaliana. *Genome research*, *15*(1), 78–91. https://doi.org/10.1101/gr.2908205

Akter, A., Islam, M. M., Mondal, S. I., Mahmud, Z., Jewel, N. A., Ferdous, S., . . . Rahman, M. M. (2014). Computational identification of miRNA and targets from expressed sequence tags of coffee (Coffea arabica). *Saudi J Biol Sci*, *21*(1), 3–12. doi: 10.1016/j.sjbs.2013.04.007

Allen, E., Xie, Z., Gustafson, A. M., Sung, G. H., Spatafora, J. W., & Carrington, J. C. (2004). Evolution of microRNA genes by inverted duplication of target gene sequences in Arabidopsis thaliana. *Nature genetics*, *36*(12), 1282–1290. https://doi.org/10.1038/ng1478

Altschul, S. F., Gertz, E. M., Agarwala, R., Schaffer, A. A., & Yu, Y. K. (2009). PSI-BLAST pseudocounts and the minimum description length principle. *Nucleic Acids Res*, *37*(3), 815–824. doi: 10.1093/nar/gkn981

Altschul, S. F., Madden, T. L., Schaffer, A. A., Zhang, J., Zhang, Z., Miller, W., & Lipman, D. J. (1997). Gapped BLAST and PSI-BLAST: a new generation of protein database search programs. *Nucleic Acids Res*, *25*(17), 3389–3402. doi: 10.1093/nar/25.17.3389

Ambros, V., Bartel, B., Bartel, D. P., Burge, C. B., Carrington, J. C., Chen, X., . . . Tuschl, T. (2003). A uniform system for microRNA annotation. *RNA*, *9*(3), 277–279. doi: 10.1261/rna.2183803

Arazi, T., Talmor-Neiman, M., Stav, R., Riese, M., Huijser, P., & Baulcombe, D. C. (2005). Cloning and characterization of micro-RNAs from moss. *The Plant journal : for cell and molecular biology*, *43*(6), 837–848. https://doi.org/10.1111/j.1365-313X.2005.02499.x

Aukerman, M. J., & Sakai, H. (2003). Regulation of flowering time and floral organ identity by a MicroRNA and its APETALA2-like target genes. *Plant Cell*, *15*(11), 2730–2741. doi: 10.1105/tpc.016238

Axtell, M. J. (2013). Classification and comparison of small RNAs from plants. *Annu Rev Plant Biol, 64*, 137–159. doi: 10.1146/annurev-arplant-050312-120043

Axtell, M. J., & Meyers, B. C. (2018). Revisiting Criteria for Plant MicroRNA Annotation in the Era of Big Data. *Plant Cell, 30*(2), 272–284. doi: 10.1105/tpc.17.00851

Bae, J. M. (2021). Coffee consumption and risk of type 2 diabetes mellitus in Asians: A meta-epidemiological study of population-based cohort studies. *World J Diabetes, 12*(6), 908–915. doi: 10.4239/wjd.v12.i6.908

Baker, C. C., Sieber, P., Wellmer, F., & Meyerowitz, E. M. (2005). The early extra petals1 mutant uncovers a role for microRNA miR164c in regulating petal number in Arabidopsis. *Current biology: CB, 15*(4), 303–315. https://doi.org/10.1016/j.cub.2005.02.017

Banks, I. R., Zhang, Y., Wiggins, B. E., Heck, G. R., & Ivashuta, S. (2012). RNA decoys: an emerging component of plant regulatory networks? *Plant Signal Behav, 7*(9), 1188–1193. doi: 10.4161/psb.21299

Barozai, M. Y. (2012). Identification and characterization of the microRNAs and their targets in Salmo salar. *Gene, 499*(1), 163–168. doi: 10.1016/j.gene.2012.03.006

Bartel, D. P. (2004). MicroRNA: Genomics, Biogenesis, Mechanism and Function. *Cell 11*(2), 281–297.

Begemann, M. B., Gray, B. N., January, E., Gordon, G. C., He, Y., Liu, H., . . . Oufattole, M. (2017). Precise insertion and guided editing of higher plant genomes using Cpf1 CRISPR nucleases. *Sci Rep, 7*(1), 11606. doi: 10.1038/s41598-017-11760-6

Bibi, F., Barozai, M. Y., & Din, M. (2017). Bioinformatics profiling and characterization of potentialmicroRNAs and their targets in the genus Coffea. *TURKISH JOURNAL OF AGRICULTURE AND FORESTRY, 41*, 191–200. doi: 10.3906/tar-1612-121

Billoud, B., De Paepe, R., Baulcombe, D., & Boccara, M. (2005). Identification of new small non-coding RNAs from tobacco and Arabidopsis. *Biochimie, 87*(9–10), 905–910. https://doi.org/10.1016/j.biochi.2005.06.001

Bonnet, E., Van de Peer, Y., & Rouze, P. (2006). The small RNA world of plants. *New Phytol, 171*(3), 451–468. doi: 10.1111/j.1469-8137.2006.01806.x

Boonphang, O., Ontawong, A., Pasachan, T., Phatsara, M., Duangjai, A., Amornlerdpison, D., . . . Srimaroeng, C. (2021). Antidiabetic and Renoprotective Effects of Coffea arabica Pulp Aqueous Extract through Preserving Organic Cation Transport System Mediated Oxidative Stress Pathway in Experimental Type 2 Diabetic Rats. *Molecules, 26*(7). doi: 10.3390/molecules26071907

Chaves, S. S., Fernandes-Brum, C. N., Silva, G. F., Ferrara-Barbosa, B. C., Paiva, L. V., Nogueira, F. T., . . . Chalfun-Junior, A. (2015). New Insights on Coffea miRNAs: Features and Evolutionary Conservation. *Appl Biochem Biotechnol, 177*(4), 879–908. doi: 10.1007/s12010-015-1785-x

Chen, H., Li, Z., & Xiong, L. (2012). A plant microRNA regulates the adaptation of roots to drought stress. *FEBS Lett, 586*(12), 1742–1747. doi: 10.1016/j.febslet.2012.05.013

Chen, X. (2008). MicroRNA metabolism in plants. *Current topics in microbiology and immunology, 320*, 117–136.

Cui, G., Zhao, M., Zhang, S., Wang, Z., Meng, M., Sun, F., . . . Xi, Y. (2020). MicroRNA and regulation of auxin and cytokinin signalling during post-mowing regeneration of winter wheat (Triticum aestivum L.). *Plant Physiol Biochem, 155*, 769–779. doi: 10.1016/j.plaphy.2020.08.032

Darling, A., Carey, L., & Feng, W. (2003). The Design, Implementation, and Evaluation of mpiBLAST. *Proc Cluster World, 2003*.

Davis, A. P., Govaerts, R., Bridson, D. M., & Stoffelen, P. (2006). An annotated taxonomic conspectus of the genus Coffea (Rubiaceae). *Botanical Journal of the Linnean Society, 152*(4), 465–512. doi: https://doi.org/10.1111/j.1095-8339.2006.00584.x

Denoeud, F., Carretero-Paulet, L., Dereeper, A., Droc, G., Guyot, R., Pietrella, M., Zheng, C., Alberti, A., Anthony, F., Aprea, G., Aury, J. M., Bento, P., Bernard, M., Bocs, S., Campa, C., Cenci, A., Combes, M. C., Crouzillat, D., Da Silva, C., Daddiego, L., . . . Lashermes, P. (2014). The coffee genome provides insight into the convergent evolution of caffeine biosynthesis. Science (New York, N.Y.), 345(6201), 1181–1184. https://doi.org/10.1126/science.1255274.

Dezulian, T., Remmert, M., Palatnik, J. F., Weigel, D., & Huson, D. H. (2006). Identification of plant microRNA homologs. *Bioinformatics, 22*(3), 359–360. doi: 10.1093/bioinformatics/bti802

Dos Santos, T. B., Soares, J., Lima, J. E., Silva, J. C., Ivamoto, S. T., Baba, V. Y., Souza, S., Lorenzetti, A., Paschoal, A. R., Meda, A. R., Nishiyama Júnior, M. Y., de Oliveira, Ú. C., Mokochinski, J. B., Guyot, R., Junqueira-de-Azevedo, I., Figueira, A., Mazzafera, P., Júnior, O. R., Vieira, L., Pereira, L., . . . Domingues, D. S. (2019). An integrated analysis of mRNA and sRNA transcriptional profiles in Coffea arabica L. roots: insights on nitrogen starvation responses. Functional & integrative genomics, 19(1), 151–169. https://doi.org/10.1007/s10142-018-0634-8

Emery, J. F., Floyd, S. K., Alvarez, J., Eshed, Y., Hawker, N. P., Izhaki, A., Baum, S. F., & Bowman, J. L. (2003). Radial patterning of Arabidopsis shoots by class III HD-ZIP and KANADI genes. *Current biology : CB*, 13(20), 1768–1774. https://doi.org/10.1016/j.cub.2003.09.035

Fei, Y., Wang, R., Li, H., Liu, S., Zhang, H., & Huang, J. (2018). DPMIND: degradome-based plant miRNA-target interaction and network database. *Bioinformatics, 34*(9), 1618–1620. doi: 10.1093/bioinformatics/btx824

Friis, E. M., Pedersen, K. R., & Crane, P. R. (2004). Araceae from the Early Cretaceous of Portugal: evidence on the emergence of monocotyledons. *Proceedings of the National Academy of Sciences of the United States of America, 101*(47), 16565–16570. https://doi.org/10.1073/pnas.0407174101

Fu, C., Sunkar, R., Zhou, C., Shen, H., Zhang, J. Y., Matts, J., Wolf, J., Mann, D. G., Stewart, C. N., Jr, Tang, Y., & Wang, Z. Y. (2012). Overexpression of miR156 in switchgrass (Panicum virgatum L.) results in various morphological alterations and leads to improved biomass production. *Plant biotechnology journal, 10*(4), 443–452. https://doi.org/10.1111/j.1467-7652.2011.00677.x

Galatenko, V. V., Galatenko, A. V., Samatov, T. R., Turchinovich, A. A., Shkurnikov, M. Y., Makarova, J. A., & Tonevitsky, A. G. (2018). Comprehensive network of miRNA-induced intergenic interactions and a biological role of its core in cancer. *Sci Rep, 8*(1), 2418. doi: 10.1038/s41598-018-20215-5

Gao, J., Yin, F., Liu, M., Luo, M., Qin, C., Yang, A., Yang, S., Zhang, Z., Shen, Y., Lin, H., & Pan, G. (2015). Identification and characterisation of tobacco microRNA transcriptome using high-throughput sequencing. *Plant biology (Stuttgart, Germany)*, *17*(3), 591–598. https://doi.org/10.1111/plb.12275

Gentile, A., Dias, L. I., Mattos, R. S., Ferreira, T. H., & Menossi, M. (2015). MicroRNAs and drought responses in sugarcane. Frontiers in plant science, 6, 58. https://doi.org/10.3389/fpls.2015.00058

Guedes, F. A. d. F., Nobres, P., Rodrigues Ferreira, D. C., Menezes-Silva, P. E., Ribeiro-Alves, M., Correa, R. L., Alves-Ferreira, M. (2018). Transcriptional memory contributes to drought tolerance in coffee (Coffea canephora) plants. *Environmental and Experimental Botany, 147*, 220–233. doi: https://doi.org/10.1016/j.envexpbot.2017.12.004

Gupta, P. (2015). *MicroRNAs and target mimics for crop improvement.*

Gustafson, A. M., Allen, E., Givan, S., Smith, D., Carrington, J. C., & Kasschau, K. D. (2005). ASRP: the Arabidopsis Small RNA Project Database. *Nucleic acids research, 33*(Database issue), D637–D640. https://doi.org/10.1093/nar/gki127

Ivamoto, S. T., Reis, O. J., Domingues, D. S., Dos Santos, T. B., de Oliveira, F. F., Pot, D., . . . Pereira, L. F. (2017). Transcriptome Analysis of Leaves, Flowers and Fruits Perisperm of Coffea arabica L. Reveals the Differential Expression of Genes Involved in Raffinose Biosynthesis. *PLoS One, 12*(1), e0169595. doi: 10.1371/journal.pone.0169595

Ivashuta, S., Banks, I. R., Wiggins, B. E., Zhang, Y., Ziegler, T. E., Roberts, J. K., & Heck, G. R. (2011). Regulation of gene expression in plants through miRNA inactivation. *PLoS One, 6*(6), e21330. doi: 10.1371/journal.pone.0021330

Jin, H., & Martin, C. (1999). Multifunctionality and diversity within the plant MYB-gene family. *Plant Mol Biol, 41*(5), 577–585. doi: 10.1023/a:1006319732410

Jones-Rhoades, M. W., & Bartel, D. P. (2004). Computational identification of plant microRNAs and their targets, including a stress-induced miRNA. *Molecular cell, 14*(6), 787–799. https://doi.org/10.1016/j.molcel.2004.05.027

Jones-Rhoades, M. W., & Bartel, D. P. (2004b). Computational identification of plant microRNA and their targets, including a stress induced miRNA. *Mol Cell, 14*, 787–799.

Kamthan, A., Chaudhuri, A., Kamthan, M., & Datta, A. (2015). Small RNAs in plants: recent development and application for crop improvement. *Front Plant Sci, 6*, 208. doi: 10.3389/fpls.2015.00208

Kasschau, K. D., Xie, Z., Allen, E., Llave, C., Chapman, E. J., Krizan, K. A., & Carrington, J. C. (2003). P1/HC-Pro, a viral suppressor of RNA silencing, interferes with Arabidopsis development and miRNA function. *Dev Cell, 4*(2), 205–217. doi: S153458070300025X [pii]

Khraiwesh, B., Ossowski, S., Weigel, D., Reski, R., & Frank, W. (2008). Specific gene silencing by artificial MicroRNAs in Physcomitrella patens: an alternative to targeted gene knockouts. *Plant Physiol, 148*(2), 684–693. doi: 10.1104/pp.108.128025

Kidner, C. A., & Martienssen, R. A. (2005). The developmental role of microRNA in plants. *Current opinion in plant biology*, *8*(1), 38–44. https://doi.org/10.1016/j.pbi.2004.11.008

Kim, J., Jung, J. H., Reyes, J. L., Kim, Y. S., Kim, S. Y., Chung, K. S., Kim, J. A., Lee, M., Lee, Y., Narry Kim, V., Chua, N. H., & Park, C. M. (2005). microRNA-directed cleavage of ATHB15 mRNA regulates vascular development in Arabidopsis inflorescence stems. *The Plant journal: for cell and molecular biology*, *42*(1), 84–94. https://doi.org/10.1111/j.1365-313X.2005.02354.x

Kim, J. Y., Huh, K. H., Park, Y. J., Fang, Y., Kang, C. M., & Kim, Y. S. (2008). Molecular mechanisms of cell death of mycophenolic acid-treated primary isolated rat islets: implication of mitogen-activated protein kinase activation. *Transplantation proceedings*, *40*(8), 2575–2577. https://doi.org/10.1016/j.transproceed.2008.07.123

Klatsky, A. L., Morton, C., Udaltsova, N., & Friedman, G. D. (2006). Coffee, cirrhosis, and transaminase enzymes. *Arch Intern Med, 166*(11), 1190–1195. doi: 10.1001/archinte.166.11.1190

Lagos-Quintana, M., Rauhut, R., Lendeckel, W., & Tuschl, T. (2001). Identification of novel genes coding for small expressed RNAs. Science (New York, N.Y.), 294(5543), 853–858. https://doi.org/10.1126/science.1064921

Lai, E. C., Tomancak, P., Williams, R. W., & Rubin, G. M. (2003). Computational identification of Drosophila microRNA genes. *Genome Biology 4*(7), R42. doi: 10.1186/gb-2003-4-7-r42

Lai, X., Wolkenhauer, O., & Vera, J. (2016). Understanding microRNA-mediated gene regulatory networks through mathematical modelling. *Nucleic Acids Res, 44*(13), 6019–6035. doi: 10.1093/nar/gkw550

Lai, Z., Wang, F., Zheng, Z., Fan, B., & Chen, Z. (2011). A critical role of autophagy in plant resistance to necrotrophic fungal pathogens. *Plant J, 66*(6), 953–968. doi: 10.1111/j.1365-313X.2011.04553.x

Lau, N. C., Lim, L. P., Weinstein, E. G., & Bartel, D. P. (2001). An abundant class of tiny RNAs with probable regulatory roles in Caenorhabditis elegans. Science (New York, N.Y.), 294(5543), 858–862. https://doi.org/10.1126/science.1065062

Lee, R. C., & Ambros, V. (2001). An extensive class of small RNAs in Caenorhabditis elegans. Science (New York, N.Y.), 294(5543), 862–864. https://doi.org/10.1126/science.1065329

Lee, R. C., Feinbaum, R. L., & Ambros, V. (1993). The *C.elegans* heterochronic gene *lin-4* encodes for small RNAs with antisense complementarity to *lin-14*. *Cell, 75*, 843–854.

Lemos, M. F., Perez, C., da Cunha, P., Filgueiras, P. R., Pereira, L. L., Almeida da Fonseca, A. F., Ifa, D. R., & Scherer, R. (2020). Chemical and sensory profile of new genotypes of Brazilian Coffea canephora. *Food chemistry*, *310*, 125850. https://doi.org/10.1016/j.foodchem.2019.125850

Lemos, S., Fonçatti, L., Guyot, R., Paschoal, A. R., & Domingues, D. S. (2020). Genome-Wide Screening and Characterization of Non-Coding RNAs in *Coffea canephora*. *Non-coding RNA*, *6*(3), 39. https://doi.org/10.3390/ncrna6030039

Li, A., & Mao, L. (2007). Evolution of plant microRNA gene families. *Cell Res, 17*(3), 212–218. doi: 10.1038/sj.cr.7310113

Li, Y., Zhang, Q., Zhang, J., Wu, L., Qi, Y., & Zhou, J. M. (2010). Identification of microRNAs involved in pathogen-associated molecular pattern-triggered plant innate immunity. Plant physiology, 152(4), 2222–2231. https://doi.org/10.1104/pp.109.151803

Libault, M., Wan, J., Czechowski, T., Udvardi, M., & Stacey, G. (2007). Identification of 118 Arabidopsis transcription factor and 30 ubiquitin-ligase genes responding to chitin, a plant-defense elicitor. *Mol Plant Microbe Interact, 20*(8), 900–911. doi: 10.1094/mpmi-20-8-0900

Lim, L. P., Lau, N. C., Weinstein, E. G., Abdelhakim, A., Yekta, S., Rhoades, M. W., Burge, C. B., & Bartel, D. P. (2003). The microRNAs of Caenorhabditis elegans. Genes & development, 17(8), 991–1008. https://doi.org/10.1101/gad.1074403

Lim, L. P., Lau, N. C., Garrett-Engele, P., Grimson, A., Schelter, J. M., Castle, J., Bartel, D. P., Linsley, P. S., & Johnson, J. M. (2005). Microarray analysis shows that some microRNAs downregulate large numbers of target mRNAs. *Nature*, *433*(7027), 769–773. https://doi.org/10.1038/nature03315

Liu, X., Wu, S., Xu, J., Sui, C., & Wei, J. (2017). Application of CRISPR/Cas9 in plant biology. *Acta Pharm Sin B, 7*(3), 292–302. doi: 10.1016/j.apsb.2017.01.002

Llave, C., Kasschau, K. D., Rector, M. A., & Carrington, J. C. (2002). Endogenous and silencing-associated small RNAs in plants. *Plant Cell, 14*(7), 1605–1619.

Loss-Morais, G., Ferreira, D. C., Margis, R., Alves-Ferreira, M., & Correa, R. L. (2014). Identification of novel and conserved microRNAs in Coffea canephora and Coffea arabica. *Genet Mol Biol, 37*(4), 671–682. doi: 10.1590/S1415-47572014005000020

Lu, S., Sun, Y. H., Shi, R., Clark, C., Li, L., & Chiang, V. L. (2005). Novel and mechanical stress-responsive MicroRNAs in Populus trichocarpa that are absent from Arabidopsis. *The Plant cell*, *17*(8), 2186–2203. https://doi.org/10.1105/tpc.105.033456

Mallory, A. C., Reinhart, B. J., Jones-Rhoades, M. W., Tang, G., Zamore, P. D., Barton, M. K., & Bartel, D. P. (2004). MicroRNA control of PHABULOSA in leaf development: importance of pairing to the microRNA 5' region. *The EMBO journal*, *23*(16), 3356–3364. https://doi.org/10.1038/sj.emboj.7600340

Maxwell, K. L. (2017). The Anti-CRISPR Story: A Battle for Survival. *Mol Cell, 68*(1), 8–14. doi: 10.1016/j.molcel.2017.09.002

McConnell, J. R., & Barton, M. K. (1998). Leaf polarity and meristem formation in Arabidopsis. *Development 125*(15), 2935–2942.

McConnell, J. R., Emery, J., Eshed, Y., Bao, N., Bowman, J., & Barton, M. K. (2001). Role of PHABULOSA and PHAVOLUTA in determining radial patterning in shoots. *Nature*, *411*(6838), 709–713. https://doi.org/10.1038/35079635

Mette, M. F., van der Winden, J., Matzke, M., & Matzke, A. J. (2002). Short RNAs can identify new candidate transposable element families in Arabidopsis. *Plant Physiol, 130*(1), 6–9. doi: 10.1104/pp.007047

Mohanan, S., & Chandrashekar, A. (2012). Computational Identification of Conserved MicroRNA and their Targets in Coffea canephora by EST Analysis. *Dynamic Biochemistry, Process Biotechnology and Molecular Biology, 6*, 70–76.

Mohanan, S., Gowda, K., Kandukuri, S. V., & Chandrashekar, A. (2013). Involvement of a novel intronic microRNA in cross regulation of N-methyltransferase genes involved in caffeine biosynthesis in Coffea canephora. *Gene, 519*(1), 107–112. doi: 10.1016/j.gene.2013.01.035

Nelson, J. M., Lane, B., & Freeling, M. (2002). Expression of a mutant maize gene in the ventral leaf epidermis is sufficient to signal a switch of the leaf's dorsoventral axis. *Development 129*(19), 4581–4589.

Noronha Fernandes-Brum, C., Marinho Rezende, P., Cherubino Ribeiro, T. H., Ricon de Oliveira, R., Cunha de Sousa Cardoso, T., Rodrigues do Amaral, L., Chalfun-Junior, A. (2017). A genome-wide analysis of the RNA-guided silencing pathway in coffee reveals insights into its regulatory mechanisms. *PLoS One, 12*(4), e0176333. doi: 10.1371/journal.pone.0176333

Palatnik, J. F., Allen, E., Wu, X., Schommer, C., Schwab, R., Carrington, J. C., & Weigel, D. (2003). Control of leaf morphogenesis by microRNAs. *Nature, 425*, 257–263.

Park, W., Li, J., Song, R., Messing, J., & Chen, X. (2002). CARPEL FACTORY, a Dicer homolog, and HEN1, a novel protein, act in microRNA metabolism in Arabidopsis thaliana. *Current Biology 12*(17), 1484–1495. doi: S0960982202010175 [pii]

Pasquinelli, A. E., Reinhart, B. J., Slack, F., Martindale, M. Q., Kuroda, M. I., Maller, B., . . . Ruvkun, G. (2000). Conservation of the sequence and temporal expression of let-7 heterochronic regulatory RNA. *Nature, 408*(6808), 86–89. doi: 10.1038/35040556

Rebijith, K. B., Asokan, R., Hande, R., V, K., & Nirmalbabu, K. (2013). In silico mining of novel microRNAs from coffee (Coffea arabica) using expressed sequence tags. *Journal of Horticultural Science and Biotechnology, 88*, 325–337. doi: 10.1080/14620316.2013.11512973

Reeder, J., Höchsmann, M., Rehmsmeier, M., Voss, B., & Giegerich, R. (2006). Beyond Mfold: recent advances in RNA bioinformatics. *Journal of biotechnology*, *124*(1), 41–55. https://doi.org/10.1016/j.jbiotec.2006.01.034

Reinhart, B. J., Slack, F. J., Basson, M., Pasquinelli, A. E., & Bettinger, J. C. (2000). The 21- nucleotide *let-7* RNA regulates developmental timing in *Caenorhabditis elegans*. *Nature, 403*, 901–906.

Reinhart, B. J., Weinstein, E. G., Rhoades, M. W., Bartel, B., & Bartel, D. P. (2002). MicroRNAs in plants. *Genes & development*, *16*(13), 1616–1626. https://doi.org/10.1101/gad.1004402

Rhoades, M. W., Reinhart, B. J., Lim, L. P., Burge, C. B., Bartel, B., & Bartel, D. P. (2002). Prediction of plant microRNA targets. *Cell, 110*(4), 513–520. doi: 10.1016/s0092-8674(02)00863-2

Robert-Seilaniantz, A., MacLean, D., Jikumaru, Y., Hill, L., Yamaguchi, S., Kamiya, Y., & Jones, J. D. (2011). The microRNA miR393 re-directs secondary metabolite biosynthesis away from camalexin and towards glucosinolates. *Plant J, 67*(2), 218–231. doi: 10.1111/j.1365-313X.2011.04591.x

Ruiz-Ferrer, V., & Voinnet, O. (2009). Roles of plant small RNAs in biotic stress responses. *Annu Rev Plant Biol, 60*, 485–510. doi: 10.1146/annurev.arplant.043008.092111

Samad, A. F. A., Sajad, M., Nazaruddin, N., Fauzi, I. A., Murad, A. M. A., Zainal, Z., & Ismail, I. (2017). MicroRNA and Transcription Factor: Key Players in Plant Regulatory Network. *Front Plant Sci, 8*, 565. doi: 10.3389/fpls.2017.00565

Schwab, R., Ossowski, S., Warthmann, N., & Weigel, D. (2010). Directed gene silencing with artificial microRNAs. *Methods Mol Biol, 592*, 71–88. doi: 10.1007/978-1-60327-005-2_6

Sharma, N., Tripathi, A., & Sanan-Mishra, N. (2015). Profiling the expression domains of a rice-specific microRNA under stress. *Frontiers in plant science*, *6*, 333. https://doi.org/10.3389/fpls.2015.00333

Sun, G., Stewart, C. N., Jr, Xiao, P., & Zhang, B. (2012). MicroRNA expression analysis in the cellulosic biofuel crop switchgrass (Panicum virgatum) under abiotic stress. *PloS one*, *7*(3), e32017. https://doi.org/10.1371/journal.pone.0032017

Sun, X., Lin, L., & Sui, N. (2019). Regulation mechanism of microRNA in plant response to abiotic stress and breeding. *Mol Biol Rep, 46*(1), 1447–1457. doi: 10.1007/s11033-018-4511-2

Sunkar, R., Chinnusamy, V., Zhu, J., & Zhu, J. K. (2007). Small RNAs as big players in plant abiotic stress responses and nutrient deprivation. *Trends in plant science*, *12*(7), 301–309. https://doi.org/10.1016/j.tplants.2007.05.001

Sunkar, R., Girke, T., Jain, P. K., & Zhu, J. K. (2005). Cloning and characterization of microRNAs from rice. *The Plant cell*, *17*(5), 1397–1411. https://doi.org/10.1105/tpc.105.031682

Sunkar, R., & Zhu, J. K. (2004). Novel and stress-regulated microRNAs and other small RNAs from Arabidopsis. *Plant Cell, 16*, 2001–2019.

Taylor, R. S., Tarver, J. E., Foroozani, A., & Donoghue, P. C. (2017). MicroRNA annotation of plant genomes - Do it right or not at all. *Bioessays, 39*(2). doi: 10.1002/bies.201600113

Unver, T., Namuth-Covert, D. M., & Budak, H. (2009). Review of current methodological approaches for characterizing microRNAs in plants. *International journal of plant genomics, 2009*, 262463–262463. doi: 10.1155/2009/262463

Vazquez, F., Vaucheret, H., Rajagopalan, R., Lepers, C., Gasciolli, V., Mallory, A. C., Hilbert, J. L., Bartel, D. P., & Crété, P. (2004). Endogenous trans-acting siRNAs regulate the accumulation of Arabidopsis mRNAs. *Molecular cell*, *16*(1), 69–79. https://doi.org/10.1016/j.molcel.2004.09.028

Voinnet, O. (2009). Origin, biogenesis, and activity of plant microRNAs. *Cell, 136*(4), 669–687. doi: 10.1016/j.cell.2009.01.046

Wang, X. J., Reyes, J. L., Chua, N. H., & Gaasterland, T. (2004). Prediction and identification of Arabidopsis thaliana microRNAs and their mRNA targets. *Genome biology*, *5*(9), R65. https://doi.org/10.1186/gb-2004-5-9-r65

Williams, L., Grigg, S. P., Xie, M., Christensen, S., & Fletcher, J. C. (2005). Regulation of Arabidopsis shoot apical meristem and lateral organ formation by microRNA miR166g and its AtHD-ZIP target genes. *Development, 132*(16), 3657–3668. doi: 10.1242/dev.01942

Xie, F., Xiao, P., Chen, D., Xu, L., & Zhang, B. (2012). miRDeepFinder: a miRNA analysis tool for deep sequencing of plant small RNAs. *Plant Mol Biol.* doi: 10.1007/s11103-012-9885-2

Xie, X., Lu, J., Kulbokas, E. J., Golub, T. R., Mootha, V., Lindblad-Toh, K., Lander, E. S., & Kellis, M. (2005). Systematic discovery of regulatory motifs in human promoters and 3' UTRs by comparison of several mammals. *Nature*, *434*(7031), 338–345. https://doi.org/10.1038/nature03441

Xie, Z., Kasschau, K. D., & Carrington, J. C. (2003). Negative feedback regulation of Dicer-Like1 in Arabidopsis by microRNA-guided mRNA degradation. *Current biology: CB*, *13*(9), 784–789. https://doi.org/10.1016/s0960-9822(03)00281-1

Yin, K., Gao, C., & Qiu, J. L. (2017). Progress and prospects in plant genome editing. *Nat Plants, 3*, 17107. doi: 10.1038/nplants.2017.107

Zaidi, S. S., Mahfouz, M. M., & Mansoor, S. (2017). CRISPR-Cpf1: A New Tool for Plant Genome Editing. *Trends Plant Sci, 22*(7), 550–553. doi: 10.1016/j.tplants.2017.05.001

Zhang, B., Pan, X., Cannon, C. H., Cobb, G. P., & Anderson, T. A. (2006). Conservation and divergence of plant microRNA genes. *Plant J, 46*(2), 243–259. doi: 10.1111/j.1365-313X.2006.02697.x

Zhang, B., Wang, Q., Wang, K., Pan, X., Liu, F., Guo, T., Cobb, G. P., & Anderson, T. A. (2007). Identification of cotton microRNAs and their targets. Gene, 397(1–2), 26–37. https://doi.org/10.1016/j.gene.2007.03.020

Zhang, B., & Wang, Q. (2015). MicroRNA-based biotechnology for plant improvement. *J Cell Physiol, 230*(1), 1–15. doi: 10.1002/jcp.24685

Zhang, P., Meng, X., Chen, H., Liu, Y., Xue, J., Zhou, Y., & Chen, M. (2017). PlantCircNet: a database for plant circRNA-miRNA-mRNA regulatory networks. *Database (Oxford), 2017*. doi: 10.1093/database/bax089

Zhang, X., Wang, J., Cheng, Q., Zheng, X., Zhao, G., & Wang, J. (2017). Multiplex gene regulation by CRISPR-ddCpf1. *Cell Discov, 3*, 17018. doi: 10.1038/celldisc.2017.18

Zhao, M., Meyers, B. C., Cai, C., Xu, W., & Ma, J. (2015). Evolutionary patterns and coevolutionary consequences of MIRNA genes and microRNA targets triggered by multiple mechanisms of genomic duplications in soybean. The Plant cell, 27(3), 546–562. https://doi.org/10.1105/tpc.15.00048

Zhao, Y., Zhang, C., Liu, W., Gao, W., Liu, C., Song, G., Xie, C. (2016). An alternative strategy for targeted gene replacement in plants using a dual-sgRNA/Cas9 design. *Sci Rep, 6*, 23890. doi: 10.1038/srep23890

Zhong, R., & Ye, Z. H. (2004). Amphivasal vascular bundle 1, a gain-of-function mutation of the IFL1/REV gene, is associated with alterations in the polarity of leaves, stems and carpels. *Plant and cell physiology 45*(4), 369–385.

Zhou, M., & Luo, H. (2013). MicroRNA-mediated gene regulation: potential applications for plant genetic engineering. *Plant Mol Biol, 83*(1–2), 59–75. doi: 10.1007/s11103-013-0089-1

Zuker, M. (2003). Mfold web server for nucleic acid folding and hybridization prediction. *Nucleic Acids Res, 31*(13), 3406–3415. doi: 10.1093/nar/gkg595

18

Current progress in tissue culture and transgenic research in coffee (Coffea *spp.)*

Manoj Kumar Mishra

CONTENTS

18.1 Introduction

Coffee is one of the most favored beverages consumed all over the world. It is estimated that during 2020–2021, around 166.63 million 60-kg bags were consumed worldwide (ICO, 2021). Coffee belongs to the genus *Coffea* of family Rubiaceae and consists of more than 125 species (Davis, 2011; Razafinarivo et al., 2013), of which only two species (*C. arabica* [arabica coffee] and *C. canephora* [robusta coffee]) are commercially exploited for beverage consumption. Coffee is cultivated in 10.2 million ha of land covering 80 countries predominantly situated in the tropical and sub-tropical climatic zones of the world, especially in Asia, Africa, and Latin America (Mishra and Slater, 2012). The global production of coffee during 2020 exceeded 175.64 million bags (of 60-kg capacity), comprising 105.325 million bags (59.96%) of arabica and 70.323 million bags (40.03%) of robusta coffee (ICO, 2021). In 2019, total world coffee exports were estimated to be worth $30.1 billion in the international market. The economies of several coffee-growing countries depend heavily on the production and export of coffee, and it is estimated that more than 125 million people worldwide were directly or indirectly

DOI: 10.1201/9781003043133-22

involved in the coffee supply chain and derive their income from the production of coffee.

Apart from its being a leading beverage, the moderate consumption of coffee has also been associated with several health benefits such as reducing diabetics, various cancers, Parkinson's disease, and Alzheimer's disease (Bidel and Tuomilehto, 2013; Wierzejska, 2017; Poole et al., 2018). Coffee brew is also considered to be an important source of nutrition (Olechno et al., 2021). The chlorogenic acids, caffeine, and other phenolic compounds in coffee are responsible for the chemoprotective effects, whereas the lipid fraction containing cafestol and kahweol act as a safeguard against some malignant cells by modulating the detoxifying enzymes (George et al., 2008).

18.2 Origin and genetic diversity

Most coffee species discovered so far naturally occur in tropical forests of Africa, Madagascar, and the Mascarene islands, while some of the species are found in the Indian subcontinent, Southeast Asia, and tropical Australasia (Davis et al., 2010, 2011; Mishra et al., 2020). *C. arabica*, which is the only self-fertile tetraploid ($2n=4x=44$) cultivated species, is believed to have originated in the highlands of southwestern Ethiopia and the Boma Plateau of South Sudan, with a wild population scattered in the Mount Marsabit forest in Kenya (Meyer, 1965). The first record of cultivation of *C. arabica* dates back 1500 years to southern Ethiopia, from where it gradually extended to Yemen during the sixth century (Anthony et al., 2002). It is widely recognized that two different genetic variants of arabica (Typica and Bourbon) spread from Yemen and formed the base of all the commercial arabica cultivars grown worldwide. On account of this, the majority of commercial arabica cultivars have a narrow genetic base (Van Der Vossen, 1985; Mishra et al., 2011), which seriously hampers the genetic improvement program and makes them vulnerable to climate change. Therefore it is imperative to identify and use novel genetic stocks using high-throughput genome technology currently available to accelerate the genetic improvement program of arabica coffee. In this regard, wild arabica coffee from the center of origin could be an important source of enriching the arabica gene pool for genetic improvement of coffee (Tornincasa et al., 2007; Mishra et al., 2014a; Jingade et al., 2019).

Unlike *C. arabica*, *C. canephora* is a diploid ($2n=2x=22$), self-incompatible species believed to be originated from the lowland tropical forest of Africa stretching from Cameroon through the Central African Republic to the Democratic Republic of Congo and then up to northern Angola, passing through Uganda and Northern Tanzania (Cubry et al., 2013). Robusta coffee was introduced to Java from Congo in 1900, from which it was introduced to India during 1903–1906 (Mishra, 2019). In the last few years, there has been a resurgence of interest in mapping the genetic diversity of *C. canephora* in countries including Brazil (Bikila et al., 2017), Vietnam and Mexico (Garavito et al., 2016), Kenya (Ogutu et al., 2016), Ecuador (Solorzano et al., 2017), Nigeria (Anagbogu et al., 2019), India (Huded et al., 2020), and Uganda (Kiwuka et al., 2021). All the recent results have suggested that genetically diverse robusta accessions should be employed in breeding programs to maximize heterosis.

18.3 Conventional breeding and its limitations

Globally, there are several constraints that affect coffee cultivation. The major factors that affect production and productivity include a number of pests and diseases, biotic and abiotic stress, and accelerated anthropogenic climatic change (Avelino et al., 2015; Bunn et al., 2014; Van der Vossen et al., 2015; DaMatta et al., 2018). For several decades, all over the world conventional breeding has been mainly used for the genetic improvement of coffee, and this has resulted in the availability of several improved cultivars for commercial exploitation. However, the continuous degradation of the environment and emergence of pests and diseases in several coffee-growing countries demands the continuous supply of improved cultivars at regular intervals. However, conventionally it is an arduous, time-consuming process and involves several steps including selection of parental material, hybridization, and progeny evaluation (Mishra, 2019). A minimum of 20–25 years is required to develop a new cultivar using conventional breeding approaches. In addition to the above, the long gestation period of the coffee tree, the high-cost involvement of conducting field trials, ploidy barriers between two commercially cultivated coffee varieties, and imprecise and uncontrollable transfer of genetic traits are some of the major bottlenecks associated with conventional coffee breeding. Further, genetic resistance to the coffee white stem borer (*Xylotrechus quadripes*) and the coffee berry borer (*Hypothenemus hampei*); tolerance to drought, salinity, heat, and cold; and herbicides are some of the features that are not readily available in the coffee gene pool or difficult to integrate using conventional breeding approaches (Mishra and Slater, 2012). Further, selection of parental material as well as hybrid identification based on phenotypic characters alone is extremely challenging considering the morphologically indistinguishable intra- and inter-varietal genotypes.

All over the world, coffee is conventionally propagated through seeds. In the case of *C. arabica*, which is an autogamous species, it is possible to obtain true to type plants by selfing the mother plant. However, in the case of *C. canephora*, which is an allogamous species and needs another partner for fecundation, vegetative propagation is considered to be an alternative approach for propagation of true to type plants. The same approach is also appropriate for arabica F1 hybrids, which tend to segregate because of meiotic recombination in the next generation. However, vegetative propagation on a large scale through sucker cuttings is impracticable due to the limited availability of suitable suckers from the elite mother plants. In the very fast-developing scenario, *in vitro* micropropagation offers a promising alternative for the rapid multiplication of elite plants in large numbers.

18.4 *In vitro* propagation

18.4.1 Explant preparation, control of phenolics, culture establishment, and micropropagation

Micropropagation offers the rapid, cost-effective propagation of the desired genotype in large numbers that are genetically identical to mother plants (Borchetia et al., 2009). In coffee, micropropagation consists of *in vitro* culture establishment of apical and auxiliary meristem culture and mass multiplication (Carneiro, 1999). However coffee is very difficult to micropropagate due to the exudation of phenolics resulting in browning of culture media, explants necrosis, high rate of latent microbial contamination, and target tissue recalcitrance. Standardization of the disinfection protocol for nodal culture and clonal propagation through apical bud culture was studied using different concentrations of benzilaminopurine (BAP) in culture medium (Ganesh and Sreenath, 1997, 1999). The maximum rate of multiplication obtained by using the apical/auxiliary meristem culture was six to nine shoots per explants (Nakamura and Sondahl, 1981; Sondahl, 1982; Carneiro and Ribeiro, 1989). Further, Nyange and Mcnicol (1993) successfully regenerated the adventitious shoots from *C. arabica* using hypocotyls as the target explants. It was observed that compared to the multiplication of micro-cuttings in a solid culture medium, culturing micro-cuttings in a temporary immersion system has resulted in a sixfold increase in multiplication rate (Berthouly et al., 1995; Teisson and Alvard 1995). In a first report, adventitious shoot buds were induced from the leaf and zygotic embryos of *C. bengalensis*, a wild coffee species using benzyladenine (BA) and thidiazuron (TDZ), respectively (Mishra and Sreenath, 2003a, 2003b). However, the frequency of shoot buds regenerated was very low. In addition to the low rate of multiplication, the initial loss of culture in high frequency due to exudation of phenolics and browning of tissue were found to be the major obstacles in commercializing the micropropagation technology in coffee (Muniswamy and Sreenath, 2004).

18.4.2 Somatic embryogenesis

Somatic embryogenesis has potential applications both in clonal multiplication and genetic transformation. Initial studies on coffee somatic embryogenesis were started almost 50 years ago (Staritsky, 1970), and since then many laboratories have refined the technology for commercial applications. In coffee, both direct and indirect somatic embryogenesis and plant regeneration have been reported using different explants. In the direct pathway, somatic embryos are formed on the explants (leaves, hypocotyls, and other parts) without intermediate callus formation. In the indirect pathway, the first callus is formed and then embryos emerge from the embryogenic callus mass (Yang and Zhang, 2010) (Figure 18.1). Several factors are shown to control somatic embryogenesis in coffee. These include the type of explants and their physiological stage, genotypic variation, media formulation for callus induction and somatic embryo induction, combination and concentration of growth regulators, the addition of growth adjuvants for induction, proliferation, maturation, and germination of somatic embryos.

18.4.3 Type of explants

In coffee, several different kinds of explants are used as the target explants for inducing somatic embryogenesis and plant regeneration. However, young leaves, which are available throughout the year, are widely used as the target explants for inducing somatic embryogenesis all over the world (Campos et al., 2017). There are several parameters that influence somatic embryogenesis using leaf explants, including physiological status of mother plants, leaf age and maturity, the growth environment of mother plants (field grown or greenhouse grown), and the month of collection (Molina et al., 2002; Santana et al., 2004). However, compared to the field-grown plants, explants collected from the greenhouse-grown plants are more conducive for somatic embryogenesis. Further, it has been observed that the percent of *in vitro* fungal and bacterial contamination was much higher in the explants collected from the plants grown in the field compared to the greenhouse (Santa Ram et al., 2002). Recently, Devasia et al. (2020) used zinc oxide nanoparticles on *in vitro* cultures of coffee and observed enhanced recovery of explants. The authors suggested that the addition of ZnO NP at 25 mg L^{-1} to the culture medium could significantly reduce microbial contamination and improve somatic embryogenesis. However, one of the most limiting factors that determines the high-frequency somatic embryogenesis in coffee is related to the genotype, and to date, no genotype-independent somatic embryogenesis protocol has

FIGURE 18.1 Steps involved in micropropagation of *Coffea* through somatic embryogenesis. (a) Mother plant; (b) explant inoculation and callus formation; (c) somatic embryo induction; (d) multiplication of somatic embryos; (e) plantlet development; (f) genetic fidelity assessment; (g) planting of *in vitro* regenerated plants; (h) flowering in tissue culture plants; (i) tissue culture plants bearing green fruits; (j) fruit maturation and ripening.

been described either for *C. arabica* or *C. canephora* genotypes (Naidu et al., 1999; Molina et al., 2002; Etienne, 2005; Rezende et al., 2011; Muniswamy et al., 2019).

18.4.4 Basal media composition

The basal media used for somatic embryogenesis in coffee consists of either full-strength or half-strength MS (Murashige and Skoog, 1962) medium (Etienne, 2005; Gatica-Arias et al., 2008; Simões-costa et al., 2009; Ganesh and Sreenath, 1999; Naidu and Srinivasan, 1999; Chaitra and Sreenath, 2010). In the majority of cases, callus appears on the cut surfaces within 3–4 weeks of incubation. These calli can be maintained as undifferentiated embryogenic calli in suspension culture for 2–4 months (Etienne, 2005; Landey et al., 2015) or made to differentiate to obtain somatic embryos (Etienne, 2005; Papanastasiou et al., 2008; Simões-costa et al., 2009). However, keeping the calli in an undifferentiated stage in the culture medium for a longer duration has a high risk of somaclonal variation.

18.4.5 Plant growth regulators and adjuvants

Even after five decades, inducing somatic embryogenesis in different coffee genotypes is still considered to be an empiric process. This is because each genotype response differently to different plant growth regulators, and therefore a right balance between the type of growth regulator and concentration is critical for somatic embryo induction (Santana-Buzzy et al., 2007). In general, two or more different media combinations are used for obtaining high-frequency somatic embryogenesis, the first media used for inducing calli and the subsequent media used for induction, maturation, and germination of somatic embryos (Etienne, 2005; Gatica-Arias et al., 2008; Menéndez-Yuffa et al., 2010; Landey et al., 2015). In the majority of coffee genotypes, the composition of callus induction medium, especially the concentration of 2,4-D, is extremely critical and decisive for somatic embryo induction (Muniswamy et al., 2019). During the last several years, attempts were made to improve the production of somatic embryos by using a temporary immersion bioreactor (Ducos et al., 2007; Etienne et al., 2018; Landey et al., 2013), which has facilitated mass production of tissue culture plants on an industrial scale.

Besides the commonly used plant growth regulators (auxins and cytokinins), several other growth regulators, amino acids, and other adjuvants including triacontanol, silver nitrate ($AgNO_3$), salicylic acid, thidiazuron, phloroglucinol, and 6-(3-methyl-3-butenylamino) purine (2ip) were used to improve somatic embryogenesis in coffee. Interestingly, salicylates in picomolar concentration were reported to enhance somatic embryogenesis in *C. arabica* (Quiroz-Figueroa et al., 2001). Similarly, triacontanol at a micromolar concentration in combination with indole 3-acetic acid (IAA), BA, or BAP improved somatic embryo production in both *C. arabica* and *C. canephora* genotypes (Giridhar et al., 2004a; Gatica-Arias et al., 2008). Additionally, TDZ also significantly improved the somatic embryogenic calli formation and differentiation of somatic embryos from the cultured leaf explants of *C. canephora* cv. CxR as well as from the leaf and root explants of *C. arabica* genotypes (Giridhar et al., 2004c; Kahia et al., 2016; Wang et al., 2018). Further, it was observed that in *C. arabica, C. canephora*, and *C. dewevrei*, the frequency of somatic embryogenesis showed a significant increase by the addition of silver nitrate to the culture medium (Giridhar et al., 2004b; Sridevi and Giridhar, 2014; Rojas-Lorz et al., 2019). It has also been observed that incorporation of polyamine biosynthetic pathway inhibitors such as DL-α-difloromethylornithine (DFMO) and DL-α-difloromethylarginine (DFMA) in the *in vitro* culture medium significantly reduced the embryogenic response in *C. canephora*, indicating the pivotal role played by polyamines in coffee somatic embryogenesis (Kumar et al., 2008). Besides the polyamines, indoleamines (melatonin and serotonin) and calcium and calcium ionophores (A23187) have also been shown to enhance somatic embryogenesis (Ramakrishna et al., 2012).

18.4.6 Environmental influence

Apart from exogenous growth hormones, the culture environment influence somatic embryogenesis in coffee. de Faria et al. (2003) evaluated the influences of two different concentrations (50% and 80%) of dissolved oxygen on the production of somatic embryos of *C. arabica* and observed that at lower concentrations, the number of somatic embryos was less but more mature compared to higher concentrations, wherein high frequency of somatic embryos were differentiated. Similarly, the effect of CO_2 on somatic embryogenesis of *C. arabica* was investigated (Barbon et al., 2008). It was observed that CO_2 concentration exerted a positive effect on the embryo growth when present in low concentrations in the medium, and at higher concentrations it became toxic and hindered embryo development. The culture density also influences somatic embryogenesis in coffee. A high density of cells promotes the multiplication of embryogenic cells instead of the development of embryos. Embryo maturation thus benefits from a low density (Santana-Buzzy et al., 2007; Etienne, 2005).

18.4.7 Gene expression and regulation during somatic embryogenesis

In coffee, the molecular events associated with somatic embryogenesis are not fully understood. Campos et al. (2016) published the proteome profile of embryogenic calli of *C. arabica* grown in suspension culture. Silva et al. (2013) identified 15 EST-contigs related to the embryogenesis process in coffee, of which five ESTs could be readily associated with coffee embryogenesis showing similarities with proteins: polygalacturonase, cysteine-proteinase, expansine, allergenine, and WUS. Silva et al. (2015) reported that a Baby Boom (BBM) gene homologs (CaBBM) showed higher expression in embryogenic calli of *C. arabica*. Similarly, Nic-Can et al. (2013) studied the expression of the Leafy Cotyledon (LEC) gene during somatic embryogenesis of coffee and observed that LEC expression is associated with somatic embryo maturation in coffee. The role of another important gene, Somatic Embryogenesis Receptor Kinase (SERK) was studied in *C. arabica* and found that SERK ortholog (CaSERK) was exclusively expressed in embryogenic tissue suggesting its role in embryogenic capacity. The Wuschel-related homeobox, a transcription factor of the family Wuschel-related homeobox, plays a specialized function in the embryo development in *Arabidopsis* (Arroyo-Herrera et al., 2008). In *C. canephora*, a heterologous of WUS capable of somatic cell induction was observed (Arroyo-Herrera et al., 2008; Silva et al., 2013), and its overexpression increased somatic embryo production by 400% (Arroyo-Herrera et al., 2008).

The crosstalk between DNA methylation and histone modifications during the initial phases of somatic embryogenesis was investigated using chromatin immunoprecipitation assays (Nic-Can et al., 2013). The authors observed that global DNA methylation increased during the early stages of somatic embryo induction and thereafter decreased during the maturation of somatic embryos. The comparison of the different stages of embryo development in SE and ZE and *in vitro* plantlets indicated that embryogenic differentiation is linked with DNA methylation. The authors have further proved that the genes LEC1, BBM1, and WUS are under epigenetic control in relation to the embryogenic capacity in *C. canephora*.

18.4.8 Field performance of tissue culture plants

Despite commercial propagation and field planting of micropropagated plants of *C. arabica* and *C. canephora* in various countries, data on agronomic performance of these micropropagated plants compared to the seedlings is unfortunately limited (Muniswamy et al., 2017). In Brazil, an experimental plot was established to evaluate the tissue culture–derived plants of Bourbon obtained through somatic embryogenesis with seedlings of the same line (Sondahl et al., 1999). In another study, micropropagated plants of five robusta clones obtained through somatic embryogenesis were planted in five coffee-producing countries (4000 plants/location): the Philippines, Thailand, Mexico, Nigeria, and Brazil. Visual assessment of 8000 field-grown plants in the Philippines revealed that all the micropropagated plants have normal vegetative and reproductive growth with normal flowering and fruit set (Sondahl et al., 2000). Further, Etienne et al. (1999) and Etienne and Bertrand (2001) evaluated 20,000 micropropagated plants of *C. arabica* planted in four Central American countries and reported that all the tissue culture–derived plants exhibit true to type agronomic characteristics. In another study, the agronomic performance of plantlets derived through somatic embryogenesis and microcuttings of axillary shoots were taken up, and no significant differences in growth and yield pattern were observed between somatic seedlings and microcutting-derived plants (Ducos et al., 2003). In India, empirical studies were undertaken, and the growth and yield pattern of micropropagated plants of *C. arabica* and *C. canephora* were compared with the respective seedlings of the same mother plants (Muniswamy et al., 2015, 2017). In *C. arabica*c.v. S.2800, a significant increase in yield was observed in somatic embryo-derived plants compared to seedlings, whereas in *C. canephora* c.v. CxR, no statistical differences in yield were recorded between seedlings and somatic embryo-derived plants.

18.4.9 Genetic fidelity of micropropagated plants

Micropropagation through somatic embryogenesis involving a prolonged callus phase is likely to produce somaclonal variants due to chromosomal rearrangements (Bhatia and Sharma 2015). Therefore it is imperative to test the genetic fidelity of the *in vitro*–derived plantlets particularly in woody perennials like coffee before being planted in the field. A number of tools are available to assess the genetic fidelity of regenerated plants, of which DNA-based molecular markers have currently gained paramount importance. Various molecular markers such as random amplification of polymorphic DNA (RAPD), inter simple sequence repeat (ISSR), amplified fragment length polymorphism (AFLP), simple sequence repeats (SSRs), sequence-related amplified polymorphism (SRAP), and start codon targeted (SCoT) are used to determine the genetic fidelity of micropropagated plants (Bhatia et al., 2011; Nookaraju and Agarwal, 2012; Kamle et al., 2013; Babu et al., 2018; Borsai et al., 2020; Sharma et al., 2019). In coffee, genetic fidelity was assessed in 30,000 somatic embryogenesis derived plants belonging to 20 clones of *C. arabica* F1 hybrids, and 3% to 10% off-type plants were observed in the population (Etienne and Bertrand, 2001). Rani et al. (2000) carried out the molecular analysis of somatic embryo-derived plants of *C. canephora* and revealed chloroplast and mitochondrial DNA polymorphism. Landey et al. (2013) analyzed *C. arabica* plants derived from embryogenic suspensions as well as secondary embryogenesis using AFLP and MSAP markers and observed high genetic and epigenetic stability of micropropagated plants. Recently Muniswamy et al. (2017) analyzed the somatic embryo-derived plants of *C. canephora*c.v. CxR using SRAP markers and observed 95% genetic similarity between the somatic embryogenesis–derived plants and mother plants (Figure 18.1). In another empirical study, two *C. arabica* F1 hybrid plants regenerated through somatic embryogenesis were critically analyzed for genetic uniformity using SRAP and SCoT molecular markers. The average genetic similarity between the tissue culture–derived plants and mother plants in both genotypes ranged from 0.966 to 0.993 using molecular markers. Sequencing of the differential fragments amplified by SRAP markers in tissue culture–derived plants revealed the occurrence of somaclonal variation in the coding region with functional attribution to zinc finger protein (Muniswamy et al., 2019).

18.5 Genetic transformation

18.5.1 Coffee cultivation challenges can be addressed by genetic engineering

Since the first success achieved to engineer crop plants using an *Agrobacterium tumefaciens* tumor-inducing (Ti) plasmid in 1983 (O'Brien, 1983), genetic transformation technology has been the subject of intense research worldwide. Currently, genetically engineered crops are grown in more than 41 countries and represent an increasing percentage of global crop production (English and Kayleen, 2020). The global importance of cultivating genetically engineered crops emanates from the fact that during June 2017, the world population has reached 7.5 billion, of whom some 800 million people are chronically undernourished, and over two billion suffer from micronutrient deficiencies (Elliott, 2017). Before the end of this millennium, there is a supreme challenge to feed 10 billion people, which require more food to grow on less land using less water, less labor, and fewer agrochemicals while we confront global climate change and reduce the agricultural footprint on the environment and further reductions in biodiversity (Elliott, 2017). Conventional and new breeding techniques (NBTs) together with advanced transformation technologies are arguably at the core of achieving global food security (Landry and Mitter, 2019). Genetic transformation, often referred to as genetic engineering, is a process by which appropriate genes are put in expression cassettes, and using single or binary vector and directed to plant cells either directly or indirectly using T-DNA followed by regeneration of transgenic plants (Newell, 2000). Plant transformation can be used for various purposes, such as the expression of one or more introduced genes (transgenes), silencing the expression of one or more endogenous genes, or modifying the activity or function of one or more endogenous genes. In the case of coffee, which is grown in a fragile ecosystem largely by small growers in poor and developing countries, sustainable cultivation is a major challenge. The major objectives for using genetic engineering technique in coffee are to introduce new traits into elite coffee genotypes and develop improved cultivars with desirable agronomic traits, which include pest and disease resistance, herbicide resistance, drought and frost tolerance, and improved quality profile, which are not feasible to incorporate using traditional breeding techniques. During the last ten years, a comprehensive review on the genetic transformation of coffee has been published (Mishra and Slater, 2012). and the current review is an addendum incorporating the new information which has emerged subsequently. During the last ten years, large-scale sequencing of coffee was undertaken, and the genome sequence of *C. canephora*, one of the progenitor species of *C. arabica*, is available in the public database (Denoeud et al., 2014). Genome sequencing initiatives of *C. arabica* accessions have been initiated by several research groups (https://coffeegenome.ucdavis.edu/12, among others), but an open-access genome assembly is not yet available in the public domain (Scalabrin et al., 2020). As early as 2016, about 35,153 unigenes of *C. arabica* and 25,574 unigenes of *C. canephora* are already available in the public database (Marraccini, 2020), which could accelerate the functional genomics and genetic improvement program in coffee. However, one of the key aspects of most functional genomics studies is the availability of an efficient genotype independent genetic transformation system that could be useful for designing strategies for gene identification, elucidation of gene functions, regulation, and interaction of genes and gene expression analysis. This will help in precisely targeting and modifying the trait of interest using various transformation tools (genes and promoters), which in turn help in sustainable genetic improvement at a reduced economic cost.

18.5.2 Candidate genes

Identification of candidate genes controlling key agronomic features and their utilization is considered to be of pivotal importance in transgenic technology. Unlike cereals, where a large number of genes have been discovered and characterized, in coffee, limited genes have been isolated and characterized. These include N-methyltransferase genes for caffeine biosynthesis (Uefuji et al., 2003; Satyanarayana et al., 2005), *p*-coumaroyl ester 3′-hydroxylase genes for chlorogenic acid biosynthesis (Mahesh et al., 2007), a sucrose synthase gene (Leroy et al., 2005), osmotic stress response genes (Hinniger et al., 2006), galactomannan biosynthesis genes (Pré et al., 2008) genes for seed oil content (Simkin et al., 2006), isoflavone reductase-like protein in *C. arabica* (Brandalise et al., 2009), monoterpene synthases genes from *C. arabica* for aroma (Del Terra et al., 2013), *C. arabica*BURP domain–containing gene for abiotic stress tolerance (Dinh and Kang, 2017), *CaSMP*, a seed maturation protein gene from *C. arabica* (Quintero et al., 2018), and several pathogen resistance genes such as *Mex-1* gene (Noir et al., 2003) and the *Ck-1* gene to CBD (Gichuru et al., 2006). However, the structural and functional expression of these intrinsic genes needs to be validated in coffee using a simple and fast transformation procedure. Some of the genes that were isolated from coffee and used in transformation experiments include a theobromine synthase gene (*CaMXMTI*) for suppressing caffeine biosynthesis (Ogita et al., 2004) and an *ACC oxidase*gene involved in ethylene biosynthesis (Ribas et al., 2005). Similarly, some of them which are obtained from the heterologous sources and genetically transferred to coffee include *cry*1Ac and *cry 10 Aa* genes from *Bacillus thuringiensis*against leaf miner and coffee berry borer, respectively (Leroy et al., 2000; Valencia-Lozano et al., 2019) and the *α-AII*gene from the common bean for imparting resistance to coffee berry borer (Cruz et al., 2004).

18.5.3 Transformation methods

Broadly, two different, direct and indirect DNA delivery systems have been successfully used to transform coffee; these are described below.

18.5.4 Electroporation

Electroporation, also known as electropermeabilization, is a technique that involves the formation of temporary pores in the cell membrane by using externally applied electrical pulses to introduce new genetic material, usually polar molecules, into cells. The technique exploits the weak interactions between phospholipid bilayers that maintain the integrity of cell membranes. Barton et al. (1991) used this technique to deliver a transgene into the protoplast of *C. arabica*, and the transformed tissue was selected on a medium containing kanamycin, but the regenerated plants did not survive due to a weak root system. In another experiment, Fernandez-Da Silva and Yuffa (2003) tested several parameters and optimized the DNA delivery system using electroporation. They have introduced pCAMBIA 3201 incorporating uidA and *bar genes* to using torpedo stage somatic embryos and obtained secondary somatic embryos. The incorporation of the transgene was verified using GUS assay and PCR analysis. Kumar et al. (2007) analyzed the expression of *uidA* driven by the N-methyltransferase (NMT) promoter in coffee endosperm by electroporation and suggested that NMT promoter could be used for determining gene expression in endosperm tissues of *C. canephora*.

18.5.5 Microprojectile bombardment

Microprojectile bombardment is a technique by which foreign genes are coated on heavy metal particles, particularly gold or tungsten, and delivered to living cells by applying specific pressure. In coffee, Van Boxtel et al. (1995) initiated the microprojectile bombardment–mediated transformation of coffee. The authors used leaf, somatic embryos, and suspension cultures of *C. arabica, C. canephora*, and Arabusta hybrid as target explants and studied the transient expression of the uidA gene driven by specific promoters (CAMV35S, EF1α, and Tntl [transposable element of tobacco]), and suggested that the EF1α promoter from *Arabidopsis thalianais* more effective than the CaMV35S promoter in driving transient *GUS* expression in the leaves of microcuttings. Rosillo et al. (2003) optimized different conditions such as variable helium pressure, target explants distance, preconditioning of target tissue in an osmotic medium using *C. arabica* suspension culture, and pCAMBIA1391Z plasmid with three different CAMV35S, α-tubulin coffee promoter and arabicin promoter adjacent to the 11S storage protein gene. It was observed that 4 hours of pretreatment of the target tissue in the medium supplemented with mannitol and sorbitol before bombardment increased the number of cells expressing the *GUS* gene without causing cell necrosis. The first report of successful regeneration of transgenic coffee using particle bombardment of embryogenic calli was demonstrated by Cunha et al. (2004) in *C. arabica*, using pBI-426 plasmid carrying the *npt*II and *uid*A genes. The selection of the transformants was performed using kanamycin antibiotics and the resistant calli were subjected to regenerate new plants. The integration of transgenes was verified by GUS assay and by PCR analysis using uidA and nptII gene specific primers.

Ribas et al. (2005) demonstrated the suitability of somatic embryos and embryogenic tissue of *C. canephora* as target explants in particle bombardment–mediated gene delivery using pCambia3301 plasmid containing the *bar* and *uidA* genes. The bombarded tissues were subjected to selection using ammonium glufosinate herbicide, and plants were regenerated. The transgenic plants were confirmed by GUS assay and PCR and Southern blot analysis. Gatica-Arias et al. (2008) bombarded the secondary somatic embryogenic suspension culture of *C. arabica* with various pCAMBIA plasmid series carrying different selectable marker genes (*bar, nptII*, and *hpt*). Although no regeneration of the transgenic plants was possible due to damage to the bombarded tissue, stable transformants were obtained with pCAMBIA 2301 plasmid carrying kanamycin-resistance gene. The successful regeneration and inheritance of transgenes was reported in *C. arabica* using bombardment of embryogenic calli followed by kanamycin selection (Albuquerque et al., 2009). The authors reported normal growth of the transgenic plants and obtained T1 progeny presenting 3:1 segregation of the *uidA* transgene. Transgenic *C. arabica* plants carrying the α-amylase inhibitor-1 gene (α-*AI1*) from the common bean, *Phaseolus vulgaris*, under control of the seed-specific phytohemagglutinin promoter (PHA-L) were regenerated using microprojectile bombardment of embryogenic calli (Barbosa et al., 2010). The presence of the α-*AI1* gene in six regenerated transgenic T1 coffee plants were identified by PCR and Southern blot. This is the first report of producing transgenic arabica coffee targeted for coffee berry borer resistance.

The first report of producing transgenic *C. arabica* incorporating *cry1ac* gene from *B. thuringiensis* for conferring resistance to leaf miner was achieved by bombarding the torpedo-stage somatic embryos without selection in antibiotic medium (De Guglielmo-Cróquer et al., 2010). The seven regenerated transformed plants were tested using PCR and Southern blot analysis, and the integration of the *cry1ac* gene was confirmed. Valencia-Lozano et al. (2019) developed an efficient protocol of the stable genetic transformation of coffee plants expressing the Cry10Aa protein of *B. thuringiensis*. Embryogenic cell lines of *C. arabica* cultivar Typica were bombarded with plasmid pMDC85 containing the *cry10Aa* gene, and the transformants were selected on hygromycin containing medium. The stable transformation was confirmed by hygromycin-resistant embryogenic lines, green fluorescent protein (GFP) expression, quantitative analyses of *Cry10Aa* by mass spectrometry, Western blot, ELISA, and Southern blot analyses. The authors have reported 16.7% transformation efficiency using the microprojectile bombardment method.

Particle bombardment is an excellent technique that could be exploited for producing marker free transgenic plants using a co-transformation approach, as demonstrated in crops like rice (Devi et al., 2012; Tang et al., 1999), maize (Klein et al., 1989), wheat (Vasil et al., 1992), and barley ((Wan and Lemaux 1994). Depending upon the integration pattern when the two genes are not genetically linked and placed in two different loci, it could be possible to obtain plants containing one of the transgenes via recombination (Devi et al., 2012).

18.5.6 *Agrobacterium tumefaciens*

Agrobacterium-mediated transformation is the most widely used indirect method for coffee transformation. *A. tumefaciens* is a pathogenic, gram-negative soil-inhabiting bacteria that usually infect dicot plants and causes tumor formation. The formation of tumor is caused by the horizontal transfer of Ti plasmid from *Agrobacterium* to plant cell and is commonly known as T-DNA transfer. The T-DNA (approx. 200 kb) forms a part of a small independent DNA molecule outside the bacterial genome and contains left and right border sequences. The Ti plasmid of *A. tumefaciens* contains several virulence genes (Vir) and genes for opine biosynthesis in plants responsible for tumor induction in the plant cell. Several different Vir genes have specific functions in T-DNA transfer, which include T-DNA excision, transfer, and the integration of the T-DNA region into the nucleus of the host cell (Gelvin, 2003). The phenolic compounds exuded from plant wounds stimulate expression of Vir genes located on the Ti plasmid and responsible for its excision, transfer, and integration into the plant genome. Therefore any foreign DNA of interest inserted between these two border sequences of T-DNA can be transferred into plants by *Agrobacterium*–mediated transformation (Gelvin, 2000, 2003). This natural phenomenon of *Agrobacterium* was exploited in plant transformation techniques by replacing some of the undesirable tumor-inducing genes with the genes of interest within the T-DNA border.

In coffee, the *A. tumefaciens*-mediated transformation technique is the method of choice for many workers and several successful reports are available in the literature (see Table 18.1 for details). Since the first report of *A. tumefaciens*-mediated transformation of coffee by Spiral and Petiard (1991) many modifications have been made in transformation protocol, some of which are described below.

18.5.7 Target explants type

The first *A. tumefaciens*-mediated transformation was carried out in *C. canephora by* co-cultivating the protoplast with various *Agrobacterium* strains carrying *nptII* and *uidA* genes (Spiral and Pétiard, 1991). Although callus tissue from the protoplast was obtained and tested positive for the GUS expression, no plant regeneration was achieved. Subsequently, protoplast was not used in *Agrobacterium*-mediated transformation. In *C. canephora*, a highly efficient *A. tumefaciens* transformation and regeneration protocol was established using hypocotyl explants as the target material (Mishra et al., 2002; Mishra and Sreenath, 2004; Sridevi et al., 2010). It was observed that the collar region of the hypocotyls was found to be more suitable for *A. tumefaciens* transformation because of their better morphogenic response (Sridevi et al., 2010). Besides hypocotyl, embryogenic calli and somatic embryos were widely used and found to be more suitable for *A. tumefaciens* transformation in both *C. arabica* and *C. canephora* genotypes Hatanaka et al., 1999; Leroy et al., 2000; Mishra et al., 2008, 2010; Ribas et al., 2011; Kumar et al., 2018). Recently, Vargas-Guevara et al. (2018) developed and optimized *A. tumefaciens* transformation protocol using ex vitro leaf explants.

18.5.8 Binary vectors

Several binary vectors, such as pMDC85, pBECKs, pGreen, pCAMBIA, pBIN, and pSB, have been designed for the transfer of transgenes using *Agrobacterium*–mediated transformation in plants (Curtis and Grossniklaus, 2003; McCormac et al., 1997; Bulgakov et al., 2006; Chung et al., 2005; Coutu et al., 2007; Gelvin, 2003; Traore and Zhao, 2011; Zeevi et al., 2012). One of the major strategies for the facile creation of binary vector constructs is to reduce the size of the plasmid vector, which allows efficient manipulation (Hellens et al., 2000). Binary vectors for *Agrobacterium*-mediated transformation of plants usually contain (1) T-DNA (with the left and right border), which carries the foreign DNA; (2) Origin of replication (ori) for multiplication in both *Escherichia coli* and *A. tumefaciens*; and (3) selectable marker gene(s) for bacterial selection (*E. coli* and *A. tumefaciens*). Different selectable marker/reporter gene either single or in combinations are incorporated in T-DNA vector for the efficient selection of transformed plants.

18.5.9 Preculture and culture conditions

Several studies have been undertaken to improve the *Agrobacterium* transformation efficiency in coffee by manipulating the culture conditions. Leroy et al. (2000) suggested that transformation efficiency is improved by incubating the somatic embryos in NaCl solution prior to *Agrobacterium* co-cultivation. However, Mishra et al. (2002) demonstrated that incubation of somatic embryos of *C. canephora* in various concentrations of NaCl for different time intervals prior to *Agrobacterium* co-cultivation did not show any positive effect. Further, it was observed that preculture of target explants for 3 days in callus induction medium prior to *Agrobacterium* co-cultivation and addition of 50 µm acetosyringone to co-cultivation medium significantly improved T-DNA delivery in *C. canephora* (Mishra et al., 2002). Canche-Moo et al. (2006) and Ribas et al. (2006) incorporated the vacuum infiltration and sonication methods during co-cultivation to increase the transformation efficiency. Recently, Kumar et al. (2018) suggested that in *C. canephora* addition of silver thiosulphate (STS) to co-cultivation medium improved the transformation potential of embryos by enhancing *A. tumefaciens* adherence to surfaces. Vargas-Guevara et al. (2018) used an optimized agroinfiltration method to deliver *Agrobacterium* to coffee leaf tissue. The authors demonstrated that agroinfiltration using one injection followed by sanding of the abaxial leaf epidermis resulted in significantly higher transient *uidA* expression.

18.5.10 *Agrobacterium* strain and co-cultivation period

The selection of appropriate *Agrobacterium* strain is critical to the success of transformation as different crop plants respond differently to various *Agrobacterium* strains. *A. tumefaciens* strains and their plasmids are classified by the function of the opine genes they carry. These opines, which are synthesized in the infected plant cells, are mainly classified as agropine, nopaline, and octopine. For coffee transformation, various *A. tumefaciens* strains such as LBA 4404 pAL4404 (octopine), C58CI pMP90 (nopaline), EHA 101 pEHA101 (agropine), EHA 105 pEHA101 (agropine), and AGL1 pTiBo542 (agropine) have been used by various workers (see Table 18.1). In an empirical study, Mishra et al. (2008) compared the transformation efficiency of four different *Agrobacterium* strains (LBA 4404, EHA101, EHA105, AGL1) in coffee using pBECKS 2000 vector constructs carrying *uidA* and *gfp* reporter genes. Based on both transient and stable transformation data, it was demonstrated that EHA 105 and EHA101 were more efficient compared to LBA4404 in T-DNA delivery and transgenic plant regeneration. By adapting this improved protocol, the authors were able to produce a large number of transgenic coffee plants in both *C. arabica* and *C. canephora*. Recently, Vargas-Guevara et al. (2018) evaluated the transformation efficiency of three *Agrobacterium* strains using uidA gene expression and observed that the expression of *uidA* was higher in GV3101-treated coffee disks than in LBA4404 and ATHV-treated samples.

The influence of the co-cultivation period on transformation efficiency was tested in *C. canephora* using *pBECKS* 400 gus intron/*hph* construct (Mishra et al., 2002). Based on the GUS assay it was determined that the transformation efficiency was very low even after 24 hours of co-cultivation but increased thereafter for 5 days. It was observed that though maximum transformation efficiency was obtained after 5 days of

TABLE 18.1

Detailed Information on Genetic Transformation Studies in *Coffea* spp.

DNA delivery method	*Coffea* species	Explants used	Binary plasmid vector	*Agrobacterium* strains	Promoter	Selectable marker genes	Target gene	Results	Reference	Country where work is carried out
Direct delivery	*C. arabica*	Protoplast	NA	-	pGA472	*nptII*	*uidA*	TE	Barton et al. (1991)	United States
Electroporation	*C. arabica*	SE	pCAMBIA 3201	-	CaMV35S	*bar*	*uidA*	TGI	Fernandez et al. (2003)	Venezuela
	C. canephora	Endosperm	pCAMBIA 1301	-	CaMV35S	*hpt*	*uidA*	TE	Kumar et al. (2007)	India
Microprojectile bombardment	*C. arabica*	Leaf	pPIGK	-	EF-1a	*bar*	*uidA*	TE	Van Boxtel et al. (1995)	France
	C. arabica	ET	pCAMBIA 2301	-	CaMV35S	*nptII*	*uidA*	TGI	Rosilo et al. (2003)	Colombia
	C. arabica	ET	pBI-426	-	CaMV35S	*nptII*	*uidA*	TGI, PR	Cunha et al. (2004)	Brazil
	C. canephora	ET	pCAMBIA 3301	-	CaMV35S	*bar*	*uidA*	TGI, PR	Ribas et al. (2005)	Brazil
	C. arabica	SE	pCAMBIA 2301	-	CaMV35S	*nptII*	*uidA*	TE	Gatica-Arias et al. (2008)	Costa Rica
	C. arabica	ET	pBI-426	-	CaMV35S	*nptII*	*uidA*	TGI, PR	Albuquerque et al. (2009)	Brazil
	C. arabica	SE	pUBC	-	Ubiquitine		*cry1ac*	PR	Guglielmo-Croqer et al. (2010)	Venezuela
	C. arabica	ET	pMDC85	-	CaMV35S	*hpt*	*cry10Aa*	PR	Valencia-Lozano et al. (2019)	Mexico
Indirect delivery	*C. arabica*	Protoplast	pGV2260	NA	CaMV35S	*hpt*	*uidA*	TE	Spiral and Petiard (1991)	France
A. tumefaciens	*C. canephora*	ET	pIGI121-Hm	EHA101		*hpt*	*uidA*	TGI, PR	Hatanaka et al. (1999)	Japan
	C. arabica and C. canephora	SE	pBIN19	LBA4404	EF-1α	*csr-1–1*	*cry1Ac*	TGI, PR	Leroy et al. (2000)	France
	C. canephora	ZE, SE, H	pBECKS 400	EHA101	CaMV35S	*hpt*	*uidA*	TGI,	Mishra et al. (2002)	India
	C. arabica and C. canephora	ET, SE	pHIBI – IG	EHA101	CaMV35S	*hpt*	*gfp & CaMXMT1*	RNAi,	Ogita et al. (2004)	Japan
	C. canephora	H	pBECKS 400	EHA 101	CaMV35S	*hpt*	*uidA*	PR	Mishra and Sreenath (2004)	India
	C. canephora	ET	pCAMBIA 3301	EHA105	CaMV35S	*ppt*	*uidA*	TGI, PR	Cruz et al. (2004)	Brazil
	C. canephora	ET, Leaf	pER10W-35SRed	C58	CaMV35S	*nptII*	*DsRFP*	TGI	Canche-Moo et al. (2006)	Mexico
	C. canephora	SE	pCAMBIA 1381	A4, EHA 101	CaMV35S	*hpt*	*NMT*	TGI, PR	Kumar et al. (2006)	India
	C. canephora	SE	pBIN19	LBA 4404	EF-1	*csr-1–1*	*cry1Ac*	RNAi	Perthuis et al. (2005)	France
	C. arabica	ET	pCAMBIA 3300	EHA 105	CaMV35S	*bar*	*ACC-oxidase*	Field test	Ribas et al. (2005)	Brazil
	C. canephora	ET	pCAMBIA 3301	EHA 105	CaMV35S	*bar*	*uidA*	AE	Ribas et al. (2006)	Brazil
	C. canephora	Embryos	pER10W-35SRed	C58C1	CaMV35S	*WUSCHEL*	*DsRFP*	TGI, PR	Arroyo-Herrera et al. (2008)	Mexico
	C. canephora and C. arabica	ET	pBECKS 2000	EHA 101, EHA 105, LBA 4404, AGL 1	CaMV35S	*hpt, visual*	*uidA, sgfp*	TGI, PR	Mishra et al. (2008)	India
	C. canephora	ET	pBECKS 2000.7	EHA 105	CaMV35S	GFP visual	*sgfp*	TGI, PR	Mishra et al. (2010)	India
	C. canephora	H	pCAMBIA 1305.2	GV3101	CaMV35S	*hpt*	*uidA*	FTP	Sridevi et al. (2010)	India
	C. arabica	ET	pMDC32	LBA1119	CaMV35S	*hpt*	*sgfp*	PR	Ribas et al. (2011)	France
	C. canephora	SE	pCAMBIA 1301	GV3101	CaMV35S	*hpt*	*uidA*	PR	Kumar et al. (2018)	India
	C. arabica	Ex vitro leaf	pCAMBIA 1303	GV3101	CaMV35S	*hpt*	*uidA*	PR TGI	Guevara et al. (2018)	Costa Rica
A. rhizozenes	*C. arabica and C. canephora*	SE	pBIN 19	A4	CaMV35S	*hpt*	*uidA*	TGI, PR	Spiral et al. (1993)	France
	C. arabica	Leaf	NA	IFO 14554	NA	NA	NA	TGI, PR	Sugiyama et al. (1995)	Japan
	C. canephora	SE	pBIN 19	A4	CaMV35S	*csr-1–1*	*cry1Ac*	TGI, PR	Gimenez et al. (1996)	France
	C. canephora	SE	pCAMBIA 1301	A4	CaMV35S	*hpt*	*uidA*	TGI, PR	Kumar et al. (2006)	India
	C. arabica	H	pBIN 19	A4	CaMV35S	visual	*uidA*	TGI, PR	Alpizar et al. (2006)	France
	C. arabica	H	pCAMBIA 2300	A4	CaMV35S	visual	*gfp*	TGI, PR	Alpizar et al. (2006)	France

Notes: ET = Embryogenic tissue; SE = somatic embryo; H = hypocotyl; TE = transient gene expression; TGI =target gene integration; PR =plant regeneration; FTP =fertile transgenic plant; RNAi = RNA interference; AE =antisense expression.

Source: Updated from Mishra et al. (2012).

co-cultivation, often it was encountered that beyond 4 days of co-cultivation lead to uncontrollable bacterial infection resulting in loss of explants. In general, 4–5 days of cocultivation is followed in *Agrobacterium*-mediated transformation of coffee irrespective of the strain used (Ribas et al., 2006; Mishra et al., 2010; Sridevi et al., 2010).

18.5.11 Plant selectable marker/reporter genes

Different types of selectable marker genes such as antibiotic or herbicide resistance (*hpt* hygromycin-R, *nptII* kanamycin-R, *csr1-1* chlor and sulfuron-R, *ppt* phosphinothricin-R) incorporated in *Agrobacterium* vector constructs are used in coffee transformation (Table 18.1). It is suggested that the type of antibiotics and their minimum concentrations need to be determined before initiating the transformation experiments. This is because one antibiotic that is efficient in selecting transformants in one crop may not be suitable in selecting transformants in another crop. Additionally, the selection of antibiotics is also critical based on the type of explants and the regeneration protocol. The fundamental principle of selecting the type and concentration of antibiotics is based upon its ability to separate transformed cells from the nontransformed cells without seriously intervening in the regeneration process.

The first work on *Agrobacterium*-mediated coffee transformation was carried out using hygromycin as the selective agent (Spiral and Pétiard, 1991), and since then hygromycin has been used as the most favored antibiotic for selecting transformants. In many transformation experiments, hygromycin at concentrations of 20–100 mg L^{-1} was used successfully for the selection of transformants (Hatanaka et al., 1999; Mishra and Sreenath, 2004; Ribas et al., 2011). Mishra et al. (2002) tested different concentration of hygromycin in both transformed and nontransformed tissue of *C. canephora*, and it was observed that 25 mg L^{-1} hygromycin completely checked nontransformed somatic embryo growth and proliferation. Canche-Moo (2006) used kanamycin (100 mg L^{-1}) and successfully regenerated plants following *Agrobacterium*-mediated transformation. However, many authors have reported the development of nontransformed somatic embryos at higher kanamycin concentrations and have attributed this to a poor capacity for transformant selection (Mishra et al., 2002; Van Boxtel et al., 1997). In addition to antibiotics, many herbicide resistance selection marker genes (chlorsulfuron [*csr1-1*], phosphinothricin [*ppt*], and ammonium glufosinate [*bar*]) were also used in coffee transformation, and transgenic plants of both *C. arabica* and *C. canephora* were regenerated (Leroy et al., 2000; Cruz et al., 2004; Ribas et al., 2006). As an alternative to negative markers such as antibiotics and herbicides, positive selection marker genes such as phosphomannose isomerase (*pmi*) and xylose isomerase (*xylA*) have also been tested for coffee transformation (Samson et al., 2004). Following transformations, selections of transformants were made in the culture medium supplemented with various concentrations of mannose and xylose, respectively, without an additional carbohydrate source. The study indicated that compared to mannose, xylose is a potential positive selective agent for coffee transformation.

18.6 Transgenic coffee regeneration without antibiotics

It has been argued that transgenic crops regenerated using antibiotic and herbicide resistance genes could have negative effects on human health and the environment, although no credential data have been presented to support this claim. However, to improve the public acceptance of transgenic plants, it is imperative to develop technology to regenerate plants without antibiotics. In coffee, Mishra et al. (2010) for the first time regenerated transgenic plants of both *C. canephora* and *C. arabica* by employing a visual selection of GFP as the sole screen following *A. tumefaciens*-mediated transformation. Further, a clean gene strategy using *Agrobacterium* vector constructs pBECKS 2000 series McCormac et al. (1999) incorporating the *cre/lox* recombination system (also known as induction-excision strategy) was also used, and transgenic coffee of *C. arabica* and *C. canephora* were regenerated (Mishra et al., 2014b).

18.6.1 *Agrobacterium rhizogenes*

Apart from a few reports, *Agrobacterium rhizogenes*-mediated transformation has not been used regularly in coffee. Spiral et al. (1993) for the first time reported *A. rhizogenes*-mediated transformation of both *C. canephora* and *C. arabica* species using hygromycin selection. Subsequently, many workers have reported successful *A. rhizogenes*-mediated transformation and plantlet regeneration in both *C. arabica* and *C. canephora* using leaf, somatic embryos, or hypocotyls explants (Gimenez et al., 1996; Kumar et al., 2006. Alpizar et al., 2006). In the majority of transformation experiments, *Agrobacterium* strain A4 was used, except in one case where bacterial strain IFO 14554 has been used (Table 18.1). Further, the binary constructs used in *A. rhizogenes* transformation are either based on a pBIN19 or pCAMBIA backbone. *A. rhizogenes* transformation could be very useful in many functional genomics studies related to root knot nematode resistance, nutrient uptake, and heavy metal tolerance studies in coffee.

18.6.2 Gene editing

Discovered only during 2012 (Jinek et al., 2012), clustered regularly interspaced short palindromic repeats (CRISPR)/CRISPR-associated (Cas9) (CRISPER/Cas9) became a powerful genome-editing tool that has been widely adopted in model organisms as well as many plants. The CRISPR/Cas9 system consists of three components: crRNA (CRISPR RNA), Cas9, and tracrRNA (trans-activating crRNA). The Cas9 is an endonuclease, which acts like a DNA sensor and cuts the double-strand break (DSB). crRNA (20–50 nucleotides) and tracrRNA (75–171 nucleotides) work in tandem and guide Cas9 to specific targets where genome editing is planned to be undertaken. Although the design and application of CRISPR/Cas9 are simpler, a lot of experiments especially the system of DNA delivery and tissue regeneration methods need to be optimized for individual species to achieve accuracy and efficiency. In *Coffea*, Breitler et al. (2018) made an in silico analysis of the whole genome sequence of *C. canephora* and identified all gRNA target sequences. In all, 8,145,748 CRISPR guides were identified corresponding to 5,338,568 different sequences. The CRISPR-associated (Cas9) gene editing was targeted at the phytoene desaturase gene (*CcPDS*) utilizing the *A. tumefaciens* transformation technique and the somatic embryogenesis plant regeneration protocol. The analysis of the RNA-guided genome-editing events revealed that 22.8% of the regenerated plants were heterozygous mutants and 7.6% were homozygous mutants, and about 30.4% mutation efficiency was estimated to be achieved at the target site. The study further demonstrated that an improved regeneration protocol other than somatic embryogenesis could be suitable for genome editing by the CRISPR/Cas9 method in coffee.

18.7 Conclusion and future perspectives

Coffee is the most important global agricultural commodity and a popular beverage consumed throughout the world. Rapid climate change has had a tremendous impact, and coffee cultivation in many countries is at a crossroads. In fact, the sustainable cultivation of coffee has practical significance not only for the stakeholders, including the producers and consumers, but also for the environment where coffee is grown under shade. The advances made in coffee *in vitro* propagation using somatic embryogenesis should be simplified so that it can be adapted with minimum infrastructure and low-cost technology. This is because a large majority of coffee growers have small holdings and do not have the resources to buy tissue culture plants at a high price. Further, empirical study should be undertaken and both seedlings and tissue culture plants should be compared for their agronomic performance. Because the cost of seedlings is much lower than tissue culture plants, the advantage of tissue culture plants over seedlings needs to be demonstrated. Another important aspect that needs to be addressed is to reduce the tissue culture–induced somaclonal variation. In coffee, genetic alterations in the coding region have already been demonstrated (Muniswamy et al., 2019), which may affect the gene functions. Therefore, maximum precaution should be taken and minimum tissue culture cycles need to be fixed for each genotype before mass propagation of a particular variety is taken up, because each genotype responds differently to culture conditions.

Transgenic technology has tremendous scope for coffee improvement because conventional breeding has several limitations. Despite significant advances made over the last 25 years, a genotype-independent transformation protocol is not yet achieved in coffee, and genetic transformation could not become a routine procedure in many laboratories. Developing a robust transformation system in coffee is now considered a priority because (1) it can be used as a tool for gene annotation and (2) the existing gene pool is inadequate for production of transgenic coffee with agronomically important traits, especially pest resistance. Further, public disquiet regarding the antibiotic resistance associated with transgenic technology needs to be addressed. In this regard, several strategies have been adopted, such as the use of positive selection markers and GFP for transformant selection and the *cre/lox* system for elimination of selectable marker genes. The development of a clean gene transgenic technology for coffee based on the *gfp*gene for visual selection and the *cre/lox* vector system is already demonstrated and needs to be pursued. Further genome editing using CRISPER/Cas9 technology has an unprecedented technological advantage, and a concerted effort to fine-tune the technology needs to be made at the earliest opportunity.

REFERENCES

Albuquerque EVS, Cunha WG, Barbosa AEAD, Costa PM, Teixeira JB, Vianna GR, Cabral GB, Fernandez D, Maria FG. *In Vitro Cell. Dev. Biol. Plant.* 2009; 45 (5): 532–539.

Alpizar E, Dechamp E, Espeout S, Royer M, Lecouls AC, Nicole M, Bertrand B, Lashermes P, Etienne H. *Plant Cell Rep.* 2006; 25: 959–967

Anagbogu CF, Bhattacharjee R, Ilori C, Tongyoo P, Dada KE, Muyiwa AA, Gepts P, Beckles M. *Genet Resour Crop Evol.* 2019; 66: 685–696.

Anthony F, Combes, M,, Astorga C, Bertrand B, Graziosi G, Lashermes P. *Theor Appl Genet.* 2002; 104: 894–900. https://doi.org/10.1007/s00122-001-0798-8

Arroyo-Herrera, Gonzalez AK, Canche Moo R, Quiroz-Figueroa FR, Loyola-Vargas VM, Rodriguez-Zapata LC, D'Hondt CB, Suárez-Solís VM, Castaño E. *Plant Cell Tiss Organ Cult.* 2008; 94: 171–180. https://doi.org/10.1007/s11240-008-9401-1

Avelino J, Cristancho M, Georgiou S, Imbach P, Aguilar L, Bornemann L, Läderach P, Anzueto F, Hruska AJ, Morales C. *Food Security.* 2015; 7: 303–321. https://doi.org/10.1007/s12571-015-0446-9.

Babu AG, Vinoth A, Ravindhran R. *Plant Cell Tiss Organ Cult.* 2018; 132. https://doi.org/10.1007/s11240-017-1319-z

Barbon R, Jiménez E, Preil W. *Plant Cell Tiss Organ Cult.* 2008; 95: 155–161. https://doi.org/10.1007/s11240-008-9427-4

Barbosa AEAD, Albuquerque EVS, Silva MCM, Souza DSL, Oliveira-Neto OB, Valencia A, Rocha TL, Maria FG. *BMC Biotechnology.* 2010; 10: 44. https://doi.org/10.1186/1472-6750-10-44

Barton CR, Adams TL, Zarowitz MA. In: *Proceedings of the 14th International Conference on Coffee Science, (ASIC)*, ASIC, San Francisco, USA. 1991; 460–464.

Berthouly M, Dufour M, Alvard D, Carasco C, Alemano L, Teisson C. In: *16th International Scientific Colloquium on Coffee (ASIC),* ASIC, Kyoto, Japan. 1995; 514–519.

Bhatia S, Sharma K. Academic Press. 2015; 361–368. https://doi.org/10.1016/B978-0-12-802221-4.00011-X.

Bhatia R, Singh KP, Sharma TR, Jhang T. *Plant Cell Tiss Organ Cult.* 2011; 104: 131–135.

Bidel S, Tuomilehto J. *European Endocrinology.* 2013; 9 (2): 99–106.

Bikila BA, Sakiyama NS, Caixeta ET. *Journal of Microbiology & Experimentation.* 2017; 5 (1): 00136.

Borchetia S, Dasa SC, Handiqueb PJ, Dasa S. *Scientia Horticulturae.* 2009; 120 (4): 544–550.

Borsai Hârţa OM, Katalin S, Kelemen C, Tripon AF, Maria-Maruni C, Clapa D. *Horticultural Science.* 2020; 47. https://doi.org/10.17221/20/2019HORTSCI

Brandalise M, Severino FE, Maluf MP, Maia IG. *Plant Cell Reports.* 2009; 28 (11): 1699–1708.

Breitler JC, Dechamp E, Campa C, Rodrigues LAZ, Guyot R, Marraccini P, *Plant Cell Tissue and Organ Culture.* 2018; 134: 383–394.

Bulgakov VP, Kiselev KV, Yakovlev KV, Zhuravlev YN, Gontcharov AA, Odintsova NA. *Journal*. 2006; 1 (4): 454–461. https://doi.org/10.1002/biot.200500045

Bunn C, Läderach P, Rivera OO, Kirschke D. *Climatic Change*. 2014; 129 (1–2): 89–101. https://doi.org/10.1007/s10584-014-1306-x

Campos N, Luciano A, Paiva V, Panis B, Carpentier SC. *Proteomics*. 2016; 16: 1001–1005.

Campos NA, Panis B, Carpentier SC. *Frontiers in Plant Science*. 2017; 8: 1460. https://doi.org/10.3389/fpls.2017.01460

Canche-Moo RLR, Ku-Gonzalez A, Burgeff C, Loyola-Vargas VM, Rodríguez-Zapata LC. Castaño. *Plant Cell Tissue and Organ Culture*. 2006; 84 (3): 373–377.

Carneiro MF. Advances in coffee biotechnology. *AgBiotechNet*. 1999; 1: 1–7.

Carneiro MF, Ribeiro TMO. *Broteria Genetica*. 1989; 85 (1989): 139–152.

Chaitra TR, Sreenath HL. *Journal of Plantation Crops*. 2010; 38 (3): 219–222.

Chung SM, Frankman EL, Tzfira T. *Trends in Plant Science*. 2005; 10: 357–361.

Coutu C, Brandle J, Brown D, Brown K, Miki B, Simmonds J, Hegedus DD. *Transgenic Research*. 2007; 16: 771–781

Cruz ARR, Paixão ALD, Machado FR. *Bol Pesq Desenvol*. 2004; 73, Embrapa Brasília.

Cubry P, Bellis FD, Pot D, Musoli P, Leroy T. *Genetic Resources and Crop Evolution*. 2013; 60 (2): 483–501.

Cunha WG, Machado FRB, Vianna GR, Teixeira JB, de Baross EVSA. Obtencao De. *Bol. Pesqui. Desenvol*. 2004; 73: 6–14.

Curtis MD, Grossniklaus U. *Plant Physiology*. 2003; 133 (2): 462–469.

DaMatta FM, Avila RT, Cardoso AA, Martins SCV, Ramalho JC. *Journal of Agricultural and Food Chemistry*. 2018; 66: 5264–5274.

Davis AP. *Nordic Journal of Botany*. 2011; 29: 471–472.

Davis AP, Rakotonasolo F, De Block P. *Nordic Journal of Botany*. 2010; 28: 134–136.

De Guglielmo-Cróquer Z, Zaidi AM Yuffá MA. *Brazilian Journal of Biology*. 2010; 70 (2): 387–393.

de Feria M, Jiménez E, Barbón R, Capote A, Chávez M, Quiala E. *Plant Cell, Tissue and Organ Culture*. 2003; 72: 1–6. https://doi.org/10.1023/A:1021202305692

Del Terra L, Lonzarich V, Asquini E, Navarini L, Graziosi G, Liverani FS, Pallavicini A. *Phytochemistry*. 2013; 89.

Denoeud F, Carretero-Paulet L, Dereeper A, Droc G, Guyot R, Pietrella M, *Science*. 2014; 345: 1181–1184.

Devasia J, Muniswamy B, Mishra MK. *International Journal of Nanoscience and Nanotechnology*. 2020; 16: 271–277.

Devi S, Mishra MK, Elliott M. *Tropical Life Science Research*. 2012; 23 (2): 39–48.

Dinh SN, Kang H. *Plant Cell Reports*. 2017; 36 (11): 1829–1839.

Ducos JP, Alenton R, Reano JF, Kanchanomai C, Deshayes A, Pétiard V. *Euphytica*. 2003; 131: 215–223.

Ducos JP, Lambot C, Pétiard V. *International Journal of Developmental Biology*. 2007; 1: 1–12.

Elliott M. Agriculture and food security: Fifth anniversary. *Agriculture and Food Security*. 2017; 6 (58).

English C, KayleenS. *LP Infographics Document the Global Growth of Agricultural Biotechnology Innovation*. https://geneticliteracyproject.org/2020/04/28/where-are-gmo-crops-grown-glp-infographics-document-the-global-growth-of-agricultural-biotechnology-innovation/

Etienne H. In: SM Jain and PK Gupta (Eds.). *Protocol for Somatic Embryogenesis in Woody Plants*. Springer, Amsterdam. 2005; 167–179.

Etienne H, Bertrand B. *Tree Physiology*. 2001; 21: 1031–1038.

Etienne, Barry-Etienne HD, Vasquez N, Berthouly M. In: B Bertrand and B Rapidel (Eds.). *Desafios de Cafecultura encentroamerica*. IICA, San Jose. 1999; 457–493.

Etienne H, Breton D, Breitler JC, Bertrand B, Déchamp E, Awada R, Marraccini P, Léran S, Alpizar E, Campa C, Courtel P, Georget F, Ducos JP. *Frontiers in Plant Science*. 2018; 871: 1–21. https://doi.org/10.3389/fpls.2018.01630

Fernandez-Da Silva SR, Yuffa AM. *Electronic Journal of Biotechnology*. 2003; 6: 29–35.

Ganesh SD, Sreenath HL. *Journal of Plantation Crops*. 1997; 25: 169–174.

Ganesh SD, Sreenath HL. *Indian Coffee*. 1999; 4: 7–9.

Garavito A, Montagnon C, Guyot R, Bertrand B. *BMC Plant Biology*. 2016; 16 (1): 242. https://doi.org/10.1186/s12870-016-0933-y

Gatica-Arias Andrés M, Arrieta-Espinoza G, Ana M, Esquivel E. *Electronic Journal of Biotechnology*. 2008; 11 (10): 1–11

Gelvin SB. *Annual Review of Plant Physiology and Plant Molecular Biology*. 2000; 51: 223–256. https://doi.org/10.1146/annurev.arplant.51.1.223

Gelvin SB. *Microbiology and Molecular Biology Reviews*. 2003; 67: 16–37.

George S, Elizabeth, Ramalakshmi K, Jagan L, Rao M. *Critical Reviews in Food Science and Nutrition*. 2008; 48: 5, 464–486. https://doi.org/10.1080/10408390701522445

Gichuru EK, Charles O, Agwanda, Combes MC, Mutitu EW, Ngugi ECK, Bertrand B, Lashermes P. In: *Proceedings of the 21st International Conference on Coffee Science (ASIC)*, ASIC, Montpellier, France. 2006; 786–793.

Gimenez C, Yuffá AM, de García E. *Phyton*. 1996; 59: 39–46.

Giridhar P, Indu EP, Chandrasekar A. *In Vitro Cellular & Developmental Biology – Plant*. 2004a; 40: 200–203. https://doi.org/10.1079/IVP2003519

Giridhar P, Indu EP, Kumar V, Chandrasekar A, Ravishankar GA. *Acta Physiologiae Plantarum*. 2004b; 26: 299–305.

Giridhar P, Kumar V, Indu EP, Ravishankar GA, Chandrasekar A. *Acta Botanica Croatica*. 2004c; 63: 25–33.

Hatanaka T, Choi YE, Kusano T, Sano H. *Plant Cell Reports*. 1999; 19: 106–110.

Hellens RP, Edwards EA, Leyland NR, Bean S, Mullineaux PM. *Plant Molecular Biology*. 2000; 42: 819–32. https://doi.org/10.1023/a:1006496308160

Hinniger C, Caillet V, Michoux F, Amor MB, Tanksley S, Lin C, Mccarthy J. *Annals of Botany*. 2006; 97: 755–765.

Huded AKC, Jingade P, Muniswamy B, Mishra MK. *International Journal of Fruit Science*. (2020); 1553–8362.

ICO; International Coffee Organization. 2021; www.ico.org/trade_statistics.asp.

Jinek M, Chylinski K, Fonfara I, Hauer M, Doudna JA, Charpentier E. *Science*. 2012; 337: 816–821.

Jingade P, Huded AKC, Kosaraju B, Mishra MK. *Journal of Crop Improvement*. 2019; 33 (3): 327–345,

Kahia J, Kirika M, Lubabali H, Mantell S. *Hortscience*. 2016; 51: 1148–1152. https://doi.org/10.21273/Hortsci10771-16

Kamle M, Kumar P, Bajpai A, Kalim S, Chandra R. *New Zealand Journal of Crop and Horticultural Science*. 2013; 42: 1–9. https://doi.org/10.1080/01140671.2013.814574.

Kiwuka C, Goudsmit E, Tournebize R, de Aquino SO, Douma JC, Bellanger L, Crouzillat D, Stoffelen P, Sumirat U, Legnaté H, Marraccini P, de Kochko A, Andrade AC, Mulumba JW, Musoli P, Anten NPR, Poncet V. *PLoS One*. 2021; 16 (2): e0245965.

Klein T, Kornstein ML, Sanford JC, Fromm ME. *Plant Physiology*. 1989; 91 (1): 440–444. https://doi.org/10.1104/pp.91.1.440.

Kumar V, Satyanarayana KV, Itty SS, Indu EP, Giridhar P, Chandrashekar A, Ravishankar GA. *Plant Cell Reports*. 2006; 25: 214–222.

Kumar V, Satyanarayana KV, Ramakrishna A, Chandrashekar A, Ravishankar GA. *Current Science*. 2007; 93 (3): 383–386.

Kumar V, Giridhar P, Chandrashekar A, Ravishankar GA. *Acta Physiologiae Plantarum*. 2008; 30 (2): 217–223. https://doi.org/10.1007/s11738-007-0110-x

Kumar A, Simmi PS, Giridhar P. *Plant Cell Tissue and Organ Culture*. 2018; 134: 369–382 https://doi.org/10.1007/s11240-018-1428-3

Landey RB, Cenci A, Georget F, Bertrand B, Camayo G, Dechamp E, Herrera JC, Santoni S, Lashermes P, Simpson J, Etienne H. *PLoS one* 2013; 8 (2): 1–15.

Landey RB, Cenci A, Guyot R, Bertrand B, Georget F, Dechamp E, Herrera JC, Aribi J, Lashermes P, Etienne H. *Plant Cell Tissue and Organ Culture*. 2015; 122: 517–531. https://doi.org/10.1007/s11240-015-0772-9

Landry MP, Mitter N. *Nature Nanotechnology*. 2019; 14: 512–514. https://doi.org/10.1038/s41565-019-0463-5

Leroy T, Henry AM, Royer M, Altosaar I, Frutos R, Duris D, Philippe R. *Plant Cell Reports*. 2000; 19: 382–389.

Leroy T, Marraccini P, Dufour M, Montagnon C, Lashermes P, Sabau X, Ferreira LP, Jourdan I, Pot D, Andrade AC, Glaszmann JC, Vieira LEG, Piffanelli P. *Theoretical and Applied Genetics*. 2005; 111: 1032–1041.

Mahesh VR, Million-Rousseau P, Ullmann N, Chabrillange J, Bustamante L, Mondolot M, Morant M, Noirot S, Hamon A, De Kochko A, Werck-Reichhart D, Campa C. *Plant Molecular Biology* 2007; 64: 145–159.

Marraccini P, In: FM Cánovas, U Lüttge, MC Risueño, and H Pretzsch (Eds). *Progress in Botany*. Springer, Cham. 2020; 82. https://doi.org/10.1007/124_2020_42

McCormac, A., Elliott M, Chen D. *Molecular Biotechnology*. 1997; 8: 199–213

McCormac A, Elliott M, Chen DF. *Molecular and General Genetics*. 1999; 261**:** 226–235. https://doi.org/10.1007/s004380050961

Menendez-Yuffa A, Dominique BE, Benoit B, Frédéric G, Hervé E. *Plant Cell Tissue and Organ Culture*. 2010; 102: 297–307. https://doi.org/10.1007/s11240-010-9734-4

Meyer FG. *Economic Botany*. 1965; 19: 136–151.

Mishra MK. In Jameel M. Al-Khayri, S. Mohan Jain and Dennis V. Johnson (Eds). *Advances in Plant Breeding strategies: Nut and Beverage Crops*. Springer Publisher, New York. 2019; 475–515. https://doi.org/10.1007/978-3-030-23112-5

Mishra MK, Sreenath HL. *Plant cell Biotechnology and Molecular Biology*. 2003a; 4: 57–62.

Mishra MK, Sreenath HL. *Journal of Plantation Crops*. 2003b; 31: 55–58.

Mishra MK, Sreenath HL. In: *Proceedings of the 19th International Conference on Coffee Science (ASIC)*, ASIC, Bangalore, India. 2004b; 792–796.

Mishra MK, Sreenath HL, Srinivasan CS In: K Sreedharan, PKV Kumar, and BM Jayarama Chulaki (Eds). *Proceedings of the 15th Plantation Crops Symposium (PLACROSYM)*, PLACROSYM, Mysore, India. 2002; 251–255.

Mishra MK, Sreenath HL, Jayarama AC, McCormac, Devi S, Elliott M. In: *Proceedings of the 22nd International Association for Coffee Science (ASIC)*, ASIC, Campinas, Brazil. 2008; 843–850.

Mishra MK, Devi S, McCormac A, Scott N, Chen DF, Elliott M, Slater A. *Biologia*. 2010; 65 (4): 639–646. https://doi.org/10.2478/s11756-010-0078-7

Mishra MK, Tornincasa P, De Nardi B, Asquini E, Dreos R, Del Terra L, Rathinavelu R, Rovelli P, Pallavicini A, Graziosi G. *Journal of Crop Science and Biotechnology*. 2011; 14 (1): 25–37. https://doi.org/10.1007/s12892-010-0035-6

Mishra MK, Slater A. *Biotechnology Research International*. 2012; 1–17. https://doi.org/10.1155/2012/580857

Mishra MK, Nishani S, Gowda M, Padmajyothi D, Suresh N, Sreenath HL, Jayarama. *Plant Breeding and Seed Science*. 2014a; 70 (1): 29–40. https://doi.org/10.1515/plass-2015-0011

Mishra MK, Sreenth HL, Cormac ACM, Elliott MC, Jayarama. *Abstracted in XXXIV Annual Meeting of PTCA (I) and National Symposium on Plant Tissue Culture and Biotechnology for Food and Nutritional Security*. PTCA (I), Mysore, India. 2014b; 57.

Mishra MK, Huded AKC, Jingade P. *Botanica*. 2020; 26 (2): 184–196.

Molina DM, Aponte ME, Cortina H, Moreno G. *Plant Cell Tissue and Organ Culture*. 2002; 71: 117–123. https://doi.org/10.1023/A:1019965621041

Muniswamy B, Sreenath HL. *Journal of Plantation Crops*. 2004; 32: 127–130.

Muniswamy B, Sreenath HL, Samuel SD, Jayarama. *Journal of Plantation Crops*. 2015; 43 (3): 225–230.

Muniswamy B, Kosaraju B, Mishra MK, Raghuramulu Y. *Open Life Science*. 2017; 12: 1–11.

Muniswamy B, Mishra MK, Jingade P, Huded AKC. *Plant Cell Tissue and Organ Culture*. 2019; 139: 91–103.

Murashige T, Skoog F. *Physiologia Plantarum*. 1962; 15: 472–497.

Naidu M, Srinivasan CS. *Journal of Coffee Research*. 1999; 27 (2): 121–126.

Naidu M, Srinivasan CS. In: Kishor PBK (Ed). *Plant Tissue Culture and Biotechnology: Emerging Trends*. 1999; 90–95.

Nakamura T, Sondahl MR. Multiplicacao *in vitro* de genes ortotropicas em *Coffea* spp. In: *Proceedings O Congresso Brasileiro Pesquisa Cafeiras*, Sao Lourenco, Brazil. 1981; 162–163.

Newell CA. *Molecular Biotechnology*.2000; 16: 53–65. https://doi.org/10.1385/MB:16:1:53

Nic-Can GI, López-Torres A, Barredo-Pool F, Wrobel K, Loyola-Vargas VM, Rojas-Herrera R. *PLoS ONE*. 2013; 8 (8): e72160. https://doi.org/10.1371/journal.pone.0072160

Noir S, et al. *Plant Pathology*. 2003; 52: 97–103.

Nookaraju A, Agarwal DC. *South African Journal of Botany*. 2012; 78: 302–306. https://doi.org/10.1016/j.sajb.2011.08.009

Nyange NE, Mcnicol RJ. In: *Proceedings of the 15th International Conference on Coffee Science (ASIC)*, ASIC, Montpellier, France. 1993; 738–740.

O'Brien. *Nature Biotechnology*. 1983; 1: 36–38. https://doi.org/10.1038/nbt0383-36

Ogita S, Uefuji H, Morimoto M, Sano H. *Plant Molecular Biology*. 2004; 54: 931–941.

Ogutu C, Fang T, Yan L, Wang L, Huang L, Wang X, Huang L, Wang X, Ma B, Deng X, Owiti A, Nyende A, Han Y. *Tree Genetics and Genomes*. 2016; 12: 54.

Olechno E, Pu ścion-Jakubik A, Socha K, Zujko MEZ *Foods*. 2021; 10 (6): 1328.

Papanastasiou I, et al. *Plant Cell Tissue and Organ Culture*. 2008; 92: 215–225. https://doi.org/10.1007/s11240-007-9326-0

PerthuisB, PradonJL, Montagnon C, Dufour M, Leroy T. *Euphytica*. 2005; 144: 321–329.

Poole R, Kennedy OJ, Roderick P, Fallowfield JA, Hayes PC, Parkes J. *BMJ*. 2018; 359. https://doi.org/10.1136/bmj.k194

Pré M. *Annals of Botany*. 2008; 102: 207–220.

Quintero FOC, Pinto LG, Barsalobres-Cavallari CF, Arcuri MDLC, Pino LE, Peres LEP. *Plant Cell Reports*. 2018; 37 (9): 1257–1268.

Quiroz-Figueroa F, Méndez-Zeel M, Larqué-Saavedra A, Loyola-Vargas V. *Plant Cell Reports*. 2001; 20: 679–684.

Ramakrishna A, *Plant Cell, Tissue and Organ Culture*. 2012; 108 (2): 267–278.

Rani V, Singh KP, Shiran B, Nandy S, Goel S, Devarumath RM. Sreenath HL, Raina SN. *Plant Cell Reports*. 2000; 19 (10): 1013–1020.

Razafinarivo NJ, Guyot R, Davis AP, Couturon E, Hamon S, Crouzillat D, Rigoreau M, Dubreuil-Tranchant C, Poncet V, De Kochko A, Rakotomalala JJ, Hamon P. *Annals of Botany*. 2013; 111: 229–248. https://doi.org/10.1093/aob/mcs283

Rezende JCD, Carvalho CSHD, Pasqua M, Santos ACR, Carvalho SMD. *Ciencia Rural*. 2011; 41: 384–389. https://doi.org/10.1590/S0103-84782011000500031

Ribas AF, Kobayashi AK, Pereira LFP, Vieira LGE. *Brazilian Journal of Plant Physiology*. 2005; 49: 493–497.

Ribas AF, Pereira LFP, Vieira LGE. *Brazilian Journal of Plant Physiology*. 2006; 18 (1): 83–94.

Ribas AF, Dechamp E, Champion A, Bertrand B, Combes MC, Verdeil JL, Lapeyre F, Lashermes P, Etienne H. *BMC Plant Biology*. 2011; 11: 92.

Rojas-Lorz L, Arrieta-Espinoza G, Valdez-Melara M, Pereira LFP, Gatica-Arias A. *BrazilianArchivesofBiologyandTechnology*. 2019; 62. https://doi.org/10.1590/1678–4324–2019180228

Rosillo AG, *Plant Cell Tissue and Organ Culture*. 2003; 74: 45–49.

Samson NP, Campa C, Noirot M, De Kochko A. In: *Proceedings of the 19th International Conference on Coffee Science (ASIC)*, ASIC, Bangalore, India. 2004; 707–713.

Santana N, González ME, Valcárcel M, Canto-Flick A, Hernández MM, Fuentes-Cerda FJ, Barahona F, Mijangos-Cortés, Loyola-Vargas VM. *In Vitro Cellular & Developmental Biology* – Plant. 2004; 40: 95–101. doi: 10.1079/IVP2003486

Santana-Buzzy N, Rojas-Herrera R, Galaz-Ávalos RM, Ku-Cauich JR, Mijangos-Cortés J, Gutiérrez-Pacheco LC, Canto A, Quiroz-Figueroa F, Loyola-Vargas VM. *In Vitro Cellular & Developmental Biology* – Plant. 2007; 43: 507–520. https://doi.org/10.1007/s11627-007-9074-1

Santa Ram A, Muniswamy B, Sreenath HL, Srinivasan CS. *Journal of Coffee Research*. 2002; 30 (1): 1–7.

Satyanarayana KV, Vinod Kumar, Chandrashekar A, Ravishankar GA. *Journal of Biotechnology*. 2005; 119: 20–25.

Scalabrin S, Toniutti L, Di Gaspero G, Scaglione D, Magris G, Vidotto M, et al. *Scientific Reports*. 2020; 10: 4642.

Sharma et al. *Physiol Mol Biol Plants*. 2019; 25 (2): 581–588.

Silva AT, Barduche D, Gabriella K do Livramento, Paiva LV. *In Vitro Cell. Dev. Biol.-Plant*. 2015; 1: 93–101.

Silva A, Paiva LV, Andrade AC Barduche D. *Genet. Mol. Res*. 2013; 12: 1698–1709.

Simkin AJ, Qian T, Caillet V, Michoux F, Ben Amor M, Lin C, Tanksley S, McCarthy J. *Journal of Plant Physiology*. 2006; 163: 691–708.

Simões-costa MC, Carapuça E, Moura IR. *Acta Horticulturae* 2009; 812: 295–300. https://doi.org/10.17660/ActaHortic.2009.812.40

Solórzano RGL, De Bellis F, Leroy T, Plaza L, Guerrero H, Subia C, Calderón D, Fernández F, Garzón I, Lopez D, Vera D. *Scientific World Journal*. 2017; 1–12. https://doi.org/10.1155/2017/1248954

Sondahl MR, In: A Fujiwara (Ed). *Plant Tissue Culture*. Maruzen, Tokyo. 1982; 417–418.

Simkin AJ, Qian T, Caillet V, Michoux F, Ben Amor M, Lin C, Tanksley S, McCarthy J. In: *IIISIBAC Symposium*, Londrina, Brazil. 1999.

Sondahl MR., Sondahl CN, Gonclaves W. *Coffee Biotechnology and Quality*. Kluwer Academic Publishers, Amsterdam, Netherlands. 2000; 143–150

Spiral J, *Comptes Rendus de l' Academie des Sciences de Paris*. 1993; 316: 1–6.

Spiral J, Pétiard V. In: *Proceedings of the 14th International Conference on Coffee Science (ASIC) Colloque Scientifique*, ASIC, San Francisco. 1991; 383–390.

Sridevi V, Giridhar P. *Plant Cell Tiss Organ Cult*. 2010; 10 (3): 339–347.

Sridevi V, Giridhar. *J. Plant Biochem. Biotechnol*. 2014 23: 112–118 (2014). https://doi.org/10.1007/s13562-012-0186-2

Staritsky G. *ActaBotanica Neerlandica* 1970; 19: 509–514.

Sugiyama M, Matsuoka C, Takagi T. In: *Transformation of coffee with Agrobacterium rhizogenes. Proceeding of the 16th International Conference on Coffee Science (ASIC)*, ASIC, Kyoto, Japan. 1995; 853–859.

Tang K, Tinjuangjun P, Xu Y, Sun X, Gatehouse JA, Ronald PC, Qi H, Lu X, Christou P, Kohli A. *Planta*. 1999; 208: 552–563.

Teisson C, Alvard D. 1995. *Current Issues in Plant. Molecular and Cellular Biology,* Terzi M, Cella R, Falavigna A, (eds) pp. 105–110. Kluwer Academic Publishers, Netherlands.

Tornincasa P, Dreos R, De Nardi B, Asquini E, Devasia J, Mishra MK, In: Proceeding of the *21st International Conference on Coffee Science*, ASIC, Montpellier, France. 2007; 778–785.

Traore SM, Zhao B. *Plant Methods* 2011; 7:42.

Uefuji H, Ogita S, Yamaguchi Y, Koizumi N, Sano H. *Plant Physiology*. 2003; 132: 372–380.

Valencia-Lozano E, Cabrera-Ponce JL, Gómez-Lim MA, Ibarra JE. *Int. J. Mol. Sci*. 2019; 20: 5334.

Van Boxtel J, Eskes A, Berthouly M. *In Vitro Cellular & Developmental Biology – Plant*. 1997; 33: 6–12. https://doi.org/10.1007/s11627-997-0033-7

Van Boxtel J, et al. *Plant Cell Rep*. 1995; 14: 748–752. https://doi.org/10.1007/BF00232915

Van der Vossen HAM. In: C Clifford, J Willson (Eds). *Coffee: Botany, Biochemistry and Production of Beans and Beverage*, Croom Herm, London. 1985; 48–96.

Van der Vossen HAM, *Euphytica*. 2015; 204 (2). https://doi.org/10.1007/s10681-015-1398-z.

Vargas-Guevara C, *3 Biotech*. 2018; 8: 471. https://doi.org/10.1007/s13205-018-1495-5

Vasil V, Castillo AM, Fromm ME, Vasil IK. *Nature Biotechnology*. 1992; 10: 667–674.

Wan Y, Lemaux PG. *Plant Physiology*. 1994; 104: 37–48.

Wang YC; Lin M, Huang B, Chung HH, Chen JT. *Acta Biologica CracoviensiaSeries Botanica*. 2018; 60 (2): 35–44. https://doi.org/10.1515/abcsb-2015

Wierzejska R. *Arch. Med. Sci*. 2017; 13 (3): 507–514. https://doi.org/10.5114/aoms.2016.63599.

Yang X, Zhang X. *Critical Reviews in Plant Sciences*. 2010; 29: 36–57. https://doi.org/10.1080/07352680903436291

Zeevi V, Liang Z, Arieli U, Tzfira T. *Plant Physiology*. 2012; 158 (1): 132–144.

Section V

Coffee consumption and human health

19

Coffee methylxanthines and human health effects

Monika Przeor and Monika Beszterda

CONTENTS

19.1 Introduction

Coffee is the most popular hot beverage in the world, and it is an integral part of our culture. In recent years, the European Union (EU) imported almost three million tons of coffee beans, about 5% more than ten years ago (International Coffee Organization, 2018). Although there are over 90 species in the *Coffea* genus, *C. arabica* and *C. canephora* are the only commercially important species (Farah and dos Santos, 2015). More than 70 countries cultivate the plant, but Brazil, Colombia, Ethiopia, and India are the leading producers.

Coffee beans contain various different bioactive phytochemicals, most notably 0.8%–2.5% caffeine but also another methylxanthines such as theophylline and theobromine, chlorogenic acids, diterpenes, tocopherols, hydroxycinnamic acids, flavonols, and more than 900 volatile compounds mainly formed during the roasting process. Interestingly, despite the fact that some reports caution against excessive coffee and caffeine consumption, research on moderate use

DOI: 10.1201/9781003043133-24

predominantly reports health-promoting effects of coffee use and a correlation of reduced mortality (Schuster and Mitchell, 2019). Caffeine has mild analgesic properties and is contained in over-the-counter medications used to manage migraines. Bitter polyphenols in coffee demonstrates sympathomimetic properties, which means that they can activate sympathetic nervous system and can increase lipolysis and thermogenesis, reduce food intake, and in this way boost energy expenditure (Westerterp-Plantenga et al., 2006; Mikołajczyk-Stecyna et al., 2020). Other components are immune active substances and antioxidants and impart several health benefits such as protecting the body from the hazardous effects of free radicals. Their effectiveness against diabetes mellitus, chronic respiratory disease, and cardiovascular disparities is well understood (Dorsey and Jones, 2017).

19.2 Methylxanthines as specific compounds found in coffee

Methylxanthines are secondary plant metabolites formed from purine and pyridine nucleotides and have evolved in nearly 100 species in 13 orders of the plant kingdom due to their pesticidal properties, being toxic for insects and many other potentially harmful pathogens (Ashihara et al., 2011). This term applies to a class of psychoactive substances that includes caffeine, theobromine, theophylline, and paraxanthine. Methylxanthines are a component of several common dietary products, such as coffee, tea, yerba mate, "energy" drinks, and chocolate. Caffeine is the most popular and consumed substance among the methylxanthines, hence most of the research on these substances focuses on caffeine (Oñatibia-Astibia et al., 2016). Despite the fact that methylxanthines are considered as safe substances, they can interact with and affect other drugs. These interactions can include drug absorption, metabolism, and clearance, but pharmacodynamic interactions between methylxanthines and other drugs have also been noted (Costa et al., 2016).

19.2.1 Main methylxanthines and pyridine alkaloid

19.2.1.1 Caffeine (1,3,7-trimethylxanthine)

The most well-known methylxanthine is caffeine (1,3,7-trimethylxanthine), which was isolated from tea and coffee in the early 1820s (Ashihara et al., 2011). The primary biosynthetic pathway is a sequence consisting of, respectively, xanthosine, 7-methylxanthosine, 7-methylxanthine, theobromine, and caffeine, where xanthosine is produced as a catabolite of purine nucleotides (Ashihara, 2015).

Caffeine is a purine derivative (Figure 19.1) with molecular weight of 194.19 g mol^{-1}. It is a white powder with a bitter taste and 21.6 g L^{-1} of water solubility. The oral LD_{50} determined for caffeine is at the level of 192 mg kg^{-1} body weight in adults and 320 mg kg^{-1} body weight in children (Dorsey and Jones, 2017). The caffeine content of seeds varies from 0.4% to 2.4% dry weight. However, the average caffeine content in commercially available coffee, which depends on the coffee plant varietals and the preparation method, ranges from 2 mg/100 mL in decaffeinated preparations to 57 mg/100 mL in a cup

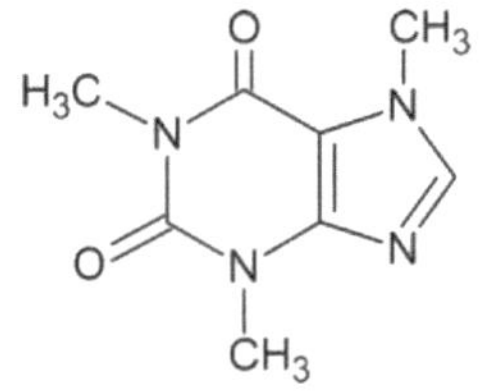

FIGURE 19.1 Chemical structure of caffeine.

made from toasted and ground coffee (Fujioka and Shibamoto, 2007; Oñatibia-Astibia et al. 2016). Arabica beans are reported to have 10–12 mg caffeine/g coffee bean (1.2%–1.4%), while Robusta contain 19–21 mg caffeine/g coffee bean (1.2%–3.3%; Fox et al., 2013; Dorsey and Jones, 2017).

In 1984, caffeine was attached to the Prohibited List of the World Anti-Doping Agency. Since 2004, caffeine has not been considered a prohibited substance; notwithstanding, it is still part of the World Anti-Doping monitoring program (Franke and Bagusat, 2015).

19.2.1.2 Trigonelline (N-methyl nicotinic acid)

The pyridine alkaloid trigonelline (Figure 19.2) is a niacin derivative found in coffee beans, at 1%–3% of dry weight. It has a molecular weight of 137.1360 g mol^{-1}. Both caffeine and trigonelline are accumulated in green coffee beans, but the first one is more well known among the public. Both are N-methyl compounds: trigonelline is N-methylnicotinic acid, and caffeine is an N-methylxanthine. Moreover, the methyl groups are both supplied by S-adenosyl L-methionine (Ashihara, 2015). However, the content of trigonelline in Arabica green coffee beans is higher than that in Robusta, hence it can be applied as a marker compound to distinguish the coffee bean species (Wei and Tanokura, 2015). The trigonelline biosynthetic pathway is closely linked to pyridine nucleotide synthesis de novo and the cycle of pyridine nucleotide. Trigonelline synthase, which catalyzes the methylation of nicotinic acid, is the key enzyme in trigonelline synthesis. Trigonelline shows different physiological activities in plants; for example, it regulates the cell cycle process and leaf movements, and acts as an osmolyte with accumulation during biotic and abiotic stress conditions (Rajasekaran et al. 2001; Mohamadi et al. 2018).

During roasting, trigonelline is converted to nicotinic acid (niacin) and serves as a source of vitamin B_3 (Trugo, 2003). It has been demonstrated that coffee roasting increases the niacin concentration of coffee beans by up to 1000% (Mazzafera, 1990). Trigonelline also results in the presence of 1-methylpyridinium and 1,2-dimethylpyridinium in proportion to the extent of roasting (Ashihara, 2015). It also contributes to

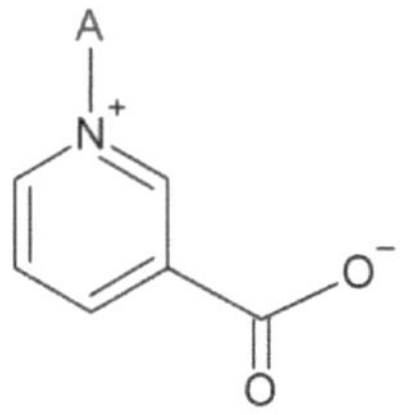

FIGURE 19.2 Chemical structure of trigonelline.

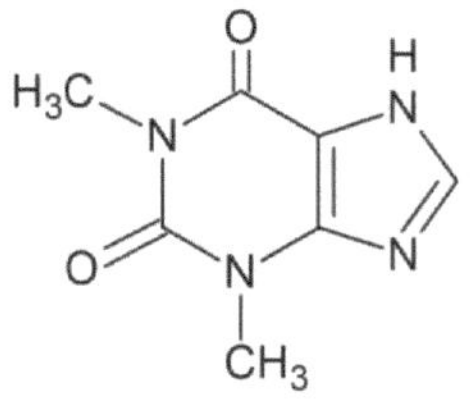

FIGURE 19.3 Chemical structure of theophylline.

the bitter taste of coffee and serves as the precursor of many volatile coffee roasting products (Dorsey and Jones, 2017).

19.2.1.3 Theophylline (1,3-dimethylxanthine)

Catabolism of caffeine starts with its conversion to theophylline (di-methylxanthine) (Figure 19.3) catalyzed by N7-demethylase. Only slight amounts of caffeine are very slowly degraded with the removal of the three methyl groups, causing the formation of xanthine in almost all caffeine-forming plant species (Ashihara et al., 2008; Ashihara et al., 2011). Simultaneously, theophylline is a methylxanthine with a long history of use as a bronchodilator. It occurs naturally in coffee and also tea, from which it may be extracted (Dasgupta and Krasowski, 2020).

19.2.2 Methylxanthine metabolism

Gastrointestinal absorption of a moderate dose (50–250 mg) of caffeine is fast (about 45 minutes) and complete, reaching almost 100% of bioavailability (Blanchard and Sawers, 1983). Once it passes the bloodstream, where caffeine binds to albumin, it is distributed in all tissues by simple diffusion or carrier-mediated transport. Caffeine metabolism is hepatic, and the cytochrome P450 enzyme CYP1A2 metabolizes over 90% of this substance, which is the responsible for the between-subject variability of caffeine plasma levels (Schuster and Mitchell, 2019). The half-life of caffeine is approximately 2.5–5.0 hours, but it can be higher during pregnancy, in infants and neonates, and in patients with hepatic diseases. The major products of the first steps in caffeine metabolism are paraxanthine (78%), theobromine (14%), theophylline (9%), and the final product, methyluric acid, are all excreted via urine (Oñatibia-Astibia et al., 2016). The subsequent caffeine metabolites in humans, 1-methylxantine and 1-methylurate, are considered to be potent antioxidants in vitro and in vivo (Gomez-Ruiz et al., 2007). These compounds also exhibited strong antioxidant properties, greater than that of caffeine (Dorsey and Jones, 2017).

In contrast to caffeine, theobromine is absorbed more slowly and it reaches peak plasma concentrations after 2.5 hours (for caffeine – approximately 30 minutes following the intake of caffeine; Fredholm, 2011). Interestingly, theobromine ingested as or with chocolate would take only 2 hours to reach peak plasma concentrations, while administration of the isolated substance would require about 3 hours. The half-life of theobromine is approximately 7–12 hours. Similar to caffeine, theobromine is distributed throughout the total body water, however caffeine penetrates the blood-brain barrier more readily than theobromine. Moreover, theobromine shows a two- to threefold lower A1 and a tenfold lower A2 adenosine receptor affinity than caffeine (Baggott et al., 2013). In contrast to caffeine, the nonselective phosphodiesterase inhibition effect of theobromine probably is stronger; for instance, moderate doses of theobromine produce vasodilation and bronchodilation in humans (Fredholm, 2011).

19.2.3 Physiological effects of methylxanthines

Four different cellular mechanisms of action have been characterized for methylxanthines: mobilization of intracellular calcium (Ca^{2+}) from the endoplasmic reticulum in both skeletal and cardiac muscle; modulation of $GABA_A$ receptors and antagonism of adenosine receptors (A1, A2A, A2B, and A3); and inhibition of phosphodiesterases (PDEs; Oñatibia-Astibia et al., 2016).

Adenosine is both a messenger molecule that has hormone-like action in the periphery and acts as a strong neuroregulator in the central nervous system (CNS) and also an intermediate metabolite (Martínez-Pinilla et al., 2015). Because of structural similarities with purine nucleosides, caffeine, theophylline, and theobromine can act as competitive inhibitors of adenosine receptors what is responsible for many of the CNS effects (Oñatibia-Astibia et al., 2016). However, theobromine has two- and threefold lower affinity compared to caffeine for the A1 and A2A receptors and is apparently less efficacious as a phosphodiesterase inhibitor (Baggott et al., 2013). In turn, theophylline inhibits the activity of phosphodiesterase and increase the intracellular level of cyclic adenosine monophosphate (cAMP). Theophylline has also been reported to increase the length and complexity of the dendritic process in melanocytes (Huang et al., 2020).

19.3 Coffee consumption and health

19.3.1 Point of tolerance

Coffee consumption affects various aspects of the human body and health, mainly due to the interaction of a number of important active substances showing a wide spectrum of action in this kind of beverage. For coffee, there was found a tolerance effect. Both experimental and clinical evidence confirmed that previous regular coffee exposure reduce future body reaction. Having this in mind, the effects observed after repetitive dosages of coffee differ from those observed at the first consumption. Tolerance for coffee intake was observed in some previous studies. Research performed that tolerance for heart rate and blood pressure, as well as renin and catecholamine levels, was achieved after 4 days of administration (Robertson et al., 1981). Similar tolerance was observed for decaffeinated coffee after 5 days (Ammon et al., 1983). However, posterior studies presented this ability to tolerance as more heterogeneous. The pharmacokinetics of caffeine in coffee have been shown to be responsible for a different level of tolerance in various consumers: only some of them got tolerance after consequent coffee intake (Lovallo et al., 2004; Smits et al., 1985b).

Based on research comparing monozygotic and dizygotic twins, it was suggested that genetic background is responsible for these differences (Kendler and Prescott, 1999). Other hypotheses suggested that the tolerance effect does not affect all possible biological effects after coffee consumption. In the studies, the increase in activity of muscle sympathetic nervous was observed as an effect of intravenous coffee (caffeinated and decaffeinated) or caffeine alone (250 mg) administration. Moreover, intravenous caffeine resulted in a slight blood pressure increase. However, in the group of regular coffee drinkers, coffee intake did not increase blood pressure (Corti et al., 2002). Therefore, it can be suspected that such tolerance differs for drinkers with different levels of sensitivity to coffee, and the regularity of coffee consumption also influences it.

19.3.2 Cardiovascular system

The impact of coffee consumption on the cardiovascular system in many areas is related to the influence on the arterial walls. It was investigated if the consumption of caffeinated and decaffeinated coffee affects endothelial function in a healthy human. Flow-mediated dilation decreased after caffeinated espresso consumption and increased after decaffeinated espresso intake (dose-dependent effect; Buscemi et al., 2009; Buscemi et al., 2010). Chlorogenic acid mediation in coffee was proposed as a hypothesis. Another randomized study detected similar results lasting for a maximum of 60 minutes – reduction in flow-mediated dilation as a result of intake of coffee containing 80 mg of caffeine, and no reduction after decaffeinated coffee intake (Papamichael et al., 2005). Investigators suggest that hydroxyhydroquinone, generated by the roasting of coffee beans, is responsible for further modulation by interfering with protective function promoted by chlorogenic acid at the vascular endothelium (Ochiai et al., 2009). The effect of coffee consumption on large arteries was observed as well. It was found that non-regular caffeinated coffee intake changed parameters of arterial stiffness: the arterial wave reflection at the aorta and the carotid-femoral pulse wave increased (Karatzis et al., 2005; Mahmud and Feely, 2001). Consumption of Ethiopian-origin Arabica coffee resulted in higher free fatty acids in serum, as well as high-density lipoprotein (HDL) cholesterol levels, in contrast to triacylglycerides in habitual drinkers. These results demonstrated a possible protective effect in the incidence of cardiovascular disease (Gebeyehu et al., 2020).

In case of the impact of coffee consumption on the cardiovascular system, caffeine is indicated as a main bioactive compound with specific action. The main directional pharmacological action of caffeine from typically drunk coffee is the antagonism of the A1 and A2A adenosine receptors (Riksen, 2011). As studies showed, the dose of 2–5 cups of coffee (300 mg of caffeine) increases the endothelial-mediated response after the use of acetylcholine and did not affect flow-mediated dilation (Umemura et al., 2006).

19.3.3 Blood pressure

The pressor effect of coffee is attributed to the caffeine. There are some studies showing that the response in human body after the ingestion of coffee can be mitigated. Chlorogenic acid, as an antioxidant improving endothelial function and limiting blood pressure increase, can be assigned with it (Zhao et al., 2012). Decaffeinated coffee had a mitigated pressor effect in one study (Smits et al., 1985a) and a similar pressor effect as caffeinated coffee in another study (Corti et al., 2002). Previously mentioned, possible interference of hydroxyhydroquinone with chlorogenic acid may partially explain these differences (Ochiai et al., 2009). In case of blood pressure, a tolerance effect was observed in habitual drinkers, despite similar to changes in sympathetic activity among nonhabitual drinkers (Corti et al., 2002). Moreover, the effect of coffee consumption on blood pressure in hypertension subjects was defined as neutral (Mesas et al. 2011). In the group of pregnant women drinking coffee, the effect of high pressure or pregnancy-induced hypertension were not achieved. However, in this group a tendency to higher birth weight, followed by coffee consumption (≥300 mg per day) was observed (van der Hoeven et al., 2017). It is also not well defined, as much caffeine level in peripheral blood after coffee consumption is responsible for the pressure effect (Smits et al., 1985b). Nonetheless, slight increase in blood pressure in non-regular coffee drinkers, at the level of 10–20 mm Hg after 1 hour, was observed years ago (Robertson et al., 1978). The reason of this seemed to be increased systemic vascular resistance, caused by increase of renin and catecholamines, combined with the action of caffeine (Hartley et al., 2004). Despite the fact that in hypertension subjects the elevation in blood pressure after a dose of 200–300 mg of caffeine was slight (Mesas et al., 2011), this kind of tolerance should be considered an individual feature (Farag et al., 2005). In line with that finding, other investigators have proposed genetic traits as the main modifying agents (polymorphisms in the α-(2)-adrenergic receptors and adenosine A2A receptors; Renda et al., 2012).

19.3.4 Lipid metabolism

The impact of coffee consumption on lipid metabolism can be considered in two ways. The first one is associated with the presence of kahweol and cafestol, diterpenes responsible for the increase in cholesterol level, in unfiltered coffee beverage (Jansen et al., 1995). The second one is about the possible antioxidant activity of polyphenols in reducing lipid oxidation.

The consumption of unfiltered coffee resulted in significant changes in lipid profile – increase in total cholesterol, triglycerides, and low-density lipoprotein (LDL) cholesterol (Cai et al., 2012). That effect in lipid metabolism was not limited only to habitual coffee drinkers but also to those drinking coffee after a 1-month abstinence. Importantly, this experiment showed a decrease in the ratio of LDL:HDL cholesterol by 8%, and apolipoprotein B:apolipoprotein A-I by 9%. In the light of these parameters, there can be concluded that great, protective balance was achieved between LDL and HDL cholesterol fractions, as well as the balance between related apolipoproteins B and A-I (Kempf et al., 2010). However, this effect vanished after a 3-week abstinence from coffee (Strandhagen and Thelle, 2003).

In the observational studies (Marventano et al., 2016), coffee consumption improved various parameters of metabolic syndrome in normal and hypercholesterolemia subjects. An 8-week ingestion of standardized coffee blends (three cups per day) resulted in a reduction of levels of triglycerides, leptin, and plasminogen activator inhibitor-1 (PAI-1), as well as glucose concentration and insulin resistance, greater in a group of hypercholesterolemic patients. In Japan, daily consumption at least one cup of coffee resulted in lower visceral obesity, lower visceral adipose tissue, and lower metabolic syndrome than green tea (Koyama et al., 2020).

19.3.5 Homocysteine

Studies showed an increase in plasma homocysteine level by 10%–20% after intake of 1000 mL unfiltered coffee per day for 2–4 weeks (Grubben et al., 2000; Urgert et al., 2000). It is caused by coffee or caffeine, independently (Verhoef et al., 2002). Folic acid is proposed to be a treatment agent controlling the reduction of high homocysteine level induced by coffee consumption (Strandhagen et al., 2003). The high level of plasma homocysteine is associated with increased risk of cardiovascular diseases (Clarke et al., 2002; Casas et al., 2005) and can result from genetic defects or nutritional insufficiency. However, reduction of homocysteine level by using vitamin B_{12} or folic acid did not change cardiovascular risk (Armitage et al., 2010).

19.3.6 Carbohydrate metabolism

Non-habitual coffee consumption has a detrimental effect on blood glucose tolerance, as well as insulin sensitivity (Beaudoin and Graham, 2011). In this case, it is similar to the pattern of caffeine observed in the studies. From the reason of impaired glucose disposal in the skeletal muscles, caffeine from coffee increases insulin resistance by 15%–30% (Beaudoin and Graham, 2011). Habitual coffee drinkers who received caffeine in a dose of 200 mg for 7 days had insulin sensitivity significantly impaired, without altering glucose levels (MacKenzie et al., 2007).

In the studies, it was detected that the areas under the curves in the oral glucose tolerance test (OGTT) and in the insulin test were lower in non-caffeine subjects after drinking decaffeinated coffee, in comparison to drinking caffeinated coffee. Moreover, decaffeinated coffee intake was followed by half lower glucose response compared with placebo (Battram et al., 2006). The decaffeinated variant of coffee should not be overlooked when considering the health effects of this drink. It turns out that decaffeinated coffee has an antagonistic effect on glucose absorption in human intestines (Johnston et al., 2003). There are no differences between decaffeinated coffee and chlorogenic acid or trigonelline in glucose and insulin areas under the curves in OGTT (Van Dijk et al., 2009). In a group of patients with diabetes mellitus who were also habitual coffee drinkers, caffeinated coffee intake resulted in slightly higher glucose response in OGTT than decaffeinated coffee or even hot water intake (Krebs et al., 2012). The reason for higher postprandial glucose level in diabetes patients was caffeine (Lane et al., 2008). In a similar group of patients, glucose and insulin responses significantly increased after enriching decaffeinated coffee with 250 mg of caffeine in mixed-meal tolerance test. For clear decaffeinated coffee, the mentioned effect was not detected (Lane et al., 2007). With this information in mind, it is possible that the development of total or even lower-magnitude coffee tolerance (caffeinated or decaffeinated), instead of tolerance of protective compounds (e.g., polyphenols), results in the protective effect of coffee consumption on carbohydrates metabolism. Protective balance after long-term consumption was associated with the impact on modulators of carbohydrate metabolism.

Korean observational studies indicated there was inverse correlation between coffee consumption (mean 2 teaspoons of powdered coffee per day in women and 2.2–2.7 teaspoons in men) and prevalence of type 2 diabetes mellitus (T2DM; Lim et al., 2019).

Quantitative markers are helpful in determining the development of diseases related to carbohydrate metabolism. In T2DM development, the molecule is interleukin-18 (IL-18), with predictive potential (Thorand et al., 2005). In oxidative stress, 8-isoprostane is recognized as such a marker (Milne et al., 2011), and in the insulin-sensitizing ability, adiponectin (Kishida et al., 2012). High doses of coffee (eight cups per day) caused an increase in adiponectin level and decrease in 8-isoprostane and IL-18 levels (Kempf et al., 2010). In the group of habitual drinkers, five cups of caffeinated coffee per day for 8 weeks resulted in adiponectin level increase, comparing with those who were drinking decaffeinated coffee or any coffee. Additionally, decaffeinated coffee consumption was followed by a decrease in the concentration of fetuin-A. Fetuin-A is known as a hepatic glycoprotein marker of T2DM (Wedick et al., 2011).

In case of carbohydrates metabolism, the same interest as on the blood pressure problem has arisen – whether polyphenols or any other compounds can modulate the mentioned metabolism. In addition, having properties of antioxidants in mind, it should be considered if the activity of these compounds results probably as protective.

On the other hand, epidemiological studies suggest that there exists a favorable balance of regular consumption of coffee on insulin sensitivity. In the cross-sectional study of 954 subjects who declared regular coffee consumption, this hypothesis was proposed (Loopstra-Masters et al., 2011). In agreement with the above are (1) a prospective study conducted for over 6 years, in which it was observed that regular coffee ingestion favored post-load glucose rather than fasting glucose (Van Dam et al., 2004); and (2) Finnish and Japanese studies that found that after coffee consumption, fasting glucose, fasting insulin, and blood glucose response after 2 hours in OGTT were inversely associated (Yamaji et al., 2004; Bidel et al., 2006).

From another point of view, in a case-control study (Goto et al., 2011), average levels of sex hormone-binding globulin, an agent inversely associated with insulin resistance and developing T2DM (Ding et al., 2009), were higher in a group of women drinking at least 4 cups of coffee than in a group of non-drinkers.

19.3.7 Oxidative stress

Due to the wide range of bioactive compounds like phenolic compounds, caffeine, trigonelline, diterpenes, and soluble fiber, coffee highlights a great antioxidant activity. This antioxidant-rich product has been suggested to be even the best source of dietary compounds with such activity. The antioxidants presented in the human diet counteract the action of reactive oxygen species (ROS) responsible for the development of oxidative stress in cells. Stress occurs when oxidant molecules produced in the cells exceed the number of antioxidants able to inactivate them, to scavenge free radicals (Martini et al., 2016). Consumption of coffee in a dose of 200 mL per day increased antioxidant capacity in the plasma of healthy subjects (Moura-Nunes et al., 2009), as well as thiol groups (SH), even higher than after tea consumption (Natella et al., 2002). On the other hand, plasma antioxidant status increased after a single dose of 400 mL of coffee, providing 420–780 mg of chlorogenic acid, but only in short-term ingestion (Agudelo-Ochoa et al., 2016). Considering the type of coffee, both medium-light roast and medium roast filtered coffee, in a dose of 150 mL per day for 4 weeks, resulted in higher total antioxidant status (about 20%–25%), and in the oxygen radical absorbance capacity (ORAC) test (only for medium light roast; Corrêa et al., 2012). The most significant markers of antioxidant stress in an organism are specific enzymes: superoxide dismutase (SOD), glutathione peroxidase (GPx), catalase (CAT), glutathione S-transferases (GSTs), and glutathione reductase (GSR). Studies are not homogenous in this case. Some indicated that 4-week consumption of light dark roasted coffee caused an increase in SOD, GPx, and CAT activities in contrast to dark roasted coffee. This effect was probably correlated with chlorogenic acid content (Kotyczka et al., 2011). On the contrary, 5-day coffee consumption of 800 mL of coffee per day had no impact on GPx and SOD activity of the lymphocytes (Mišík et al., 2010). GSR blood level increased in a 4-week consumption of brewed coffee (750 mL per day; Bakuradze et al., 2011). Moreover, intake of 1000 mL of coffee per day in a 5-day intervention resulted in the increase in plasma activity of GST (Steinkellner et al., 2005).

19.3.8 Immune modulator

Many results show that caffeine is an immune active compound with therapeutically accessible dosages to the average coffee consumer. Caffeine may have a pronounced effect on cAMP, which is a strong immune modulator with the capacity to suppress immunocompetent and inflammatory and cells (Dorsey and Jones, 2017; Shirvani et al., 2020). According to a review by Horrigan et al. (2006), a large number of studies have indicated that caffeine has immune modulating effects on innate and adaptive immunity, natural killer cell function, cytokine production, free radical production, lymphocyte proliferation antibody production, histamine release, leukocyte counts, immune organ weights, leukocyte chemotaxis, myeloperoxidase production, immune cell apoptosis, and in vivo hypersensitivity reactions (Dorsey and Jones, 2017).

19.3.9 Respiratory tract diseases

19.3.9.1 Apnea of prematurity

Apnea of prematurity (AOP) is a common complication of premature neonates that affects upward of 80% of neonates with a birth weight less than 1000 g. Episodes of AOP and its associated hypoxia, bradycardia, and cyanosis are characterized by a pause in breathing for greater than 20 seconds, accompanied by a heart rate less than 67% from baseline for more than 4 seconds in neonates less than 37 weeks' gestation, and an oxygen saturation (SpO_2) less than 80% for more than 4 seconds (Schoen et al., 2014; Oñatibia-Astibia et al., 2016).

Methylxanthine therapies, including caffeine citrate and theophylline, are the key pharmaceutical agents used in the treatment and prevention of AOP (Theobald et al., 2000; Zhao et al., 2011). As a nonspecific antagonist of adenosine receptors (A1, A2A, A2B, and A3), caffeine may improve minute ventilation, CO_2 sensitivity, contraction of the diaphragmatic muscles, and neural respiratory regulation (Martin et al., 2004). Furthermore, methylxanthines are often prescribed to obviate apneic episodes around the periextubation period and to facilitate weaning off of mechanical ventilation (Aranda et al., 1983). These benefits of methylxanthines in reducing the frequency of apneic episodes in infants were first proven in the late 1970s (Aranda et al., 1977). Caffeine citrate can be administrated orally or intramuscularly with a recommended initial dose of 20 mg kg^{-1}, followed by a daily maintenance dose of 5 mg kg^{-1}, while the standard theophylline regimen involves an oral loading dose of 1.5 mg kg^{-1} and 4 mg kg^{-1} in preterm neonates and neonates, respectively (Arora et al. 2012; Steer et al. 2004), followed by a maintenance dose of 1–3 mg kg^{-1}/12 h (Tracy et al., 2010). It is worth noting that caffeine is the recommended first-line pharmaceutical therapy, as it is associated with less adverse events and has a wider therapeutic window than theophylline (Schoen et al., 2014).

19.3.9.2 Asthma

Asthma is a chronic respiratory disease that usually presents the first symptoms at a young age. The major symptoms are serious coughing periods and wheezing that severely affect the ability to breathe. Although asthma was identified about 2000 years ago, anti-inflammatory drugs were not used until the 1960s (Oñatibia-Astibia et al., 2016). As early as the eighteenth century, physicians considered that strong brewed coffee was effective at improving symptoms in asthmatics. Already in 1860, medical textbooks recommended coffee consumption for the treatment of dyspnea in bronchial asthma (Tilley, 2011).

It is known that caffeine and theobromine correct lung function and induce bronchodilation in asthma patients (Simons et al., 1985; Bara et al. 2001). Theophylline has a bronchodilator action and can reduce inflammation underlying asthma (Pauwels, 1987). Moreover, it also induces a relaxation in airways and of vascular smooth muscle, thus lowering the mean pulmonary arterial pressure (Rabe et al., 1995; Walker et al., 2014). As above, theophylline is one of the most frequent prescribed medications for the treatment of asthma worldwide. Guidelines for the treatment of asthma, however, recommend the use of inhaled corticosteroids and β-agonists as a first-line agent.

19.3.10 Neurological and neurodegenerative disease

Taking into account that methylxanthines have the ability to cross the blood-brain barrier and that the adenosinergic system is omnipresent in the CNS, these compounds are considered to have a marked overall influence on neuronal function (Monteiro et al., 2019). One mechanism is the inhibiting the activity of cyclic nucleotide phosphodiesterase through inhibition of depletion of cAMP. This results in accumulation of cAMP, a second messenger that causes prolonged effects of epinephrine, strengthening its actions. The next effect acts through adenosine; caffeine acts antagonistically on adenosine receptors A1 and A2A with stressed effects in the striatal area (Nehlig et al., 1992; Ferre et al., 2008). The another mechanism of action is the mobilization of intracellular Ca^{2+}. These three directions of action are well known. However, other mechanisms of action are considered, including antagonistic effects on γ-aminobutyric acid receptors (Taketo et al., 2004; Franke and Bagusat, 2015).

As above, it is clear that caffeine has applied as a therapeutic in a large scope of applications. Caffeine has been shown to exhibit significant beneficial influence through the improvement of motor activity for Parkinson's disease patients. These effects were mediated via neurorestoration through tropic proteins (TGF-β, neurturin, bone morphogenic proteins, and glial cell–derived neurotropic factor) and neuroprotective effects (Chen and Chern, 2011; Postuma et al., 2012). Beside, caffeine appears to have apply in Alzheimer's disease prevention and reversal symptoms of Alzheimer's disease through inhibition and reducing generation of amyloid-β plaque formations (Cupino and Zabel, 2013; Gahr et al., 2013; Dorsey and Jones, 2017). Another beneficial effects demonstrated for caffeine in the context of the nervous system include reduced neuronal damage after ischemic insult and neuroprotective effects against spinal cord injury (Rivera-Oliver and Dıaz-Rios, 2014; Rudolphi et al. 1989). In anesthetic practice, caffeine has been applied in the temporary symptomatic relief of post-dural puncture headaches. Reduction in cerebral blood volume and cerebral arterial vasoconstriction are probably responsible for any symptomatic relief (Warner et al., 2018; Premnath et al., 2020). Moreover, applying the methylxanthines, mainly caffeine as an analgesic, especially as an analgesic adjuvant has been reported to significantly increase pain relief when combined with other common analgesics (aspirin, paracetamol, or ibuprofen; Derry et al., 2014).

Other than the aforementioned psychostimulant activity, cognitive-enhancing effects in humans have also been consistently associated with methylxanthines. Main effects of caffeine on cognition are increased vigilance, attention, reaction time, acute and chronic memory effects, executive function, as well as judgment and risk-taking behavior and concentration (McLellan et al., 2016). Taking 150–600 mg of caffeine in normal healthy, unimpaired subjects results in slight reduction of reaction time in simple psychomotor tests through slightly increased attention and vigilance. Moreover, subjective self-evaluation is slightly increased (Franke and Bagusat, 2015). Caffeine has been also assumed as a pertinent candidate for counteracting memory loss, and a significant number of prospective studies indicated protective effects against cognitive decline elicited by frequent coffee or caffeine consumption (Cunha et al., 2008; Arab et al., 2011; Vercambre et al., 2013). In comparison to well-rested individuals, caffeine has substantial procognitive effects in tired or sleep-deprived individuals. Procognitive effects in sleep-deprived subjects on higher cognitive domains like memory, are probably secondary effects mediated via increased concentration, vigilance, and attention (Loke, 1988; Rogers et al., 2008; Ruxton, 2008; Franke and Bagusat, 2015). It has been also proven essential effect of caffeine on vehicle control (both longitudinal and lateral parameters of car driving tasks) in sleep-restricted individuals (Irwin et al., 2020).

Many of the studies have shown that the interaction of some of the bioactive compounds results in greater improvement in neurocognitive function than from caffeine alone. For instance, the combination of L-theanine and caffeine is among the most studied, with notable effects on attention via increases subjective alertness (measured with the Bond-Lader test; Owen et al., 2008; Giesbrecht et al., 2010). Earlier research by Mumford et al. (1994) demonstrated behavioral effects from theobromine, such as an increase in alertness, motivation, and energy but also sleepiness in some subjects. Probably, this exceptional properties of theobromine could lie in its synergistic effect with caffeine (Schuster and Mitchell, 2019). In Table 19.1, the effect of coffee consumption on status of selected diseases is shown.

19.4 Coffee consumption and the risk of diseases

19.4.1 Coronary heart disease

Some case-control studies suggest that coronary risk is associated with coffee consumption (Greenland, 1993; Kawachi et al., 1994). On the other hand, in cohort studies, it was problematic to find the adverse effect of coffee intake on coronary heart disease (CHD). Moreover, in women, a more beneficial effect was observed with moderate consumption of coffee (Wu et al., 2009). Myocardial infarction in history did not increase the risk of CHD disease from coffee consumption (Mukamal et al., 2009). The same was concluded for regular coffee drinkers with previous hypertension (Mesas et al., 2011). Some coffee components have acute effects on the cardiovascular system. Considering the risk of CHD, frequency of coffee intake should be taken into account: in more sensitive patients or in immediate time to coffee consumption, the risk is probably higher. It is higher 1 hour after ingestion in nonhabitual coffee consumers or those from a high-risk group (Baylin et al., 2006).

19.4.2 Stroke

The risk of stroke following coffee consumption has not been investigated well. It is possible that coffee consumption is inversely associated with the risk of stroke, a 10%–20% reduction (Larsson and Orsini, 2011). The ischemic stroke risk increase in the first hour after coffee intake in a group of irregular drinkers (Mostofsky et al., 2010).

19.4.3 Arrhythmia

Higher frequency of ventricular arrhythmia in most experimental models and clinical studies was associated with high

TABLE 19.1
Effect of Each Dose of Coffee on Status of Selected Diseases

	Coffee (CO)/caffeine (CA) consumption per day	↑ or ↓ or ↔	Reference
Cardiovascular system	80 mg CA	↓ flow-mediated dilation	Papamichael et al., 2005
	habitual consumption	↑↑ free fatty acids ↑↑ HDL-cholesterol ↓↓ triacylglycerides	Gebeyehu et al., 2020
	2–5 cups/d	↑↑ endothelial-mediated response	Umemura et al., 2006
Homocysteine	1000 mL d^{-1} for 2–4 weeks	↑↑ plasma homocysteine	Grubben et al., 2000; Urgert et al., 2000
Blood pressure	1 cup/d (unhabitual)	↑ 10–20 mm Hg in 1 h	Robertson et al., 1978
	200–300 mg CA	↑ blood pressure (hypertension patients)	Mesas et al., 2011
Carbohydrate metabolism	200 mg CA/d for 7 days	↓ insulin sensitivity (diabetes)	MacKenzie et al., 2007
	250 mg CA	↑↑ glucose response ↑↑ insulin response (diabetes)	Lane et al., 2007
	habitual consumption	↓↓ fasting glucose ↓↓ fasting insulin ↓↓ blood glucose response	Yamaji et al., 2005; Bidel et al., 2006
T2DM	2 teaspoons CO/d (women) 2.2–2.7 teaspoons CO/d (men)	↓ incidence	Lim et al., 2019
	≥ 4 cups CO/d	↑↑ sex hormone-binding globulin	Goto et al., 2011
	8 cups CO/d	↑↑ adiponectine ↓↓ 8-isopropane	Kempf et al., 2010
	5 cups CO/d for 8 weeks	↑↑ adiponectine	Wedick et al., 2011
Lipid metabolism	3 cups CO/d for 8 weeks	↓↓ triglycerides ↓↓ leptin ↓↓ plasminogen activator inhibitor-1 ↓↓ glucose concentration ↓↓ insulin resistance	Marventano et al., 2016
	≥ 1 cup CO/d	↓↓ visceral obesity ↓↓ visceral adipose tissue ↓↓metabolic syndrome	Koyama et al., 2020
Oxidative stress	200 mL CO/d	↑↑ antioxidant capacity (healthy)	Moura-Nunes et al., 2009
	400 mL CO/d	↑ plasma antioxidant status (short-term)	Agudelo-Ochoa et al., 2016
	150 mL CO/d for 4 weeks	↑↑ antioxidant status ↑ ORAC	Corrêa et al., 2012
	800 mL CO/d for 5 days	↔ GPx ↔ SOD	Mišík et al., 2010
	750 mL d^{-1} for 4 weeks	↑↑ GSR	Bakuradze et al., 2011
	1000 mL d^{-1} for 5 days	↑↑ plasma GST	Steinkellner et al., 2005
Neurological and neurodegenerative disease	150–600 mg CA/d	↓ reaction time (healthy) ↑ self-evaluation (healthy)	Franke and Bagusat, 2015

Notes: ↑↑ = increase; ↑ = slight increase; ↔ = no change; ↓ = slight decrease; ↓↓ = decrease.

dosages of caffeine. Catecholamine seemed to be a mediator in this effect (Pelchovitz and Goldberger, 2011). The direct impact of coffee consumption on the higher risk of occurrence of arrhythmia was not well indicated. Additionally, consumption of coffee was not associated with risk for atrial fibrillation (Shen et al., 2011). Also, the mentioned relation between coffee ingestion and the risk of ventricular arrhythmia was not confirmed in long-term observational studies (Zhang et al., 2011a). In lack of evidence, caution should be imposed in concluding that coffee does not affect the increased risk of arrhythmia, especially in prone subjects.

19.4.4 Heart insufficiency

In case of heart insufficiency, moderate coffee consumption was inversely associated with the risk of heart failure (Mostofsky et al., 2012).

19.4.5 Liver diseases

A fat-, fructose-, and cholesterol-rich diet coupled with intake of decaffeinated coffee (6 g kg^{-1} of body weight) resulted in lower parameters of insulin resistance and non-alcoholic steatohepatitis in mice (Brandt et al., 2019). Regular coffee consumption prevents damage of hepatocytes. This protection was observed in a reduced number of abnormal liver function tests (hepatocellular carcinoma and cirrhosis). The risk of cirrhosis, caused by tissue remodeling after hepatitis inflammation, also decreased (Masterton and Hayes, 2010). Moreover, coffee consumption reduced levels of gamma-glutamyltransferase (GGT; Nakanishi et al., 2000), aspartate aminotransferase (AST; Jang et al., 2012), and alanine aminotransferase (ALN; Ruhl and Everhart, 2005). There are some suggestions for coffee as a protective agent of the liver, but the mechanism is unclear. Caffeine, in the range of the usually consumed amount of coffee, is known as an antagonist of A2 adenosine receptors. This inhibition in immune cells is followed by exacerbating acute inflammation in the liver (Ohta and Sitkovsky, 2011). However, some kind of protection in the effect of coffee consumption is regularly observed in clinical studies. Having this in mind, again, there should be considered a counterbalancing action, detrimental of caffeine and protective of other active compounds.

19.4.6 Diabetes

In case of diabetes mellitus, relation of coffee consumption is repeatedly inverse in many studies. The ingestion of over 6 cups of coffee per day affects the relative risk of T2DM at the range of 0.65 (95% CI), and 4–6 cups of coffee per day at 0.72, in comparison with lower consumption (<2 cups of coffee per day; Van Dam and Hu, 2005). There was also observed a logarithmic-linear correlation in an analyzed group of over 450,000 subjects, with a specific conclusion: the risk of T2DM was reduced by 7% with every additional cup. For decaffeinated coffee, similar results were established (Huxley et al., 2009). Interestingly, in a group of Native Americans, known to have a relatively high incidence of diabetes mellitus, high (>12 cups per day) and regular consumption of coffee reduced the risk of T2DM by 67%. Mentioned cohort studies were conducted over 7 years (Zhang et al., 2011b). The confirmation of the aforementioned relationship was the reduction of risk of diabetes mellitus in the range of 25% as an effect of over 4 cups of coffee intake, which was observed in another prospective study (Fuhrman et al., 2009). It was concluded that to achieve protection against diabetes mellitus, moderate or high consumption of coffee is recommended, and in susceptible populations as well.

19.4.7 Osteoporosis

The debate about the relation between bone fracture and bone density has been going on for years. Coffee increases the acidic load and, probably due to this mechanism, increases urinary calcium output. Bones, being a reservoir of calcium salts, are a kind of buffer for the organism and in acidosis constitute a great physiological counterbalance (Lemann et al., 1966). A few studies found a relation between coffee consumption and fractures of the forearm (Hernandez-Avila et al., 1991) and hip (Kiel et al., 1990), but another study contradicted this conclusion (Cooper et al., 1992). In any case, however, this kind of harm following by drinking over 2–3 cups of coffee per day (Kiel et al., 1990; Hasling et al., 1992), as well as in susceptible patients (e.g., women with low Ca^{2+} intake or elderly people; Cooper et al., 1992). The bone loss (Rapuri et al., 2001), lower density (Korpelainen et al., 2003), fracture (Hallström et al., 2006), as a result of high dosage of coffee in daily diet were observed.

On the other hand, no associations between consumption and bone status or fracture were identified in the Mediterranean region (Johnell et al., 1995; Lloyd et al., 1997), nor any coffee-correlated determinant of bone loss (Sakamoto et al., 2001). Importantly, coffee was not included to the list of fracture risk factors by the World Health Organization.

19.4.8 Cancer

Studies about cancer are methodologically limited. According to the occurrence of cancer, there were some epidemiological studies conducted. In the vast majority, coffee consumption neutrally affected cancers of the bladder (Zhou et al., 2012), colon/rectum (Zhang et al., 2010), stomach (Botelho et al., 2006), pancreas (Turati et al., 2012), prostate (Park et al., 2010), lung (Tang et al., 2010), breast (Tang et al., 2009), and ovary (Braem et al., 2012). Unclear information was presented in the case of bladder cancer – either a positive correlation (Zeegers et al., 2001) or inconclusive balance (Zhou et al., 2012). In pancreatic cancer, coffee intake was identified as a neutral (Turati et al., 2012) or even protective factor (Dong et al., 2011). There are studies indicating liver cancer protection following coffee consumption (Bravi et al., 2007; Leung et al., 2011). Protection against oral cavity–pharyngeal cancer (Hildebrand et al., 2013) and endometrial cancer (Je and Giovannucci, 2012) were also suggested. The effect on esophageal squamous cell carcinoma, esophageal adenocarcinoma, and laryngeal cancer was neutral in contrast (Hildebrand et al., 2013). The mechanism of coffee action against cancer is supposed to be the ability of its caffeine to antagonist potential of the A2 receptors. This leads to a situation where caffeine acts as a pro-inflammatory agent and behaves as factors in immunotherapy in cancer treatment. In this way, alkaloid causes inflammation; a similar approach is used in cancer vaccines.

Due to some discrepancies, further epidemiological studies on the association of cancer with coffee intake are needed.

19.5 Conclusion

Because a cup of coffee accompanies most of us every day, it is worth considering its impact on our health and well-being. Coffee is a complex beverage containing many valuable ingredients.

Caffeine (1,3,7-trimethylxanthine) is often considered to be the main active compound of coffee. Because this ingredient easily crosses intracellular barriers, including blood-brain and placenta, it is easily detectable in body fluids (i.e., urine, saliva, bile, milk, semen, and sweat) and exhibits various biological effects such as stimulatory effects on the CNS, a positive effect on long-term memory, relaxation of human airway smooth muscle, active vasodilator action on the coronary vessels, and even stimulation of thermogenesis and appetite modulation. As stated above, although the most known component of the coffee is caffeine, the balance between caffeine and other active compounds seems to determine its final biological properties. Studies discussing the effect of coffee consumption on the human organism do not clearly determine which effect is more significant – caffeine itself or rather coffee (beverage) containing caffeine and other compounds simultaneously (synergistic effect). Moreover, the consumed dose of coffee should be considered. Detrimental effects have been attributed to typical amounts of caffeine consumption, such as increased nervousness, anxiety, increase in blood pressure, and tachycardia. Yet there is increasing evidence that other constituent ingredients in caffeine-containing drinks (e.g., polyphenols) can antagonize some of the adverse effects of caffeine.

19.6 Acknowledgment

We thank the editor, Magdalena Jeszka-Skowron, for valuable tips in writing the chapter.

REFERENCES

Agudelo-Ochoa G.M., Pulgarín-Zapata I.C., Velásquez-Rodriguez C.M., Duque-Ramírez M., Naranjo-Cano M., Quintero-Ortiz M.M. *J Nutr.* 2016; 146(3): 524–531.

Ammon H., Bieck P., Mandalaz D, Verspohl E. *Br J Clin Pharmacol.* 1983; 15(6): 701–706.

Arab L., Biggs M.L., O'Meara E.S., Longstreth W.T., Crane P.K., Fitzpatrick A.L. *J Alzheimers Dis.* 2011; 27: 553–66.

Aranda J.V., Turmen T., Davis J., Trippenbach T., Grondin D., Zinman R. *J Pediatr.* 1983; 103(6): 975–8.

Aranda J.V., Gorman W., Bergsteinsson H., Gunn T. *J Pediatr.* 1977; 90: 467–72.

Armitage J.M., Bowman L., Clarke R.J., Wallendszus K., Bulbulia R., Rahimi K. *J Am Med Assoc.* 2010; 303(24): 2486–94.

Arora P. *J Neonat Biol.* 2012;1: 1–2.

Ashihara H. *Nat Prod Commun.* 2008; 3(9): 1423–8.

Ashihara H. Coffee in Health and Disease Prevention. Academic Press, 2015.

Ashihara H., Kato M., Crozier A. Methylxanthines. Springer, 2011.

Baggott M.J., Childs E., Hart A.B., de Bruin E., Palmer A.A., Wilkinson J.E. *Psychopharmacol.* 2013; 228(1): 109–18.

Bakuradze T., Boehm N., Janzowski C., Lang R., Hofmann T., Stockis J.P. *Mol Nutr Food Res.* 2011; 55(5): 793–7.

Bara A.I., Barley E.A. *Cochrane Database Syst Rev.* 2001; 4:CD001112. doi: 10.1002/14651858.CD001112.

Battram D.S., Arthur R., Weekes A., Graham T.E. *J Nutr.* 2006; 136(5): 1276–80.

Baylin A., Hernandez-Diaz S., Kabagambe E.K., Siles X., Campos H. *Epidemiol.* 2006; 17(5): 506–11.

Beaudoin M.S., Graham T.E. Methylxanthines. Handbook of Experimental Pharmacology. Springer, 2011.

Bidel S., Hu G., Sundvall J., Kaprio J., Tuomilehto J. *Horm Metab Res.* 2006; 38(1): 38–43.

Blanchard J., Sawers S.J.A. *Eur J Clin Pharmacol.* 1983; 24(1): 93–8.

Botelho F., Lunet N., Barros H. *Revisão.* 2006; 22(5): 889–900.

Braem M.G.M., Onland-Moret N.C., Schouten L.J., Tjønneland A., Hansen L., Dahm C.C. *Am J Clin Nutr.* 2012; 95(5): 1172–81.

Brandt A., Nier A., Jun Jin C., Baumann A., Jung F., Ribas V. *Redox Biol.* 2019; 21: 101092. doi:10.1016/j.redox.2018.101092.

Bravi F., Bosetti C., Tavani A., Bagnardi V., Gallus S., Negri E. *Hepatology.* 2007; 46(2): 430–5.

Buscemi S., Verga S., Batsis J.A., Donatelli M., Tranchina M.R., Belmonte S. *Eur J Clin Nutr.* 2010; 64(5): 483–9.

Buscemi S., Verga S., Batsis J.A., Tranchina M.R., Belmonte S., Mattina A. *Eur J Clin Nutr.* 2009; 63(10): 1200–05.

Cai L., Ma D., Zhang Y., Liu Z., Wang P. *Eur J Clin Nutr.* 2012; 66(8): 872–7.

Casas J.P., Bautista L.E., Smeeth L., Sharma P., Hingorani A.D. *Lancet.* 2005; 365(9455): 224–32.

Chen JF, Chern Y. Methylxanthines. Handbook of Experimental Pharmacology. Springer, 2011.

Clarke R., Collins R., Lewington S., Donald A., Alfthan G., Tuomilehto J. *J Am Med Assoc.* 2002; 288(16): 2015–22.

Cooper C., Atkinson E.J., Wahner H.W., O'Fallon W.M., Riggs B.L., Judd H.L. *J Bone Min Res.* 1992; 7(4): 465–71.

Corrêa T.A.F., Monteiro M.P., Mendes T.M.N., de Oliveira D.M., Rogero M.M., Benites C.I., de Matos Vinagre C.G.C., Mioto B.M., Tarasoutchi D., Tuda V.L., Machado César L.A., da Silva Torres E.A.F. *Plant Foods for Hum Nutr.* 2012; 67(3): 277–82.

Corti R., Binggeli C., Sudano I., Spieker L., Hänseler E., Ruschitzka F., Chaplin W.F., Lüschner T.F., Noll G. *Circulation.* 2002; 106(23): 2935–40.

Costa G., Morelli M., Simola N. *Neuropathology of drug addictions and substance misuse.* Academic Press, 2016.

Cunha R.A. *Med Clin.* 2008; 131: 790–95. doi:10.1016/S0025-7753(08)75506-4.

Cupino T.L., Zabel M.K. *Transl Stroke Res.* 2013; 5(3): 330–37.

Dasgupta A, Krasowski MD. Therapeutic Drug Monitoring Data (Fourth Edition). Academic Press, 2020.

Derry C.J., Derry S., Moore R.A. *Cochrane Database Sys Rev.* 2014; 12: 1–62.

Ding E.L., Song Y., Manson J.E., Hunter D.J., Lee C.C., Rifai N., Buring J.E., Gaziano J.M., Liu S. *N Eng J Med.* 2009; 361(12): 1152–63.

Dong J., Zou J., Yu X.F. *World J Gastroenterol.* 2011; 17(9): 1204–10.

Dorsey B.M., Jones M.A. Handbook of coffee processing by-products: sustainable applications. Academic Press, 2017

Farag N.H., Vincent A.S., McKey B.S., Whitsett T.L., Lovallo W.R. *Am J Cardiol.* 2005; 95(11): 1389–92.

Farah A., dos Santos T.F. *Coffee in health and disease prevention.* Academic Press, 2015.

Ferre S., Ciruela F., Borycz J. *Front Biosci.* 2008; 13: 2391–9.

Fox G.P., Wu A., Yiran L., Force L. *J Agricul Food Chem.* 2013; 61: 10772–78.

Franke A.G., Bagusat C. Coffee in Health and Disease Prevention. Academic Press, 2015.
Fredholm B.B. Methylxanthines. Springer, 2011.
Fuhrman B.J., Smit E., Crespo C.J., Garcia-Palmieri M.R. *Public Health Nutr.* 2009; 12(6): 842–8.
Fujioka K., Shibamoto T. *Food Chem.* 2007; 106: 217–221.
Gahr M., Nowak D.A., Connemann B.J., Schönfeldt-Lecuona C. *Brain Res.* 2013; 26(1519): 19–30.
Gebeyehu G.M., Feleke D.G., Molla M.D., Admasu T.D. *Heliyon.* 2020; 6(9): e04886. doi: 10.1016/j.heliyon.2020.e04886.
Giesbrecht T., Rycroft J.A., Rowson M.J., De Bruin E.A. *Nutr Neurosci.* 2010; 13(6): 283–90.
Gomez-Ruiz J.A., Leake D.S., Ames J.M. *J Agri Food Chem.* 2007; 55: 6962–9.
Goto A., Song Y., Chen B.H., Manson J.E., Buring J.E., Liu S. *Diabetes.* 2011; 60(1): 269–75.
Greenland S. *Epidemiol.* 1993; 4: 366–74.
Grubben M.J., Boers G.H., Blom H.J., Broekhuizen R., De Jong R., Van Rijt L., de Ruijter E., Swinkels D.W., Nagengast F.M., Katan M.B. *Am J Clin Nutr.* 2000; 71(2): 480–4.
Hallström H., Wolk A., Glynn A., Michaëlsson K. *Osteoporosis Int.* 2006; 17(7): 1055–64.
Hartley T.R., Lovallo W.R., Whitsett T.L. *Am J Cardiol.* 2004; 93(8): 1022–6.
Hasling C., Sondergaard K., Charles P., Mosekilde L. *J Nutr.* 1992; 122: 1119–26.
Hernandez-Avila M., Colditz G.A., Stampfer M.J., Rosner B., Speizer F.E., Willett W.C. *Am J Clin Nutr.* 1991; 54(1): 157–63.
Hildebrand J.S., Patel A.V., McCullough M.L., Gaudet M.M., Chen A.Y., Hayes R.B., Gapstur S.M. *Am J Epidemiol.* 2013; 177(1): 50–8.
Horrigan L.A., Kelly J.P., Connor T.J. *Pharmacol Ther.* 2006; 111(3): 877–89.
Huang H.C., Yen H., Lu J.Y., Chang T.M., Hii C.H. *Food Chem Toxicol.* 2020; 137: 111165. doi:10.1016/j.fct.2020.111165.
Huxley R., Lee C.M.Y., Barzi F., Timmermeister L., Czernichow S., Perkovic V., Grobbe D.E., Batty D., Woodward M. *Arch Intern Med.* 2009; 169(22): 2053–63.
International Coffee Organization. 2018. http://www.ico.org/documents/cy2017-18/annual-review-website-e.pdf
Irwin C., Khalesi S., Desbrow B., McCartney D. *Neurosci Biobehav Rev.* 2020; 108: 877–88.
Jang E.S., Jeong S.H., Hwang S.H., Kim H.Y., Ahn S.Y., Lee J., Lee S.H., Park Y.S., Hwang J.H., Kim J-W., Kim N., Lee D.H. *BMC Gastroenterol.* 2012; 12(1): 145. doi: 10.1186/1471-230X-12-145.
Jansen D.F., Nedeljkovic S., Feskens E.J.M., Ostojic M.C., Grujic M.Z., Bloemberg B.P.M., Kromhout D. *Arterioscler Thromb Vasc Biol.* 1995; 15(11): 1793–7.
Je Y., Giovannucci E. *Int J Cancer* 131(7): 1700–10.
Johnell O., Gullberg B., Kanis J.A., Allander E., Elffors L., Dequeker J., Dilsen G., Gennari C., Lopes Vaz A., Lyritis G., Mazzuoli G., Miravet L., Passeri M., Perez Cano R., Rapado A., Ribot C. J *Bone Miner Res.* 1995; 10(11): 1802–15.
Johnston K.L., Clifford M.N., Morgan L.M. *Am J Clin Nutr.* 2003; 78(4): 728–33.
Karatzis E., Papaioannou T.G., Aznaouridis K., Karatzi K., Stamatelopoulos K., Zampelas A. *Int J Cardiol.* 2005; 98(3): 425–30.
Kawachi I., Colditz G.A., Stone C.B. *Heart.* 1994; 72(3): 269–7.
Kempf K., Herder C., Erlund I., Kolb H., Martin S., Carstensen M. *Am J Clin Nutr.* 2010; 91(4): 950–7.
Kendler K.S., Prescott C.A. *Am J Psychiatry.* 1999; 156(2): 223–8.
Kiel D.P., Felson D.T., Hannan M.T., Anderson J.J., Wilson P.W.F. *Am J Epidemiol* 132(4): 675–84.
Kishida K., Funahashi T., Shimomura I. *Endocr Metab Immune Disord - Drug Targets.* 2012; 12(2): 118–31.
Koyama T., Maekawa M., Ozaki E., Kuriyama N., Uehara R. *Nutrients.* 2020; 12(10): 3090. doi: 10.3390/nu12103090.
Korpelainen R., Korpelainen J., Heikkinen J., Väänänen K., Keinänen-Kiukaanniemi S. *Osteoporosis Inter.* 2003; 14(1): 34–43.
Kotyczka C., Boettler U., Lang R., Stiebitz H., Bytof G., Lantz I. *Mol Nutr Food Res.* 2011; 55(10): 1582–6.
Krebs J.D., Parry-Strong A., Weatherall M., Carroll R.W., Downie M. *Metab Clin Exp.* 2012; 61(9): 1231–7.
Lane J.D., Feinglos M.N., Surwit R.S. *Diabetes Care.* 2008; 31(2): 221–2.
Lane J.D., Hwang A.L., Feinglos M.N., Surwit R.S. *Endocr Pract.* 2007; 13(3): 239–43.
Larsson S.C., Orsini N. *Am J Epidemiol.* 2011; 174(9): 993–1001.
Lemann J., Litzow J.R., Lennon E.J. *J Clin Investig.* 1966; 45(10): 1608–14.
Leung W.W.M., Ho S.C., Chan H.L.Y., Wong V., Yeo W., Mok T.S.K. *J Epidemiol Community Health.* 2011; 65(6): 556–8.
Lim Y., Park Y., Choi S.K., Ohn J.H. *Nutrients.* 2019; 11(10): 2377. doi: 10.3390/nu11102377.
Lloyd T., Rollings N., Eggli D.F., Kieselhorst K., Chinchilli V.M. *Am J Clin Nutr.* 1997; 65(6): 1826–30.
Loke W.H. *Physiol Behav.* 1988; 44(3): 367–72.
Loopstra-Masters R.C., Liese A.D., Haffner S.M., Wagenknecht L.E., Hanley A.J. *Diabetologia.* 2011; 54(2): 320–8.
Lovallo W.R., Wilson M.F., Vincent A.S., Sung B.H., McKey B.S., Whitsett T.L. *Hypertension.* 2004; 43(4): 760–5.
MacKenzie T., Comi R., Sluss P., Keisari R., Manwar S., Kim J. *Metab ClinExp.* 2007; 56(12): 1694–8.
Mahmud A, Feely J. *Hypertension.* 2001; 38(2): 227–31.
Martin R.J., Abu-Shaweesh J.M., Baird T.M. *Paediatrc Respir Rev.* 2004; 5: 77–82.
Martínez-Pinilla E., Oñatibia-Astibia A., Franco R. *Front Pharmacol.* 2015; 6:30. doi: 10.3389/fphar.2015.00030.
Martini D., Del Bo' C., Tassotti M., Riso P., Del Rio D., Brighenti F., Porrini M. *Molecules.* 2016; 21(8): 979. doi: 10.3390/molecules21080979.
Marventano S., Salomone F., Godos J., Pluchinotta F., Del Rio D., Mistretta A. *Clin Nutr.* 2016; 35(6): 1269–81.
Masterton G.S., Hayes P.C. *Eur J Gastroenterol Hepatol.* 2010; 22(11): 1277–83.
Mazzafera P. *Phytochem.* 1990; 30(7): 2309–10.
McLellan T.M., Caldwell J.A., Lieberman H.R. *Neurosci Biobehav Rev.* 2016; 71: 294–312.
Mesas A.E., Leon-Muñoz L.M., Rodriguez-Artalejo F., Lopez-Garcia E. *Am J Clin Nutr.* 2011; 94(4): 1113–26.
Mikołajczyk-Stecyna J., Malinowska A.M., Mlodzik-Czyzewska M., Chmurzynska A. *Food Qual Prefer.* 2020; 104115. https://doi.org/10.1016/j.foodqual.2020.104115
Milne G.L., Yin H., Hardy K.D., Davies S.S., Roberts L.J. *Chem Rev.* 2011; 111(10): 5973–96.
Mišík M., Hoelzl C., Wagner K.H., Cavin C., Moser B., Kundi M. *Mutat Res - Fund Mol M* 692(1–2): 42–8.

Mostofsky E., Schlaug G., Mukamal K.J., Rosamond W.D., Mittleman M.A. *Neurology*. 2010; 75(18): 1583–8.
Mohamadi N., Sharififar F., Pournamdari M., Ansari M. *J Diet Suppl*. 2018; 15(2): 207–22.
Monteiro J., Alves M.G., Oliveira P.F., Silva B.M. *Crit Rev Food Sci Nutr*. 2019; 59(16): 2597–625.
Mostofsky E., Rice M.S., Levitan E.B., Mittleman M.A. *Circ Heart Fail*.- 2012; 5(4): 401–5.
Moura-Nunes N., Perrone D., Farah A., Donangelo C.M. *Int J Food Sci Nutr*. 2009; 60(6): 173–81.
Mukamal K.J., Hallqvist J., Hammar N., Ljung R., Gémes K., Ahlbom A. *AmHeart J*. 2009; 157(3): 495–501.
Nakanishi N., Nakamura K., Nakajima K., Suzuki K., Tatara K. *Eur J Epidemiol*. 2000; 16(5): 419–23.
Natella F., Nardini M., Giannetti I., Dattilo C., Scaccini C. *J Agri Food Chem*. 2002; 50(21): 6211–6.
Nehlig A., Daval J.L., Debry G. *Brain Res Rev*. 1992; 17(2): 139–70.
Ochiai R., Chikama A., Kataoka K., Tokimitsu I., Maekawa Y., Ohishi M. *Hypertension Res*. 2009; 32(11): 969–74.
Ohta A., Sitkovsky M. Methylxanthines. Handbook of Experimental Pharmacology. Springer, 2011.
Oñatibia-Astibia A., Martínez-Pinilla E., Franco R. *Respir Med*. 2016; 112: 1–9.
Owen G.N., Parnell H., De Bruin E.A., Rycroft J.A. *Nutri Neurosci*. 2008; 11(4): 193–8.
Papamichael C.M., Aznaouridis K.A., Karatzis E.N., Karatzi K.N., Stamatelopoulos K.S., Vamvakou G. *Clin Sci*. 2005; 109(1): 55–60.
Park C.H.., Myung. SK, Kim T.Y., Seo H.G., Jeon Y.J., Kim Y. *BJU Int*. 2010; 106(6): 762–9.
Pauwels R. *Chest*. 1987; 92: 32S-37S.
Pelchovitz D.J., Goldberger J.J. *Am J Med*. 2011; 124(4): 284–9.
Postuma R.B., Lang, A.E., Munhoz, R.P., Charland, K., Pelletier, A., Moscovich, M. *Neurology*. 2012; 79(7): 651–8.
Premnath S., Alalshaikh G., Alfotawi R., Philip M. *Cureus*. 2020; 12(4): e7820. Doi:10.7759/cureus.7820.
Rabe K.F., Magnussen H., Dent G. *Eur Respir J*. 1995; 8: 637–42.
Rajasekaran L.R., Aspinall D., Jones G., Paleg L. *Can J Plant Sci*. 2001; 81(3): 487–98.
Rapuri P.B., Gallagher J.C., Kinyamu H.K., Ryschon K.L. *Am J Clin Nutr*. 2001; 74(5): 694–700.
Renda G., Zimarino M., Antonucci I., Tatasciore A., Ruggieri B., Bucciarelli T. *Am J Clin Nutr*. 2012; 95(1): 241–8.
Riksen N.P., Smits P., Rongen G.A. Methylxanthines. Handbook of Experimental Pharmacology. Springer, 2011.
Rivera-Oliver M., Dıaz-Rıos M. *Life Sci*. 2014; 101: 1–9. doi:10.1016/j.lfs.2014.01.083.
Robertson D., Frölich J.C., Carr R.K., Watson J.T., Hollifield J.W., Shand D.G., Oates J.A. *N Eng J Med*. 1978; 298(4): 181–6.
Robertson D., Wade D., Workman R., Woosley R.L., Oates J.A. *J Clin Invest*. 1981; 67(4): 1111–7.
Rogers P.J., Smith J.E., Heatherley S.V., Pleydell-Pearce C.W. *Psychopharmacol*. (Berl). 2008;195(4): 569–77.
Rudolphi K.A., Keil M., Fastbom J., Fredholm B.B. *Neurosci Lett*. 1989; 103: 275–80. doi:10.1016/0304-3940(89)90112-2.
Ruhl C.E., Everhart J.E. *Gastroenterol*. 2005; 128(1): 24–32.
Ruxton C.H.S. *Nutr Bull*. 2008; 33(1): 15–25.
Sakamoto W., Nishihira J., Fujie K., Iizuka T., Handa H., Ozaki M., Yukawa S. *Bone*. 2001; 28(3): 332–6.
Schoen K., Yu T., Stockmann C., Spigarelli M.G., Sherwin C.M. *Pediatr Drugs*. 2014; 16(2): 169–77.
Schuster J., Mitchell E.S. *Prog Neuropsychopharmacol. Biol Psychiatry*. 2019; 89: 263–74.
Shen J., Johnson V.M., Sullivan L.M., Jacques P.F., Magnani J.W., Lubitz S.A. *Am J Clin Nutr*. 2011; 93(2): 261–266.
Shirvani H, Arabzadeh E, Akbari J. *Sci Sports*. 2020; 35(3): 65–74.
Simons F.E., Becker A.B., Simons K.J., Gillespie C.A. *J Allergy Clin Immunol*. 1985; 76: 703–707.
Smits P., Thien T., van't Laar A. *Bri J Clin Pharmacol*. 1985a; 19(6): 852–854.
Smits P., Thien T., van't Laar A. *Am J Cardiol*. 1985b; 56(15): 958–963.
Steer P., Flenady V., Shearman A., Charles B., Gray P., Henderson-Smart D. *Arch Dis Child*. 2004; 89: 499–503.
Steinkellner H., Hoelzl C., Uhl M., Cavin C., Haidinger G., Gsur A. *Mutat Res*. 2005; 591(1–2): 264–275.
Strandhagen E., Landaas S., Thelle D.S. *Eur J Clin Nutr*. 2003; 57(11): 1411–1417.
Strandhagen E., Thelle D.S. *Eur J Clin Nutr*. 2003; 57(9): 1164–1168.
Taketo M., Matsuda H., Yoshioka T. *Brain Res*. 2004; 1016(2): 229–239.
Tang N., Wu Y., Ma J., Wang B., Yu R. *Lung Cancer*. 2010; 67(1): 17–22.
Tang N., Zhou B., Wang B., Yu R. *Am J Obstet Gynecol*. 2009; 200(3): 290.e1–290.e9. doi: 10.1016/j.ajog.2008.10.019.
Theobald K., Botwinski C., Albanna S., McWilliam P. *Neonatal Netw*. 2000; 19(6): 17–24.
Thorand B., Kolb H., Baumert J., Koenig W., Chambless L., Meisinger C. *Diabetes*. 2005; 54(10): 2932–2938.
Tilley S.L. Methylxanthines. Springer, 2011.
Tracy M.B., Klimek J., Hinder M., Ponnampalam G., Tracy S.K. *Acta Paediatr*. 2010; 99(9): 1319–1323.
Trugo L.C. Encyclopedia of Food Sciences and Nutrition (Second Edition). Academic Press, 2003.
Turati F., Galeone C., Edefonti V., Ferraroni M., Lagiou P., La Vecchia C. *Ann Oncol*. 2012; 23(2): 311–318.
Umemura T., Ueda K., Nishioka K., Hidaka T., Takemoto H., Nakamura S. *Am J Cardiol*. 2006. 98(11): 1538–1541.
Urgert R., Van Vliet T., Zock P.L., Katan M.B. *Am J Clin Nutr*. 2000; 72(5): 1107–1110.
Van Dam R.M., Dekker J.M., Nijpels G., Stehouwer C.D.A., Bouter L.M., Heine R.J *Diabetologia*. 2004; 47(12): 2152–2159.
Van Dam R.M., Hu F.B. *JAMA*. 2005; 294(1): 97–104.
Van Der Hoeven T., Browne J.L., Uiterwaal C.S.P.M., van der Ent C.K., Grobbee D.E., Dalmeijer G.W. *PLoS One*. 2017; 12(5): e0177619. doi: 10.1371/journal.pone.0177619.
Van Dijk A.E., Olthof M.R., Meeuse J.C., Seebus E., Heine R.J., Van Dam R.M. *Diabetes Care*. 2009; 32(6): 1023–1025.
Vercambre M.N., Berr C., Ritchie K., Kang J.H. *J Alzheimer's Dis*. 2013; 35(2): 413–421. doi:10.3233/JAD-122371.
Verhoef P., Pasman W.J., Van Vliet T., Urgert R., Katan M.B. *Am J Clin Nutr*. 2002; 76(6): 1244–1248.
Walker M.L., Holt K.E., Anderson G.P., Teo S.M., Sly P.D., Holt P.G. *Front Immunol*. 2014; 5: 447. doi:10.3389/fimmu.2014.00447.
Warner N.S., Warner M.A., Schroeder D.R., Sprung J., Weingarten T.N. *Bosnian J Bas Med Sci*. 2018; 18(1): 101–104.

Wedick N.M., Brennan A.M., Sun Q., Hu F.B., Mantzoros C.S., Van Dam R.M. *Nutr J.* 2011; 10(1): 93. doi: 10.1186/1475-2891-10-93.

Wei F., Tanokura M. Coffee in health and disease prevention. Academic Press, 2015.

Westerterp-Plantenga M., Diepvens K., Joosen A.M.C.P., Bérubé-Parent S., Tremblay A. *Physiol Behav.* 2006; 89(1): 85–91.

Wu J., Ho S.C., Zhou C., Ling W., Chen W., Wang C. *Int J Cardiol.* 2009; 137(3): 216–225.

Yamaji T., Mizoue T., Tabata S., Ogawa S., Yamaguchi K., Shimizu E. *Diabetologia.* 2005; 47(12): 2145–2151.

Zeegers M.P.A., Tan F.E.S., Goldbohm R..A, Van Den Brandt P.A. *Int J Epidemiol.* 2001; 30(2): 353–362.

Zhang X., Albanes D., Beeson W.L., van den Brandt P.A., Buring J.E., Flood A. *J Nat Cancer Inst.* 2010; 102(11): 771–783.

Zhang Y., Lee E.T., Cowan L.D., Fabsitz R.R., Howard B.V. *Nutr Metab Cardiovasc Dis.* 2011b; 21(6): 418–423.

Zhang Y., Post W.S., Dalal D., Blasco-Colmenares E., Tomaselli G.F., Guallar E. *PLoS ONE.* 2011a; 6(2): e17584. doi: 10.1371/journal.pone.0017584.

Zhao J., Gonzalez F., Mu D. *Eur J Pediatr.* 2011; 170(9): 1097–1105.

Zhao Y., Wang J., Ballevre O., Luo H., Zhang W. *Hypertens Res.* 2012; 35(4): 370–374.

Zhou Y., Tian C., Jia C. *Prev Med.* 2012; 55(1): 14–22.

20

Current understandings of health benefits of coffee on cardiovascular and metabolic syndromes

Prakash Narayana Reddy, Harish Babu Kolla, Hanumohan Reddy Konatham, Parvatam Giridhar, and Vijaya R. Dirisala

CONTENTS

20.1 Introduction

Coffee is the most commonly consumed beverage next to water across the globe. It is bitter in taste and has slightly acidic pH. Coffee is cultivated in nearly 70 countries around the world. Different varieties of coffee are cultivated, of which the commonly consumed types are Arabica, Robusta, and Liberica (Garriguet, 2008). The coffee beverage is a complex mixture whose main constituents are caffeine (1,3,7-trimethylxanthine), chlorogenic acid (CGA), and the diterpenes kahweol and cafestol (Cano-Marquina et al., 2013; Gonzalez de Mejia and Ramirez-Mares, 2014). Several other components are also present in coffee, such as potassium, magnesium, niacin, and other biologically active compounds (Gómez-Ruiz et al., 2007). Tocopherols present in coffee beans are well known for their antioxidant nature. Bean variety, grinding, roasting, and type of brew affect the key metabolic constituents in the beverage. Coffee undergoes chemical metamorphosis during industrial processing (Cano-Marquina et al., 2013; Gonzalez de Mejia and Ramirez-Mares, 2014). Of all the biochemical constituents, caffeine is the most discussed and investigated metabolite. Caffeine is a psychoactive alkaloid molecule that is found in coffee, tea, and other consumable beverages such as cola drinks. The concentration of caffeine in coffee varies depending on the variety and agronomical environments during the cultivation stage (Zhang et al., 2014). Along with caffeine, phenolic CGA, caffeic acid, diterpenes, lactones, precursor of vitamin B_3, trigonelline, niacin, kahweol, and cafestol are present in the metabolic composition of coffee. Coffee is also an important source for other phytochemicals such as polyphenols:phenolic glycosides and other flavonoids, saponins, isoflavonoids, and lignans are transformed by gut bacteria into estrogen-like compounds that show wide range of antioxidant properties (Blaut and Clavel, 2007). The important polyphenol 5-caffeoylquinic acid (5-CQA) found in coffee shows a broad range of antioxidant activity (Nicoli et al., 1997). The numerous bioactive metabolites in coffee interact and provide unique health benefits. The bioavailability of coffee is determined by the genetic background and gut microbiota composition of the host. Roasted beans exhibit higher antioxidant activity than non-roasted beans. Upon roasting, the levels and concentration of the complex metabolites in coffee beans come to their optimum levels. CGA levels are brought to a minimum by roasting, showing greater antioxidant properties. CGAs occur as caffeoylquinic acid, quinic acid, p-coumaroylquinic acid, and feruoylquinic acids in the phenolic fraction of green coffee. Apart from antioxidative properties, CGAs also show their hypoglycemic, hepatoprotective, and antiviral properties on human health. CGAs in coffee could modulate the gut microbiome and regulate host immunity and metabolism. Coffee consumption significantly increases the numbers of *Bifidobacterium* spp. and the

DOI: 10.1201/9781003043133-25

metabolic activity of host cells (Jaquet et al., 2009). Being a key component of coffee, caffeine stimulates the central nervous system (CNS) and regulates gastrointestinal and cardiac metabolism (Chou and Benowitz, 1994; Mehta et al., 1997). Caffeine is a naturally occurring methylxanthine class of alkaloid that is metabolized by cytochrome P450 enzymes in hepatocytes (Pardo et al., 2007). Caffeine, along with the other metabolites such as CGA, shows some beneficial effects against the development of Alzheimer's disease (Lindsay et al., 2002; Cano-Marquina et al., 2013). A hypothesis made by Ascherio and colleagues stated that daily consumption of coffee could show a protective effect against the development of Parkinson's disease, heart stroke, and cancer (Ascherio et al., 2001). Phytochemicals like polyphenols, phenolic acids, flavonoids, and terpenoids in coffee beans were found to be more effective with few or lesser toxic effects and play a major role in scavenging free radicals from several drug discovery studies for various types of cancers and metabolic diseases (Singh et al., 2016). Correlation between coffee consumption and cardiovascular health is a debatable topic for research because of old scientific research methodologies and lack of advanced infrastructure (Rebello and van Dam, 2013). Studies following those methodologies and case studies reported for a conclusion of coffee consumption with cardiovascular and metabolic risks. Recent advancements in metabolomics and analytical techniques have become very helpful in monitoring the activity of coffee metabolites in vitro and in vivo. Literature surveys and conclusions drawn from such studies represented the experimental and meta-analysis data showing the positive effects of coffee consumption toward a beneficial and neutralizing potential of coffee metabolites (Sofi et al., 2007). In this chapter, we review the existing data and reports on scientific studies illustrating the metabolic properties of coffee constituents along with their biological significance and chemical activity. Further, the regulatory effects of metabolites on body metabolism during cardiovascular and related metabolic risks are discussed briefly.

20.2 Key constituents of coffee and their metabolic effects

Coffee includes a wide variety of chemical compounds such as caffeine, phenols, and diterpenes. Out of thousands of available molecules in coffee, only a few have been investigated thoroughly for their regulatory functions on health and well-being. Among them caffeine, CGA, kahweol, and cafestol are of primary importance in the metabolomics of coffee (Figure 20.1). Kahweol and cafestol are diterpenes that are present in unfiltered coffee. They are associated with a rise in cholesterol levels in the body. There was a misconception that the coffee consumption is associated with cardiovascular risks. Contrary to these statements, most prospective scientific cohort studies and meta-analysis data unfolded the controversy behind the topic of coffee consumption and explained in detail about the metabolites in coffee and their regulatory effect of cardiovascular health. In this section, we discuss the complex metabolites and their regulative biochemical properties.

20.2.1 Caffeine

Caffeine is a naturally occurring alkaloid and is the most investigated and studied metabolite in coffee. Being a psychoactive molecule, caffeine shows stimulatory effects on the body. Chemically, caffeine is 1,3,7-trimethylxanthine, a purine alkaloid that occurs naturally in the seeds, fruits, and leaves of several plants and acts as a natural pesticide. The load of caffeine is highest in coffee when compared to other consumable beverages such as tea, cola drinks, and so forth. Concentration of caffeine is affected by the method of preparation and extent of roasting of the coffee beans. Caffeine metabolizes into several metabolites such as paraxanthine, theobromine, and theophylline (Etherton and Kochar, 1993), but the final product of caffeine metabolism is paraxanthine. Caffeine metabolism is illustrated by five pathways. The first

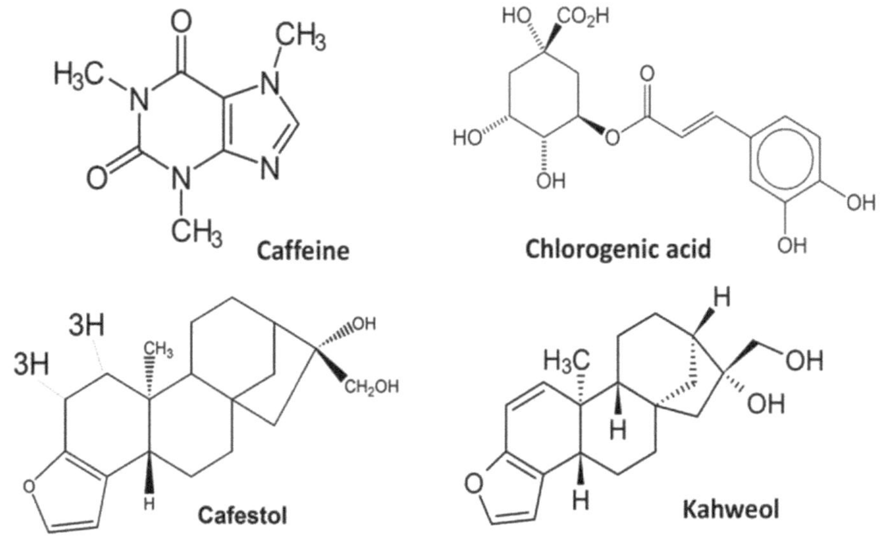

FIGURE 20.1 Major bioactive compounds of coffee.

three pathways include demethylation reactions to form theophylline and paraxanthine. These reactions influence vasodilation, mediating increased and relaxed blood flow. Coffee has a half-life of 4–6 hours. It is metabolized by isoforms of cytochrome P450 CYP1A2 in the liver contributing to 95% of metabolism. CYP1A2 catalyzes the demethylation of caffeine to paraxanthine. CYP1A2 demethylates paraxanthine to 1-methylxanthine, which is further oxidized by xanthine oxidase to 1-methyluric acid. The fourth and fifth pathways account for uracil metabolite formation and renal elimination of caffeine (Pardo et al., 2007). Caffeine exerts its antagonistic effects on A_1, A_{2A}, A_{2B}, and A_3 adenosine receptors (Fredholm et al., 2001). These receptors are expressed in most of the tissue systems. Caffeine metabolism results in a wide range of potential metabolic modifications and inflammatory responses affecting the cardiovascular and metabolic mechanisms.

20.2.2 Chlorogenic acid

CGA is a phenolic acid found in a variety of fruits, vegetables, and oils. CGA is thermally unstable and readily decomposes into caffeic acid and quinic acid. CGA is a major component of caffeine, accounting for nearly 10%. CGA strongly influences the taste and aroma of coffee. CGA is mostly investigated for its antioxidant and anti-inflammatory activities (Veljkovic et al., 2018). It could protect the brain from oxidative stress and neurodegeneration by microglial activation. Experimental evidences show that the green color of coffee extracts when they are treated with ferric chloride is due to the presence of CGA. Feruloylquinic, dicaffeoylquinic, and caffeoylquinic acids are the isomers of quinic acid that are present in coffee beans (Trugo, 2003). CGA acts as an antagonist to peroxisome proliferator-activated alpha (PPARa) receptors and boosts fat burning. Sometimes caffeine induces the release of fatty acids from stored fat in the body, and CGA in turn assists the liver for the processing of free fatty acids contributing to weight loss. Through its antioxidant nature, it further represses the proliferation of new fatty acids. CGA with caffeine is involved in maintaining homeostasis in the body by controlling fatty acid levels. CGA also mediates anti-obesity activities. Recent studies also proved the anti-tumor property of CGA. It helps kidney cells recover from necrosis and inflammation. Virological investigations on CGA revealed its anti-viral nature. CGA successfully exhibits its anti-viral potential toward H1N1 influenza and hepatitis B viruses (Garg et al., 2016). One-third of CGAs are absorbed in the upper gastrointestinal tract post-ingestion, where they are further metabolized into hydroxycinnamic acids (HCA). HCAs exist in several conjugated forms including amides (conjugated with mono- or polyamines, amino acids, or peptides) and esters (esters of hydroxy acids, such as tartaric acid and sugar derivatives, and glycosides; Teixeira et al. 2013). Subsequently they are released into circulation with several structural modifications, such as glucuronidation, sulfation, and methylation.

20.2.3 Diterpenes

Kahweol and cafestol are the diterpenes that are present in coffee beans. Both of them are lipophilic and have the same molecular formula ($C_{20}H_{30}O_3$). They are present at higher concentrations in the unfiltered fraction of coffee (Cruchten et al., 2010). Both kahweol and cafestol are structurally similar, with kahweol having an extra double bond to cafestol (Halvorsen et al., 1998). Cafestol has higher melting point than kahweol. Nearly 70% of the diterpenes are absorbed in the small intestine. They are metabolized and accumulated in the enterohepatic system, further triggering their pharmacological properties changing the structural and functional aspects of metabolism (Cruchten et al., 2010). The two molecules cafestol and kahweol demonstrate their biological and pharmacological properties such as hepatoprotection, tumor suppression, anti-inflammatory and anti-diabetic responses. Kahweol and cafestol are nearly similar in their metabolic properties with slight variations. Pioneer studies have confirmed their property of rising triacylglycerol and low-density lipoproteins (LDLs) in blood circulation, which is a severe risk in the progression and pathogenesis of cardiovascular diseases (Roos et al., 2000). Contrary to this conclusion, several observations on animal model and human volunteer studies suggest that both the diterpene molecules impart their little-known effects like tumor suppression activities and anti-inflammatory properties, which are positive outcomes that encourage identifying the target sites and drug evaluation process (Ren et al., 2019). Disproving the previous statements on kahweol and cafestol regarding their cardiac issues, studies from Urgert and colleagues made a strong report stating that kahweol and cafestol could protect consumers from cardiovascular-related diseases by lowering the serum levels of LDL and triacylglycerol on chronic consumption of coffee (Urgert and Katan, 1997; Urgert et al., 1997). Apart from the cardiovascular metabolism, both metabolites are reported to provide several health benefits such as detoxification, tumor suppression, and anti-inflammatory response. Of both molecules, kahweol is found to have more antioxidation potential. It blocks the expression of pro-inflammatory cytokines and induces several detoxifying enzymes (Ren et al., 2019). Both molecules are involved in certain metabolic pathways and regulate the flow of metabolic flux. But the mechanistic involvement of diterpene molecules in metabolic and physiologic therapy is still unknown but hypothetic. Directions with cutting-edge discoveries in metabolic engineering and metabolomics research are essential for drug discovery and mechanotherapy with coffee diterpene molecules and evaluation of coffee's role as a functional food.

20.3 Coffee consumption and cardiovascular health

The topic of coffee consumption and cardiovascular health is quite interesting but also contentious. Studies focusing on both the beneficial and risky effects of coffee consumption provided data enough adequate to support their investigations. Most reports studying the risk factors of coffee consumption associated with cardiovascular health are older. These studies are mostly argued and challenged for their observations by today's research works on animal and human studies by simulating the coffee beverage preparation process by ostracizing the risk factor metabolites in the coffee beans during process through decaffeination and filtration. Although the mechanisms behind the coffee consumption and its correlation with cardiac diseases are unknown, recently published data suggests that moderate consumption of coffee (2–3 cups daily) could protect us from cardiac related issues. The main risk and development factors of cardiovascular diseases in coffee are diterpenes. They can raise the levels of triacylglycerols and LDLs in blood, which are potential progenitors for the pathogenesis of cardiovascular diseases (De Roos et al., 2000). Diterpenes such as cafestol and kahweol are present at higher rates in the unfiltered fraction of coffee. Coffee is considered as safe if it is filtered, since the concentrations of diterpenes are reduced to their optimum. Coffee is produced industrially for consumption through three main processes: boiling, filtration and decaffeination. It is advised by doctors and many physiologists that consumption of filtered coffee could be safe and protective. Caffeine is the most investigated metabolite for its stress relief benefits and neurostimulation properties, but chances of an increase in blood pressure (BP) are associated with high caffeine in coffee. A temporary increase in BP is observed in coffee-naïve individuals (Ammar et al., 2001). But chronic consumption of 3–4 cups of coffee daily does not affect cardiovascular health (Bohn et al., 2012; Eilat-Adar et al., 2013). Chronic consumption of filtered coffee with moderate caffeine content is safe and protective toward cardio metabolism. It is also advised to consume decaffeinated coffee, since the caffeine concentration is brought to its optimum.

Another prominent metabolically active and protective fraction of coffee is polyphenolics. CGA, a polyphenolic constituent of coffee, is well known for its antioxidant properties and protection against atherosclerosis. Studies conducted by Anderson et al in the early 2000s have stated that moderate consumption of coffee reduces the risk of cardiovascular diseases (Anderson et al., 2006). Inflammation is the most claimed condition in the body for the development of several chronic diseases like diabetes, cancer, and atherosclerosis. Inflammation promotes the increase of reactive oxygen and nitrogen species to damage the cells. Coffee metabolites such as CGA are well known antioxidants. Hence the practice of coffee consumption in beverage forms induces protection toward inflammatory responses, thereby reducing the risk of cardiovascular diseases. Endothelial dysfunction is a severe attribute

TABLE 20.1

Positive Aspects of Coffee Consumption over Cardiovascular Diseases

Supplement	Study design	Studied disease	Inference	Reference
Coffee beverage	Meta-analysis of21 cohort studies	CHD	Moderate coffee consumption could reduce the risk of CHD	Wu et al., 2009
Coffee beverage	Randomized controlled trial on 103 acute MI patients	MI	Coffee consumption is safe and confers protection toward MI	Richardson et al., 2009
Caffeine dose	Randomized, double-blind, placebo-controlled, cross-over study	CAD	Caffeine ingestion reduced the levels of CRP, a CVD biomarker	Shechter et al., 2011
Coffee beverage	130,054 adult Californians	Arrhythmia	Consuming 4 cups of coffee daily is associated with less incidence of arrhythmias and AF Coffee consumption is inversely correlated with arrhythmias	Klatsky et al., 2011
Caffeine dose	Placebo control trials with caffeine dose of 500 mg d^{-1}	Arrhythmia	Caffeine ingestion reduced the frequency of ventricular arrhythmias	Myers, 1991
Caffeine dose	Electrophysiological ingestion of 275 mg of caffeine by22 patients with ahistory of ventricular arrhythmias	Arrhythmia	Caffeine did not alter ventricular arrhythmias	Chelsky et al., 1990
Caffeine dose	Case-control study	Arrhythmia	Consumption of >6 cups of coffee daily predispose QT abnormalities and arrhythmias	Zhang et al., 2011
Caffeine dose	Health study on 33,638 women of more than 45 years free of CVDs	AF	Caffeine consumption is not associated with a rise in the risk of AF	Conen et al., 2010
Caffeine consumption in coffee, cola and tea	Danish Diet, Cancer and Health Study of47,949 volunteers aged50–64 years	AF	Caffeine consumption is not associated with the risk of AF	Frost and Vestergaard, 2005
Coffee beverage	Framingham Heart Study of4,526 human volunteers	AF	No significant relation between coffee consumption and AF is registered	Shen, 2011
Coffee beverage	Swedish Mammography Cohort study of34,670 women with no history of CVDs	Heart stroke	Coffee consumption may significantly reduce the risk of stroke	Larsson, 2011
Coffee beverage	Longitudinal study on 50,739 women	Depression	Consumption of ≥4 cups of coffee daily reduce depression levels	Lucas et al., 2011

Notes: AF = atrial fibrillation; CAD = coronary artery disease; CHD = coronary heart disease; CRP = C-reactive protein; CVD = cardiovascular disease; MI = myocardial infarction.

to metabolic syndrome and atherosclerosis. CGA, a phenolic constituent of coffee, shows its protective nature towards endothelial functioning by its antioxidant and anti-inflammatory properties. CGA is involved in scavenging the reactive oxygen species (ROS) and enhances the availability of nitric oxide and reduces oxidative stress, which is a contributing factor for heart failure (Suzuki et al., 2008). Protective effects of CGA studied in the rat model demonstrated that CGA lowers the plasma concentration of triacylglycerol and LDLs, ensuring cardiovascular protection (Yukawa et al., 2004). CGA inhibits the activity of NADPH and lipid peroxidase activity (Clifford, 1999). CGA ensure its anti-inflammatory nature by repressing the expression of pro-inflammatory cytokines such as IL-β, IL-6, and TNF-α (Kotyczka et al., 2011). Some of the interesting findings of research works on coffee correlation with cardiovascular diseases are summarized in Table 20.1. Experimental data provided by current investigations over a decade with insights into metabolomics and molecular nutrition on coffee consumption and its regulatory effects on cardiovascular health clearly suggest that consumption of decaffeinated and filtered coffee with optimal levels of both caffeine and diterpenes having active polyphenolics such as CGA could impact cardiac health, with protective and beneficial effects on cardiovascular and metabolic mechanisms.

20.4 Coffee consumption over metabolic issues

Advances in metabolomics and nutraceutical research supports deep investigation in discovering the potential health benefits of coffee consumption and its regulatory effect on metabolic health (Figure 20.2). Recent research findings from animal and volunteer case control studies provide enough data supporting coffee consumption as a beneficial practice. The effect of the coffee beverage on diseases of metabolic syndrome such as oxidative stress, type 2 diabetes (T2DM), hypertension, obesity, and non-alcoholic fatty liver disease (NAFLD) is briefly discussed in the following sections.

20.4.1 Oxidative stress

Coffee, being a rich source of dietary antioxidant compounds (Vinson, 2005), is thought to thwart the action of ROS. Oxidative stress occurs when the cellular production of oxidant molecules surpasses the accessibility of antioxidants able to rout these insults. The coffee bioactives caffeine, phenolic compounds, trigonelline, diterpenes, and soluble fiber can play a significant role against this condition through the scavenging of free radicals. Once ingested, caffeine readily absorbs in the upper gastrointestinal tract, with a peak value within 60 minutes of ingestion (Lang, 2013). The intake of caffeine has been connected with a high number of biological effects, mainly concerning the stimulation of the central and sympathetic nervous systems, typically associated with a feeling of alertness after coffee consumption (Godos et al., 2014; Owens et al., 2014; Cappelletti et al., 2015). Recently, the intake of caffeine has been also reported to exert ergogenic effects; in 2011, the request of a health claim for caffeine in this direction received a positive opinion from the European Food Safety Authority (EFSA NDA Panel, 2011). Research on the potential protective effects of coffee and its bioactives (caffeine, CGAs, and diterpenes) against oxidative stress and related chronic disease risk has been increasing in recent years.

20.4.2 Hypertension

Prospective studies suggest a protective effect of high coffee intake (≥4 cups per day) against hypertension, mainly in women, and studies have been devoted to effects of caffeine on hypertension factors (Ammon et al., 1983). However, other bioactives such as polyphenols, soluble fiber, and potassium could exert a beneficial effect in the cardiovascular system (Andersen et al., 2006). The relation between coffee consumption and BP is not

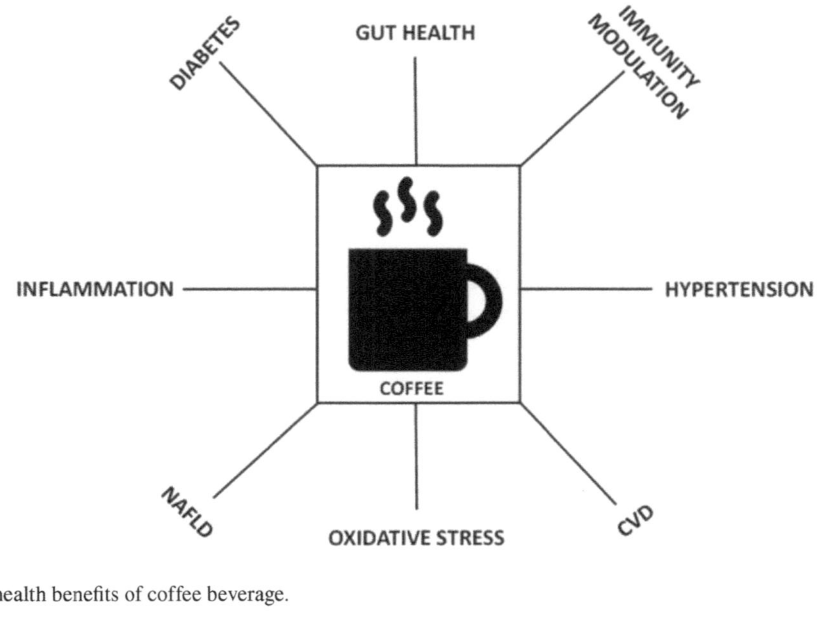

FIGURE 20.2 Potential health benefits of coffee beverage.

known. It is opined that as such coffee does not cause any rise of BP. Sporadic reports stated increase in systolic BP and diastolic BP at rest and during mental and physical stress (Bonita et al., 2007). This pressor effect is due to the elevation by caffeine of peripheral vascular resistance rather than enhancement of cardiac output (Nurminen et al., 1999). The capability of caffeine to upsurge vascular resistance nurtures the question of its consequence in hypertension development. Daily intake of 5 cups of decaffeinated coffee for a period of 6 weeks suggestively reduced ambulant BP by 1.5 mmHg systolic and 1.0 mmHg diastolic compared to a similar amount of regular coffee. A number of mechanisms have been proposed by which caffeine could raise BP, including sympathetic overactivation, antagonism of adenosine receptors, increased norepinephrine release via direct effects on the adrenal medulla, renal effects, and activation of the renin-angiotensin system (Gilbert, 1984).

20.4.3 Non-alcoholic fatty liver disease (NAFLD)

Coffee drinking is inversely related with the risk of NAFLD. A gap in the literature still exists regarding the intestinal mechanisms that are tangled in the protective effect of coffee consumption towards NAFLD. Epidemiological data illustrates that its intake is protective towards several diseases including liver fibrosis, cirrhosis, chronic liver disease, and liver cancer (Poole et al., 2017) that substantiated by marker-based studies converse associations between coffee drinking and markers of liver disease risk, such as serum levels of liver enzymes, that were reported for both caffeinated and decaffeinated coffee (Xiao et al., 2014). The increase of caffeinated and decaffeinated coffee consumption by 2 cups per day reduced the risk of hepatocellular carcinoma by 27% and 14%, respectively (Kennedyet al., 2017). Components in coffee other than caffeine may be critical in eliciting the health benefits of the beverage (Salomone et al., 2017; Alferinket al., 2018). There is experimental evidence that coffee with high CGA concentrations can modulate glucose intolerance and improve/decrease NAFLD development in obese rats (Panchal et al., 2012). The effect of coffee on the evolution of liver disease has also been attributed to its antifibrotic effects. In a rat model, coffee has been shown to attenuate thioacetamide-induced liver inflammation and fibrosis (Furtado et al., 2012). Animal studies have shown that coffee decreases expression of transforming growth factor-β and connective tissue growth factor, thus contributing to reduced fibrosis (Arauz et al., 2013). Furthermore, in rat models of alcohol-induced liver injury, caffeine has been shown to be protective against alcohol-induced liver fibrosis by dampening the cAMP/PKA/CREB pathway in rat hepatic stellate cells (Wang et al., 2015). Coffee is helpful for health in general and mainly for patients with liver disease. Coffee consumption of ≥2 cups/day protects against the progression of almost all forms of liver disease. The occurrence of progressive fibrosis and cirrhosis and hepatocellular carcinoma is inferior among coffee drinkers. The shielding effects are irrespective of the etiology of liver disease and more pronounced in alcohol-related liver disease. However, more interventional trials are required before coffee finds its way into the regular prescription of liver disease patients.

20.4.4 Type 2 diabetes

Caffeine can improve vinsulin sensitivity significantly, but it is not clear whether tolerance for this effect develops after long-term regular intake. Furthermore, it is plausible that the effects of coffee are different from those of caffeine. Generally, high consumption of caffeinated and decaffeinated coffee was associated with a substantially lower risk of T2DM, suggesting non-caffeine components of coffee may be responsible. A study on the bioactives of coffee reported that CGA, a strong antioxidant, might help regulate blood sugar levels and decrease the risk of diabetes. CGA inhibits intestinal glucose absorption and improves insulin sensitivity. Coffee exerts functional and beneficial effects on human health, as determined in various meta-analysis studies (Hitman, 2014). Drinking 3–4 cups of coffee per day leads to a 25% lower risk of developing T2DM compared to consuming no coffee or fewer than 2 cups per day (Park et al., 2010; Kim et al., 2013). The relative risk of T2DM for the highest level of coffee intake (>6 cups/day) was 0.71 (0.67–0.76) for caffeinated coffee and 0.79 (0.69–0.91) for intake of decaffeinated coffee. People who increase their consumption of coffee (+1 cup/day) had a decrease of 11% on T2DM risk in the subsequent 4 years. On the other hand, people who decreased their coffee intake by more than 1 cup/day had a 17% higher risk for T2DM in the next 4 years (Dórea and da Costa, 2005). Dose-response analysis suggested that the incidence of T2DM decreased by 12% (0.88 (0.86–0.90)) for intake of every 2 cups/day of caffeinated coffee and 11% (0.89 (0.82–0.98)) for intake of every 2 cups/day of decaffeinated coffee (Huxley et al., 2009). Also, there is indication that habitual consumers develop a tolerance to the effects caused by caffeine in insulin sensitivity and glucose tolerance (Ludwig et al., 2014). The reduction of T2DM risk found in epidemiological studies may be explained by the long-term effects (in years) of the bioactive compounds that downmodulate the oxidative stress and chronic inflammation, thus favoring an improvised glucose metabolism.

20.5 Conclusion

Advancements in analytical techniques and metabolomics tools helped to investigate the beneficial properties of coffee metabolites and their dynamic nature. Coffee in beverage form is considered a functional food because of its biological and metabolic activity. Recent studies revealed the potentiality of coffee consumption practice in immune boosting and metabolic improvement. Studies targeting metabolic syndrome in animal models and human volunteers over a decade proved the efficacy of coffee drinks as functional foods with health benefits.

20.6 Acknowledgments

Prakash Narayana Reddy thanks the INSPIRE division of Department of Science and Technology, Government of India, for conferring the INSPIRE faculty award and research grant [DST/INSPIRE/04/2017/000565] during the study period.

Authors also thank the administration and management of Vignan's University for providing necessary support and facilities during the period of study.

REFERENCES

Alferink LJM, et al. *Semin Liver Dis*. 2018; 38(3): 193–214.
Ammar R, et al. *Pharmacotherapy*. 2001; 21(4): 437–442.
Ammon HP, et al. *Br J Clin Pharmacol*. 1983; 15(6): 701–706.
Andersen LF, et al. *Am J Clin Nutr*. 2006; 83(5):1039–1046.
Arauz J, et al. *J Appl Toxicol*. 2013; 33(9):970–979.
Ascherio A, et al. *Ann Neurol*. 2001; 50(1):56–63.
Blaut M, Clavel T. *J Nutr*. 2007; 137(3 Suppl 2): 751S–5S.
Bøhn SK, et al. *Food Funct*. 2012; 3(6): 575–591.
Bonita JS, et al. *Pharmacol Res*. 2007; 55(3): 187–198.
Cano-Marquina A, et al. *Maturitas*. 2013; 75(1): 7–21.
Cappelletti S, et al. *Curr Neuropharmacol*. 2015; 13(1): 71–88.
Chelsky LB, et al. *JAMA*. 1990; 264(17):2236–2240.
Chou TM, Benowitz. *Comp Biochem Physiol C Pharmacol Toxicol Endocrinol*. 1994; 109(2): 173–189.
Clifford MN. *J Sci Food Agric*. 1999; 79(3):362–372.
Conen D, et al. *Am J Clin Nutr*. 2010; 92(3):509–514.
De Roos B, et al. *J Intern Med*. 2000;248(3):211–216.
Dórea JG, da Costa TH. *Br J Nutr*. 2005; 93(6): 773–782.
EFSA Panel on Dietetic Products, Nutrition and Allergies (NDA). *EFSA Journal* 2011; 9(4):2054.
Eilat-Adar S, et al. *Nutrients*. 2013; 5(9): 3646–3683.
Etherton GM, Kochar MS. *Arch Fam Med*. 1993; 2(3): 317–322.
Fredholm BB, et al. *Pharmacol Rev*. 2001; 53(4): 527–552.
Frost L, Vestergaard P. *Am J Clin Nutr*. 2005; 81(3): 578–582.
Furtado KS, et al. *Basic Clin Pharmacol Toxicol*. 2012; 111(5): 339–347.
Garg SK. In *Nutraceuticals*. Academic Press, 2016, pp. 653–667.
Garriguet D. *Health Rep*. 2008; 19(4): 23–29.
Gilbert RM. *Prog Clin Biol Res*. 1984; 158: 185–213.
Godos J, et al. *Int J Food Sci Nutr*. 2014; 65(8): 925–936.
Gómez-Ruiz JÁ, *J Agric Food Chem*. 2007 Aug 22; 55(17): 6962–6969.
Gonzalez de Mejia E, Ramirez-Mares. *Trends Endocrinol Metab*. 2014;25(10): 489–492.
Halvorsen B, *J Lipid Res*. 1998; 39(4): 901–912.
Hitman, G. A. *Diabetic Medicine* 2014; 7(31): 763.
Huxley R, et al. *Arch Intern Med*. 2009; 169(22): 2053–2063.
Jaquet M, Ro et al. *Int J Food Microbiol*. 2009; 130(2): 117–121.
Kennedy OJ, et al. *BMJ Open*. 2017; 7(5): e013739.
Kim EK, et al. *Korean J Community Nutr*. 2013; 18(20): 134–141.
Klatsky AL *Perm J*. 2011; 15(3): 19–25.
Kotyczka C, et al. *Mol Nutr Food Res*. 2011; 55(10): 1582–1586.
Lang R, *Anal Bioanal Chem*. 2013; 405(26): 8487–8503.
Larsson SC, *J Am Coll Cardiol*. 2011; 58(17): 1828–1829.
Lindsay J, et al. *Am J Epidemiol*. 2002; 156(5): 445–453.
Lucas M, et al. *Arch Intern Med*. 2011; 171(17): 1571–1578.
Ludwig IA, et al. *Food Funct*. 2014; 5(8): 1718–1726.
Mehta A, et al. *Acta Cardiol*. 1997; 52(3): 273–283.
Myers MG. *Ann Intern Med*. 1991; 114(2): 147–150.
Nicoli MC, et al. *LWT-Food Sci Technol*. 1997; 30(3): 292–297.
Nurminen ML, et al. *Eur J Clin Nutr*. 1999; 53(11): 831–839.
Owens JA, et al. *Nutr Rev*. 2014; 72(1): 65–71.
Panchal SK, et al. *J Nutr*. 2012; 142(4): 690–697.
Pardo Lozano R, et al. *Adicciones*. 2007; 19(3): 225–238.
Park JY, et al. *Korean J Nutr*. 2010; 43: 374–381.
Poole R, et al. *BMJ*. 2017; 359: j5024.
Rebello SA, van Dam RM. *Curr Cardiol Rep*. 2013; 15(10): 403.
Ren Y, et al. *Int J Mol Sci*. 2019; 20(17): 4238.
Richardson T, et al. *QJM*. 2009; 102(8): 555–561.
Salomone F, et al. *Nutrients*. 2017; 9(1): 85.
Shechter M, et al. *Am J Cardiol*. 2011; 107(9): 1255–1261.
Shen J. *Am J Clin Nutr*. 2011; 93(2): 261–266.
Singh S, et al. *Front Plant Sci*. 2016; 7: 1667.
Sofi F, et al. *Nutr Metab Cardiovasc Dis*. 2007; 17(3): 209–223.
Suzuki A, et al. *Am J Hypertens*. 2008; 21(1): 23–27.
Teixeira J, et al. *Biomed Res Int*. 2013; 251754.
Trugo LC. "COFFEE| Analysis of Coffee Products." In: *Encyclopedia of Food Sciences and Nutrition*. Academic Press, 2003, pp. 1498–506.
Urgert R, Katan MB. *Annu Rev Nutr*. 1997; 17: 305–324.
Urgert R, et al. *Eur J Clin Nutr*. 1997; 51(7): 431–436.
Veljkovic E, et al. *Nicotine and Other Tobacco Compounds in Neurodegenerative and Psychiatric Diseases: Overview of Epidemiological Data on Smoking and Preclinical and Clinical Data on Nicotine*. Academic Press, 2018.
Vinson JA. In: *Abstracts of Papers of the American Chemical Society*, vol. 230, pp. U7–U7.1155. American Chemical Society, 2005.
Wang Q, et al. *Int Immunopharmacol*. 2015; 25(2): 340–352.
Wu JN, et al. *Int J Cardiol*. 2009; 137(3): 216–225.
Xiao Q, et al. *Hepatology*. 2014; 60(6): 2091–2098.
Yukawa GS, et al. *Biochemistry (Mosc)*. 2004; 69(1): 70–74.
Zhang LF, et al. *Drug Des Devel Ther*. 2014; 9: 241–272.
Zhang Y, et al. *PLoS One*. 2011; 6(2): e17584.

21

Coffee diterpenes: Significance and biological functions

Vaddadi Sridevi, Anusree Poyyakara Veetil, and Parvatam Giridhar

CONTENTS

21.1 Introduction

Coffee beans are a rich source for a wide range of bioactive compounds, which contribute to the quality of coffee brew and are also reported to have a good number of functional attributes in humans. Among the major bioactives, diterpenes of coffee brew have significance like other important metabolites (namely, caffeine, chlorogenic acids, trigonelline, and nicotinic acid; Mullen et al., 2011; Jeszka-Skowron et al., 2015). Diterpenes present in the lipid fraction of coffee beans are important in plant defense mechanisms besides nutraceutical and sensual properties, which have gradually gained consideration due to their physiological properties in human health. Lipids are an important component of coffee beverage flavor and aroma. A good number of investigations on the positive properties of coffee in humans have considerable scientific support, mostly based on observations (Cadden et al., 2007). Various biological functions such as antioxidant, restraining inflammatory response, anti-mutagenic, and inhibition of cancer cell growth were attributed to coffee (George et al., 2008). Coffee oil is rich in diterpenes of cafestol ($C_{20}H_{28}O_3$) and kahweol ($C_{20}H_{26}O_3$), which are mainly pentacyclic diterpene alcohols based on the kaurene skeleton of 20 kaurene carbons that are found exclusively in coffee plants (Garattini et al., 1993; Urgert and Katan 1997). A decent number of reports indicate their hypercholesteremic effects. During the last two decades, significant contributions from researchers mainly addressed various aspects of diterpenes of coffee that include both bioactivities and pharmacological properties. These diterpenes are reported as antinutritional factors due to their specified role in the increase in serum cholesterol, low-density lipoproteins (LDLs) and triglycerides, which cause cardiac difficulties and impairment of liver cells in humans (Urgert et al., 1995a; Speer and Kölling-Speer 2001).

21.2 Diterpenes cafestol and kahweol in coffee

The content of the diterpenes cafestol and kahweol in coffee beans varies with the species, variety, climatic factors, and post-harvest processing conditions. A significant distinction between cafestol and kahweol is the level of saturation of the double bond connection between carbons 1 and 2, prompting a range with the greatest peak absorption at different wavelengths. Both cafestol and kahweol are delicate to acids, temperature, and light; particularly kahweol is weak and unstable in its purified state. In 1989, 16-O-methylcafestol (16-OMC) was identified from Robusta espresso beans and its structure was elucidated by synthesis (Speer and Mischnick, 1989). As indicated by subsequent investigations, cafestol and kahweol are found in both Arabica (*Coffea arabica*) and Robusta coffees (*C. canephora*; Kölling-Speer et al., 1999). Later, 16-O-methylkahweol was reported in Robusta coffee beans by Speer and Kölling-Speer (2001).

DOI: 10.1201/9781003043133-26

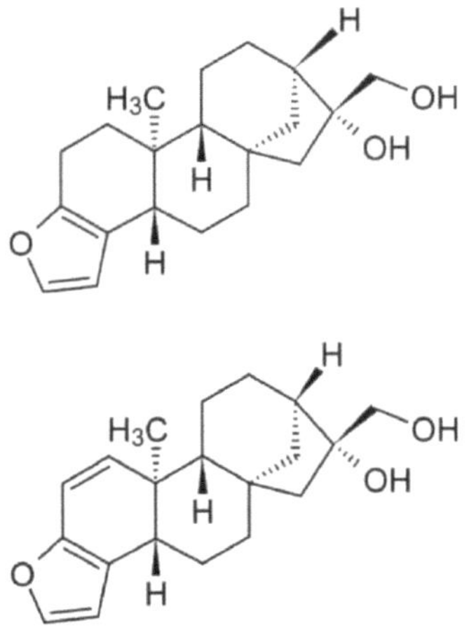

FIGURE 21.1 Chemical structure of coffee diterpenes (a) Cafestol and (b) kahweol.

21.2.1 Coffee lipid and diterpene content

Diterpenes present in the lipid fraction of coffee beans are important in plant defense mechanisms besides nutraceutical and sensual properties. These diterpenes gained the attention of researchers owing to their physiological effects in human health (Urgert et al., 1995a, 1995b). The lipid contents range from 7% to 17%, with an average of 15% in Arabica coffee and 10% in Robusta coffee. A large portion of the lipids as coffee oil are situated in the endosperm of green coffee beans (Speer and Kolling-Speer, 2006). The significant amount of lipid is present in the fluid structure inside the cells. Around 1%–2% of lipids are coffee wax that covers the bean as a thin adhering film. The fatty oils (75%) are the primary segment of coffee lipids, with diterpenes and unsaturated fat esters summing up to 20%. The combined total of diterpenes in *C. arabica* is 1.3%–1.9%, and that of *C. canephora* is 0.2%–1.5% (De Roos et al., 1997; Sridevi et al., 2010). Cafestol and kahweol are the distinctive diterpenes present in the unsaponifiable lipid portion of crude coffee principally esterified to unsaturated fats at the C-17 position. Only a small portion (1%–3%) is present in the free form in coffee. Cafestol levels range between 0.15% and 0.37% d.m. of beans in Robusta and 0.27%–0.67% d.m. of beans in Arabica. Kahweol levels are of 0.11%–0.35% d.m. in Arabica beans and <0.1% d.m. in Robusta beans (Kurzrock et al., 1998). Cafestol and kahweol are in low concentrations in *Coffea* that cultivated in western and central African forests, while species starting from eastern Africa indicated more significant levels of kahweol (De Roos et al., 1997). The content of cafestol and kahweol from both Arabica and Robusta coffee samples varies with the geographical origin and distribution of *Coffea* species (Sridevi et al., 2010).

21.2.2 Cafestol and kahweol biosynthetic pathway

Earlier reports pointed at the existence of great diversity for diterpene content between the species (De Roos et al., 1997; Mazzafera et al., 1998) and within the species (Speer and Kölling-Speer, 2001), which specifies the indication of genetic polymorphism of the respective enzymes directing the cafestol biosynthesis pathways. There is no information about actual biosynthesis of cafestol and kahweol in plants except sporadic reports that linked to the gibberellic acid (GA) pathway at the enter kaurene step diversion (De Castro and Marraccini 2006). The final steps of cafestol and kahweol biosynthesis in coffee plants have not yet been elucidated, but in plants, most terpenoids are synthesized by members of the cytochrome P450s gene family (CYPs) in order to understand genetic and biochemical aspects of kahweol biosynthesis (Suzana et al., 2012; Ivamoto et al., 2013). For both the diterpenes cafestol and kahweol, the entkauronic acid (EK) is the immediate precursor, but the actual key enzyme is not yet reported. The same EK is also precursor for GA biosynthesis (Liu et al., 2005), but no reports are available about the actual enzymes involved in cafestol and kahweol production from EK. Recently, cafestol and kahweol contents were quantified in flowers and fruit perisperm (30–210 days after flowering [DAF]), which was further supported by transcriptional analysis of eight CYP genes in Arabica coffee. Out of five genes observed, a similar pattern between diterpene concentration and gene transcription was reported. Three CYPs had transcriptional patterns similar to cafestol accumulation, whereas those of the two CYPs were related with kahweol accumulation (Suzana et al., 2012; Ivamoto et al., 2013).

21.3 Influence of various factors on coffee diterpenes

Coffee is a popular non-alcoholic beverage that holds intricate taste, aroma, and flavor properties that are swayed by several parameters such as coffee variety, origin, roasting, and preparation methods (Sridevi et al., 2011). Following extraction of brew, various physicochemical modifications occur and alter the organoleptic qualities of coffee (Laurent et al., 2012). The inherent quality of coffee is an important factor for the price determination at market. The main quality changes of coffee beans during the storage period are associated with fluctuations in temperature and relative humidity of the storage environment that affect the intensity and speed of the grain deterioration. If the oxygen, nitrogen, and reactive substances are predominant, they react with other proteins, carbohydrates, and lipids, which causes changes in the mechanism of cell homeostasis that leading to loss of functionality (Reni et al., 2012). Post-harvest processing methods, especially the pulped natural/semi-washed process, helps to improve bean quality.

Reports of extreme weather encountered by coffee farmers are becoming common, which ultimately affects sustainable coffee production (Baker 2012). Similarly, high temperature and dry weather conditions 3 months after blooming are reported to hamper the quality of the coffee fruits by developing a physiological disorder in coffee beans called *Coracao negro* (black heart), as reported in Brazil, but this is associated with a specific environmental condition and vanishes after weather conditions return to normal (Anonymous, 2014).

The influence of the environment plays a crucial role in coffee cultivation, as this plant is cultivated at different elevations; climate poses a profound influence on caffeine (Sridevi and Giridhar, 2014a), trigonelline (Sridevi and Giridhar, 2013), and diterpenes (Sridevi and Giridhar, 2016). Coffee bean composition is affected by genetic characteristics, edaphoclimatic conditions (Sridevi et al., 2010), post-harvest and roasting processes, and the final preparation of coffee brews (Sridevi et al., 2011). Hydrosoluble (nicotinic and chlorogenic acid trigonelline and caffeine) and liposoluble (cafestol and kahweol) compounds and melanoidins stand out for their contribution to health, mainly as antioxidants (Cintia et al., 2012; Sridevi and Giridhar, 2015). Various biotechnological approaches have been investigated by researchers to downregulate caffeine in coffee (Sridevi and Giridhar, 2008; Mohanan et al., 2014) and also to develop plants with reduced levels of diterpenes (Sridevi and Giridhar, 2014b).

21.4 Significances of cafestol and kahweol with respect to biological functions

As of now, the available literature reveals the various potential bioactivities and pharmacological properties such as anti-inflammation, anti-angiogenesis, anti-carcinogenic, anti-diabetic capability, hepatoprotective, cardioprotective, chemoprotective, cytoprotective, anti-nociceptive activity and anti-osteoclastogenesis activities of coffee extracts.

21.4.1 Hypercholesterolemic effect

The cafestol and kahweol are considered very active molecules, as intake of 10 mg each of these substances (equivalent to three cups of boiled coffee) leads to an elevated levels of total cholesterols (Table 21.1), especially high LDLs and

TABLE 21.1

Studies on Coffee Diterpenes on Hypercholesteremic Effect

Experimental animal/cell lines	Dose and duration	Results	Reference
Sprague-Dawley, 42 days old	Cafestol and kahweol in coffee beans, 10% coffee beans in diet for 10 d	A 1.7-fold increase in GST activity of liver (ratio of test to control) and 2.1-fold of small intestine observed.	Wattenberg and Lam (1984)
NMRI, age n.p.	Cafestol palmitate, 2 mg g^{-1} in diet for 4 d	Significant increase in GST activity of the cytosolic fraction of the liver recorded and also ~30% increase in weight of the liver noted.	Di Simplicio et al. (1989)
Hamsters, Syrian, "adult"	Coffee solution comprising cafestol and kahweol given ad libitum with high-fat diet	Elevated both total cholesterol and LDL cholesterol lipid fraction coffee control.	Sanders and Sandaradura (1992)
Cebus and Rhesus monkeys,	Cafestol and kahweol in 0.5% coffee oil. Cebus: 5.13 mg cafestol and 6.21 mg kahweol in the feed per kg b.w. Rhesus: 5.70 mg cafestol and 6.90 mg m.mol kahweol in the feed per kg b.w.	No effect was observed on plasma cholesterol or triglyceride concentrations. Similarly, no effect on ALT activity was detected.	Terpstra et al. (1995)
Human skin fibroblasts	99% pure cafestol (0.5–50 μg mL^{-1}) for 18 h; combination of cafestol and kahweol (48:47:5 isokahweol), 98% pure at 1–50 μg mL^{-1} incubated up to 24 h	Consequence of cafestol on cell-association of 125I-TC-LDL: after 6 h, a 9% and 18% lessening with 10 and 20 μg mL^{-1}, respectively. After 18 h, a dose-dependent cut ensued (17%–66% decrease with growing amounts of 10–50 μg mL^{-1}.	Halvorsen et al. (1998)
HepG2 cells	Cafestol and a blend of cafestol and kahweol, 20 μg mL^{-1} (63 μM) for 18 h pre-incubation	Cafestol reduced the uptake of 125I-TC-LDL by 15%–20% and the degradation by 20%–30% after pre-incubation, with all-out effects after 6 h and 10 h, correspondingly. A 6-h pre-incubation reinforced 17% drop of precise binding of LDL. No influence on cholesterol synthesis detected for both cafestol and mix.	Rustan et al. (1997)
29-year-old and 30-year-old male and female	900 mL filtered coffee brew administered containing cafestol (39 mg) and kahweol (33 mg) per day for 24 wk	Elevated activity of ALT and LDL cholesterol content by up to 80% and 9%–14%, respectively. Triglyceride concentrations were increased by 26% within 2 wk.	Urgert et al. (1995b)
24-year-old male and 26-year-old female	Ground coffee containing cafestol (39 mg) and kahweol (49 mg) per d for 21 d	Average serum cholesterol levels augmented by 25 mg dL^{-1}; triglyceride level and ALT activity improved by 27 mg dL^{-1}.	Urgert et al. (1995a)
64-year-old male and female	Cafestol (68 mg) and kahweol (85 mg) in Arabica and 29 mg cafestol and 1 mg kahweol in Robusta oils per kg	Arabica oil elevated serum total cholesterol by 44.1 mg dL^{-1} and plasma triglycerides by 72 mg dL^{-1}. An average serum ALT activity of 17.9 U L^{-1} was observed for Robusta oil treatment; other effects insignificant.	Van Rooij et al. (1995)
Hamsters, Syrian, 4 weeks old	Cafestol and kahweol in boiled coffee (44.1 mg kg^{-1} cafestol and 45.4 mg kg^{-1} kahweol in unfiltered coffee)	No influence on serum cholesterol and triacylglycerol levels was seen.	Beynen et al. (1996)
Male and female Wistar rats, 7–8 weeks old	Kahweol (2000 mg kg^{-1})	Subacute and acute studies in rats inferred that green coffee oil could be discovered clinically to develop hypercholesteremic and hypoglycemic agents.	De Oliveira et al. (2020b)

very-low-density lipoproteins (VLDLs) that have ill effects on health (Weusten-Van der Wouw et al., 1994; Ratnayake et al., 1995; Urgert et al., 1995a, 1995b). The underlying mechanism behind the ability of coffee diterpene to affect lipid metabolism in the human body is unfamiliar. These diterpenes are reported to elevate the content of total cholesterol in humans along with LDL cholesterol and triglycerides, which are known to increase the risk of coronary diseases, and it is also known to increases alanine aminotransferase (ALT) activity, which induces damage to hepatocytes (Table 21.2). However, other studies have shown that cafestol promotes uptake and degradation of LDL (Ranheim et al., 1995).

Sanders and Sandaradura (1992) explained how the cafestol and kahweol could elevate both total cholesterol and LDL, when hamsters were fed a high-fat diet coupled with diterpenes. Both plasma cholesterol and triglyceride levels could not alter much. Also, no effect on ALT was recorded (Terpstra et al., 1995). Similar observations were found when boiled coffee containing cafestol and kahweol and unfiltered coffee were tested against 4-week-old hamsters, where no effect on serum cholesterol and diglycerol levels was seen (Beynen et al., 1996). A drastic rise in serum cholesterol levels was demonstrated in 64-year-old males and females by Van Rooij et al. (1995). In their study, both Arabica and Robusta coffee oils comprising 68 mg and 85 mg and 29 mg and 1 mg of cafestol and kahweol, respectively, per kilogram of oil were tested (Table 21.1). Arabica oil was found to elevate serum cholesterol levels by 44.1 mg dL^{-1} and plasma triglycerides by 72 mg dL^{-1}. However, ground coffee containing 39 mg and 49 mg of cafestol and kahweol when given for 21 days to young volunteers leads to moderate levels of increase in serum cholesterol and triglycerides along with a rise in ALT activity (Urgert et al., 1995a). Similarly, filtered coffee brew containing 39 mg and 33 mg of cafestol and kahweol per day, when given for 24 week to young male and female volunteers, showed elevated ALT activity and LDL levels up to 80% and 14%, respectively (Urgert et al., 1995b). Studies in human hepatoma cells (HepG2) showed decreased uptake of 125I-TC-LDL by 15%–25% in the presence of a cafestol and kahweol mixture (Rustan et al., 1997). Similar studies with 99% pure cafestol and a mixture of cafestol, kahweol, and isokahweol were pursued by Halvorsen et al. (1998) in human fibroblasts; they found a concentration-dependent decrease in total cholesterol and LDL after 18 hours. Recent studies of De Oliveira et al. (2020a) in male and female Wistar rats showed green coffee oil rich in kahweol has a role as an agent in hypercholesteremic effect and hypoglycemia. Different human cell line and animal studies have uncovered that cafestol and kahweol display chemoprotective and cytoprotective effects.

21.4.2 Induction of mammary tumors

Early studies on the diet containing 10% and 20% of green coffee beans containing both diterpenes was investigated in Sprague-Dawley rats for 14 days, revealing that 75% and 44% of the animals' mammary tumors were detected at 10% and 20% dosage, respectively (Wattenberg, 1983). Even instant coffee containing cafestol and kahweol palmitates was reported to induce mammary tumors in these rats (Wattenberg and Lam, 1984). Similarly, occasional tumors of the pouch and lip were reported in Syrian golden hamsters (Table 21.2) that were under the treatment of green coffee beans containing diterpene palmitates (Miller et al., 1988). Sporadic studies against pathogenic bacteria such as *Salmonella typhimurium* TM677 revealed that cafestol acetate and kahweol acetate and their palmitate esters showed a dose-dependent inhibition of bacteria (Pezzuto et al., 1989).

21.4.3 Hepatoprotective, anti-carcinogenic, and antioxidant activity

Post et al. (2000) explained the ability of a cafestol- and kahweol-containing diet to inhibit bile acid synthesis by downregulation of both neutral and acidic pathways, leading to a decrease in expression of LDLR and an elevated secretion of cholesteryl esters in VLDSL, as shown in female mice. The inhibitory effect of cafestol and kahweol on B[a]P-DNA adduct formation in primary rat hepatocytes was reported, which relied on the concentration of cafestol and kahweol used (Cavin et al., 2003). Similarly, they affect hepatic activity elevation of O-6-methylglucanine – DNA methyltransferase as shown in male F344 rats (Huber et al., 2003). Nitric oxide inhibition and iNOS gene transcription suppression potential of kahweol in mouse macrophage cell RAW 264.7 was documented, where this particular response was found due to the inhibition of NF-kB activation in LPS-activated macrophages (Kim et al., 2004). Kim et al. (2006b) showed TNFα-induced VCAM-1 and ICAM-1 expression by kahweol in human endothelial cell line ECV304. They hypothesized that it could happen via a mechanism involving the inhibition of the JAK2-P13K/Akt-NF-kB signaling pathway in response to TNFα. Other studies with male ICR mice have shown the hepatoprotective and antioxidant activity of cafestol and kahweol treatment could forestall the liver injuries brought about via carbon tetrachloride in mice. The cancer prevention agent impact and superoxide radical scavenging activities of cafestol and kahweol was considered utilizing mouse liver homogenate treated with $FeCl_2$-ascorbate, which incites lipid peroxidation in the cell (Lee et al., 2007a). Anti-carcinogenic activity by induction of placental glutathione S-transferase (GST-P) activity was reported (Schilter et al., 1996). In an interesting study, Cavin et al. (2002) demonstrated the ability of coffee diterpenes on inhibition of AFB1 adduct formation in human hepatocytes. Anti-carcinogenic properties of C&K have been seen in human cell lines as well as in experimental animal models. Iwamoto et al. (2019) saw that treatment of human prostate malignancy cell lines PC-3, DU145, and LNCaP with kahweol acetic acid and cafestol at a 30-μM dose for 48 hours hindered the multiplication and movement of prostate disease cells by inciting apoptosis and restraint of epithelial mesenchymal progress (EMT). The protective role of kahweol on human dopaminergic neurons from 6-hydroxydopamine-derived oxidative stress was demonstrated by Hwang and Jeong (2008).

Auxiliary studies have shown apoptosis inducing the ability of kahweol in human leukemia U937 cells (Oh et al., 2009). Kahweol probably utilizes its favorable to apoptotic properties by hindering the declaration of HSP70 in a fixation subordinate way. Kim et al. (2006a) revealed that kahweol in human

TABLE 21.2

Some Important Investigations on Effects of Cafestol and Kahweol of Coffee on Humans, Animals, and Studies Pertaining to Anti-carcinogenicity and Genotoxicity

Sex and age	Dose and duration	Results	Reference
Anti-carcinogenicity and Genotoxicity			
Sprague-Dawley, 34 days old	Diet containing 10% or 20% of green coffee beans containing cafestol and kahweol for 14 d	Mammary tumors noticed in 75% and 44% of the animals for 10% and 20% dose.	Wattenberg (1983)
Sprague-Dawley, 34 days old	Instant coffee comprising cafestol and kahweol palmitates, diet containing of 10% instant coffee served for 14 days *prior to* DMBA administration (p.o.; 12 mg in 1 mL olive oil on day 15)	Mammary tumors seen in 75% of animals against 100% in the control group.	Wattenberg and Lam (1984)
Salmonella typhimurium strain TM677	Cafestol, cafestol acetate, and kahweol acetate: 0.31, 0.62, 1.25, 2.5, and 5.0 mg mL^{-1} kahweol: 0.1–1.6 mg mL^{-1} palmitic acid esters of cafestol and kahweol: 0.06–5.0 mg mL^{-1}	Approximately 50% of bacteria lived at the highest concentration for kahweol+S9: ~ – S9 treatment; a dose-reliant bacterial existence detected (with range 0.015–1.0 mg mL^{-1}, survival was 91% and 3.2%, respectively.	Pezzuto et al. (1989)
Syrian golden	Green coffee beans containing cafestol and kahweol palmitates: Group 1, normal chow; Group 2, 20% green coffee beans in chow	All animals in Group 1 were observed to possess multiple tumors. But in Group 2 animals, occasional tumors of the pouch and lip were observed. Gross tumors were found in 75% of in Group 1 and 22% in Group 2 animals.	Miller et al. (1988)
Human studies			
Human hepatocytes THLE cell line	Kahweol and cafestol (0–4 µg mL^{-1}) for 24, 48 or 72 h	Inhibition of AFB1 adduct formation in human hepatocytes by cafestol and kahweol reported. Stimulation of detoxifying enzyme and inhibition of AFB1-activating enzyme expression are the mechanisms that underwrite the protective effect.	Cavin et al. (2003)
Human endothelial cell line ECV304	Kahweol (10 µM) for 1 h	The TNFα-induced VCAM-1 and ICAM-1 expression was inhibited by kahweol. Also, it thwarts the adhesion of monocytes to endothelial cells via a mechanism including the inhibition of the JAK2-PI3K/Akt-NF-κB signaling pathways in retort to TNFα.	Kim et al. (2006b)
Human neuroblastoma SH-SY5Y	Kahweol (10 µM)	Cytoprotection against oxidative injury was evident from the kahweol induced HO-1 expression in dopaminergic SH-SY5Y neuronal cells. Kahweol also tempts Nrf2 nuclear translocation, which is upstream of kahweol-triggered HO-1 expression, and activates PI3K/Akt and p38 phosphorylation. The PI3K/Akt and p38 pathways are involved in kahweol-induced Nrf2 nuclear translocation, HO-1 expression, and cytoprotection.	Hwang and Jeong (2008)
Human leukemia U937 cells	kahweol (1.0–10 µM) for 24 h	Apoptosis by Akt dephosphorylation, JNK instigation, and downregulation of anti-apoptotic proteins was detected. Because of the apoptosis persuading capacity of kahweol could make it a perhaps active/deterrent and/or therapeutic agent against leukemia.	Oh et al. (2009)
The human lung cancer cell line A549	Kahweol 20 µM	Inhibition of A549 cell growth and apoptosis via downregulation of the STAT3 signaling pathway was observed in upon kahweol treatment.	Kim et al. (2006a)
Human umbilical endothelial cells (HUVEC)	Cafestol (0–20 µM)	Cafestol decreased the phosphorylation of Akt. Subsequently inhibited the migration, tube formation, and nitric oxide production of the VEGF-stimulated HUVECs. These consequences offer that cafestol disturbs the angiogenesis of HUVECs by regulating the phosphorylation of FAK and Akt.	Wang et al. (2012)
MSTO-211H and H28 cells, a type of human mesothelioma cells	82.07 µM cafestol and 56.00 µM khaweol 48 h incubation	Malignant pleural mesothelioma was influenced by both cafestol and kahweol that attributed to their chemoprotective ability. They were observed to regulate sp1 target protein, ensuing in apoptosis by the suppression of sp1 level in MSTO-211H cells.	Lee et al. (2012)

(*Continued*)

TABLE 21.2 (Continued)

Sex and age	Dose and duration	Results	Reference
Human microvascular endothelial cells (HMVECs)	Cafestol palmitate and kahweol palmitate (50, 75, and 100 mM) for 24 h	Several steps involved in angiogenesis were inhibited by cafestol palmitate and kahweol palmitate. Preventing the release of VEGF to the HMVEC culture medium via the repression of the Akt signaling pathway besides its inhibitory properties on phosphorylates (active) VEGFR-2 was obviously showed.	Wang et al. (2012)
Human oral squamous cancer cell (OSCC) lines HN22 and HSC4	Khaweol (0–60 μM) 48-h treatment	Cell growth hang-up action and persuades apoptosis induction by Sp1 in existence of kahweol displayed, after that regulation of the Sp1 target proteins and apoptosis connected proteins shown.	Chae et al. (2014)
Human renal carcinoma Caki cells	Cafestol (30 μM)	Cafestol acts as a robust enhancer of ABT-737-induced apoptosis in cancer cells, which have Mcl-1 facilitated resistance to ABT-737. Amazingly reduced tumor formation and induced apoptosis in a xenograft model that established with collective treatment with cafestol and ABT-737. The main mechanism of cafestol-mediated ABT-737-induced apoptosis may be downregulation of Mcl-1 expression and upregulation of Bim expression.	Woo et al. (2014)
The head and neck squamous cell carcinoma cell lines (SCC25, CAL27, FaDu)	Cafestol (50–100 μM) for 48 and 72 h	Treatment with cafestol resulted in increased apoptosis. Cafestol with irradiation caused to a notable fall of clonogenic survival in the cell lines SCC25 and CAL27.	Kotowski et al. (2015)
KHYG-1 NK leukemia cells	Cafestol and khaweol (30 μM) for 72 h	Together, cafestol and kahweol caused intensification the level of granzyme B by activating the transcription factors that regulate the granzyme B gene promoter. The better levels of granzyme B might add to the higher cytolytic activity of KHYG-1 cells.	Saito et al. (2015)
HT-29 human colorectal cancer cells	Kahweol (10–200 μM) for 24 h	Colorectal tumor cell growth inhibition by promoting apoptosis was observed, presumably by suppressing HSP70 expression.	Choi et al. (2015)
Human colon cancer cell lines, HCT116, SW480	Kahweol (0, 12.5, 25, 50 μM) for 24 and 48 h	Strikingly inhibited the multiplying of human colorectal cancer cell lines HCT116 and SW480. Reduced cyclin D1 protein level in both these cells.	Park et al. (2016)
Non-small cell lung cancer (NSCLC) cell lines NCI-H358 and NCI-H1299	Kahweol (30, 60 and 90 μM) 24 or 48 h	The kahweol-treated cells exhibited diminished cell viability, superior nuclear condensation, and a better number of Annexin V-positive NSCLC cells.	Jeon et al. (2016)
Human colorectal cancer cell lines (HCT116, SW480, LoVo, HT-29)	Kahweol (50 μM)	Increased apoptosis in human colorectal cancer cells observed. Similarly observed augmented ATF3 expression through the transcriptional activity. ATF3 overexpression increased kahweol-mediated cleaved PARP. Inhibition of ERK1/2 and GSK3β blocked kahweol-mediated ATF3 expression.	Park et al. (2017)
Human prostate cancer cell lines (PC-3, DU145, LNCaP)	Kahweol acetate and cafestol (30 μM) for 48 h	Both these diterpenes synergistically repressed the multiplication and passage of prostate cancer cells in a dose-dependent manner.	Iwamoto et al. (2019)
Endothelial cells (HUVECs)	Cafestol (3 and 10 μM) for 12 h	Investigations in vascular endothelial cells revealed the potential of cafestol in attenuation of cyclic-strain-stimulated intercellular adhesion molecule-1 (ICAM-1), monocyte chemoattractant protein-(MCP-)1, and interleukin-(IL-)8 secretion. Also, inhibition of the cyclic-strain-induced phosphorylation of extracellular signal-regulated kinase and p38 MAPK by cafestol verified.	Hao et al. (2018)
Human neuroblastoma SH-SY5Y cell line	Kahweol (0.1–10 μM) for 12 h	Kahweol at 10 μM exhibited mitochondrial protection in methylglyoxal-treated human neuroblastoma SH-SY5Y cells.	De Olivera et al. (2020b)
Jurkat cells	Kahweol (0, 10, 20, 40 mM) for 10 min,	Evaluated in PHA/PMA activated lymphocytic Jurkat cells via suppressing the phosphorylation of ERK and c-Fos.	Park et al. (2020)

Sex and age	Dose and duration	Results	Reference
Animal studies			
Female apoE*3-Leiden, female LDLR1/2, WT C57BL/6 mice	0.05% (wt/wt) cafestol and 0.025% (wt/wt) kahweol (high-cafestol diet), 0.01% (wt/wt) cafestol and 0.005% (wt/wt) kahweol (low-cafestol diet)	Delays bile acid production by downregulation of together the neutral and the acidic pathway, leading to a fall in expression of the LDLR and an elevated up secretion of cholesteryl esters in VLDL.	Post et al. (2000)
Sprague-Dawley rat primary hepatocytes	Kahweol and cafestol (0–4 µg mL^{-1}) for 24, 48 or 72 h	Cafestol and kahweol constrain B[a]P–DNA adduct formation in primary rat hepatocytes.	Cavin et al. (2003)
male F344 rats	Kahweol and cafestol	Kahweol and cafestol could raise the hepatic activity of O-6-methylguanine -DNA methyltransferase in a dose-dependent manner.	Huber et al. (2003)
Mouse macrophages cell RAW 264.7	Kahweol (5 µM) for 24 h	Nitric oxide production repressed and suppression of iNOS gene transcription via the inhibition of NF-kB activation in LPS-activated macrophages displayed.	Kim et al. (2004)
Male ICR mice	Kahweol and cafestol (100 mg kg^{-1}), 3 d	Kahweol and cafestol have a compelling hepatoprotective action upon CCl4-induced hepatic damage in mice. The hepatoprotective effects of kahweol and cafestol attributed to their ability to block the bioactivation of CCl4 by inhibiting CYP2E1 activity.	Lee et al. (2007b)
RAW264.7 cells	Kahweol (0–100 µM) for 1 h	Inhibited NO and PGE2 production in LPS-treated RAW264.7 cells at the transcriptional level. Kahweol downregulated the phosphorylation of nuclear NF-kB and STAT-1 but not c-Jun and c-Fos. This kahweol suppressed the phosphorylation of Akt and JAK2, which are upstream signaling components for the activation of NF-k B and STAT-1.	Shen et al. (2010)
Male Wistar rats	Kahweol (40–80 µg)	Exerted antinociceptive activity, and could also secondarily bring peripheral antinociception by freeing endogenous opioid peptides that leads to the instigation of opioid receptors on the terminals of primary afferent neurons.	Guzzo et al. (2015)
KKAy mice	Cafestol	Fasting plasma glucose was 28%–30% lower, fasting glucagon was 20% lower, and insulin sensitivity improved by 42% in the high-cafestol group compared with the control group KKAy mice. Improvement in insulin secretion from isolated islets by 75%–87% observed.	Mellbye et al. (2017)
3T3-L1 cell line	Kahweol (25 µg mL^{-1})	Decreased lipid buildup and expression levels of adipogenesis and lipid accrual-related factors detected. Kahweol suggestively improved AMP activated protein kinase (AMPK) instigation.	Baek et al. (2017)
Twenty 9-week-old male, specific pathogen-free C57BL/6J mice	Dialyzed coffee extract	In vivo protein expression changes detailed C57BL/6J in mouse livers treated with dialyzed coffee extract, which approves that DCE-2.5 and DCE-5 were safer and more favorable to murine hepatocytes than DCE-10.	Yoon et al. (2018)
C57B/6 primary Kupffer cells (KC) and primary hepatocytes (HC)	Kahweol	Kahweol displays a defensive outcome by constraining liver inflammation in primary Kupffer cells and primary hepatocytes cultures along with their co-cultures. Kahweol reduced the LPS-induced production of IL-1α, IL-1β, IL-6, and TNFα.	Seo et al. (2018)
Male C57BL/6 wild-type (WT) mice, 8–10 weeks old	Kahweol (5 mg kg^{-1})	Kahweol protects against traumatic brain injury (TBI) by plummeting immune responses and may serve as a potential therapeutic intervention for TBI patients.	Lee et al. (2019)
Primary cultures of rat neonatal ventricular myocytes	cafestol (3, 10 µM) for 12 h	Cafestol (3–10 uM) pre-treatment to neonatal rat cardiomycetes pointedly disallowed U-II-induced cardiomycyte hypertrophy through Nrf2/HO-1 activation and inhibition of redox signaling, ensuing in cardioprotective effects.	Hao et al. (2019)
Male Balb/c mice 96-week-old	Cafestol (20 mg kg^{-1} and 40 mg kg^{-1})	Alleviated serum liver enzyme and pathological injury, but curbed the discharge of inflammation mediators, comprising TNFα and IL-1β, and hepatocellular apoptosis and autophagy. The hepatic protective effects of cafestol on HIRI partly was owing to the inhibition of ERK/PPARγ pathway.	Ji et al. (2020)

lung adenocarcinoma A549 cells could prompt apoptosis by downregulation of the STAT3 signals pathway. The exact system of apoptotic impact was additionally concentrated on human colorectal disease cells by Lee et al. (2007a). Their investigations uncovered that kahweol initiates apoptosis through upregulation of ERK1/2 OR GSK3β-subordinate ATF3 expression. Also, kahweol found to build the cleavage of PARP in colorectal disease cell lines. Kahweol inhibited NO and PGE2 production in LPS-treated RAW264.7 cells at the transcription level (Shen et al., 2010).

Investigations on human microvascular endothelial cells (HMVECs) confirmed the inhibitory role of cafestol and kahweol on various steps involved in angiogenesis (Wang et al., 2012). Cafestol affects the angiogenesis of HUVECs by regulating the phosphorylation of FAK shown in human umbilical endothelial cells (Wang et al., 2012). Likewise, chemotherapeutic effect of cafestol and kahweol was evident due to its influence on malignant pleural mesothelioma (Lee et al., 2012), wherein cafestol and kahweol regulate sp1 target protein, resulting in apoptosis by the suppression of sp1 level in MSTO-211H cells. Woo et al. (2014) explained the main mechanism of cafestol-mediated ABT-737-induced apoptosis may be due to downregulation of Mc1–1 expression and upregulation of Bim expression. Treatment with cafestol resulted in increased apoptosis in the head and neck squamous cell carcinoma cell lines (SCC25, CAL27, FaDu), wherein a noteworthy lessening of clonogenic existence in the cell lines SCC25 and CAL27 was recorded (Kotowski et al., 2015). It has been affirmed that cafestol and kahweol increase the degree of granzyme B, which can upgrade cytotoxic action that was demonstrated in KHYG-1 NK leukemia cells, wherein cafestol and kahweol has been found in activating the transcription factors that control the granzyme B gene promoter (Saito et al., 2015).

21.4.4 Antinociceptive activity and colorectal cancer inhibition

Kahweol is able to downregulate the phosphorylation of nuclear NF-kB and SAT-1 but not c-Jun and c-Fos. The antinociceptive activity of kahweol was shown in male Wistar rats by Guzzo et al. (2015). Choi et al. (2015) exhibited the impact of kahweol in HT-29 human colorectal cancer growth cells. It was discovered that kahweol could anticipate the improvement of colorectal tumor cells by inducing apoptosis. Decreased cell viability and enhanced nuclear condensation was another feature displayed by kahweol-treated non-small cell lung cancer cell lines NCI-HI299 (Jeon et al., 2016). Kahweol at different concentrations (0–50 μM) when administered to human colon cancer cell lines HCT116 and SW480, followed by observations at 24 and 48 hours, revealed a striking inhibitory effect on the proliferation of these cells (Park et al., 2016). In addition, kahweol-interceded cyclin D1 degradation may add to the hindrance of the multiplication in human colorectal cancer cells (Park et al., 2016). Further studies on the effect of kahweol at 50 μM established increased apoptosis in human colorectal cancer cell lines HCT116, SW480, LoVo, and HT-29. It was observed that inhibition of ERK1/2, JNK, and GSK3β blocked kahweol-mediated ATF3 expression in these cells (Park et al., 2017).

21.4.5 Anti-adipogenesis and antidiabetic potential

Kahweol supports to less lipid buildup and expression levels of adipogenesis and lipid-accumulation factors. Studies in this regard in 3T3-L1 cell line by using kahweol at 25 μg mL^{-1} showed augmented AMP-activated kinase activation (Baek et al., 2017).

Elevated glucose uptake in muscle cells under 10^{-8} M cafestol treatment was reported and also improved insulin secretion in clonal rat insulinoma cell line INS-1E was reported (Mellbye et al. 2015); however, studies pertain to underlying mechanism are lacking. Cafestol at low (0.4 mg d^{-1}) to high dose (1.2 mg d^{-1}) improved diabetic rats biochemical profile after 10 weeks, wherein a notable slump in fasting blood glucose and fasting glucagon, along with improvement in secretion of insulin and rise in sensitivity was recorded in a T2D murine model KKAY mice (Mellbye et al., 2017). Anti-adipogenesis effect of kahweol (25 μg mL^{-1}) reduced lipid accumulation in 3T3-L1 cells was found to be through downregulation of PPARγ (Kim et al. 2018). Adenosine 5′-monophosphate-activated protein kinase (AMPK) activation contributes to improved glucose uptake and reduced lipid accumulation. Kahweol, being the activator of AMPK, could be a choice for metabolic disorders such as diabetes (Zhang et al., 2009); however, further studies and clinical trials in this regard are required to substantiate their exact role.

21.4.6 The anti-inflammatory effect of coffee diterpenes

The anti-inflammatory effect of kahweol was explored by Seo et al. (2018) in C57B/6 primary Kupfer cells and hepatocytes, wherein kahweol could decline the LPS-stimulated inflammatory cytokines, which are connected to the downregulation of NFkB, STAT3, and p38 MAPK. Cafestol fundamentally could lessen the total expression of inflammatory molecules in endothelial cells (HUVECs). The discharge of ICAM-1, MCP-1, and IL-8 was subdued by cafestol (Hao et al., 2018). Additionally, in cyclic-strain-treated HUVECs, cafestol restrained phosphorylation of ERK and p38 MAPK. The mechanism of activity is obviously identified with HO-1 and Sirt1 upregulation (Hao et al., 2018). Dialyzed coffee extracts at lower concentrations are reported to be beneficial to murine hepatocytes when the same were tested against 9-week-old male pathogen-free C57BL/6J mice (Yoon et al., 2018). Cardioprotective effects of cafestol were observed in primary cultures of rat neonatal ventricular myocytes, wherein cafestol significantly prevented U-II-induced cardiomyocyte hypertrophy through Nrf2/HO-I activation and inhibition of redox singling (Hao et al., 2019). Kahweol able to show protective response against traumatic brain injury (TBI) by reducing immune responses as shown in mice (Lee et al., 2019). Further studies have shown cytoprotective effect of kahweol in human neuroblastoma cells SH-SY5Y (Ribas et al., 2019). So on the basis of the aforementioned existing literature, it is understood that cafestol and kahweol functional attributes are mainly due to various mechanisms pertain to downregulation of inflammatory mediators through apoptosis of tumor cells and increase in glutathione.

Recent studies in experimental animal models, such as 96-week-old male Balb/c mice when given cafestol at 20- and 40 mg kg^{-1} doses, alleviate serum liver enzyme, a pathological injury, but at the same time could suppress the release of inflammation mediators that include TNFα and IL-1β and hepatocellular apoptosis and autophagy (Ji et al., 2020). De Olivera et al. (2020a) saw that kahweol exhibited a cancer prevention agent impact on the layers of mitochondria obtained from the methylglyoxal-treated SH-SY5Y cell lines. In general, the methylglyoxal incites mitochondria-related redox and bioenergetics weakness. Kahweol pretreatment at 10 μM altogether repressed the methylglyoxal-induced impacts through a mechanism associated with activation of the signaling pathways PI3K/Akt and p38 MAPK/Nrf2. At this point these leads are preliminary and further examinations are needed to affirm the impact of kahweol in vivo. A reduction in fat accumulation and improvement in locomotor activity by 20% and 38% was noted in *Caenorhabditis elegans* in presence of cafestol (Farias-Pereira et al., 2020). It was demonstrated that, cafestol could increase fat oxidation and energy expenditure via DAF-12 dependent pathway. Park et al. (2020) explored the impact of C&K on IL-2 cytokine contrasted with other coffee mixes in PHA/PMA-treated T-lymphocytic Jurkat cells. IL-2 production was not curbed by trigonelline, niacin, or chlorogenic acids. Conversely, cafestol and kahweol could hinder IL-2 item at an exceptionally low fixation (10 mM). It was discovered that ERK phosphorylation just as c-Fos phosphorylation in PHA/PMA-treated Jurkat cells was repressed by kahweol, which demonstrates that the IL-2 articulation in the Jurkat cells might be connected to the ERK and c-Fos pathways.

21.4.7 Other aspects of diterpenes

Overall, cafestol and kahweol display correlated biological actions but not exactly the same, and the reason could be owing to the occurrence of one conjugated double bond on the furan ring of the kahweol.

In humans, cafestol and kahweol were recovered from the feces but not from urine. However, cafestol and kahweol as simple conjugates of glucuronic or sulfuric acid in those treated with β-glucuronidase (Raymond, 1999) were recorded in urine as excretes, which accounts for up to 6% of ingested cafestol and 3% of kahweol. Many applications have been patented for cafestol and kahweol coffee bean oil, which contains cafestol and kahweol in cosmetic compositions (Bertholet, 1987; Karl and Claire, 2012). In combination with suitable carrier molecules that appropriate to respective formulation pertain to cosmetics or pharmacy, various topical compositions comprising cafestol can be used in the treatment of skin related problems.

21.5 Conclusion

The existing literature on the effect of coffee diterpenes influence on humans is a dose-dependent response and is also contingent on the content of these diterpenes in coffee brew, which in turn varies with the method of coffee brew preparation. As most of the investigations in human volunteers are devoted on looking at hypercholesteremic effects, other studies on cell lines and experimental animals indicate the anti-carcinogenic potential of coffee diterpenes. In-depth studies on these aspects at the mechanism level further elucidate the actual role of cafestol and kahweol at the cellular level.

REFERENCES

Anonymous. Most Expensive Coffee. November 17, 2011.

Baek JH, Kim NJ, Song JK, Chun KH. Kahweol Inhibits Lipid Accumulation and Induces Glucose-Uptake through Activation of AMP-Activated Protein Kinase (AMPK). *BMB Rep.* 2017; 50 (11): 566–71. https://doi.org/10.5483/BMBRep.2017.50.11.031

Baker PS. *The Changing Climate for Sustainable Coffee.* TW209TY, UK. ASIC 2012, 24th International Conference on Coffee Science, November 11–16, 2012. Costa Rica, pp-69. https://www.researchgate.net/publication/260596801

Bertholet R. Preparation of Cafestol. U.S. Patent No. 4,692,534. U.S. Patent and Trademark Office. Patent Full Text and Image Database 1987. Filed on April 3, 1986. Internet last accessed on August 8, 1999. https://patents.google.com/patent/US4692534A/en?oq=US4692534A

Beynen AC, Marianne PME. Weusten-Van Der Wouw, De Roos B, Katan MB. Boiled Coffee Fails to Raise Serum Cholesterol in Hamsters and Rats. *British J of Nutri.* 1996; 76 (5): 755–64. https://doi.org/10.1079/bjn1996008.

Cadden, ISH, Partovi N, Yoshida EM. Review Article: Possible Beneficial Effects of Coffee on Liver Disease and Function. *Alimentary Pharmacol and Ther.* 2007; 26 (1): 1–8. https://doi.org/10.1111/j.1365-2036.2007.03319.x.

Cavin CD, Bezencon C, Guignard G, Schilter B. Coffee Diterpenes Prevent Benzo[a]Pyrene Genotoxicity in Rat and Human Culture Systems. *Biochem and Biophysic Res Commun.* 2003; 306 (2): 488–95. f https://doi.org/10.1016/S0006-291X(03)00994-X.

Cavin CD, Scharf GH, Constable A, Huber WW, Schilter B. Cafestol and Kahweol, Two Coffee Specific Diterpenes with Anticarcinogenic Activity. *Food and Chem Toxicol.* 2002; 40 (8): 1155–63. https://doi.org/10.1016/S0278

Chae J-Il, Jeon Y-J, Shim J-H. Anti-Proliferative Properties of Kahweol in Oral Squamous Cancer through the Regulation Specificity Protein 1. *Phytotherapy Res.* 2014; 28 (12): 1879–86. https://doi.org/10.1002/ptr.5217.

Choi DW, Lim MS, Lee JW, Chun W, Lee SH, Nam YH, Park JM, et al. The Cytotoxicity of Kahweol in HT-29 Human Colorectal Cancer Cells Is Mediated by Apoptosis and Suppression of Heat Shock //Protein 70 Expression. *Biomolecules and Therap.* 2015; 23 (2): 128–33. https://doi.org/10.4062/biomolther.2014.133.

Cintia KSG, Maria BSS, Joao S.BGD, Luiz FP, Marta TB. *Composition and Antioxidant Activity of Roasted Arabica Coffee from Different Cultivars Growing in the Same Edapho-climatic Condition.* Departments de Fisiologia Vegetal-Laboratorio de Ecofisiologia, institute Agronomico do parana-lapar Londrina Brazil centro technologic Cocari, Mandaguari, PR, Brazil. ASIC, 24th International Conference on Coffee Science, November 11–16, 2012; Costa Rica, p. 106.

De Castro RD, Marraccini P. Cytology, Biochemistry and Molecular Changes during Coffee Fruit Development. *Brazilian J of Plant Physiol.* 2006; 18 (1): 175–99. https://doi.org/10.1590/S1677-04202006000100013.

De Oliveira NA, Sandini TM, Cornelio-Santiago HP, Martinelli ECL, Raspantini LER, Raspantini PC, Momo C, de Oliveira AL, Fukumasu H. Acute and Subacute (28 Days) Toxicity of Green Coffee Oil Enriched with Diterpenes Cafestol and Kahweol in Rats. *Regulatory Toxicol and Pharmacol*, 2020a; 110 (November 2019): 104517. https://doi.org/10.1016/j.yrtph.2019.104517.

De Oliveira RM, Custódio de Souza IC, Fürstenau CR. Mitochondrial Protection Promoted by the Coffee Diterpene Kahweol in Methylglyoxal-Treated Human Neuroblastoma SH-SY5Y Cells. *Neurotoxicity Res*. 2020b; 37 (1): 100–10. https://doi.org/10.1007/s12640-019-00107-w.

De Roos B, Van Der Weg G, Urgert R, Van De Bovenkamp P, Charrier A, Katan MB. Levels of Cafestol, Kahweol, and Related Diterpenoids in Wild Species of the Coffee Plant *Coffea*. *Journal of Agr and Food Chem*. 1997; 45 (8): 3065–69. https://doi.org/10.1021/jf9700900

Di Simplicio P, Jensson H, Mannervik B. Effects of Inducers of Drug Metabolism Mouse Glutathione Transferase Basic Hepatic Forms Of Mouse Glutathion Transferase. *Biochemical Journal*. 1989; 263 (3): 679–85. https://doi.org/10.1042/bj2630679.

Farias-Pereira R, Kim E, Park Y. Cafestol Increases Fat Oxidation and Energy Expenditure in Caenorhabditis Elegans via DAF-12-Dependent Pathway. *Food Chem*. 2020; 307: 125537. https://doi.org/10.1016/j.foodchem.2019.125537.

Garattini ES, Parker D, Gregoriadis G, Poste G. 1993 Monographs Negri. Institute for Pharmaceutical Research, n.d., pp. 143–44.

George SE, Ramalakshmi K, Rao LJM. A Perception on Health Benefits of Coffee. *Critical Rev in Food Sci Nutri*. 2008; 48 (5): 464–86. https://doi.org/10.1080/10408390701522445.

Guzzo LS, Romero TRL. Queiroz CM, Caliari MV, Azevedo AO, Perez AC, Duarte IDG. Involvement of Endogenous Opioid Peptides in the Peripheral Antinociceptive Effect Induced by the Coffee Specific Diterpene Kahweol. *Pharmacological Reports*. 2015; 67 (5): 1010–15. https://doi.org/10.1016/j.pharep.2015.02.009.

Halvorsen B, Ranheim T, Nenseter MS, Huggett AC, Drevon CA. Effect of a Coffee Lipid (Cafestol) on Cholesterol Metabolism in Human Skin Fibroblasts. *Journal of Lipid Res*. 1998; 39 (4): 901–12. https://doi.org/10.1016/s0022-2275(20)32576-1.

Hao WR, Sung LC, Chen CC, Chen PY, Cheng TH, Chao HH, Liu JC, Chen JJ. Cafestol Inhibits Cyclic-Strain-Induced Interleukin-8, Intercellular Adhesion Molecule-1, and Monocyte Chemoattractant Protein-1 Production in Vascular Endothelial Cells. *Oxidative Med and Cellular Long*. 2018; 2018. https://doi.org/10.1155/2018/7861518.

Hao WR, Sung LC, Chen CC, Hong HJ, Liu JC, Chen JJ. Cafestol Activates Nuclear Factor Erythroid-2 Related Factor 2 and Inhibits Urotensin II-Induced Cardiomyocyte Hypertrophy. *American J of Chinese Med*. 2019; 47 (2): 337–50. https://doi.org/10.1142/S0192415X19500162.

Huber WW, Scharf G, Nagel G, Prustomersky S, Schulte-Hermann R, Kaina B. Coffee and Its Chemopreventive Components Kahweol and Cafestol Increase the Activity of O6-Methylguanine-DNA Methyltransferase in Rat Liver – Comparison with Phase II Xenobiotic Metabolism. *Mutation Research – Fundamental and Molecular Mechanisms of Mutagenesis*. 2003; 522 (1–2): 57–68. https://doi.org/10.1016/S0027-5107(02)00264-6.

Hwang YP, Jeong HG. The Coffee Diterpene Kahweol Induces Heme Oxygenase-1 via the PI3K and P38/Nrf2 Pathway to Protect Human Dopaminergic Neurons from 6-Hydroxydopamine-Derived Oxidative Stress. *FEBS Letters*. 2008; 582 (17): 2655–62. https://doi.org/10.1016/j.febslet.2008.06.045.

Ivamoto ST, Pagiatto NF, Kitzberger CSG, Schola MBS, Vieira LGE, Domingues DS, Pereira LFP. Transcriptional Pattern of CYPS Candidate Genes and Biochemical Aspects of Cafestol and Kahweol Biosynthesis in Coffee Arabica. *In Simposio Brasileiro De Genetica Molecular De Plantas*. 2013, Brazilian Society of Genetics (SBG), Riberirao Preto, SP, Brazil.

Iwamoto H, Izumi K, Natsagdorj A, Naito R, Makino T, Kadomoto S, Hiratsuka K, et al. Coffee Diterpenes Kahweol Acetate and Cafestol Synergistically Inhibit the Proliferation and Migration of Prostate Cancer Cells. *Prostate*. 2019; 79 (5): 468–79. https://doi.org/10.1002/pros.23753.

Jeon YJ, Bang W, Cho JH, Lee RH, Kim SH, Kim MS, Park SM, et al. Kahweol Induces Apoptosis by Suppressing BTF3 Expression through the ERK Signaling Pathway in Non-Small Cell Lung Cancer Cells. *Int J of Oncol*. 2016; 49 (6): 2294–302. https://doi.org/10.3892/ijo.2016.3727.

Jeszka-Skowron M, Zgoła-Grześkowiak A, Grześkowiak T. Analytical Methods Applied for the Characterization and the Determination of Bioactive Compounds in Coffee. *Eur Food Res Technol*. 2015; 240: 19–31.

Ji J, Wu L, Feng J, Mo W, Wu J, Yu Q, Li S, et al. Cafestol Preconditioning Attenuates Apoptosis and Autophagy during Hepatic Ischemia-Reperfusion Injury by Inhibiting ERK/PPARγ Pathway. *Int Immunopharmacol*. 2020; 84 (April). https://doi.org/10.1016/j.intimp.2020.106529.

Karl I, Claire MC. Slimming Cosmetic Composition Comprising Cafestol or Kahweol. Patent no. ep 1 648 403 B1. 2012. https://data.epo.org/publication-server/rest/v1.0/publicationdates/20120815/patents/EP1648403NWB1/document.html

Kim HG, Hwang YP, Jeong HG. Kahweol Blocks STAT3 Phosphorylation and Induces Apoptosis in Human Lung Adenocarcinoma A549 Cells. *Toxicology Lett*. 2006a; 187 (1): 28–34. https://doi.org/10.1016/j.toxlet.2009.01.022.

Kim HG, Kim JY, Hwang YP, Lee KJ, Lee KY, Kim DH, Kim DH, Jeong HG. The Coffee Diterpene Kahweol Inhibits Tumor Necrosis Factor-α-Induced Expression of Cell Adhesion Molecules in Human Endothelial Cells. *Toxicol Appl Pharmacol*. 2006b; 217 (3): 332–41. https://doi.org/10.1016/j.taap.2006.09.013.

Kim JS, Lee SG, Kang YG, Kwon TK, Nam JO. Kahweol Inhibits Adipogenesis of 3T3-L1 Adipocytes through Downregulation of PPARγ. *Natural Product Res*. 2018; 32 (10): 1216–19. https://doi.org/10.1080/14786419.2017.1326039.

Kim JY, Jung KS, Jeong HG. Suppressive Effects of the Kahweol and Cafestol on Cyclooxygenase-2 Expression in Macrophages. *FEBS Lett*. 2004; 569 (1–3): 321–26. https://doi.org/10.1016/j.febslet.2004.05.070.

Kölling-Speer I, Strohschneider S, Speer K. Determination of Free Diterpenes in Green and Roasted Coffees. *HRC J of High-Res Chrom*. 1999; 22 (1): 43–46. https://doi.org/10.1002/(SICI)1521-4168(19990101)22:1<43::AID-JHRC43>3.0.CO;2-P.

Kotowski U, Heiduschka G, Seemann R, Eckl-Dorna J, Schmid R, Kranebitter V, Isabella Stanisz I, Markus Brunner M,

Lill C, Thurnher D. Effect of the Coffee Ingredient Cafestol on Head and Neck Squamous Cell Carcinoma Cell Libes. *Strahlentherapie und Onkologie*. 2015; 191 (6): 511–17. https://doi.org/10.1007/s00066-014-0807-x.

Kurzrock T, Kolling-Speer I, Speer K. Identification of Dehydrocafestol Fatty Acid Esters in Coffee. In *Proceeding 20th in Symposium on Capillary Chromotogr*, edited by P. Sandra, AJ Rackstraw. No. 27. Naxos Software Solution, M. Schaefer, Schriesheim, Germany. Dissertation, Technical University of Dresden, 1998.

Laurent B, Jean-Claude M, Noel D, Vincent C, Jacques L, Celine A, Didier M. Evolution of Physicochemical and Organoleptic Properties of Brewed Coffee during Storage Related to Green Coffee Variety and Roasting Degree. *UMR Qualisud (CIRAD), Universite Monitpeller* (1): 34095. Montipeller Cedex 5, France. ** CAFÉ RICHARD. 106 rue du Fosse Blanc-92230 Gennevilliers-France. ASIC 2012, 24th International Conference on Coffee Science, November 11–16, 2012, Costa Rica, p. 132.

Lee HF, Lin JS, Chang CF. Acute Kahweol Treatment Attenuates Traumatic Brain Injury Neuroinflammation and Functional Deficits. *Nutrients*. 2019; 11 (10): 1–15. https://doi.org/10.3390/nu11102301

Lee KA, Chae JI, Shim JH. Natural Diterpenes from Coffee, Cafestol and Kahweol Induce Apoptosis through Regulation of Specificity Protein 1 Expression in Human Malignant Pleural Mesothelioma. *Journal of Biomed Sci*. 2012; 19 (1): 1–10. https://doi.org/10.1186/1423-0127-19-60.

Lee KJ, Choi JH, Jeong HG. Hepatoprotective and Antioxidant Effects of the Coffee Diterpenes Kahweol and Cafestol on Carbon Tetrachloride-Induced Liver Damage in Mice. *Food and Chem Toxicol*. 2007b; 45 (11): 2118–25. https://doi.org/10.1016/j.fct.2007.05.010

Lee KJ, Inoue M, Otani T, Iwasaki M, Sasazuki S, Tsugane S. Coffee Consumption and Risk of Colorectal Cancer in a Population-Based Prospective Cohort of Japanese Men and Women. *International Journal of Cancer*. 2007a; 121 (6): 1312–18. https://doi.org/10.1002/ijc.22778.

Liu Y, Wang H, Ye HC, Li GF. Advances in the Plant Isoprenoid Biosynthesis Pathway and Its Metabolic Engineering. *Journal of Integrative Plant Biol*. 2005; 47 (7): 769–82. https://doi.org/10.1111/j.1744-7909.2005.00111.x.

Mazzafera P, Soave D, Zullo MAT, Guerreiro Filho O. Oil Content of Green Beans from Some Coffee Species. *Bragantia*. 1998; 57: 45–48. http://dx.doi.org/10.1590/S0006-87051998000100006

Mellbye FB, Jeppesen PB, Hermansen K, Gregersen S. 2015. Cafestol, a Bioactive Substance in Coffee, Stimulates Insulin Secretion and Increases Glucose Uptake in Muscle Cells: Studies in Vitro. *Journal of Nat Prod*. 78 (10): 2447–51. https://doi.org/10.1021/acs.jnatprod.5b00481.

Mellbye FB, Jeppesen PB, Shokouh P, Laustsen C, Hermansen K, Gregersen S. Cafestol a Bioactive Substance in Coffee, Has Antidiabetic Properties in KKAy Mice. *J Nat Prod*. 2017; 80 (8): 2353–59.

Miller EG, Formby WA, Rivera-Hidalgo F, Wright JM. Inhibition of Hamster Buccal Pouch Carcinogenesis by Green Coffee Beans. *Oral Surgery Oral Med Oral Pathol*. 1988; 65 (6): 745–49. https://doi.org/10.1016/0030-4220(88)90022-9

Mohanan S, Satyanarayana KV, Sridevi V, Kalpashree, Giridhar P, Arun Chandrashekar A, Ravishankar GA. Evaluating the Effect and Effectiveness of Different Constructs with a Conserved Sequence for Silencing of *Coffea Canephora* N-Methyltransferases. *Journal of Plant Biochem and Biotechnol*. 2014; 3 (4): 399–409. https://doi.org/10.1007/s13562-013-0224-8.

Mullen WB, Nemzer B, Ou A. Stalmach J, Hunter MN, Clifford, Combet E. The Antioxidant and Chlorogenic Acid Profiles of Whole Coffee Fruits Are Influenced by the Extraction Procedures. *Journal of Agri Food Chem*. 2011; 59 (8): 3754–62. https://doi.org/10.1021/jf200122.

Oh JH, Lee JT, Yang ES, Chang JS, Lee DS, Kim SH, Choi YH, Park JW, Kwon TK. The Coffee Diterpene Kahweol Induces Apoptosis in Human Leukemia U937 Cells through Down-Regulation of Akt Phosphorylation and Activation of JNK. *Apoptosis*. 2009; 14 (11): 1378–86. https://doi.org/10.1007/s10495-009-0407-x.

Park GH, Song HM, Jeong JB. The Coffee Diterpene Kahweol Suppresses the Cell Proliferation by Inducing Cyclin D1 Proteasomal Degradation via ERK1/2, JNK and GKS3β-Dependent Threonine-286 Phosphorylation in Human Colorectal Cancer Cells. *Food and Chemical Toxicol*. 2016; 95: 142–48. https://doi.org/10.1016/j.fct.2016.07.008

Park GH, Song HM, Jeong JB. Kahweol from Coffee Induces Apoptosis by Upregulating Activating Transcription Factor 3 in Human Colorectal Cancer Cells. *Biomolecules and Ther*. 2017; 25 (3): 337–43. https://doi.org/10.4062/biomolther.2016.114.

Park JB. Kahweol Found in Coffee Inhibits IL-2 Production Via Suppressing the Phosphorylations of ERK and c-Fos in Lymphocytic Jurkat Cells. *Journal of Dietary Supplements*. 2020; (June 25): 1–11. https://doi.org/10.1080/19390211.2020.1784347.

Pezzuto JM, Dhammika Nanayakkara NP, Compadre CM, Swanson SM, Kinghorn AD, Guenthner TM, Sparnins VL, Lam LKT. Characterization of Bacterial Mutagenicity Mediated by 13-Hydroxy-Ent-Kaurenoic Acid (Steviol) and Several Structurally-Related Derivatives and Evaluation of Potential to Induce Glutathione S-Transferase in Mice. *Mutation Research/Genetic Toxicol*. 1989; 169 (3): 93–103. https://doi.org/10.1016/0165-1218(86)90088-1

Post SM, De Roos B, Vermeulen M, Afman L, Jong MC, Dahlmans VEH, Havekes LM, Katan FSMB, Princen HMG. Cafestol Increases Serum Cholesterol Levels in Apolipoprotein E*3- Leiden Transgenic Mice by Suppression of Bile Acid Synthesis. *Arteriosclerosis, Thrombosis, and Vascular Biol*. 2000; 20 (6): 1551–56. https://doi.org/10.1161/01.atv.20.6.1551.

Ranheim T, Halvorsen B, Huggett AC, Blomhoff R, Drevon CA. Effect of a Coffee Lipid (Cafestol) on Regulation of Lipid Metabolism in CaCo-2 Cells. *J Lipid Res*. 1995; 36 (10): 2079–89.

Ratnayake WMN, Pelletier G, Hollywood R, Malcolm S, Stavric B. Investigation of the Effect of Coffee Lipids on Serum Cholesterol in Hamsters. *Food and Chem Toxicol*. 1995; 33 (3): 195–201. https://doi.org/10.1016/0278-6915(94)00131-7.

Raymond T. Cafestol, Kahweol, Review of Toxicological Literature. Prepared for Scott Masten, Ph.D. 1999; National Institute of Environmental Health Sciences P.O. Box 12233 Research Triangle Park, North Carolina 27709 Contract No. N01-ES-65402. https://ntp.niehs.nih.gov/ntp/htdocs/chem_background/exsumpdf/cafestol_508.pdf

Reni S, Gerson GS, Flavio BM, Rosa AM, Sttela DVF, Fernando B. *Sensory Quality Alterations and Activity of Lea*

Proteins during the Coffee Beans Storage. Agronomic Institute (IAC) Campinas, SP, Brazil, Federal University of Lavras (UFLA) Larvas, MG, and Brazil. San Paulo State University (UNESP), Botucatu, SP, Brazil, ASIC 2012, 24th International Conference on Coffee Science, November 11–16, 2012, Costa Rica, p. 137.

Ribas FC, Custódio de Souza IC, Roberto de Oliveira M. The Effects of Kahweol, a Diterpene Present in Coffee, on the Mitochondria of the Human Neuroblastoma SH-SY5Y Cells Exposed to Hydrogen Peroxide. *Toxicology in Vitro*. 2019; 61 (July): 104601. https://doi.org/10.1016/j.tiv.2019.104601

Rustan AC, Halvorsen B, Huggett AC, Ranheim T, Drevon CA. Effect of Coffee Lipids (Cafestol and Kahweol) on Regulation of Cholesterol Metabolism in HepG2 Cells. *Arteriosclerosis, Thrombosis, and Vascular Biol*. 1997; 17 (10): 2140–49. https://doi.org/10.1161/01.ATV.17.10.2140.

Saito T, Abe D, Nogata Y. Coffee Diterpenes Potentiate the Cytolytic Activity of KHYG-1 NK Leukemia Cells. *Food Sci and Technol Res*. 2015; 21 (2): 281–84. https://doi.org/10.3136/fstr.21.281.

Sanders, TAB, Sandaradura S. The Cholesterol-Raising Effect of Coffee in the Syrian Hamster. *British Journal of Nutr*. 1992; 68 (2): 431–34. https://doi.org/10.1079/bjn1992010034.

Schilter B, Perrin I, Cavin C, Huggett AC. Placental Glutathione S-Transferase (GST-P) Induction as a Potential Mechanism for the Anti-Carcinogenic Effect of the Coffee- Specific Components Cafestol and Kahweol. *Carcinogenesis*. 1996; 17 (11): 2377–84. https://doi.org/10.1093/carcin/17.11.2377

Seo HY, Kim MK, Lee So-H, Hwang JS, Park KG, Jang BK. Kahweol Ameliorates the Liver Inflammation through the Inhibition of NF-κB and STAT3 Activation in Primary Kupffer Cells and Primary Hepatocytes. *Nutrients*. 2018; 10 (7): 863. http://doi.org/10.3390/nu10070863

Shen T, Park YC, Kim SH, Lee J, Cho JY. Nuclear Factor-KB/ Signal Transducers and Activators of Transcription-1-Mediated Inflammatory Responses in Lipopolysaccharide-Activated Macrophages Are a Major Inhibitory Target of Kahweol, a Coffee Diterpene. *Biological and Pharm Bull*. 2010; 33 (7): 1159–64. https://doi.org/10.1248/bpb.33.1159

Speer K, Kölling-Speer I. Lipids. In *Coffee Recent Developments*, edited by RJ Clarke, OG Vitzthum. 2001, pp. 331–49.

Speer K, Kolling-Speer I. The Lipid Fraction of the Coffee Bean. *Brazilian J of Plant Physiol*. 2006; 18 (1): 201–16. https://doi.10.1590/s1677-04202006000100014

Speer K, Mischnick P. 16-O-Methylcafestol - Ein Neues Diterpen Im Kaffee Entdeckung Und Identifizierung. *Zeitschrift Für Lebensmittel-Untersuchung Und -Forschung*. 1989; 189 (3): 219–22. https://doi.org/10.1007/BF01028067.

Sridevi V, Giridhar P. Recent Trends in Coffee Biotechnology towards Quality Improvement—A Review. *Indian J Bot Res*. 2008; 4: 5–12.

Sridevi V, Giridhar P. Influence of Altitude Variation on Trigonelline Content during Ontogeny of *Coffea Canephora* Fruit. *Journal of Food Stud*. 2013; 2 (1): 62–74. https://doi.org/10.5296/jfs.v2i1.3747.

Sridevi V, Giridhar P. Changes in Caffeine Content during Fruit Development in *Coffea Canephora* P. Ex. Fr. Grown at Different Elevations. *Journal of Biol Earth Sci*. 2014a; 4 (2): 168–75.

Sridevi V, Giridhar P. Establishment of Somaclonal Variants of Robusta Coffee with Reduced Levels of Cafestol and Kahweol. *In Vitro Cellular and Developmental Biol. Plant*. 2014b; 50 (5): 618–26. https://doi.org/10.1007/s11627-014-9613-5.

Sridevi V, Giridhar P. Antioxidant Potential of Free Diterpenes Cafestol and Kahweol Rich Extractives of Coffee Beans. *Canadian J of Basic and Appl Sci*. 2015; 03 (03): 78–84.

Sridevi V, Giridhar P. Variations in Diterpenes-Cafestol and Kahweol Content in Beans of Robusta Coffee Grown at Different Altitudes. *Proceedings of the National Academy of Sciences India Section B - Biological Sciences*. 2016; 86 (2): 291–97. https://doi.org/10.1007/s40011-014-0429-1.

Sridevi V, Giridhar P, Ravishankar GA. Free Diterpenes Cafestol and Kahweol in Beans and in vitro Cultures of *Coffea* Species. *Current Sci*. 2010; 99 (8): 1101–4. http://www.jstor.org/stable/24066120.

Sridevi V, Giridhar P, Ravishankar GA. Evaluation of Roasting and Brewing Effect on Antinutritional Diterpenes-Cafestol and Kahweol in Coffee. *Global J of Medical Res*. 2011; 11 (5): 1–7.

Suzana TI, David P, Pierre C, Pierre M, Lucia FP, Douglas S, Luiz VGE, Luiz PFP. *Diterpenes in Coffee Arabica: Biochemical Aspects and Transcriptional Analysis of Candidate CYPS Involved in Cafestol and Kahweol Biosynthesis*. ASIC 2012, 24th International Conference on Coffee Science, November 11–16, 2012; Costa Rica, p. 216.

Terpstra AHM, Katan MB, Weusten-Van der Wouw MPME, Nicolosi RJ, Beynen AC. Coffee Oil Consumption Does Not Affect Serum Cholesterol in Rhesus and Cebus Monkeys. *Journal of Nutr*. 1995; 125 (9): 2301–6. https://doi.org/10.1093/jn/125.9.2301

Urgert R, Katan MB. The Cholesterol-Raising Factor from Coffee Beans. *Annual Review of Nutr*. 1997; 17: 305–24. https://doi.org/10.1146/annurev.nutr.17.1.305.

Urgert R, Schulz GM, Angela, Katan MB. Effects of Cafestol and Kahweol from Coffee Grounds on Serum Lipids and Serum Liver Enzymes in Humans. *American Journal of Clinical Nutr*. 1995a; 61 (1): 149–54. https://doi.org10.1093/ajcn/61.1.149

Urgert R, van der Weg G, Kosmeijer-Schuil TG, van de Bovenkamp P, Hovenier R, Katan MB. Levels of the Cholesterol-Elevating Diterpenes Cafestol and Kahweol in Various Coffee Brews. *Journal of Agr and Food Chem*. 1995b. 43 (8): 2167–72. https://doi.org/10.1021/jf00056a039

Van Rooij J, Van der Stegen GHD, Schoemaker RC, Kroon C, Burggraaf J, Hollaar L, Vroon TFFP, Smelt AHM, Cohen AF. A Placebo-Controlled Parallel Study of the Effect of Two Types of Coffee Oil on Serum Lipids and Transaminases: Identification of Chemical Substances Involved in the Cholesterol-Raising Effect of Coffee. *American Journal of Clin Nutr*. 1995; 61 (6): 1277–83. https://doi.org/10.1093/ajcn/61.6.1277.

Wang S, Yeo CY, Mi JS, Haeng JH, Jae HP. Antiangiogenic Properties of Cafestol, a Coffee Diterpene, in Human Umbilical Vein Endothelial Cells. *Biochemical and Biophysical Res. Commun*. 2012; 421 (3): 567–71. https://doi.org/10.1016/j.bbrc.2012.04.046.

Wattenberg LW. Inhibition of Neoplasia by Minor Dietary Constituents. *Cancer Res*. 1983; 43 (5): 2448s–24553s.

Wattenberg LW, Lam LKT. Protective Effects of Coffee Constituents on Carcinogenesis in Experimental Animals. In *Banbury Report No. 17*, edited by B Mac Mahon, T

Sugimura. Coffee and Health; Symposium, October 1983. Cold Spring Harbor Laboratory, Cold Spring Harbor, NY, 1984, pp. 137–45.

Weusten-Van der Wouw MPME, Katan MB, Viani R, Huggett AC, Liardon R, Lund-Larsen PG, Thelle DS et al. Identity of the Cholesterol-Raising Factor from Boiled Coffee and Its Effects on Liver Function Enzymes. *Journal of Lipid Res.* 1994; 35 (4): 721–33. https://doi.org/10.1016/s0022-2275(20)41169-1

Woo SM, Min KJ, Seo BR, Nam JO, Choi KS, Yoo YH, Kwon TK. Cafestol Overcomes ABT-737 Resistance in Mcl-1-Overexpressed Renal Carcinoma Caki Cells through Downregulation of Mcl-1 Expression and Upregulation of Bim Expression. *Cell Death and Dis.* 2014; 5 (11). https://doi.org/10.1038/cddis.2014.472.

Yoon CS, Kim MK, Kim YS, Lee SK. In Vivo Protein Expression Changes in Mouse Livers Treated with Dialyzed Coffee Extract as Determined by IP-HPLC. *Maxillofacial Plastic and Reconstructive Surg.* 2018; 40 (1). https://doi.org/10.1186/s40902-018-0183-z.

Zhang BB, Zhou G, Li C. AMPK: An Emerging Drug Target for Diabetes and the Metabolic Syndrome. *Cell Metabol.* 2009; 9 (5): 407–16. https://doi.org/10.1016/j.cmet.2009.03.012.

22

Indoleamines and biogenic amines profiles in Coffea *species and human health effects*

Akula Ramakrishna, Soumya Mukherjee, Sridevi Vaddadi, and Parvatam Giridhar

CONTENTS

22.1 Introduction

Coffee is one of the commonly traded commodities in the transnational market. Commercial coffee production relies mainly on either *Coffea arabica* (Arabica) or *C. canephora* (Robusta; Vinod Kumar et al. 2006; Butt and Tauseef 2011). Robusta coffee beans are important for the production of soluble coffee. Indoleamines, namely, melatonin (*N*-acetyl-5-methoxytryptamine, MEL) and serotonin (5-hydroxytryptamine, SER), are neurotransmitters in animals and they have also been reported in plants (Badria 2002; Roshchina 2001). The existence of MEL and SER in plants has been established qualitatively by GC-MS or LC-MS studies (Cao et al. 2006; Kolar et al. 1997; Huang and Mazza 2011; Ramakrishna et al. 2011; Soumya Mukherjee and Ramakrishna 2018). Both MEL and serotonin are reported to be in abundant quantities in herbs and higher plants, especially in edible parts such as seeds (Rayne et al. 2010; Hano et al. 2017; Meng et al. 2017; Meng et al. 2018; Briguglio et al. 2018). Therefore, MEL- and SER-rich foods are considered as nutraceuticals (Ferrari et al. 2004; Iriti et al. 2006; Delgado et al. 2013; Sae-Teaw et al. 2013; Gomes Domingos et al. 2017; Briguglio et al. 2018; Salehi et al. 2019). Eating the palatable plant parts that constitute the MEL is reported to increase the circulating MEL level (Reiter et al. 2005), and this study has created a wide interest to screen various plant species for their presence. In view of the presence of MEL and SER in several edible plants, medicinal herbs, and seeds, their antioxidant activity is related to their therapeutic value for dealing with many human ailments, such as cardiovascular damage (Reiter and Tan 2003), Alzheimer's disease (Pappolla et al. 2000), and cancer (Sauer et al. 2001). Intake of tryptophan-rich foods can result in an increase of MEL levels both in tissue and in blood (Hardeland et al. 2005).

In the 1970s, melatonin was considered a by-product of the coffee extraction process. Subsequent reports revealed its presence in both lower and higher plants. The presence of SER was reported in leaves and seeds of *Coffea* (Strance 1997) and in green and roasted beans (Casal et al. 2004) and coffee wax (Kele and Ohmacht 1996). Being a neurological mediator, the SER content of *Coffea* appears to have significance from a health standpoint (Casal et al. 2004). The content of SER is also important and considered as chemical marker to distinguish Arabica and Robusta coffee samples. Bioactive amines profiles in instant coffee were reported (Da Silveira et al. 2007). Caffeine, the major metabolite in coffee, has been widely studied. Given the levels of SER and MEL in coffee and the importance of these bioactive molecules to food quality and well-being, it is commendable to explore the SER and MEL levels in different organs of *Coffea* ssp. This chapter describes the indoleamines profiles in *Coffea* ssp. and possible human health effects.

22.2 Immunolocalization of serotonin in different tissues of *C. canephora*

SER in various in vitro cultures of coffee and also in immature *C. canephora* fruit was localized by adopting the method of Perrot-Rechenmann and Gadal (1986). Such

DOI: 10.1201/9781003043133-27

immunolocalization techniques helps to identify SER efficiently in the vascular tissues of stem, root, somatic embryos, and the endocarp region (husk) of the immature fruits of coffee. These immunolocalization studies serve as a tool to interpret the role of SER in the cellular signaling mechanism in plants (Ramakrishna et al. 2012a).

22.3 Analysis of melatonin and serotonin in in vitro and ex vitro tissues of *C. canephora* by LC-MS-ESI

SER and MEL were analyzed in *C. canephora* LC-MS-ESI, wherein MEL was confirmed by liquid chromatography (HPLC)/tandem mass spectrometry and the ESI+ scan spectrum of a MEL standard was reported to exhibit one major ion at m/z 233 corresponding to [M]+. Analysis of indoleamines profiles using HPLC in *C. canephora* fruit tissues (i.e., zygotic embryos and endosperm tissues), seedling parts, and in vitro established cultures showed significant differences in their levels (Ramakrishna et al. 2012b).

22.4 Indoleamines profiles in in vitro cultures of *C. canephora*

It is quite interesting to see very little difference in the profiles of SER and MEL in 2-week-old ex vitro (8.25 ± 1 and 6.04 ± 1 μg g^{-1} fresh weight [FW]) and in vitro (9.10 ± 1 and 4.07 ± 1 μg g^{-1} FW) plant leaves. Callus cultures (4 weeks) and in vitro developed plant stalk (6 weeks) were found to possess 75.74 and 38.25 μg g^{-1} FW, but in vitro roots had low content (16.08 μg g^{-1}). But, the SER quantity was more in globular stage somatic embryos (31.8 μg g^{-1}) compared to callus and in vitro plant stalk and roots. Seedling roots of *C. canephora* contains high content of MEL (15.04 μg g^{-1} FW) compared to shoot tip (6.25 ± 2 μg g^{-1}) and first leaves (4.00 μg g^{-1}). The cotyledonary leaves were observed to have moderately good MEL content (46.23 ± 4 μg g^{-1} FW). Surprisingly, the levels of SER are fourfold higher compared to MEL in these first leaves of seedlings (Ramakrishna et al. 2011).

22.5 Role of indoleamines profiles in seed physiology

Freshly harvested green beans of *C. canephora* contain a high content of MEL (115.25±6 μg g^{-1}) and SER (98.5 μg g^{-1}). Similarly, substantial content of SER and MEL recorded in zygotic embryos (65.25 and 96.54 μg g^{-1}) and endosperm (34.08 and 51.08 μg g^{-1} FW) of ripened fruits in contrast to immature green fruits. It is curious to know a perfect differentiation in both SER and MEL among ex vitro grown seedling parts, wherein great levels of MEL content were noted in cotyledons (46.23 ± 4), which was somewhat more than that of hypocotyls (44.25 ± 4 μg g^{-1} FW). In the initial stages of fruit ontogeny, the indoleamine profile in seeds was little to modest trailed by advancement with the fruit growth progression (Ramakrishna et al. 2012b). However, in sweet cherry, MEL accumulation was more in young fruits and leaves and then reduced gradually with maturation (Xia et al. 2020). The presence of SER during the differentiation stage of embryos in *Coffea* was further supported by its presence in other tissues, such as seed wax, which contains 4.1 mg g^{-1} of SER (Kele and Ohmacht 1996), and also in polished coffee called Astra, which contained 20–40 mg SER/100 g. (Strance 1997). Such high levels of MEL and SER in seeds could aid in seed germination and variability, as reported in *Mimosa pudica* L. (Ramakrishna et al. 2009). In general, such high MEL content in seeds could protect the delicate lipid-rich tissues of the embryo from oxidative stress (Manchester et al. 2000). Nevertheless, the exciting thing to be examined is the high quantity of SER (up to 4.1 mg g^{-1}) in coffee wax and its implication, if any, in offering protection to the green beans from damage during storage.

22.6 Occurrence of indoleamines in coffee beans

The green beans of *C. canephora* are found to contain more SER (10.5±1.2 μg g^{-1} dry weight) compared to roasted beans (7.3±1.5 μg g^{-1} dry weight). In contrast to this, MEL content was less in green beans (5.8±0.8 μg g^{-1} dry weight) than to roasted beans (8.0±0.9 μg g^{-1} dry weight). Even coffee brew holds good content of MEL (3.0±0.6 μg/50 mL). A comparable trend in indoleamine profiles of *C. arabica* was detected, wherein green beans comprise 12.58±0.84 μg g^{-1} dry weight of SER and roasted beans hold 9.60±0.8 μg g^{-1} MEL (Ramakrishna et al. 2011). Green coffee prepared from Robusta beans possess high tryptamine levels; consequently the use of Robusta beans at a higher proportion in the blend could deliver a coffee with greater tryptamine content (Casal et al. 2002).

22.7 Indoleamines in processed coffee

Both MEL (3.9±0.2 μg/50 mL) and SER (7.3±0.6 μg/50 mL) were identified in coffee brew. In regular coffee, the predominant indole compound is serotonin followed by cadaverine, spermidine, and tryptamine (Da Silveira et al. 2007). According to Lima et al. (1999), intake of four to five 50-mL cups of coffee beverage provide 24.5 μg tryptamine and 39 μg SER daily. Similarly one cup of coffee contains 4–5 g of roasted coffee and contains ~40 μg of melatonin. Consuming a single cup of coffee is hypothetically sufficient to raise blood melatonin levels. Ursing et al. (2003) found an upsurge in circulating melatonin in human studies upon consuming coffee. It is estimated that those individuals who profited from drinking three or six cups of coffee daily also had at least 120 μg or 240 μg of melatonin, respectively, in the consumed beverage.

Studies on melatonin concentrations in Chinese green tea (Shiya tea) and black tea have also been measured (Chen et al. 2003). The significance of high levels of melatonin in popular

TABLE 22.1

Melatonin and Serotonin Levels in Different Coffee Beans of *C. canephora* and *C. arabica*

Plant part/product	Melatonin content	Serotonin content	Reference
C. canephora			
Green beans	5.8±0.8 μg/g dw	10.5±0.6 μg/g dw	Ramakrishna et al. 2012c
Roasted beans	8.0±0.9 μg/g dw	7.3±0.5 μg/g dw	Ramakrishna et al. 2012c
Decoction (brew)	3.0±0.6 (μg/50 mL)	4.0±0.7 (μg/50 mL)	Ramakrishna et al. 2012c
C. arabica			
Green beans	6.8±0.4 μg/g dw	12.5±0.8 (μg/g dw)	Ramakrishna et al. 2012c
Roasted beans	9.6±0.8 μg/g dw	8.7±0.4 (μg/g dw)	Ramakrishna et al. 2012c
Decoction (brew)	3.9±0.25 (μg/50 mL)	7.3±0.6 (μg/50 mL)	Ramakrishna et al. 2012b
Green coffee	–	1.13±0.15 mg/100 g	Cirilo et al. 2003
American roasted	–	0.16±0.05 mg/100 g	Cirilo et al. 2003
French roasted coffee	–	0.29±0.22 mg/100 g	Cirilo et al. 2003
Dry instant coffee	–	0.04–1.80 mg/100 g	Cirilo et al. 2003

Note: dw = dry weight.

beverages could prove beneficial for some of these conditions. Roasting temperature is found to influence SER and MEL profiles. Beans subjected to light roast had low levels of SER (Cirilo et al. 2003; Casal et al. 2004). Serotonin was detected in American roasted (0.16±0.05 mg/100 g), French roasted (0.29±0.22 mg/100 g), and dry instant coffee (0.04–1.80 mg/100 g), as shown in Table 22.1.

22.8 Biogenic amines in coffee

Amorim et al. (1977) were the first to study the levels of free amines in coffee. Biogenic amines (put, cadaverine, serotonin (5-Hydroxy tryptamine, SER) tyramine, spd, and spm) were reported in Robusta and Arabica coffees (Casal et al. 2004). These three amines are considered the most dominant amines present in roasted beans, with mean ranges of 50%–64% (putrescine), 23%–29% (spermidine), and 12%–21% (spermine; Amorim et al. 1977; Cirilo et al. 2003). In contrast to conjugated polyamines, free polyamines have been widely investigated in food products comprising scanty reports on coffee (Cirilo et al. 2003). Free biogenic amine levels in Arabica and Robusta coffee samples have been used as chemical markers for the two coffee species (Casal et al. 2002). The effect of roasting on the levels of amines in high (soft) and low (rio) quality coffees was investigated by Oliveira et al. (2005), wherein higher putrescine content for low-quality coffee in comparison to high-quality coffee was detected. Yet this roasting does not had any noticeable correlation for spermidine and spermine levels for cup quality. Other minor amines such as histamine, cadaverine, tryptamine, and tyramine is also associated with low-quality coffee (Oliveira et al. 2005).

Biogenic amines such as polyamines are well studied in various plants, and being an essential compounds for growth and development in plants, their presence in various edible parts of plants has triggered interest in researchers. In this regard, the significance of the endogenous polyamines content in various parts of *Coffea arabica* and *C. canephora* during the fruit formation stages was reported (Sridevi et al. 2009). Putrescine (Put), spermine (Spm) and spermidine (Spd) are mainly detected during the ontogeny of fruit, and a rise in their levels relates to that of the progression of fruit development, whereas in the initial stages of flower and fruit development, Spm levels are at a low. Similarly, the higher levels of cadavarine (Cad) and other biogenic amines (namely, tyramine [Tyr] and tryptamine [Try]) were also documented during the ontogeny of fruit in *C. canephora* (Sridevi et al. 2009). The predominance of Put, Spd, and Spm in zygotic embryos than in somatic embryos was recorded (Sridevi et al. 2009). In the same investigation, the endogenous polyamines profiles in *Coffea* tissues and beans has been addressed. Apart from this, the presence of polyamines in the pulp of fruit could be useful in further interpreting their beneficial attributes from a physiological point of view and also as quality traits of *Coffea* (Sridevi et al. 2009). In a recent study, Macheiner et al. (2021) reported the richness of monoamines (MA), diamines (DA), and polyamines (PA) as well as free amino acids (fAA) in green coffee and interpreted green bean derivatives as use as nutritional supplements and infusions. Giridhar et al. (2010) investigated the correlation between indoleamines and polyamines in *C. canephora* and its possible influence on the caffeine alkaloids pathway.

It is incidental to find that the dissimilarity in MEL and SER in green to roasted beans is important. The levels of SER is high in the wax layer of coffee beans compared to roasted beans, since robusta green beans, being a source of high tryptamine content, could offer specialty blends (Casal et al. 2002). Degradation of free amines during roasting was observed. Nevertheless, roasted coffee contained mostly serotonin, which appears to be the most resistant amine to the roasting process (Da Silveira et al. 2007). Biogenic amines levels in green and roasted coffee are depicted in Table 22.2.

TABLE 22.2

Biogenic Amine Levels in Green and Roasted Coffee

Amines	Green coffee (mg/100 g)	American roasted coffee	French roasted coffee	Dry instant coffee (mg/100 g) (minimum-maximum)
Putrescine	1.03 ± 0.09	nd	nd	0.04–0.53
Cadaverine	–	nd	nd	0.04–0.81
Histamine	–	–	–	0.04–0.14
Serotonin	1.13 ± 0.15	0.16 ± 0.05	0.29 ± 0.22	0.04–1.80
Spermidine	0.60 ± 0.07	0.12±0.02c	0.20±0.01b	0.04–0.77
Spermine	0.44 ± 0.08	–	–	0.05–0.41
Tryptamine	–	–	–	0.04–0.55
Agmatine	nd	nd	0.12 ± 0.05	–

Note: nd = no data.

Source: Adapted from Cirilo et al. (2003).

22.9 Effect of roasting temperature on biogenic amines

During the roasting process, most of the free amines present in green coffee were degraded due to pyrolysis that has an adverse effect on the SER due to oxidation and breakage of the carbon chain. Da Silveira et al. (2007) reported that during roasting of coffee, there is a significant decrease in the levels of amines. Also, there was a report of large variation in biogenic amines in *Coffea* (Sridevi et al. 2009). There is a lot of variation in biogenic amines in *Coffea*, especially because they are prone to processing conditions. Özdestan (2014) profiled the biogenic amines along with minerals in Turkish coffee and assessed the daily intake through its consumption. According to this report, the total amine levels in the brewed coffee was in the range of 5.679–48.88 mg L^{-1} compared to 126.0–352.2 mg kg^{-1} of the ground coffee. Similarly, Restuccia et al. (2015) analyzed the effect of various brewing methods on biogenic amines of brew, which revealed that putrescine is prevalent among various biogenic amines, followed by PUT, followed by SPD, HIS, TYR, CAD, SPM, PHE, and SER. However, upon roasting, reduction in their content was recorded. Interestingly, in coffee brew, substantially very low content of these biogenics in amines in the order of PUT, SPM, TYR, CAD, SPD, PHE, HIS, and SER were observed (Restuccia et al. 2015).

22.10 Human health effects

Coffee quality not only depends on organoleptic properties but also on its positive health attributes. Accordingly, much interest has been evoked on various factors that are responsible for coffee quality. The presence of indoleamines in roasted beans that are transferred to the coffee brew may be useful for consumer health. Consumption of edible plant tissues that contain the SER and MEL would be useful, as their antioxidant activity has medicinal value for treatment of human diseases such as cardiovascular damage (Reiter and Tan 2003), Alzheimer's disease (Pappolla et al. 2000), and cancer (Sauer et al. 2001). Intake of four to five 50-mL cups of coffee beverage provide 24.5 μg tryptamine and 39 μg SER daily (Lima and Gloria 1999). Moreover, SER serves as key substrate for feruloylserotonin and 4-coumaroylserotonin, which have high antioxidative activity, chemotherapeutic effects, and antimicrobial activity against plant pathogens (Koyama et al. 2006). Moreover, a fatty acid derivative of serotonin, called eicosanoyl-5-hydroxytryptamide (EHT), was found in coffee. A combination of EHT and caffeine may be able to slow or stop the progression of Parkinson's disease and dementia with Lewy bodies (DLB; Yan et al. 2018).

Melatonin, being a molecule with antioxidant activity, may contribute to the health of coffee beverage consumers. Numerous published articles have proven that melatonin effectively protects against the hepatopathologies mentioned above (Cruz et al. 2005; Wang et al. 2005, 2010; Solis-Munoz et al. 2011). Elevated levels of circulating melatonin in coffee consumers is ascribed to the inhibitory effect of coffee on the hepatic melatonin metabolizing enzyme CYP1A2 (Ursing et al. 2003). The effects of coffee consumption on the rate of melatonin secretion, as reflected by urinary excretion of 6-sulphoxymelatonin (6-SMT) and sleep quality as assessed by actigraphy. Caffeine intake ensued in reduced sleep quality and secretion of melatonin, the main hormone that regulates sleep (Shilo et al. 2002).

22.11 Conclusion

Coffee beans have considerable quantities of MEL, as part of a healthy diet, and hence may be beneficial in counteracting free radical damage. Therefore, they would be of nutraceutical value apart from their role as neurotransmitters. As the content of SER and MEL in seeds is high compared to all other parts of the plant, further investigations are necessary for understanding their physiological role. The prevalence of SER and MEL in coffee beans, both green and roasted and brew, suggest their importance as nutraceuticals. As indoleamine bioactives are stated to have a health beneficial role, their further relevance to food, nutritional sciences, and human health was of interest to explore their levels in coffee, an important universal beverage. The mechanisms behind the melatonin secretion in presence of caffeine are yet to be elucidated (Shilo et al. 2002). Further studies are necessary to define the significance of these indoleamines in human health.

REFERENCES

Amorim et al, J Agri Food Chem. 1977; 25:957–958.
Badria FA. J Med Food 2002; 5:153–157
Briguglio et al, Nutrients 2018; 10:591.
Butt MS, Tauseef M, Crit. Rev. Food Sci. Nutr 2011; 51:363–373.
Cao et al, J Chromatogr A. 2006; 1134:333–337
Casal et al, J Liq Chromatogra & Rel Tech 2002; 25:2535–2549.
Casal et al, J Agric Food Chem 2004; 52:6188–6192
Chen et al, Life Sci 2003; 73:19–26
Cirilo et al, Food Chem. 2003; 82:397–402.
Cruz et al, J Pineal Res. 2005; 39:143–150.
Da Silveira et al, J Food Comp Anal 2007; 20:451–457.
Delgado et al, J Anim Physiol Anim Nutr 2013; 97:137–145.
Ferrari et al, Biogerontology, 2004; 5:275–289.
Giridhar et al, 2010, In: Proceedings of the 23rd International conference on coffee science, ASIC, pp. 633–637.
Gomes Domingos et al, J. Crit Rev Food Sci Nutr 2017; 59:133–140.
Hano et al, Sci Hortic 2017; 214:07–113
Hardeland et al, Nut Met 2005; 2:1–15.
Huang X, Mazza G, Crit Rev Food Sci Nutr 2011; 51:269–284
Iriti et al, J Sci Food Agric 2006; 86:1432–1438
Kele M, Ohmacht R, J Chromatogra A 1996; 730:59–62
Kolar et al, Phytochem 1997; 8:1407–1413
Koyama et al, J Agric Food Chem 2006; 54:4970–4976.
Lima, Gloria, Bol Soc Brasil Cien Tecnol Aliment 1999; 33:70–79
Macheiner et al, Eur Food Res Technol. 2021; 247:85–99.
Manchester et al, Life Sci. 2000; 67:3023–3029.
Meng et al, Nutrients. 2017; 9:367
Meng et al, Nutrients 2018; 10:591.
Oliveira et al, Food Chem 2005; 287–291.
Özdestan Ö. Food Res Int. 2014; 61:67–175
Pappolla et al, J Neural Transmis 2000; 107:203–231
Perrot-Rechenmann C, Gadal P, In: Wang TL, ed. Immunology in plant science. Cambridge University Press, Cambridge, 1986, pp. 59–88.
Ramakrishna et al, Plant Signaling & Behavior 2009; 12:1136–1141.
Ramakrishna et al, Indian J Exp Biol 2011; 49:234–240.
Ramakrishna et al, Plant Cell, Tissue and Organ Culture. 2012a; 108:267–278.
Ramakrishna et al, Acta Phy. Plant. 2012b; 34:393–396
Ramakrishna et al, J Pineal Res 2012c; 52:470–476
Rayne S. Nature Proceedings. 2010; 1–3.
Reiter, R.J. and Tan, D.X. Cardiovascular Research, 2003; 58:10–19.
Reiter et al, Nutrition 2005; 21:920–924.
Restuccia et al, Food Chem. 2015; 175:143–150
Roshchina VV. Science Publishers, Enfield, 2001, 4–81
Sae-Teaw et al, J Pineal Res 2013; 55:58–64.
Salehi et al, Cells 2019; 8:681.
Sauer et al, Biochem Pharmacol 2001; 61:1455–1462
Shilo et al, Sleep Med. 2002; 3:271–273.
Solis-Munoz et al, J Pineal Res 2011; 51:113–123.
Soumya Mukherjee, Ramakrishna. Neurotransmitters in Plants: Perspectives and Applications 2018. Taylor & Francis, CRC Press, New York.
Sridevi et al, Acta Physiologiae Plantarum 2009; 31:757–764
Strance A. Przemysl spozywezy 1997; 47:76–77
Tahan et al, Canadian J. Surgery 2010; 53:313–318.
Ursing et al, J. Endocrin. Investig. 2003; 26:403–406.
Vinod Kumar et al, Plant Cell Tissue Organ Cult 2006; 87:49–65.
Wang et al, Life Sciences 2005; 77:1902–1915.
Xia et al, Molecules. 2020; 25:753.
Yan et al, Proc Natl Acad Sci U S A. 2018;18(115):E12053–E12062.

23

Bioactive components in coffee and their health aspects

Palakkil Mavilavalappil Sandeep and Iychettira Machaiah Mandappa

CONTENTS

23.1 Introduction

Coffee is a popular beverage consumed worldwide (Oliveira et al., 2009; Salinardi et al., 2010). Since its introduction into Arabia from Ethiopia many centuries ago, its cultivation has continued in three regions of our globe, namely, Africa, Latin America, and Asia. *Coffea arabica* and *C. canephora* are the two most widely cultivated species (Simon-Gruita et al., 2019). Green coffee beans are mainly composed of carbohydrates, lipids, proteins, phenols, minerals, fatty acids, caffeine, trigonelline, and free amino acids. Roasting of coffee beans results in a decrease in the levels of carbohydrates, proteins, phenols, and free amino acids without bringing a notable change in the levels of lipids, minerals, fatty acids, caffeine, and trigonelline. The process of coffee roasting is accompanied by the Maillard reaction, resulting in complex and diverse coffee melanoidins, which account for 29% of the total weight of roasted coffee. The lipid part of coffee consists mainly of triacylglycerols and esters of diterpene alcohols, and contains a small amount of esters of steroids, diterpene alcohols, sterols, tocopherols, phosphatides, and tryptamine derivatives. So far, most of the diterpenes found in coffee are present in the form of fatty acid esters, with the highest content of cafestol, kahweol, and 16-O-methoxycafestol (16-OMC; Hu et al., 2019).

23.2 Biological activities of constituents in coffee

Coffee drinking is closely related to human health (Bae et al., 2014). Previous research studies shows that the bioactivity of coffee is closely related to caffeine, chlorogenic acids (CGAs), trigonelline, diterpenoids (especially cafestol and kahweol; Bułdak et al., 2018), and coffee melanoidins (Delgado-Andrade et al., 2005). The present research understanding of the chemical constituents and physiological functions of coffee and the diseases associated with its consumption are mainly based on observational studies and meta-analyses using crude coffee extracts (Hu et al., 2019). In this chapter, we attempt to summarize the biological activity of the main active constituents in coffee and their health effects based on existing research data. A brief overview of beneficial and adverse effects of major bioactive components of coffee is given in Tables 23.1 and 23.2, respectively.

DOI: 10.1201/9781003043133-28

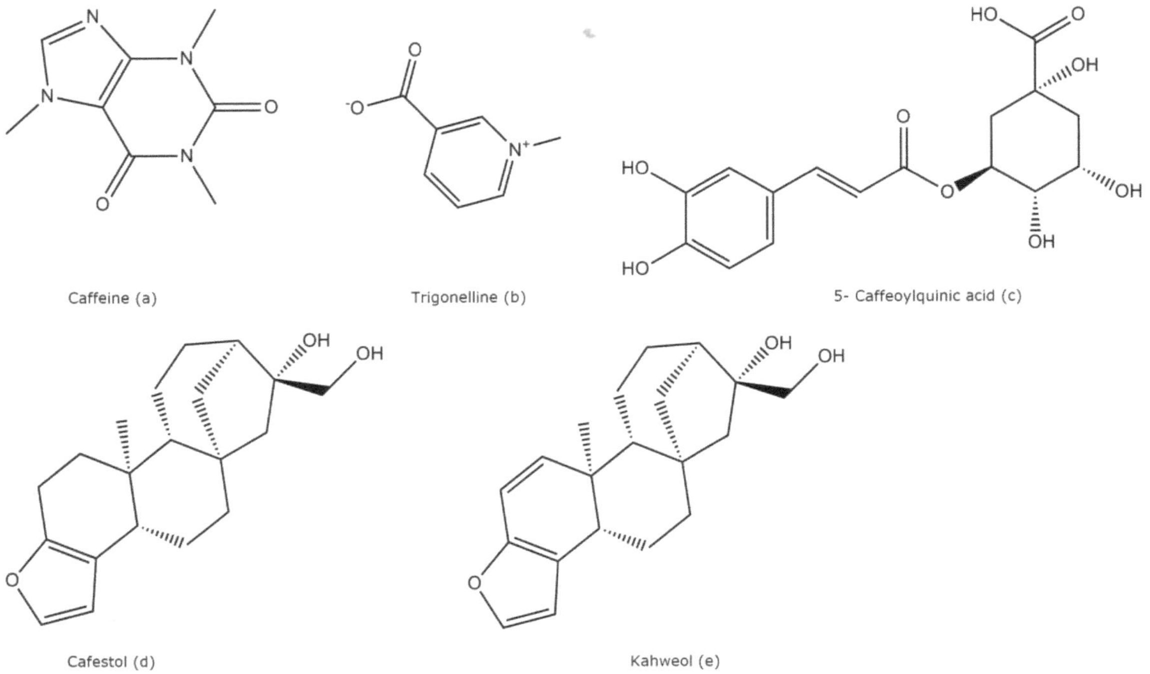

FIGURE 23.1 Chemical structure of (a) caffeine (PubChem CID: 2519), (b) trigonelline (PubChem CID: 5570), (c) 5-caffeoylquinic acid (PubChem CID: 12310830), (d) cafestol (PubChem CID: 108052), and (e) kahweol (PubChem CID: 114778).

TABLE 23.1

Beneficial Effects of Major Bioactive Compounds in Coffee

Bioactive compounds in coffee	Beneficial effects	References
Caffeine	Neuroprotective activity	Volkow et al., 2015; Manalo et al., 2018; López-Cruz et al., 2018
	Anti-cancer activity	Miwa et al., 2012; Fehrholz et al., 2014; Sugimoto et al., 2019
	Anti-inflammatory and antioxidant activities	Ikram et al., 2019
	Anti-diabetic activity	Johnston et al., 2003; Arti et al., 2012; Shearer et al., 2007; van Dijk et al., 2009
	Immunomodulatory activity	Chavez Valdez et al., 2011
	Increases psychomotor activity and alertness	Volkow et al., 2015
	Allows DNA repair systems to correct replication errors	Bode et al., 2007; Dubrez et al., 2001
	Regulates epileptic disorders	Isokawa et al., 2016; Alasmari, 2020
	Influences gut microbiota	Bazzaz et al., 2019; Kang et al., 2012
Trigonelline	Neuroprotective activity	Farid et al., 2020
	Hypoglycemic and hypolipidemic activity	Zhou et al., 2013; Yoshinari et al., 2009
	Anti-cancer activity	Hirakawa et al., 2005; Arlt et al., 2013
	Estrogenic activity	Allred et al., 2009
	Antibacterial activity	Antonio et al., 2010

Bioactive compounds in coffee	Beneficial effects	References
Chlorogenic acids	Antioxidant activity	Hu et al., 2019
	Anti-inflammatory activity	Xu et al., 2010; Chauhan et al., 2012; Hwang et al., 2014
	Anti-diabetic activity	Bassoli et al., 2008; Johnston et al., 2003; Meng et al., 2013
	Anti-cancer activity	Luo et al., 2019; Bułdak et al., 2018; Feng et al., 2005; Kang et al., 2011
	Hypolipidemic activity	Rodriguez de Sotillo et al., 2002; Yukawa et al., 2004
	Anti-microbial activity	Wang et al., 2009; Urushisaki et al., 2011; Ma et al., 2010; Lou et al., 2011
	Antihypertensive and neuroprotective activities	Hu et al., 2019; Kwon et al., 2010
Cafestol and kahweol	Anti-inflammatory properties	Kim et al., 2004; Shen et al., 2010
	Anti-angiogenic and anti-tumorigenic properties	Kotowski et al., 2015; Woo et al., 2014; Lee et al., 2012; Choi et al., 2011
	Potential drug for degenerative bone treatment	Fumimoto et al., 2012; Fukuma et al., 2015
	Reduces DNA damage and prevents tumor initiation	Huber et al., 2008; Lee et al., 2007; Cavin et al., 2002; Cavin et al., 1998; Higgins et al., 2008; Tao et al., 2008
	Reduces fasting blood glucose and fasting glucagon along with improving insulin secretion and sensitivity	Zhang et al., 2009; Mellbye et al., 2015
Coffee melanoidins	Antioxidant and antimicrobial activities	Rufián-Henares et al., 2009
	Anti-cancer and antihypertensive properties	Rufián-Henares and Morales, 2007

TABLE 23.2

Adverse Effects of Major Bioactive Compounds in Coffee

Bioactive compounds in coffee	Adverse effects	References
Caffeine	Mostly due to higher doses. Mild effects like anxiety, restlessness, fidgeting, insomnia, facial flushing, increased urination, muscle twitches or tremors, irritability, agitation, elevated or irregular heart rate, and gastrointestinal problems. Severe effects like disorientation, hallucinations, psychosis, seizure, arrhythmias, ischemia, and rhabdomyolysis.	Evans et al., 2021
Trigonelline	Intensifies osteoporotic changes in diabetes-induced rat models.	Folwarczna et al., 2016
Cafestol and kahweol	Deleterious effects on lipids by increasing triacylglycerol and low-density lipoprotein (LDL) levels and their adverse effects on liver enzymes.	Godos et al., 2014; De Roos et al., 2000
Coffee Melanoidins	Inhibitory activity on some proteolytic enzymes in the alimentary tract, including trypsin.	Ibarz et al., 2009; Oste et al., 1984
Chlorogenic acids	Reports not available.	

23.2.1 Caffeine

Coffee is one of the major sources of caffeine in food (Nawrot et al., 2003). The presence of caffeine in green coffee can range from 1.45% to 2.38% in *C. arabica* and *C. canephora*, respectively (Bicho et al., 2013). Caffeine content in roasted coffee is much higher in *C. canephora* (1.7%–4.0%) than in *C. arabica* (0.8%–1.4%; Gaibor et al., 2020). Caffeine (Figure 23.1a), an alkaloid that is chemically known as 1,3,7-trimethylxanthine (or 1,3,7-trimethyl-1H-purine-2, 6 [3H, 7H]-dione), containing two fused rings that are related to purines, is among the most commonly consumed stimulants worldwide.

23.2.1.1 Bioavailability of caffeine

After consumption, caffeine is rapidly and completely absorbed into the bloodstream from the gastrointestinal tract. The peak plasma concentration was observed to be 4–5 mg kg^{-1} with 30–120 minutes of administration with half-lives ranging from 2.5 to 5.0 hours (Arnaud, 1993). Caffeine is distributed to all body fluids such as plasma, saliva, bile, cerebrospinal fluid, breast milk, semen, and umbilical cord blood and to all tissue organs. Caffeine's lipophilic properties enable it to cross the placental and blood-brain barriers (Callahan et al., 1982; Grosso et al., 2008; Sachse et al., 2008; Scott et al., 1984). The absorbed caffeine then undergoes hepatic metabolism to form paraxanthine (1,7-dimethylxanthine), theobromine (3,7-dimethylxanthine), and theophylline (1,3-dimethylxanthine; Tang-Liu et al., 1983). In adults, only 1%–5% of caffeine ingested is recovered unchanged in the urine; whereas infants 8 to 9 months old excrete about 85% of caffeine ingested unchanged (Stavric et al., 1990).

23.2.1.2 Caffeine consumption and health aspects

Coffee is considered as one of the best known psychostimulants worldwide because of its caffeine content (Butt et al., 2011; Higdon et al., 2006; Fredholm et al., 1999). About 430–440 mL (one cup) of brewed coffee consumed will deliver an average of 188 mg of caffeine to the body (McCusker et al., 2003). The primary mechanism of action is by regulating the multiple physiological activities mediated through endogenous adenosine and adenosine receptors (AR). ARs include A1, A2A, A2B, and A3 subtypes where A1 subtypes are mainly distributed in the central nervous and peripheral nervous systems, and they are also distributed in the kidneys, lungs, bladder, and heart. The A2A receptor is mainly distributed in the dopamine-rich region of brain tissue, whereas the A2B receptor is mainly distributed in the digestive system and the A3 receptor is widely distributed in the spleen, lungs, heart, kidneys, other organs, and the surface of inflammatory cells. Consumption of caffeine has shown many beneficial effects in various animal and human studies. Studies have shown that caffeine have neuroprotective, anti-cancer, anti-inflammatory, and antioxidant properties. Moderate intake of caffeine (3–5 cups per day) is associated with reducing fatigue, revised cognitive, and improved alertness, leading to better yield in psychomotor activities (Smith, 2002; van Gelder et al., 2007). Caffeine may induce changes in dopaminergic systems in humans and animals through modulation of adenosine A receptors (Volkow et al., 2015; Manalo et al., 2018; López-Cruz et al., 2018). It is reported that by inducing protection effects against dopaminergic neuronal loss, caffeine may have therapeutic activity against neurodegenerative diseases, including Parkinson's disease (Manalo et al., 2018). Caffeine has also been found to induce various effects on γ-amino butyric acid (GABA) receptors, thereby suggesting its role in the regulation of epileptic disorders (Isokawa, 2016; Alasmari, 2020). Volkow et al. (2015) suggest the involvement of caffeine-induced increase in post-synaptic D2/D3 receptor availability in the ventral striatum and thereby increasing the level of alertness. Caffeine is reported to have antioxidant activity due to its impact against endogenous antioxidant regulators, such as nuclear factor erythroid 2-related factor 2 (Nrf-2) and heme oxygenase-1 (HO-1) and by regulating the level of lipid peroxidation (LPO) and reactive oxygen species (ROS; Ikram et al., 2019). The interaction of caffeine with signaling pathways and thereby suppressing transforming growth factor β (Fehrholz et al., 2014), inactivation of phosphoinositide-3-kinase/AKT/mammalian target of rapamycin (Sugimoto et al., 2019), and inhibition of mitogen-activated protein kinase pathways (Miwa et al., 2012) could play a vital role in inhibiting the progression of colorectal cancer. In addition, caffeine can modulate the immune response with its response to cytokines (Chavez Valdez et al., 2011), influence gut bacteria (Bazzaz et al., 2019; Kang et al., 2012), and regulate cell cycle checkpoint pathways allowing DNA repair systems to correct replication errors (Bode et al., 2007; Dubrez et al., 2001). The effect of caffeine in glucose metabolism has been studied by many researchers worldwide (Johnston et al., 2003; Arti et al., 2012; Shearer et al., 2007; van Dijk et al., 2009). The studies suggest that caffeine can increase insulin sensitivity and decrease glucose storage, thereby reducing the extent of type 2 diabetes mellitus (T2DM).

The adverse effects of caffeine range from mild to severe depending on the dose consumed and the individual's sensitivity. Mild effects include anxiety, restlessness, fidgeting, insomnia, facial flushing, increased urination, muscle twitches or tremors, irritability, agitation, elevated or irregular heart rate, and gastrointestinal problems. Severe effects include disorientation, hallucinations, psychosis, seizure, arrhythmias, ischemia, and rhabdomyolysis (Evans et al., 2021). If habitual users abruptly stop consumption, caffeine can cause withdrawal symptoms. The symptoms may start within 12–24 hours from last consumption, peak in 1–2 days, and may persist for up to 1 week (Juliano et al., 2004).

23.2.2 Trigonelline

After caffeine, trigonelline is the second most abundant alkaloid compound in raw coffee beans (Viani, 1974). Trigonelline content is demonstrated to be in the range of 0.87%–0.9% in *C. arabica* and 0.74% in *C. canephora*, of which 50%–65% degrades on roasting (Kim et al., 2006). Trigonelline or N-methyl nicotinic acid (Figure 23.1b) is a pyridine alkaloid that is derived from the methylation of the nitrogen atom of nicotinic acid (niacin). During roasting, trigonelline was heated above 180°C, yielding a nutritionally significant amount of nicotinic acid, vitamin B_3, or nicotinate (Taguchi et al., 1985). Roasting of green trigonelline will also yield several pyridine derivatives, including 1-methylpyridinium and 1,2-dimethylpyridinium (Viani, 1974; Stadler et al., 2002a; Stadler et al., 2002b).

23.2.2.1 Bioavailability of trigonelline

Bioavailability of trigonelline in human and animals has been investigated in several studies. Initial absorption of trigonelline occurs in the stomach, where it was detected 15 minutes after ingestion of coffee. The study showed that one cup of coffee containing 808 μM of trigonelline attained its peak plasma concentration of 6 μM in 3 hours and had an elimination half-life of 5 hours. Within 0–8 hours post-ingestion of coffee, urinary secretion of trigonelline was 57.4% of intake for males and 46.2% for females, suggesting that a reasonable amount passed through without getting subjected to phase II metabolism. The coffee also contained 1-methylpyridinium which reached its peak plasma concentration of 0.8 μM and an elimination half-life of 2.3 hours. The urinary excretion of 1-methylpyridinium was more than 60% (Lang et al., 2010).

23.2.2.2 Trigonelline consumption and health aspects

Trigonelline has been shown to possess neuroprotective, hypoglycemic, hypolipidemic, anti-cancer, estrogenic, and antibacterial activities. A study conducted by Farid et al. (2020) showed the potential of trigonelline as a new drug candidate in memory recovery signaling by penetrating the brain and may activate creatine kinase B-type, leading to axonal formation. Trigonelline exerts its action on key enzymes of glucose and lipid metabolism, such as glucokinase, glucose-6-phosphatase, fatty acid synthase, and carnitine palmitoyl transferase in diabetic rats, which leads to lowering of blood sugar, cholesterol, and blood lipid (Zhou et al., 2013; Yoshinari et al., 2009). It has also shown beneficial effects in streptozotocin- and high-carbohydrate/high-fat diet-induced diabetic rats displaying peripheral neuropathy, a condition for which there is no effective drug for its treatment. The study suggested that, through the glucagon-like peptide-1 receptor/p38 mitogen-activated protein kinases signaling pathway, nerve conduction velocity, antioxidant enzyme activity, improving micropathological changes of sciatic nerve and decreasing lipid peroxidation trigonelline improved diabetic peripheral neuropathy (Zhou et al., 2012). The anti-invasive activity of trigonelline and niacin was demonstrated in the hepatoma cell line of AH109A previously cultured with a reactive oxygen species (ROS)–generating system (Hirakawa et al., 2005). Trigonelline was also reported to inhibit Nrf2 transcription factors, thereby decreasing proteasomal gene expression and proteasome activity, rendering pancreatic cancer cells more susceptible to apoptosis (Arlt et al., 2013). The activation of estrogen receptor (ER) by trigonelline stimulated the expression of ER target genes (pS2, progesterone receptor, and cyclin D1) similar to 17β-estradiol (E2) in the MCF-7 human breast cancer cell clearly indicating its phytoestrogenic activity (Allred et al., 2009). Trigonelline content of *C. canephora* extracts at a minimum inhibitory concentration (MIC) of 0.8 mg mL^{-1} inhibited biofilm formation by *Streptococcus mutans*. However, an inverse correlation was observed between bacterial colony forming units and roasting degree. Light and medium roast showed higher inhibitory activity than darker roast (Antonio et al., 2010). However, trigonelline was reported to have an unfavorable effect on the skeletal system of streptozotocin-induced diabetic rats, intensifying the osteoporotic changes, while favorably affecting bones in the non-hyperglycemic (nicotinamide/streptozotocin-treated) rats (Folwarczna et al., 2016).

23.2.3 Chlorogenic acids

Chlorogenic acids (CGA) are phenolic compounds formed by the esterification of cinnamic acids, such as caffeic, ferulic, and p-coumaric acids, with (–) quinic acid. Out of total of 82 CGAs detected in green coffee beans (Narita et al., 2015), the main types of chlorogenic acids in green coffee are caffeoylquinic acids (CQA), dicaffeoylquinic acids (diCQA), feruloylquinic acids (FQA), and less abundantly, p-coumaroylquinic acids (CoQA) and caffeoyl-feruloylquinic acids (CFQA) with their several isomers (Farah et al., 2008). The most abundant CGA is 5-CQA (76%–84%; Figure 23.1c; Lu et al., 2020). The CGAs of *C. arabica* and *C. canephora* are in the range of 3.40%–7.24% and 5.17%–14.4% w/w dry matter, respectively, whereas the contents may decrease with roasting of green coffee beans, and the extent of this decrease depends on the degree of roasting. The average contents of total CGAs, caffeoylquinic acids, feruloylquinic acids, and dicaffeoylquinic acids in 12 commercial roasted coffee beans were reported to be 2.66%, 2.26%, 0.21%, and 0.19% w/w dry matter, respectively (Narita et al., 2015).

23.2.3.1 Bioavailability of chlorogenic acids

A portion of the ingested CGA is absorbed early in the stomach. The remaining amount of CGA proceeds toward the intestine, where it will cross the ileum and jejunum cells and enter the hepatic portal circulation, which leads directly to the liver, where they undergo a series of enzymatic reactions. In addition, they will undergo one or more reactions such as hydrolysis and/or conjugation with sulfate or glucuronic acid for elimination or later absorption in the colon by enzymes located in the intestinal mucosa cells and in the lumen. Studies on the bioavailability of CGAs indicates that they are differentially absorbed and metabolized in humans, with a large inter-individual variation. The CGAs metabolites recovered from plasma of adults after brewed coffee consumption includes three CQA isomers and three diCQA isomers. However, the metabolites identified in urine were dihydrocaffeic, gallic, isoferulic, ferulic, vanillic, caffeic, 5-CQA, sinapic, rho-hydroxybenzoic, and rho-coumaric acids (Monteiro et al., 2007; Mubarak et al., 2012; Farah et al., 2008; Renouf et al., 2010).

23.2.3.2 Consumption of chlorogenic acids and health aspects

Many beneficial pharmacological activities have been attributed to CGAs. These includes antioxidant, anti-inflammatory, anti-diabetic, anti-cancer, hypolipidemic, anti-microbial, anti-hypertensive, and neuroprotective activities. For instance, CGAs prevent diseases caused by free radical damage, such as free radical–induced cardiovascular damage, nervous system damage, liver damage, and tumors (Hu et al., 2019). CGAs can control inflammatory response by suppression of pro-inflammatory substances like IL-1b, TNF-α, IL-6 (Xu et al., 2010; Chauhan et al., 2012), and chemokine CXCL1 (Hwang et al., 2014). Anti-diabetic and anti- obesity effects of CGAs were studied by researchers. Bassoli et al. (2008) demonstrated that CGA can enhance the reduction in the plasma glucose peak by attenuating intestinal glucose absorption, indicating its effect as a glycemic index–lowering agent reducing the liability of T2DM. CGA is found to stimulate insulin secretion (Johnston et al., 2003) and improve glucose tolerance and insulin resistance (Meng et al., 2013). The important mechanisms behind these effects are the inhibition of the glucose-6 phosphate system and subsequent delay of glucose uptake in the intestine, inhibition of the activity of α-glucosidase, alteration of glucose-dependent insulinotropic polypeptide (GIP) concentrations, activation of 5′-adenosine monophosphate-activated protein kinase (AMPK), inhibition of HMG CoA reductase, strengthening the activity of carnitine palmitoyl transferase, and upregulation of hepatic peroxisome proliferator-activated receptor (PPAR)-α. CGAs are reported to suppress migration and invasion and induce apoptosis in numerous cancer cell lines including colon, breast, and lung (Luo et al., 2019; Bułdak et al., 2018). For instance, CGA was demonstrated to inhibit lung cancer cell proliferation by suppressing the NF-κB and MAPK signaling pathways (Feng et al., 2005). Kang et al. (2011) revealed that CGA derived from coffee, impaired colorectal cancer cell metastasis by suppressing NF-κB, MEK, and T-LAK cell-originated protein kinase. CGAs affects lipid metabolism by significantly reducing the plasma cholesterol, triacylglycerol, and liver triacylglycerol concentrations (Rodriguez de Sotillo et al., 2002) by reducing low-density lipoprotein (LDL) oxidation susceptibility and decreasing LDL cholesterol and malondialdehyde (MDA) levels (Yukawa et al., 2004) by inhibiting fat absorption and activating fat metabolism in liver. The study conducted in mice models suggests the role of chlorogenic acids in improving obesity hormone levels. CGAs significantly inhibited fatty acid synthase, 3-hydroxy-3-methylglutaryl CoA reductase, and acyl-CoA cholesterol acyltransferase activities, while they increased fatty acid beta-oxidation activity and PPAR-α expression in the liver compared to the high-fat group (Yukawa et al., 2004). CGA have shown neuroprotective function via the inhibition of acetyl cholinesterase when administered to mice under scopolamine-induced amnesia (Kwon et al., 2010). Several microbial entities are susceptible to CGA's antimicrobial activities. CGA has shown anti-hepatitis B virus activity in a duckling model (Wang et al., 2009) and anti-H1N1 influenza virus activities (Urushisaki et al., 2011). CGA analogs synthesized by Ma et al. (2010) showed appreciable anti-HIV activities. CGA has also shown antibacterial and antifungal functions. CGA significantly increased the outer and plasma membrane permeability, resulting in the loss of the barrier function and leakage of cytoplasmic contents inhibiting the bacterial pathogens tested (Lou et al., 2011). A study conducted by Sung and Lee (2010) suggested that chlorogenic acid can exert antifungal activity against *Candida albicans* by disrupting the structure of the cell membrane.

Despite promising data on health benefits of CGA, the side effects of both short- and long-term consumption of CGA have not yet been investigated thoroughly.

23.2.4 Diterpenes: cafestol and kahweol

Diterpenes are the main class of coffee oils after triglycerides and belong to the kauren family, as assimilation of isoprene units forms the skeleton of 20 kauren carbons. The total amount of diterpenes in green beans of *C. arabica* ranges from 1.3% to 1.9% (w/w), while their content in green beans of *C. canephora* ranges from 0.2% to 1.5% (w/w). So far, nearly 100 diterpenes with different structures have been identified in coffee, of which the most dominant representatives are cafestol ($C_{20}H_{28}O_3$), kahweol ($C_{20}H_{26}O_3$), and 16-OMC ($C_{21}H_{30}O_3$). Out of these three compounds, cafestol (Figure 23.1d) and kahweol (Figure 23.1e) are known to have biological activity (Ren et al., 2019). These compounds are mainly esterified with fatty acids and rarely exist as free forms (Guercia et al., 2016; Kitzberger et al., 2013; Hu et al., 2019; Wermelinger et al., 2011). The content of cafestol in green Arabica and Robusta coffees varied from 478 to 658 mg/100 g and from 156 to 280 mg/100 g, respectively. The content of kahweol in green Arabica was about 312–1082 mg/100 g and in green Robusta was 0–20 mg/100 g. Other diterpenes like 16-OMC was found predominantly in green Robusta as well as in trace amounts in Arabica coffees (Campanha et al., 2010; Kitzberger et al., 2013; Mori et al., 2016; Scholz et al., 2014; De Roos et al., 1997; Gunning et al., 2018). The roasting temperature and time influences the diterpenes profiles of roasted beans. The full-city roasted Arabica showed 56% and 61% reduction of cafestol and kahweol, respectively, when compared to the normal roast. However, Robusta showed a lesser reduction in cafestol and kahweol content, indicating the higher sensitivity of Arabica coffee to roasting degree (Sridevi et al., 2011). The 16-OMC content of Robusta showed only a slight reduction (from 27 mg g^{-1} to 24.9 mg g^{-1} coffee oil) in green and roasted Robusta samples, indicating its heat stability (Moeenfard, 2016). Since Robusta coffee had lower levels of cafestol and an absence or negligible amount of kahweol, when compared to Arabica, the kahweol/cafestol ratio can be used as a tool for identifying species in different blends of roasted coffee. Specifically, the kahweol/cafestol ratio above 1.0 represented pure Arabica coffee (Souza et al., 2012), and the kahweol/cafestol ratio was observed to be 0.0–0.06 in Robusta (Mori et al., 2016).

23.2.4.1 *Bioavailability of cafestol and kahweol*

The content of cafestol and kahweol in Arabica or Robusta Indian filter brew was approximately 1.2 ± 0.4 (mg/150 mL) and 0.8 ± 0.3 (mg/150 mL), respectively (Sridevi et al., 2011). About 70% of the cafestol and kahweol consumed can be absorbed in the small intestine (Urgert et al., 1996). Theoretically, the major pathways for the xenobiotic transformation in mammalian species are glucuronidation and sulphation, which occurs mostly in the liver to form water-soluble products that can be excreted through urine. However, previous studies suggest that only about 1% or less of the consumption was detected to be excreted as conjugate of glucuronic or sulfuric acid in urine (De Roos et al., 1998; Urgert et al., 1996).

23.2.4.2 *Cafestol and kahweol consumption and health aspects*

In addition to the deleterious effects on lipids by increasing triacylglycerol and LDL levels (Godos et al., 2014; De Roos et al., 2000) and its adverse effects on liver enzymes in some cases, extensive studies have demonstrated that cafestol and kahweol exhibit a wide variety of beneficial activities, including anti-inflammatory, anti-angiogenic, and anti-tumorigenic properties. Cafestol and kahweol can significantly inhibit prostaglandin E2 (PGE2) and TNF-α synthesis (Kim et al., 2004) and inhibit the activation of inhibitor of nuclear factor kappa-B kinase (IKK) (Shen et al., 2010) in lipopolysaccharide (LPS)-activated macrophages in a dose-dependent manner, indicating the mechanism of anti-inflammatory activity. It was also observed that cafestol and kahweol act differently for generating the anti-inflammatory response. Cafestol directly reduces the activity of extracellular regulated protein kinases-2 (ERK2) enzyme, thereby reducing PEG2 production (Shen et al., 2010), while kahweol downregulates phosphorylation of signal transducers and activators of transcription 1 (STAT1) without altering its total level (Villarino et al., 2017). Cafestol is reported to inhibit osteoclast differentiation and promote osteoblast differentiation through its anti-inflammatory and antioxidant activities mentioned above. However, the role of kahweol is limited only to inhibition of osteoclasts. The inhibitory action of cafestol and kahweol on osteoclastogenesis indicates their potential as a drug source in degenerative bone treatment (Fumimoto et al., 2012; Fukuma et al., 2015). Several studies have demonstrated that the combination of cafestol and kahweol can downregulate the expression of cytochrome P450s (CYP450s) or directly inhibit their activity, thereby inhibiting the formation of electrophilic or oxidant intermediate and stimulating its detoxification. This action reduces DNA damage and prevents tumor initiation (Huber et al., 2008; Lee et al., 2007; Cavin et al., 2002; Cavin et al., 1998; Higgins et al., 2008; Tao et al., 2008). The study conducted by Higgins et al. (2008) has confirmed that cafestol and kahweol can intervene in Keapl/Nrf2/ARE signaling pathways and induce the expression of phase II detoxifying enzymes, antioxidant proteins, and Nrf2, thus blocking several carcinogenic stages. Recent studies also indicated that cafestol and kahweol can significantly regulate cyclins, apoptosis-related proteins, and inhibit tumor cell activity through multiple targets, thereby blocking tumor cell proliferation and inducing apoptosis (Kotowski et al., 2015; Woo et al., 2014; Lee et al., 2012; Choi et al., 2011). Previous studies on the antidiabetic potential of cafestol and kahweol indicates that they have varying efficacy and mechanism of action. A study carried out by Mellbye et al. (2015) showed that cafestol acutely stimulated the clonal rat insulinoma cell line INS-1E to secrete insulin and also increased glucose uptake in muscle cells. They confirmed the results in vivo using T2DM mouse models, wherein significant reductions in fasting blood glucose and fasting glucagon were observed along with improved insulin secretion and sensitivity. In contrast, kahweol activates 5′-adenosine monophosphate–activated protein kinase (AMPK), thereby inhibiting lipid accumulation and stimulating glucose uptake (Zhang et al., 2009).

23.2.5 Coffee melanoidins

Melanoidins are heterogeneous, nitrogen-containing brown polymers formed from the Maillard reaction, which takes place during thermal processing of coffee. They are formed in a multistage reaction between reducing sugars and compounds possessing free amino groups (Langner and Rzeski, 2014). They constitute about 25% of the coffee brew dry matter and are responsible for part of the specific sensory properties of coffee as well as viscosity of the brew (Nunes et al., 2007; Argirova, 2005; Bekedam et al., 2008). The composition of melanoidins depends on polysaccharides (galactomannans and arabinogalactans), amino acids, proteins, and phenolic compounds (chlorogenic, caffeic, or ferulic acids) found in coffee. Some of the phenolic compounds are linked to melanoidins' core with covalent bonds; others are linked through non-covalent interactions (Rufián-Henares et al., 2007; Nunes et al., 2007). Low-molecular-weight (LMW) melanoidins constitutes about 41% of all coffee melanoidins, whereas 59% are high-molecular-weight (HMW) melanoidins (Bekedam et al., 2008). The roasting time of coffee beans plays a major role in their formation, as longer roasting time leads to increased HMW melanoidins formation and the level of LMW melanoidins remain unchanged (Nunes et al., 2007).

23.2.5.1 Bioavailability of coffee melanoidins

The information about melanoidins absorption and bioavailability are scarce. However, they are generally considered as poorly absorbable and bioavailable compounds. Melanoidins, both soluble and insoluble, can act as dietary fibers that quench the radical species in the gastrointestinal tract due to their antioxidant activity (Finot et al., 1981). They escape digestion and pass through the upper gastrointestinal tract, where they can interact with the different microbial species present in the hindgut (Gniechwitz et al., 2008), and part of these compounds are fermented by microbes in the gut and exhibit the properties of soluble dietary fibers (Gniechwitz et al., 2008; Reichardt et al., 2009).

23.2.5.2 Coffee melanoidin consumption and health aspects

Several biological activities, such as antioxidant, antimicrobial, anti-carcinogenic, and antihypertensive activities, have been attributed to coffee melanoidins. Due to their diversity, complexity, difficulties in purification and identification, and poor solubility in water and organic solvents, assigning definite properties to individual melanoidins is difficult. The negative effect of melanoidins was linked to its inhibitory activity on some proteolytic enzymes in the alimentary tract, including trypsin (Ibarz et al., 2009; Oste et al., 1984). Melanoidins are reported to exert an antioxidant effect due to their ability to scavenge free radicals (hydroxyl, superoxide, and peroxyl radicals) and chelate metal cations, especially iron. Melanoidins exhibits antimicrobial activity through chelation of metal ions. Three mechanisms found responsible include (1) ion chelation at low concentration from the median; (2) chelation with siderophores-Fe^{3+} complex; and (3) removal of Mg^{2+} from the outer membrane, thereby promoting cell destruction and release of intracellular molecules (Rufián-Henares et al., 2009). The antihypertensive activity of melanoidins is attributed to its modulatory activity on angiotensin-converting enzyme (ACE). They are also demonstrated to inhibit matrix metalloproteases (MMPs), a class of zinc-containing endopeptidases, thereby playing an important role in the prevention of colon cancer (Rufián-Henares and Morales, 2007).

23.3 Conclusion

Coffee is one of the most frequently consumed beverages worldwide. Hence, it is interesting to study the potential beneficial and adverse biological effects of coffee and its active constituents. In addition to the scientific literatures available on coffee and its health aspects, more scientific data correlating the in vitro and in vivo studies are essential to have a better understanding on coffee and its different forms of beverages.

23.4 Acknowledgments

The authors would like to thank Dr. K. Basavaraj and Dr. M. Ramya of Coffee Quality Division, Coffee Board, Bengaluru, India for their guidance and support throughout the preparation of this chapter.

REFERENCES

Alasmari F. *Saudi Pharmaceutical Journal: SPJ: The Official Publication of the Saudi Pharmaceutical Society*. 2020;28: 445–451.

Allred KF, Yackley KM, Vanamala J, Allred CD. *Journal of Nutrition*. 2009;139:1833–1838.

Antonio AG, Moraes RS, Perrone D, Maia LC, Santos KRN, Iório NLP, Farah A. *Food Chemistry*. 2010;118:782–788.

Argirova MD. *Journal of Agricultural and Food Chemistry*. 2005; 53:1210–1214.

Arlt A, Sebens S, Krebs S, Geismann C, Grossmann M, Kruse ML, Schreiber S, Schäfer H. *Oncogene*. 2013;32:4825–4835.

Arnaud M. *Caffeine, Coffee and Health*. 1993:43–95.

Arti M, Prasad M, Monali S. *Current Diabetes Reviews*. 2012;8:162–168.

Bae J-H, Park J-H, Im S-S, Song D-K. *Integrative Medicine Research*. 2014;3:189–191.

Bassoli BK, Cassolla P, Borba-Murad GR, Constantin J, Salgueiro-Pagadigorria CL, Bazotte RB, da Silva RS, de Souza HM. *Cell Biochemistry and Function*. 2008;26:320–328.

Bazzaz BSF, Fakori M, Khameneh B, Hosseinzadeh H. *Journal of Pharmacopuncture*. 2019;22:49–54.

Bekedam EK, Loots MJ, Schols HA, Van Boekel MA, Smit G. *Journal of Agricultural and Food Chemistry*. 2008;56:7138–7145.

Bicho NC, Leitão AE, Ramalho JC, de Alvarenga NB, Lidon FC. *International Journal of Food Properties*. 2013;16: 895–904.

Bode AM, Dong Z. *Cancer Letters*. 2007;247:26–39.

Bułdak RJ, Hejmo T, Osowski M, Bułdak Ł, Kukla M, Polaniak R, Birkner E. *Molecules* (Basel, Switzerland). 2018;23(12):3309.

Butt MS, Sultan MT. *Critical Reviews in Food Science and Nutrition*. 2011;51(4):363–373.

Callahan MM, Robertson RS, Arnaud MJ, Branfman AR, McComish MF, Yesair DW. *Drug Metabolism and Disposition*. 1982;10:417–423.

Campanha FG, Dias RCE, Benassi MT. *Coffee Science*. 2010;5:87–96.

Cavin C, Holzhäuser D, Constable A, Huggett AC, Schilter B. *Carcinogenesis*. 1998;19:1369–1375.

Cavin C, Holzhaeuser D, Scharf G, Constable A, Huber WW, Schilter B. *Food and Chemical Toxicology*. 2002;40:1155–1163.

Chauhan PS, Satti NK, Sharma P, Sharma VK, Suri KA, Bani S. *Phytotherapy Research*. 2012;26:1156–1165.

Chavez Valdez R, Ahlawat R, Wills-Karp M, Nathan A, Ezell T, Gauda EB. *Journal of Pediatrics* 2011;158(1):57–64, e1.

Choi MJ, Park EJ, Oh JH, Min KJ, Yang ES, Kim YH, Lee TJ, Kim SH, Choi YH, Park JW, Kwon TK. *Chemico-biological Interactions*. 2011;190:102–108.

Delgado-Andrade C, Morales FJ. *Journal of Agricultural and Food Chemistry*. 2005;53:1403–1407.

De Roos B, Van der weg G, Urgert R, Van de Bovenkamp P, Charrier A, Katan MB. *Journal of Agricultural and Food Chemistry*. 1997;45:3065–3069.

De Roos B, Meyboom S, Kosmeijer-Schuil TG, Katan MB. *Journal of Internal Medicine*. 1998;244:451–460.

De Roos B, Van Tol A, Urgert R, Scheek LM, Van Gent T, Buytenhek R, Princen HM, Katan MB. *Journal of Internal Medicine*. 2000;248:211–216.

Dubrez L. Coll JL, Hurbin A, Solary E, Favrot MC. *Journal of Biological Chemistry*. 2001;276:38980–38987.

Evans J, Richards JR, Battisti AS. *StatPearls*. Treasure Island, FL: © 2021, StatPearls Publishing LLC.; 2021.

Farah A, Monteiro M, Donangelo CM, Lafay S. *The Journal of Nutrition*. 2008;138(12):2309–2315.

Farid MM, Yang X, Kuboyama T, Tohda C. *Scientific Reports*. 2020;10(1):16424.

Fehrholz M. Speer CP. Kunzmann S. *PLoS One*. 2014;9(5):e97357.

Feng R. Lu, Y. Bowman LL. Qian Y. Castranova V. Ding M. *Journal of Biological Chemistry*. 2005;280(30):27888–27895.

Finot PA, Magnenat E. *Progress in Food & Nutrition Science*. 1981;5(1–6):193–207.

Folwarczna J, Janas A, Pytlik M, Cegieła U, Śliwiński L, Krivošíková Z, Štefíková K, Gajdoš M. *Nutrients*. 2016;8(3):133.

Fredholm BB, Bättig K, Holmén J, Nehlig A, Zvartau EE. *Pharmacological Reviews*. 1999;51(1):83–133.

Fukuma Y, Sakai E, Nishishita K, Okamoto K, Tsukuba T. *BioFactors*. 2015;41(4):222–231.

Fumimoto R, Sakai E, Yamaguchi Y, Sakamoto H, Fukuma Y, Nishishita K, Okamoto K, Tsukuba T. *Journal of Pharmacological Sciences*. 2012;118:479–486.

Gaibor J, Morales D, Carrillo Terán W. *Asian Journal of Crop Science*. 2020;12:90–96.

Gniechwitz D, Reichardt N, Meiss E, Ralph J, Steinhart H, Blaut M, Bunzel M. *Journal of Agricultural and Food Chemistry*. 2008;56:5960–5969.

Godos J, Pluchinotta FR, Marventano S, Buscemi S, Li Volti G, Galvano F, Grosso G. *International Journal of Food Sciences and Nutrition*. 2014;65(8):925–936.

Grosso LM, Triche E, Benowitz NL, Bracken MB. *Annals of Epidemiology*. 2008;18(3):172–178.

Guercia E, Berti F, Navarini L, Demitri N, Forzato C. *Tetrahedron: Asymmetry*. 2016;27(14):649–656.

Gunning Y, Defernez M, Watson AD, Beadman N, Colquhoun IJ, Le Gall G, Philo M, Garwood H, Williamson D, Davis AP, Kemsley EK. *Food Chemistry*. 2018;248:52–60.

Higdon JV, Frei B. *Critical Reviews in Food Science and Nutrition*. 2006;46(2):101–123.

Higgins LG, Cavin C, Itoh K, Yamamoto M, Hayes JD. *Toxicology and Applied Pharmacology*. 2008;226(3):328–337.

Hirakawa N, Okauchi R, Miura Y, Yagasaki K. *Bioscience, Biotechnology, and Biochemistry*. 2005;69(3):653–658.

Hu GL, Wang X, Zhang L, Qiu MH *Food Function*. 2019;10(6): 3113–3126.

Huber WW, Rossmanith W, Grusch M, Haslinger E, Prustomersky S, Peter-Vörösmarty B, Parzefall W, Scharf G, Schulte-Hermann R. *Food and Chemical Toxicology*. 2008;46(4):1230–1238.

Hwang SJ, Kim YW, Park Y, Lee HJ, Kim KW. *Inflammation Research: Official Journal of the European Histamine Research Society*. 2014;63(1):81–90.

Ibarz A, Garvín A, Garza S, Pagán J. *Food and Chemical Toxicology*. 2009;47(8):2071–2075.

Ikram M, Saeed K, Khan A, Muhammad T, Khan MS, Jo MG, Rehman SU, Kim MO. *Nutrients*. 2019;11(5):1082

Isokawa M. *Neural Plasticity*. 2016: 1–7.

Johnston KL, Clifford MN, Morgan LM. *American Journal of Clinical Nutrition*. 2003;78(4):728–733.

Juliano LM, Griffiths RR. *Psychopharmacology*. 2004;176(1):1–29.

Kang NJ, Lee KW, Kim BH, Bode AM, Lee HJ, Heo YS, Boardman L, Limburg P, Lee HJ, Dong Z. *Carcinogenesis*. 2011;32(6):921–928.

Kang TM, Yuan J, Nguyen A, Becket E, Yang H, Miller JH. *Antimicrobial Agents and Chemotherapy*. 2012;56(6):3216–3223.

Kim JY, Jung KS, Jeong HG. *FEBS Letters*. 2004;569(1–3):321–326.

Kim K-J, Park S-K. *Korean Journal of Food Science and Technology*. 2006;38.

Kitzberger CSG, Scholz MBdS, Pereira LFP, Vieira LGE, Sera T, Silva JBGD, Benassi MdT. *Journal of Food Composition and Analysis*. 2013;30(1):52–57.

Kotowski U, Heiduschka G, Seemann R, Eckl-Dorna J, Schmid R, Kranebitter V, Stanisz I, Brunner M, Lill C, Thurnher D. *Strahlentherapie und Onkologie: Organ der Deutschen Rontgengesellschaft [et al]*. 2015;191(6):511–517.

Kwon SH, Lee HK, Kim JA, Hong SI, Kim HC, Jo TH, Park YI, Lee CK, Kim YB, Lee SY, Jang CG. *European Journal of Pharmacology*. 2010;649:210–217.

Lang R, Wahl A, Skurk T, Yagar EF, Schmiech L, Eggers R, Hauner H, Hofmann T. *Analytical Chemistry*. 2010;82(4):1486–1497.

Langner E, Rzeski W. *International Journal of Food Properties*. 2014;17(2):344–353.

Lee KJ, Choi JH, Jeong HG. *Food and Chemical Toxicology*. 2007;45(11):2118–2125.

Lee KA, Chae JI, Shim JH. *Journal of Biomedical Science*. 2012;19(1):60.

López-Cruz L, Salamone JD, Correa M. *Frontiers in Pharmacology*. 2018;9:526.

Lou Z, Wang H, Zhu S, Ma C, Wang Z. *Journal of Food Science*. 2011;76(6):M398–403.

Lu H, Tian Z, Cui Y, Liu Z, Ma X. *Comprehensive Reviews in Food Science and Food Safety*. 2020;19:3130–3158.

Luo C, Xu X, Wei X, Feng W, Huang H, Liu H, Xu R, Lin J, Han L, Zhang D. *Pharmacological Research*. 2019;148:104409.

Ma CM, Kawahata T, Hattori M, Otake T, Wang L, Daneshtalab M. *Bioorganic & Medicinal Chemistry*. 2010;18(2):863–869.

Manalo RVM, Medina PMB. *Frontiers in Neuroscience*. 2018;12:137.

McCusker RR, Goldberger BA, Cone EJ. *Journal of Analytical Toxicology*. 2003;27(7):520–522.

Mellbye FB, Jeppesen PB, Hermansen K, Gregersen S. *Journal of Natural Products*. 2015;78(10):2447–2451.

Meng S, Cao J, Feng Q, Peng J, Hu Y. *Evidence-based Complementary and Alternative Medicine: eCAM*. 2013;2013:801457.

Miwa S, Sugimoto N, Yamamoto N, Shirai T, Nishida H, Hayashi K, Kimura H, Takeuchi A, Igarashi K, Yachie A, Tsuchiya H. *Anticancer Research*. 2012;32(9):3643–3649.

Moeenfard, M. Coffee Diterpenes: Quantification and Anti-angiogenic Effects. 2016.

Monteiro M, Farah A, Perrone D, Trugo LC, Donangelo C. *Journal of Nutrition*. 2007;137(10):2196–2201.

Mori ALB, Kalschne DL, Ferrão MAG, Fonseca AFAd, Ferrão RG, Benassi MDT. *Journal of Food Composition and Analysis*. 2016;52:52–57.

Mubarak A, Bondonno CP, Liu AH, Considine MJ, Rich L, Mas E, Croft KD, Hodgson JM. *Journal of Agricultural and Food Chemistry*. 2012;60(36):9130–9136.

Narita Y, Inouye K. Chlorogenic Acids from Coffee. In: Preedy VR, editor. *Coffee in Health and Disease Prevention*. San Diego: Academic Press; 2015.

Nawrot P, Jordan S, Eastwood J, Rotstein J, Hugenholtz A, Feeley M. *Food Additives and Contaminants*. 2003;20:1–30.

Nunes FM, Coimbra MA. *Journal of Agricultural and Food Chemistry*. 2007;55(10):3967–3977.

Oliveira A, Cabral F, Eberlin M, Bolini H. *Food Science and Technology (Campinas)*. 2009;29:76–80.

Oste R, Sjödin P. *Journal of Nutrition*. 1984;114(12):2228–2234.

PubChem. Cafestol. 2021/07/04/03:55:43 [cited; Available from: https://pubchem.ncbi.nlm.nih.gov/compound/Cafestol]

PubChem. Caffeine, CID: 2519.2021/06/24/18:33:25 [cited; Available from: https://pubchem.ncbi.nlm.nih.gov/compound/Caffeine]

PubChem. Chlorogensaure. 2021/07/04/03:49:03 [cited; Available from: https://pubchem.ncbi.nlm.nih.gov/compound/Chlorogensaure]

PubChem. Kahweol. 2021/07/04/04:01:28 [cited; Available from: https://pubchem.ncbi.nlm.nih.gov/compound/Kahweol]

PubChem. Trigonelline. 2021/07/03/17:58:17 [cited; 2021: [Available from: https://pubchem.ncbi.nlm.nih.gov/compound/Trigonelline]

Reichardt N, Gniechwitz D, Steinhart H, Bunzel M, Blaut M. *Mol Nutr Food Res*. 2009;53(2):287–299.

Ren Y, Wang C, Xu J, Wang S. *International Journal of Molecular Sciences*. 2019;20:4238.

Renouf M, Guy P, Marmet C, Longet K, Fraering AL, Moulin J, Barron D, Dionisi F, Cavin C, Steiling H, Williamson G. *British Journal of Nutrition*. 2010;104(11):1635–1640.

Rodriguez de Sotillo DV, Hadley M. *Journal of Nutritional Biochemistry*. 2002;13(12):717–726.

Rufián-Henares JA, Morales FJ. *Journal of Agricultural and Food Chemistry*. 2007;55(4):1480–1485.

Rufián-Henares JA, Morales FJ. *Journal of Agricultural and Food Chemistry*. 2007;55(24):10016–10021.

Rufián-Henares JA, de la Cueva SP. *Journal of Agricultural and Food Chemistry*. 2009;57(2):432–438.

Sachse KT, Jackson EK, Wisniewski SR, Gillespie DG, Puccio AM, Clark RS, Dixon CE, Kochanek PM. *Journal of Cerebral Blood Flow and Metabolism: Official Journal of the International Society of Cerebral Blood Flow and Metabolism*. 2008;28:395–401.

Salinardi TC, Rubin KH, Black RM, St-Onge MP. *Journal of Nutrition*. 2010;140(11):1943–1948.

Scholz MBS, Pagiatto NF, Kitzberger CSG, Pereira LFP, Davrieux F, Charmetant P, Leroy T. *Food Research International*. 2014;61:176–182.

Scott NR, Chakraborty J, Marks V. *Annals of Clinical Biochemistry*. 1984;21 (Pt 2):120–124.

Shearer J, Sellars EA, Farah A, Graham TE, Wasserman DH. *Canadian Journal of Physiology and Pharmacology*. 2007;85:823–830.

Shen T, Lee J, Lee E, Kim SH, Kim TW, Cho JY. *Biological and Pharmaceutical Bulletin*. 2010;33:128–132.

Simon-Gruita A, Pojoga MD, Constantin N, Duta-Cornescu G. In: Grumezescu AM, Holban AM, editors. *Caffeinated and Cocoa Based Beverages*: Woodhead Publishing; 2019.

Smith A. *Food and Chemical Toxicology*. 2002;40:1243–1255.

Souza R, Benassi M. *Journal of the Brazilian Chemical Society*. 2012;23:1347–1354.

Sridevi V, Giridhar P, Gokare R. *Global Journal of Medical Research*. 2011;11:1–7.

Stadler RH, Varga N, Hau J, Vera FA, Welti DH. *Journal of Agricultural and Food Chemistry*. 2002a;50(5):1192–1199.

Stadler RH, Varga N, Milo C, Schilter B, Vera FA, Welti DH. *Journal of Agricultural and Food Chemistry*. 2002b;50:1200–1206.

Stavric B, Gilbert SG. *Acta Pharmaceutica Jugoslavica*. 1990;40:475–489.

Sugimoto N, Katakura M, Matsuzaki K, Sumiyoshi E, Yachie A, Shido O. *Basic & Clinical Pharmacology & Toxicology*. 2019;124:575–581.

Sung WS, Lee DG. *Pure and Applied Chemistry*. 2010;82:219–226.

Taguchi H, Sakaguchi M, Shimabayashi Y. *Agricultural and Biological Chemistry*. 1985;49(12):3467–3471.

Tang-Liu DD, Williams RL, Riegelman S. *Journal of Pharmacology and Experimental Therapeutics*. 1983;224(1):180–185.

Tao KS, Wang W, Wang L, Cao DY, Li YQ, Wu SX, Dou KF. *Medical Hypotheses*. 2008;71(5):730–736.

Urgert R, Kosmeijer-Schuil TG, Katan MB. *Biochemical Society Transactions*. 1996;24(3):800–806.

Urushisaki T, Takemura T, Tazawa S, Fukuoka M, Hosokawa-Muto J, Araki Y, Kuwata K. *Evidence-based Complementary and Alternative Medicine: eCAM*. 2011;2011:254914.

van Dijk AE, Olthof MR, Meeuse JC, Seebus E, Heine RJ, van Dam RM. *Diabetes Care*. 2009;32(6):1023–1025.

van Gelder BM, Buijsse B, Tijhuis M, Kalmijn S, Giampaoli S, Nissinen A, Kromhout D. *European Journal of Clinical Nutrition*. 2007;61(2):226–232.

Viani R. *Journal of Food Science*. 1974;39(6):1216–1217.

Villarino AV, Kanno Y, O'Shea JJ. *Nature Immunology*. 2017;18(4):374–384.

Volkow ND, Wang GJ, Logan J, Alexoff D, Fowler JS, Thanos PK, Wong C, Casado V, Ferre S, Tomasi D. *Translational Psychiatry*. 2015;5(4):e549.

Wang GF, Shi LP, Ren YD, Liu QF, Liu HF, Zhang RJ, Li Z, Zhu FH, He PL, Tang W, Tao PZ, Li C, Zhao WM, Zuo JP. *Antiviral Research*. 2009;83(2):186–190.

Wermelinger T, D'Ambrosio L, Klopprogge B, Yeretzian C. *Journal of Agricultural and Food Chemistry*. 2011;59(17):9074–9079.

Woo SM, Min KJ, Seo BR, Nam JO, Choi KS, Yoo YH, Kwon TK. *Cell Death & Disease*. 2014;5(11):e1514.

Xu Y, Chen J, Yu X, Tao W, Jiang F, Yin Z, Liu C. *Inflammation Research: Official Journal of the European Histamine Research Society [et al]*. 2010;59(10):871–877.

Yoshinari O, Sato H, Igarashi K. *Bioscience, Biotechnology, and Biochemistry*. 2009;73(5):1033–1041.

Yukawa GS, Mune M, Otani H, Tone Y, Liang XM, Iwahashi H, Sakamoto W. *Biochemistry Biokhimiia*. 2004;69(1):70–74.

Zhang BB, Zhou G, Li C. *Cell Metabolism*. 2009;9(5):407–416.

Zhou JY, Zhou SW. *Evidence-based Complementary and Alternative Medicine: eCAM*. 2012;2012:164219.

Zhou J, Zhou S, Zeng S. *Fundamental & Clinical Pharmacology*. 2013;27(3):279–287.

Index

U

V

W

X

Y